Novel Thermal and Non-Thermal Technologies for Fluid Foods

Food Science and Technology

International Series

A complete list of books in this series appears at the end of this volume.

Novel Thermal and Non-Thermal Technologies for Fluid Foods

Edited by

P.J. Cullen
School of Food Science & Environmental Health,
Dublin Institute of Technology,
Dublin, Ireland

Brijesh K. Tiwari
Department of Food and Consumer Technology,
Manchester Metropolitan University,
Manchester, UK

Vasilis P. Valdramidis
Biosystems Engineering,
School of Agriculture,
Food Science and Veterinary Medicine,
University College Dublin,
Dublin, Ireland

ELSEVIER

AMSTERDAM • BOSTON • HEIDELBERG • LONDON NEW YORK • OXFORD
PARIS • SAN DIEGO • SAN FRANCISCO • SINGAPORE • SYDNEY • TOKYO

Academic Press is an imprint of Elsevier

Academic Press is an imprint of Elsevier
32 Jamestown Road, London NW1 7BY, UK
225 Wyman Street, Waltham, MA 02451, USA
525 B Street, Suite 1800, San Diego, CA 92101-4495, USA

First edition 2012

British Library Cataloguing-in-Publication Data
A catalogue record for this book is available from the British Library

Library of Congress Cataloging-in-Publication Data
A catalog record for this book is available from the Library of Congress

ISBN : 978-0-12-381470-8

For information on all Academic Press publications
visit our website at www.elsevierdirect.com

Typeset by MPS Limited, a Macmillan Company, Chennai, India
www.macmillansolutions.com

Printed and bound in United States of America

10 11 12 13 14 15   10 9 8 7 6 5 4 3 2 1

Contents

Contributors xiii
Foreword xv

CHAPTER 1 Status and Trends of Novel Thermal and Non-Thermal Technologies for Fluid Foods 1
- **1.1** What are Novel Thermal and Non-Thermal Technologies? 1
- **1.2** Fluid Foods and Process Validation 2
- **1.3** Technology Drivers 2
  - 1.3.1 Regulation 2
  - 1.3.2 Shelf-Life Extension 3
  - 1.3.3 Nutritional and Sensory Aspects 3
  - 1.3.4 Novel Functional and Organoleptic Properties 4
  - 1.3.5 Consumer Acceptability 4
  - 1.3.6 Environmental Impact 5
- **1.4** Hurdle Concept 5
- **1.5** Book Objectives 5

CHAPTER 2 Fluid Dynamics in Novel Thermal and Non-Thermal Processes 7
- **2.1** Introduction 7
- **2.2** Some Basic Considerations on Fluid Mechanics 8
- **2.3** High-Pressure Processing 10
- **2.4** Pulsed Electric Fields 17
- **2.5** Ultrasound 25
- **2.6** Light Pulses 26
- **2.7** Ohmic Heating 27
- **2.8** Conclusion 28

CHAPTER 3 Fluid Rheology in Novel Thermal and Non-Thermal Processes 35
- **3.1** Introduction 35
- **3.2** Fluid Rheology 36
- **3.3** Influence of Rheology on Process Efficacy 39
  - 3.3.1 Continuous Flow-Through Systems 40
  - 3.3.2 Batch Systems 43
  - 3.3.3 Solid–Liquid Mixtures 44
- **3.4** Effects of Treatment on Fluid Rheology 45
  - 3.4.1 Pulsed Electric Fields 45
  - 3.4.2 High-Pressure Processing 47
  - 3.4.3 Power Ultrasound 51

3.4.4 Irradiation 52
3.4.5 Ultraviolet and Pulsed White Light 54
3.4.6 Ozone 55
3.4.7 Dense-Phase Carbon Dioxide 56
3.4.8 Ohmic Heating 57
3.5 Conclusions 57

CHAPTER 4 Pulsed Electric Field Processing of Fluid Foods 63
4.1 Introduction 63
4.2 Principles of Operation 64
4.3 Process Control Parameters 66
4.4 Treatment Parameters 66
4.5 Microorganism, Enzyme and Quality- and Health-Related Compound Characteristics 68
4.6 Product Parameters 69
4.7 Microbial Inactivation Mechanism 71
4.7.1 Pulsed Electric Field Inactivation Mechanism 71
4.7.2 Effects of Pulsed Electric Fields on Microorganisms 71
4.8 Pulsed Electric Field in Combination with Other Technologies 77
4.9 Effects on Fluid Food Nutritional and Quality Parameters 80
4.9.1 Effects on Food-Quality Related Enzymes 80
4.9.2 Effects on Quality Parameters 85
4.9.3 Effects on Nutritional Parameters 87
4.10 Shelf-Life of Pulsed Electric Field-Processed Fluid Foods 91
4.11 Current Status and Future Trends 95

CHAPTER 5 High-Pressure Processing of Fluid Foods 109
5.1 Introduction 109
5.2 Technology Governing Principles 110
5.2.1 Process Equipment 110
5.2.3 Pressure-Transmitting Fluid 112
5.2.4 Food Material Response to Combined Temperature–Pressure Treatment 112
5.2.5 pH Shift Under Pressure 117
5.3 Microbiological Safety 117
5.4 Impact of Combined Pressure–Heat Treatments on Quality Attributes of Foods 118
5.5 Combined Pressure–Heat Treatment Effects on Nutrient Content of Foods 126
5.6 Regulatory Aspects 128
5.7 Conclusions 128

CHAPTER 6 Ultrasound Processing of Fluid Foods.................... 135
6.1 Introduction.................... 135
6.2 Generation of Power Ultrasound.................... 136
6.3 The Measurement of Ultrasonic Energy (Dosimetry).................... 137
6.3.1 Calorimetry.................... 139
6.3.2 Chemical Dosimetry.................... 139
6.4 Ultrasound in Liquid Food Sterilization.................... 140
6.4.1 Microbial Inactivation in Liquid Foods.................... 141
6.4.2 Enzyme Inactivation.................... 147
6.4.3 Fruit Juices.................... 147
6.4.4 Milk.................... 151
6.5 Effect on Quality Parameters.................... 152
6.6 Current Status and Challenges.................... 156
6.7 Conclusions.................... 157

CHAPTER 7 Irradiation of Fluid Foods.................... 167
7.1 Introduction.................... 167
7.2 Irradiation Technologies.................... 168
7.3 General Mode of Action.................... 171
7.4 Food Safety and Shelf-Life of Irradiated Liquids.................... 172
7.5 Sensory and Nutritive Properties.................... 173
7.6 Operational Parameter Case Study: Container Orientation.................... 174
7.7 Regulations and Consumer Attitude of Food Irradiation: China.................... 177
7.8 Case Study: Irradiated Wine, China.................... 180
7.9 Conclusions.................... 181

CHAPTER 8 Ultraviolet and Pulsed Light Processing of Fluid Foods.................... 185
8.1 Introduction.................... 185
8.2 Principles of Operation.................... 187
8.2.1 Mechanisms of Ultraviolet Light Generation.................... 187
8.2.2 Gas Discharge.................... 188
8.2.3 Pulsed Power Energization.................... 189
8.2.4 Design Parameters.................... 189
8.3 Process Control Parameters.................... 193
8.3.1 Fluid Food Parameters.................... 193
8.3.2 Ultraviolet Light Delivery.................... 196
8.3.3 Fluids Flow Dynamics.................... 197
8.3.4 Utraviolet Light Sources.................... 197
8.4 Microbial Inactivation Mechanism.................... 199
8.4.1 Photochemical Effect.................... 199

8.4.2 Other Microbial Inactivation Mechanisms Specific to Pulsed Light 203
8.4.3 Kinetics of Microbial Inactivation 204

8.5 Antimicrobial Effects of Ultraviolet and Pulsed Light in Specific Products 206
8.5.1 Ultraviolet Inactivation of Microorganisms in Fresh Juice 206
8.5.2 Ultraviolet Inactivation Case Studies in Dairy Products: Milk, Cheese Brines and Whey Protein 210
8.5.3 Antimicrobial Effects of Pulsed Light 210

8.6 Effects on Fluid Food Nutritional and Quality Parameters 211

8.7 Industrial Scale-Up and Challenges 214
8.7.1 Designs of Ultraviolet Reactors for Liquid Foods 214
8.7.2 Challenges and Future Perspectives 218

8.8 Conclusions 219

CHAPTER 9 Ozone Processing of Fluid Foods 225

9.1 Introduction 225

9.2 Generation of Ozone 226
9.2.1 Electrical (Corona) Discharge Method 226
9.2.2 Electrochemical (Cold Plasma) Method 227
9.2.3 Radiochemical Ozone Generation (RCOG) 227
9.2.4 Ultraviolet Method 227

9.3 Ozone in Liquid Food Processing 227

9.4 Factors Affecting Efficacy of Ozone Processing 228
9.4.1 Extrinsic Parameters 229
9.4.2 Intrinsic Parameters 233

9.5 Microbial Inactivation 237

9.6 Microbial Inactivation Mechanism 243

9.7 Mycotoxins 246

9.8 Effect on Fluid Food Quality 247
9.8.1 Physical Quality 248
9.8.2 Nutritional Quality 249
9.8.3 Enzyme Inactivation 250

9.9 Current Status and Future Trends 251

CHAPTER 10 Dense-Phase Carbon Dioxide Processing of Fluid Foods 263

10.1 Principles 263
10.1.1 Introduction to Dense-Phase Carbon Dioxide Technology 263
10.1.2 Carbon Dioxide Properties 264
10.1.3 Engineering Aspects of Dense-Phase Carbon Dioxide Technology 266

10.2 Process Control Parameters 270
10.2.1 Pressure and Temperature 270
10.2.2 CO_2 Physical States 276
10.2.3 CO_2 Solubility 277
10.2.4 Mixing Speed 280
10.2.5 Depressurization Rate 280
10.2.6 Pressure Cycling 281
10.3 Microbial Inactivation Mechanisms 282
10.4 Effects on Fluid Food Nutritional and Quality Parameters 285
10.4.1 Effect on Microbial Inactivation and Stability 288
10.4.2 Effect on Enzyme Inactivation 289
10.4.3 Effect on the Physical, Chemical, Sensory and Nutritional Attributes 293
10.5 Current Status and Future Trends 295
10.5.1 Commercialization Efforts 295

CHAPTER 11 Ohmic Heating of Fluid Foods 305
11.1 Introduction 305
11.2 Principles 306
11.3 Process Control Parameters 311
11.3.1 Voltage, Current and Power Applied 311
11.3.2 Electrical Conductivity 312
11.3.3 Temperature 320
11.3.4 Electrode Types and Electrochemical Reactions 321
11.3.5 Food Properties 324
11.3.6 Flow Properties 325
11.4 Microbial Inactivation Mechanism 327
11.5 Effects on Nutritional and Quality Characteristics of Fluid Food 333
11.5.1 Soluble Solid Content and Acidity 334
11.5.2 Proteins 335
11.5.3 Vitamins 340
11.5.4 Phenolic Compounds and Flavor Characteristics 344
11.5.5 Color Properties 345
11.5.6 Enzyme Inactivation 348
11.5.7 Rheological Properties 353
11.5.8 Sensory Properties 358
11.6 Current Status and Future Trends 359

CHAPTER 12 Microwave Heating of Fluid Foods 369
12.1 Introduction 369
12.2 General Principles of Microwave Heating 371
12.3 Role of Dielectric Properties in Microwave Heating 372
12.3.1 Electrical/Dielectric Properties 373
12.3.2 Factors Influencing Dielectric Properties of Foods 374

12.4 Interactions of Microwave with Food Components 377
12.4.1 Estimation of Heat Generation 378
12.4.2 The Depth of Penetration of Microwaves 378
12.5 Dielectric Properties of Selected Foods 378
12.6 Microwave Processing of Foods 380
12.6.1 Microwave Processing of Liquid Foods 382
12.6.2 Microwave Pasteurization: Thermal and Non-Thermal Effects 395
12.7 Microwave Processing and Equipment 396
12.7.1 Microwave Systems 396
12.8 Challenges in Microwave Processing 401
12.8.1 Properties of Material 401
12.8.2 Addition of Ingredients/Change in Food Compositions 402
12.8.3 Microwave Processing Equipment 402
12.9 Conclusions 404

CHAPTER 13 Infrared Heating of Fluid Foods 411
13.1 Introduction 411
13.2 Basic Principles Governing Infrared Radiation 413
13.2.1 Planck's Law 417
13.2.2 Wien's Displacement Law 418
13.2.3 Stefan–Boltzmann's Law 418
13.3 Opportunities for the Infrared Processing of Liquid Foods 418
13.3.1 Microbes in Suspension 418
13.3.2 Orange Juice 420
13.3.3 Milk 420
13.3.4 Honey 421
13.3.5 Beer 424
13.4 Equipment for Infrared Processing of Liquid Foods 425
13.5 Conclusions and Suggestions for Future Work 425

CHAPTER 14 Modeling the Kinetics of Microbial and Quality Attributes of Fluid Food During Novel Thermal and Non-Thermal Processes 433
14.1 Introduction 433
14.2 *AD HOC* Hypotheses for Kinetic Studies 435
14.2.1 Are the Applied Processes Really Static? 435
14.2.2 The Case of Microbial Non-Linearities: Experimental Artefact or A Challenging Research Objective? 435
14.3 Parameters of Interest for Novel Thermal and Non-Thermal Processes 436
14.4 Experimental Design 436

14.4.1 Microbial Data for Developing Sound Modeling Approaches 439
14.4.2 Enzymatic and Chemical Data for Kinetic Studies 440
14.4.3 Collection of Data Under Dynamic Conditions 441
14.5 Data (Pre)Processing 441
14.6 Model Structure Characterization of Microbial Dynamics 441
14.6.1 Primary Models 442
14.6.2 Assessing the Achieved Log-Reduction 445
14.6.3 Secondary Models 447
14.6.4 Probabilistic Type of Models 448
14.7 Model Structure Characterization of Chemical Dynamics 450
14.7.1 Primary Models 450
14.7.2 Secondary Models 452
14.8 Model Parameter Estimation 456
14.9 Model Validation 457
14.10 Applications on the Design and Optimization of Thermal and Non-Thermal Technologies 459
14.11 Conclusion and Future 461

CHAPTER 15 Regulatory and Legislative Issues for Non-Thermal Technologies: An EU Perspective 473
15.1 Introduction 473
15.2 An Overview of How Law is Created in the EU 474
15.2.1 The Institutions of the European Union 474
15.2.2 Types of European Legislation 476
15.2.3 The Supremacy of the European Union 476
15.3 Overview of the Regulation of Food in the European Union 477
15.4 The Novel Food Regime 478
15.4.1 The Expedited Process: Concept of Substantial Equivalence 481
15.4.2 Proposal for New Novel Food Regulation 482
15.5 How Does this Apply to Specific Emerging Non-Thermal Technologies? 484
15.5.1 Irradiation 484
15.5.2 High Pressure 487
15.6 Other Existing Regulatory Requirements 488
15.6.1 Specific Requirements for Particular Fluid Foods 488
15.7 It's a Small World, but is One Set of Global Rules Realistic? 490
15.8 Conclusions 491

Food Science and Technology International Series 495
Index 499

Contributors

Murat O. Balaban
Fishery Industrial Technology Center, University of Alaska, Kodiak. USA

V.M. Balasubramaniam
Department of Food Science and Technology and Department of Food Agricultural Enginneering, The Ohio State University, Columbus, USA

Paula Bourke
School of Food Science and Environmental Health, Dublin Institute of Technology, Dublin, Ireland

P.J. Cullen
School of Food Science & Environmental Health, Dublin Institute of Technology, Dublin, Ireland

Antonio Delgado
Friedrich-Alexander-Universität Erlangen-Nürnberg, Department of Chemical and Bioengineering, Institute of Fluid Mechanics, Germany

Pedro Elez-Martínez
Food Technology Department, University of Lleida, Lleida, Spain

Giovanna Ferrentino
Department of Materials Engineering and Industrial Technologies, University of Trento, Trento. Italy

Meixu Gao
Chinese Academy of Agricultural Sciences, Institute for Application of Atomic Energy, Beijing, China

Vicente M. Gómez-López
Instituto de Ciencia y Tecnología de Alimentos, Facultad de Ciencias, Universidad Central de Venezuela

Rockendra Gupta
Department of Food Science and Technology, The Ohio State University, Columbus, USA

H.U. Hebbar
Department of Food Engineering, Central Food Technological Research Institute, Council of Scientific and Industrial Research, Mysore, India

Filiz Icier
Food Engineering Department, Engineering Faculty, Ege University, Bornova, Turkey

Tatiana Koutchma
Agriculture and Agri-Food Canada, Guelph, Ontario, Canada

Leszek Kulisiewicz
Friedrich-Alexander-Universität Erlangen-Nürnberg, Department of Chemical and Bioengineering, Institute of Fluid Mechanics, Germany

Karl Linden
Civil, Environmental, and Architectural Engineering, University of Colorado at Boulder, Boulder, USA

Olga Martín-Belloso
Food Technology Department, University of Lleida, Lleida, Spain

T.J. Mason
Sonochemistry Centre, School of Science and the Environment, Coventry University, Coventry, UK

Brian McKenna
Emeritus Professor of Food Science, University College, Dublin. Scientific Co-ordinator, European Technology Platform Food for Life

Brendan A. Niemira
U.S. Department of Agriculture, Agricultural Research Service, Eastern Regional Research Center, Wyndmoor, USA

Sonal Patil
School of Food Science and Environmental Health, Dublin Institute of Technology, Dublin, Ireland

Navin K. Rastogi
Department of Food Engineering, Central Food Technological Research Institute, Council of Scientific and Industrial Research, Mysore, India

Cornelia Rauh
Friedrich-Alexander-Universität Erlangen-Nürnberg, Department of Chemical and Bioengineering, Institute of Fluid Mechanics, Germany

Ángel Sobrino-López
Food Technology Department, University of Lleida, Lleida, Spain

Robert Soliva-Fortuny
Food Technology Department, University of Lleida, Lleida, Spain

Nikolaos G. Stoforos
Agricultural University of Athens, Department of Food Science and Technology, Athens, Greece

Petros S. Taoukis
National Technical University of Athens, School of Chemical Engineering, Laboratory of Food Chemistry and Technology, 5, Iroon Polytechniou, Athens, Greece

Brijesh K. Tiwari
Department of Food and Consumer Technology, Manchester Metropolitan University, Manchester, UK

Vasilis P. Valdramidis
Biosystems Engineering, School of Agriculture, Food Science and Veterinary Medicine, University College Dublin, Dublin, Ireland

Jan F.M. Van Impe
Katholieke Universiteit Leuven - Department of Chemical Engineering, BioTeC - Chemical and Biochemical Process Technology and Control, Leuven (Heverlee), Belgium

Dominic Watkins
Bond Pearce LLP, London, UK

Andreas Wierschem
Friedrich-Alexander-Universität Erlangen-Nürnberg, Department of Chemical and Bioengineering, Institute of Fluid Mechanics, Germany

Foreword

This timely book could easily be divided into two sub-sections. One section would cover some general scientific issues covering all novel process development such as fluid dynamics, rheology, modeling, and regulatory issues, while the other section would consist of 10 chapters on specific novel processes.

While this book on novel processing of fluid foods includes both thermal and non-thermal technologies, any study of process development in recent years will show that the vast majority of new processing technologies fall into the non-thermal category. This book is no exception to this trend, with the specific processes covered in seven non-thermal chapters and just three thermal chapters.

Such an emphasis on non-thermal processing of foods is a timely topic. Already, non-thermal processing commands its own annual international workshop, jointly sponsored by IFT (Institute of Food Technologists) and EFFoST (European Federation of Food Science & Technology) which attracts 200–300 participants. While thermal processing is now at least 200 years old (Nicolas Appert won his prize for sterilization in 1809), most of the more recent new processing technologies can be categorized as non-thermal. Even where some heating effects are present, they are normally just side-effects, with the primary functionality of the process being non-thermal.

However, despite this emphasis, it is significant that several chapters are devoted to thermal processing. Despite its long history, it is still the most significant method of food preservation. A quick survey in any supermarket in the developed world will reveal a vast array of products, both traditional and novel, that have been processed using thermal technologies. It is, therefore, gratifying to see that research is still continuing in this well-established field, with novel methods of generating and applying heat constantly being brought from laboratory research to industrial application. Significant chapters are presented on Ohmic heating, Microwave heating, and Infrared heating. While the emphasis here is on their application to fluid foods, all of these processes are also applicable to solid foods but the "novel" aspect is the processing of liquids with these methodologies. While all three processes are at different stages of development, all have either achieved or are about to achieve significant application potential.

However, I am jumping a little too far into the book with this description. Before specific technologies are considered, several chapters are devoted to the underlying sciences behind liquid processing. Following a general introductory chapter, specific chapters are devoted to fluid dynamics and rheology as applied in the processing of liquid foods, whether by thermal or non-thermal treatments.

Following these scientific introductions, the largest segment of seven non-thermal processing chapters can be found. All of the current "hot topics" in non-thermal processing are covered including Pulsed electric field, High-pressure processing, Ultrasound processing, Irradiation, Ultraviolet and pulse-light technology, Ozone processing, and Dense-phase carbon dioxide processing. Of course, some of these technologies are closer to industrial application than others. Pulsed electric field processing could become as common as domestic microwave heating if the concepts could be properly communicated to the consumer. Some researchers have suggested recently that micropulse technology might be a more appropriate title. High-pressure processing on the other hand is quite well developed. Here, the main constraints to be overcome are the batch nature of the process and the current volume limitations of the processing vessels. Irradiation was once considered as a technology lost to the food industry. Thankfully, food scientists are still developing the technology, although its acceptance by the consumer is always going to pose a significant problem. Ultraviolet and pulse-light technology have promising futures. The mild processing implied by the technologies will be attractive for the consumer and I would expect to see significant application of the technology in the coming decades. Similar consumer attitudes to Ultrasound should ensure that this promising process enjoys similar significant application in the future. Developments in power ultrasound should ensure its future. Ozone processing is at an earlier stage in its development compared to the above processes. However, if problems such as inducing sensory change can be overcome, then its future is also bright. The final chapter in the non-thermal series is devoted to inactivation of microorganisms and enzymes by dense-phase carbon dioxide. Here, again, the probability of industrial application is high.

The common factor running through all the processes, whether thermal or non-thermal, is microbial inactivation and changes in product quality induced by the process. It is very relevant, therefore, that a chapter is devoted to modeling methods for predicting such effects in any of the foregoing processes.

Finally, no novel technology will see successful industrial application without complying with the relevant safety legislation and regulation in the geographical region in which the technologies are being applied. The final chapter of the book covers regulatory aspects and compares the processes in Europe and the USA.

Thankfully, food engineers will continue to develop and apply new technologies for more efficient, milder processing and the development of new food products. Consequently, the coming decades will see not only the widespread application of the technologies outlined in this book, but also the development of a new generation of potential applications.

Brian M. McKenna
Professor
Emeritus Professor of Food Science,
University College Dublin Scientific Co-ordinator,
European Technology Platform *Food for Life*
March 2011

CHAPTER

Status and Trends of Novel Thermal and Non-Thermal Technologies for Fluid Foods

1

P.J. Cullen, Brijesh K. Tiwari and Vasilis P. Valdramidis

1.1 WHAT ARE NOVEL THERMAL AND NON-THERMAL TECHNOLOGIES?

Traditional thermal treatments are a cornerstone of the food industry providing required safety profiles and extensions of shelf-life. However, such treatments may lead to losses of desired organoleptic properties and damage to temperature-labile nutrients and vitamins. Consequently, the food industry has long sought alternative or synergistic approaches to provide the treatment objectives. Novel thermal and non-thermal technologies have been designed to meet the required food product safety or shelf-life demands while minimizing the effects on its nutritional and quality attributes.

The potential of novel thermal electromagnetic technologies for processing have gained increased industrial interest and have potential to replace, at least partially, the traditional well-established preservation processes (Vicente and Castro, 2007). Technologies including ohmic and microwave heating, are promising alternatives to conventional methods of heat processing. Such technologies are regarded as volumetric forms of heating, in which thermal energy is generated directly inside the food. Advantages over conventional methods include not having to rely on limiting heat transfer coefficients and the requirement of high wall temperatures. Other advantages include improved retention of quality and nutritional parameters, shorter processing times and higher yields.

The term "non-thermal processing" is often used to designate technologies that are effective at ambient or sublethal temperatures. Such approaches may offer the potential for improved retention of desired quality and nutritional parameters. Although the term "non-thermal" is used to classify these technologies, many of these approaches may result in temperature rises and efficacy is commonly found to have a synergistic effect with temperature. Consequently, temperature rises may be expected or even desired to be industrially practical. There is no universal temperature at which significant detrimental effects on nutritional or quality

Novel Thermal and Non-Thermal Technologies for Fluid Foods. DOI: 10.1016/B978-0-12-381470-8.00001-3

profiles are observed. Each treatment will be governed by the food commodity, its volatile components, physics, consumer expectations, shelf-life requirements, practical treatment times, etc.

Given the vast range and increased complexities of foods consumed, there is no universal technology available for processors. However, there have been significant advances in recent years in the development of approaches which show commercial promise. The important issue is not the classification or definition of the treatment process but its validation and control of the identified critical control parameters, be it temperature, voltage, dose, etc., or combination thereof.

1.2 FLUID FOODS AND PROCESS VALIDATION

Most novel thermal and non-thermal technologies may be employed for either solid or fluid foods. However, it is likely that the most practical adoption of these approaches will be initially for fluid foods. This is due to the mode of operation of many of these technologies and the ease of integration to continuous processing streams. Nevertheless, consideration of the scale-dependent inhomogeneities and transport phenomena is paramount to ensure process validation for complex fluid systems. Coupling of microbial inactivation and quality parameter kinetics with fluid dynamics can lead to process validation, mitigation of over- or under-processing and ultimately industry adoption.

Validation may be defined as a documented program that provides a high degree of assurance that a specific process, method, or system will consistently produce a result meeting predetermined acceptance criteria. Thermal processing of fluid foods which display complex flows such as purees, pastes, soups, and jams has always been a challenge for the food industry. To improve treatment homogeneity, various forced convection approaches such as rotary impellers, static mixers, and scraped-surface heat exchangers are typically employed. Nevertheless, over-processing is common within the food industry to ensure safety requirements are met, leading to avoidable nutritional and sensory losses. Although certain novel thermal and non-thermal technologies may overcome some of the difficulties arising with processing such fluids, treatment homogeneity still remains a significant challenge. Knowledge of the spatial treatment within the confines of any process facilitates treatment homogeneity and ultimately process adoption.

1.3 TECHNOLOGY DRIVERS

1.3.1 Regulation

While food safety assurance is a global concern, approaches to regulation may differ throughout the world. The recent Food and Drug Administration (FDA) requirement for US fruit juice processors to achieve a 5-log reduction in the numbers of the most resistant pathogens in their finished products may be seen as a driver for processors to seek alternative approaches to traditional pasteurisation. Comparably, the lack of such a ruling in the EU and a trend toward minimal processing of fruit juices could also lead processors to adopt technologies which provide an extension of shelf-life with minimal degradation of nutrition and sensory profiles. There is a general consensus among the regulatory agencies that novel technologies are of benefit to both food processors and consumers alike. Given the complexities of food matrices and the range of foods produced, validating processes is a challenge for industry. However more expedited validation processes are likely with validation of comparable products for a given technology.

1.3.2 Shelf-Life Extension

Thermal and non-thermal technologies may be employed to extend shelf-life, where mild processing can provide a significant benefit for short shelf-life products but with minimal effects on quality parameters particularly for products valued for their taste and nutritional profiles. There are numerous examples of novel technologies showing shelf-life extension with fluids of various complexities, as outlined throughout this book. An example of non-thermal technology for such an approach was shown by Walkling-Ribeiro *et al.* (2010) where an improved microbiological shelf-life was achieved in a fruit juice smoothie-type beverage using a combination of moderate heat and pulsed electric fields compared to mild thermal pasteurization with shelf-life of the products expiring after 28 and 21 days, respectively.

1.3.3 Nutritional and Sensory Aspects

Taste and sensory properties are consistently rated as the most important factors that drive consumption and repeat purchase of food products. The principal driver for industrial adoption of any processing technology is to meet consumers' demands for improved taste and nutrition. Interestingly, even for such key quality parameters, consumer variation between countries is evident with respect to novel technologies. A recent survey based upon 3000 personal interviews in the United Kingdom, Germany, and France highlighted the importance of improved flavor. However, this was found to differ with culture. About half of French consumers indicate they would purchase a product processed by high pressure for improved quality while only 4% of British consumers said they would purchase for this benefit (Butz *et al.*, 2003). In contrast, almost 40% of German consumers indicated they would purchase a product processed by high pressure for enhanced nutritional parameters while this was an important influence to only 18% of French consumers (Butz *et al.*, 2003).

1.3.4 Novel Functional and Organoleptic Properties

Consumers are increasingly demanding foods that are not only good from a nutritional or taste perspective, but are health promoting, i.e. functional food or neutraceutical. Therefore, manufacturers are beginning to address this need by exploring how novel food processing and preservation methods like those described in this book could help them meet this consumer demand.

High-pressure processing in particular has unique effects on enzyme stability and catalytic activity. It is leading to new product development and/or opening a new dimension in food processing (Duvetter *et al.*, 2009) as it offers exciting possibilities for restructuring food proteins and promoting protein–polysaccharide interactions (Hendrickx and Knorr, 2001) . Additionally, it can induce starch gelatinization (Bauer *et al.*, 2005), affect protein conformation (e.g., soy, meat, egg (Messens *et al.*, 1997)) and lead to protein denaturation, aggregation or gelation depending on the food protein system.

Other approaches including pulsed electric fields (PEF) and power ultrasound have also been identified as technologies which can create novel functional characteristics in foods. Factors like coagulation, foaming and emulsification kinetics have been altered through the application of some of these processes. Food functional changes induced by such technologies offers significant potential for novel food product development.

1.3.5 Consumer Acceptability

Consumers are not only concerned about the ingredients within the foods they consume but also the processes which are employed along the "farm to fork" food chain. Paradoxically, consumers are demanding foods which are minimally processed, meet their nutritional and taste desires, yet require minimal preparation. Understanding and addressing consumer issues related to novel food processes is one of the most important challenges facing the developers of innovative food products. Research suggests that acceptance of new technologies is based to a great extent on public perceptions of the associated risks, and that perceptions of risk are influenced by trust in information and the source which provides it. Several consumer research studies have consistently shown that consumers have poor knowledge and awareness levels towards most novel food-processing techniques, which serves as a major impediment to their acceptance. Thus, effective communication regarding details of the technologies and their benefits will be essential for successful marketing of these products.

If a novel technology allows the introduction of new products with tangible benefits, consumers are most likely to accept it. Several investigators have examined the role of product and benefit information on the acceptance and likely consumption of foods processed by novel technologies (Cardello *et al.*, 2007). In these studies, the information has often focused on the benefits to be accrued from the new technology as quantified by scientific, governmental, or other food industry organizations.

1.3.6 Environmental Impact

To achieve the full benefits from commercial exploitation of such technologies, issues related to environmental impacts, such as wastewater and gas emissions, the conservation of non-renewable resources and energy consumption are increasingly attracting the attention of the food processors since they can represent significant reductions in processing costs (Pereira and Vicente, 2010). The food industry is a significant consumer of energy with the principal type of energy used for traditional thermal processing being fossil fuel. Pereira and Vicente (2010) recently reviewed emerging thermal and non-thermal technologies with respect to environmental impacts and concluded they hold benefits, be it by improving the overall energy efficiency of the process, or by reducing the use of non-renewable resources. A number of studies have proposed significant reductions in energy consumptions for PEF and ohmic heating approaches over traditional pasteurization.

1.4 HURDLE CONCEPT

Combining a non-thermal processes with conventional preservation or other non-thermal technology may enhance the overall antimicrobial effect so that lower process intensities can be employed. This approach, known as "hurdle technology", has already been applied successfully using traditional techniques of food preservation. Combining non-thermal methods with other food preservation techniques can (1) enhance the lethal effects of non-thermal processing, (2) reduce the severity of non-thermal treatment needed to obtain a given level of microbial inactivation, and/or (3) prevent the proliferation of survivors following treatment (Barbosa-Cánovas and Rasoa, 2003). The choice of hurdles, i.e. several combinations of either novel thermal, novel non-thermal or conventional processing technologies is generally made to maximize the synergistic effect on the microbial inactivation kinetics. Food preservation using combined methods involves successive or simultaneous applications of various individual treatments. Combined treatments are advantageous, principally because many individual treatments alone are not adequate to ensure food safety or stability.

1.5 BOOK OBJECTIVES

The objective of this book is to provide a comprehensive overview of recent innovations in thermal and non-thermal technologies for fluid foods. This book brings together information on fluid behavior and biochemical dynamics for various products including fruit juice, alcoholic beverages, milk, etc. This book will help readers to better assess and understand the benefits and challenges of adopting novel technologies for fluid food processing. Such scientific knowledge can be exploited in different multidisciplinary areas of risk assessment, food engineering, product development, etc. It is hoped that this text will provide the basis for the processing of safer and more competitive products which meet current consumers' demands for high quality processed fluid foods.

The book can be used as a professional reference for those working in food-processing research areas in universities and the food industry. In addition, it can serve as a good practical source for people working in the areas of quality assurance/ quality control, product development, technical services, and regulatory affairs in food processing operations.

References

Bauer, B.A., Knorr, D., 2005. The impact of pressure, temperature and treatment time on starches: pressure-induced starch gelatinisation as pressure time temperature indicator for high hydrostatic pressure processing. J. Food Eng. 68 (3), 329–334.

Barbosa-Cánovas, G.V., Rasoa, J., 2003. Nonthermal preservation of foods using combined processing techniques. Crit. Rev. Food Sci. Nutr. 43 (3), 265–285.

Butz, P., Needs, E.C., Baron, A., Bayer, O., Geisel, B., Gupta, B., et al., 2003. Consumer attitudes to high pressure food processing. Food, Agri. Environ. 1, 30–34.

Cardello, A.V., Schutz, H.G., Lesher, L.L., 2007. Consumer perceptions of foods processed by innovative and emerging technologies: a conjoint analytic study. Innovat. Food Sci. Emerg. Tech. 8, 73–83.

Hendrickx, M.E.G., Knorr, D., 2001. Ultra High Pressure Treatments of Food. Kluwer Academic/Plenum Publisher, New York.

Hirsch, A.R., Knauss (née Resch), A., Carle, R., Neidhart, S., 2009. Pectins in processed fruit and vegetables: Part I—stability and catalytic activity of pectinases. Compr. Rev. in Food Sci. Food Saf. 8 (2), 75–85.

Messens, W., Van Camp, J., Huyghebaert, A., 1997. The use of high pressure to modify the functionality of food proteins. Trends Food Sci. Tech. 8 (4), 107–112.

Pereira, R.N., Vicente, A.A., 2010. Environmental impact of novel thermal and non-thermal technologies in food processing. Food Res. Int. 43, 1936–1943.

Vicente, A., Castro, I., 2007. Novel thermal processing technologies. In: Tewari, G., Juneja, V. (Eds.), Advances in Thermal and Non-Thermal Food Preservation. Blackwell Publishing, Oxford, UK.

Walkling-Ribeiro, M., Noci, F., Cronin, D.A., Lyng, J.G., Morgan, D.J., 2010. Shelf life and sensory attributes of a fruit smoothie-type beverage processed with moderate heat and pulsed electric fields. LWT 43 (7), 1067–1073.

CHAPTER

2 Fluid Dynamics in Novel Thermal and Non-Thermal Processes

Antonio Delgado, Leszek Kulisiewicz, Cornelia Rauh and Andreas Wierschem

Friedrich-Alexander-Universität Erlangen—Nürnberg, Department of Chemical and Bioengineering, Institute of Fluid Mechanics, Germany

2.1 INTRODUCTION

Highly competitive markets in industrial nations and well-developed countries are generating an urgent need for developing novel food-processing technologies. More than economical reasons, process ecology as well as product quality are the chief driving forces for any novel technology. Regarding product quality, consumers demand food with high levels of organoleptic and nutritional quality but free of any health risks. The development of such kinds of products is usually connected with a reduction in the processing temperature (compared with traditional technologies), since thermal treatment often leads to loss of the desired organoleptic properties of fresh products and damage to temperature-labile nutrients and vitamins. Therefore, the general tendency in development of novel food technologies consists in non-thermal processing or mild thermal processing, where the impact of temperature is aimed to be reduced by replacing pure thermal treatment by methods such as elevated pressure, electrical or acoustic fields, and light pulses or by combining methods. Such innovative technologies allow exploitation of additive or synergistic effects of temperature and other process parameters, bringing high rates of inactivation of unwanted microorganisms or enzymes and preserving flavor, color and also nutritional and functional values of foods. This is the leading thought of the hurdle concept, which postulates that preservation technologies combined at lower individual intensities display synergistic antimicrobial effects, while their impact on sensory and nutritive properties remains low and the required energy input is reduced (Leistner, 1992; Ross *et al.*, 2003).

This chapter deals with the novel treatment technologies applicable for food capable of flowing, which can be considered as fluid in a general sense. In the following, the major fluid-dynamical phenomena connected with novel food

Novel Thermal and Non-Thermal Technologies for Fluid Foods. DOI: 10.1016/B978-0-12-381470-8.00002-5

treatment concepts are discussed. In particular the non-thermal technologies: high-pressure processing (HPP), pulsed electric fields (PEF), power ultrasound, and pulsed light, and also thermal treatment making use of ohmic heating are addressed, being the most promising ones at present. The quality of food processing is usually inherently correlated to the fluid-dynamical or, more generally, to the thermofluid-dynamical features of the particular process. Although this fact has become increasingly recognized and accepted in the professional circles devoted to the investigation and production of food in the last decades, the literature dedicated to describing phenomena, mechanisms and structures related to fluid mechanical effects in food capable of flowing is relatively sparse. This situation is more crucial when considering novel thermal and non-thermal technologies. Due to the novel character of these technologies and the complexity of modeling all effects related to the capability of flowing, only a few publications are available so far (see, for example, Delgado *et al.* (2007), Otero *et al.* (2002), de Heij *et al.* (2002), Delgado *et al.* (2008), Salih *et al.* (2009), Gerlach *et al.* (2008), Krauss *et al.* (2010)).

Principally, the analysis of a food treatment process aimed at inactivation of particular unwanted biological systems (e.g., microorganisms, enzymes) or another quality-related biochemical reaction always requires consideration of the effects of temperature and other lethal agents such as pressure, ultrasound intensity, electrical field intensity, and light intensity. In food capable of flowing, this requires the application of fundamental laws of conservation of mass, momentum, and energy, coupled with kinetic models describing biological reactions to inactivating agents (temperature and others involved in the particular processes). As a result, the reaction of the biotic matter dispersed in a fluid continuum can be resolved in time and space based on the mass, momentum, and energy transport (Hartmann and Delgado, 2005; Rauh *et al.*, 2009; Baars *et al.*, 2007). Only in this way can a sufficiently exact figure of the effect of processing on the biomatter be achieved. The mere analysis of kinetic (only time-dependent) aspects of biochemical reactions in foods without consideration of the scale-dependent inhomogeneities and transport phenomena, leads to unacceptable simplifications and errors in evaluation of the impact of technological process on product quality. As a result, the non-uniform spatiotemporal distribution of processing parameters resulting in onset of over- and under-processed fractions is neglected.

2.2 SOME BASIC CONSIDERATIONS ON FLUID MECHANICS

There is a large gap in the literature regarding a general overview of fluid mechanical, i.e. thermofluid-dynamical, effects on novel thermal and non-thermal processes. Particular processes such as high-pressure treatment and PEF are described in Gerlach *et al.* (2008), Krauss *et al.* (2010), Hartmann and Delgado (2005), Rauh *et al.* (2009), Baars *et al.* (2007), Otero *et al.* (2002), Kowalczyk *et al.* (2004), and Kowalczyk and Delgado (2007). The basis for the mathematical model applied for the analysis of transport phenomena in fluid matter are governing equations, which are usually formulated as partial differential conservation equations of mass, momentum, and energy, defined here for a Newtonian fluid with viscosity $\eta(T, p)$ and density $\rho(T, p)$. The conservation equation of mass is

$$\frac{\partial \rho}{\partial t} + \nabla \cdot (\rho U) = 0, \tag{2.1}$$

the conservation equation of momentum is

$$\rho \frac{DU}{Dt} = -\nabla \rho + \nabla \cdot \tau + \rho g, \tag{2.2}$$

where τ is the viscous stress tensor, for vanishing bulk viscosity defined as

$$\tau = -\frac{2}{3}\eta(\nabla \cdot U)\delta + \eta(\nabla U + (\nabla U)^T), \tag{2.3}$$

where, U is the velocity, t is the time, p is the pressure. The conservation equation of energy for a fluid with specific thermal capacity c_p, thermal expansion coefficient α, and heat conductivity λ can be written as

$$c_p \rho \frac{DT}{Dt} = \alpha T \frac{Dp}{Dt} + \nabla \cdot (\lambda \nabla T) + \eta \Phi + Q, \tag{2.4}$$

where T is the temperature and Φ is the dissipation function determining the thermal energy generated by viscous heating (see, for example, Kowalczyk and Delgado (2007)). The last term on the right-hand side represents energy input to the fluid due to the action of external energy sources, for example electric or acoustic fields, light irradiation, etc.

Strictly speaking, non-thermal treatment of food proves to be possible only in the academic case of isothermal treatment. This can be concluded directly from a rough analysis of the above governing equations, particularly of Equation 2.4. As a rule, any treatment is connected to an exchange of mechanical or thermal energy. Furthermore, real processes can not be protected from dissipative effects. In this regard, as soon as the fluid is set in motion, the dissipation of kinetic energy would cause a change of the inner energy by viscous effects (see term $\eta\Phi$ in Equation 2.4), being incrementally significant at increasing fluid deformation rates. Furthermore, it is clear that if large pressure changes are involved, the first term on the right-hand side of Equation 2.4, standing for compression work, would become significant. Thus, at least one term on the right-hand side of Equation 2.4 must deviate from zero, and must be consequently balanced by temperature changes. This leads to the conclusion that in the strict sense every process that is carried out on the liquid food system must be considered with the inclusion of temperature. In this respect, the differentiation between thermal and non-thermal processes is ambiguous when fluid-dynamical effects are involved and should be considered only in relation to the mechanism of the particular biochemical reaction caused by the concerned process.

Detailed analysis of flow phenomena requires solving the set of governing Equations 2.1–2.4, which is not an easy task, since analytical solutions can be provided only in a few special cases (Gerlach *et al.*, 2008; Krauss *et al.*, 2010). Therefore, approximate solutions are pursued instead, which provide sufficient accuracy for most applications. In order to obtain an approximate numerical solution, a discretization method has to be used, which approximates the differential equations by a system of algebraic equations. These can then be solved on a computer for a given boundary and initial conditions (Ferziger and Peric, 2002). The numerical solution of the discretized equations the provides results at discrete points in space, at the elements or the control volumes of the numerical grid, and in time. This approach can be considered as an alternative to experiments or at least a complementary method in such cases, when experiments are difficult or impossible to perform. Numerical simulations would, for example, be advantageous, when the measuring equipment itself disturbs the process under consideration. The substitution of parts of the experiments by numerical simulations also becomes interesting when the experimental development of a process is too expensive and time consuming or when the scale-up is considered.

In the following, fluid-dynamic effects in novel technologies for fluid foods are discussed in more detail, referring to the most important processes in respect to both experimental and numerical analyses.

2.3 HIGH-PRESSURE PROCESSING

The feature of pressure treatment that has attracted the attention of food technologists is the ability to kill pathogenic microorganisms, even at low temperatures. Bacteria, yeasts, moulds, and viruses can be inactivated by means of high-pressure treatment. Well-suited overviews on high-pressure processing are given by Knorr and Mathys (2008), Guerrero-Beltrán *et al.* (2005), Ludikhuyze *et al.* (2003), Cheftel (1995), Balny (2004), Smeller (2002), and Northrop (2002). Thermofluid-dynamical aspects are considered systematically in Delgado *et al.* (2007), Delgado *et al.* (2008), Kowalczyk and Delgado (2007), and Delgado and Hartmann (2002). Obviously, the process parameters such as pressure and temperature, rate of pressure change, treatment time, and mode of pressurization (single or repeated cycles) influence inactivation. Besides the reduction of viable microorganism cells, HPP enables destabilization and inactivation of certain enzymes present in foods (Rauh *et al.*, 2009; Knorr and Mathys, 2008; Guerrero-Beltrán *et al.*, 2005; Ludikhuyze *et al.*, 2003; Balny, 2004; Smeller, 2002; Northrop, 2002). In this way the stability and shelf-life of the product can be extended and unwanted changes in taste or appearance catalyzed by enzymes can be avoided.

Reaction of the biological system to HPP depends not only on the hydrostatic pressure but also on a number of other parameters such as pressure-holding time, temperature, pH-value, and other interactions with the ambient media.

These dependencies may significantly influence the results in industrial applications. Heterogeneous distribution of temperature or pH-value may induce non-uniform spatiotemporal enzyme inactivation, which is an undesired effect, since it results in over- and under-processed fractions.

Industrial inactivation is generally carried out as a batch process. The treated food is often packaged and then placed into an autoclave made from steel. After filling the autoclave with a liquid, generally water, which serves as a pressure-transmitting medium, the autoclave is pressurized. Subsequently, the process continues with a phase of constant high pressure or periodic pressure cycles. Finally, the pressure is released, the product recovered, and a new treatment may start.

As shown by Delgado and Hartmann, the high-pressure treatment in autoclaves is intrinsically an inhomogeneous process (Delgado and Hartmann, 2002) under any non-academic conditions. As mentioned above, the effect of compression must be balanced by the other field and transport quantities as expressed by Equation 2.4. The most prominent effect is an increase in enthalpy and, therefore, in temperature, often denominated as compression heating in literature. For calculating this temperature increase, Equation 2.4 can be strongly simplified by assuming the fluid food to be at rest. Furthermore, when adiabatic conditions are postulated the temperature increase is:

$$\frac{\partial T_{ad}}{\partial p} = \frac{\alpha_p}{\rho c_p} T, \tag{2.5}$$

where α_p, ρ, c_p, and the subscript *ad* indicate isobaric thermal expansion coefficient, density, specific heat capacity and adiabatic process, respectively. Since α_p, ρ, and c_p are material parameters, the temperature increase differs for the materials involved. This has far-reaching consequences. The filled pressure vessel represents a non-uniform system consisting of pressure-transmitting medium, steel vessel wall, product and packaging material. Hence, a heterogeneous temperature field can arise in the vessel during pressurizing and pressure-holding time, despite an initial uniform temperature distribution (Hartmann and Delgado, 2005; Kowalczyk *et al.*, 2004; Denys *et al.*, 1997; Sanz *et al.*, 1997; Denys *et al.*, 2000; Otero *et al.*, 2000; Hartmann, 2002; Hartmann *et al.*, 2003; Ghani and Farid, 2007; Knoerzer *et al.*, 2007; Khurana and Karwe, 2009; Pehl and Delgado, 1999; Pehl *et al.*, 2000). Subsequently heat-transfer processes set in. Due to the pressure–temperature dependency of microorganisms and enzyme inactivation, heterogeneous temperature distributions can influence the uniformity of the process. This holds even in the basic system consisting of the autoclave and a pressure-transmitting medium. The temperature increase of the autoclave walls made from stainless steel is weaker than that of the pressure-transmitting medium; this creates temperature differences between the container walls and the pressure-transmitting liquid (Ting *et al.*, 2002). Heat conduction through the container walls cools down the neighboring liquid and causes thermal heterogeneities within the vessel as pointed out by Delgado and Hartmann (2002).

Various groups have studied velocity fields and temperature distributions in the pressure-transmitting medium in autoclaves (Delgado and Hartmann, 2002;

Pehl and Delgado, 1999; Pehl *et al.*, 2000; Ting *et al.*, 2002; Rauh, 2008; Chen *et al.*, 2007; Rauh *et al.*, 2006). Delgado and collaborators showed numerically that the horizontal temperature gradient yields an unstable density configuration in the neighborhood of the cool side walls (Delgado and Hartmann, 2002). Due to gravity, the cooler liquid flows downwards in the vicinity of the side walls. In large industrial containers, this results in a hydrodynamic boundary layer (Rauh, 2008). The maximum downflow velocity in such a case as determined numerically is shown in Figure 2.1. Rauh (2008) showed that in water and water–saccharose solutions the corresponding thermal boundary layer is thinner than the hydrodynamic one. Hence, hotter liquid close to the cool near-wall liquid is also dragged downward, seen as hot fingers in the lower part of the insets depicted in Figure 2.1. As time proceeds during pressure holding, the same figure shows that natural convection produces a stable temperature and density stratification in the pressure vessel. Traces of the stratification may remain even after pressure release (Rauh, 2008).

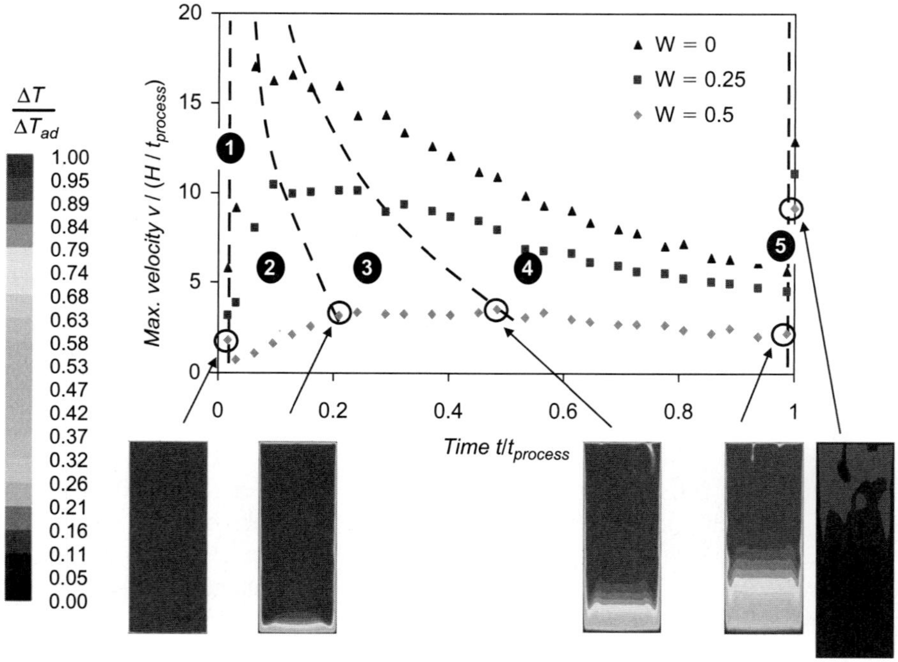

FIGURE 2.1

Maximum velocity in a cylinder-piston system as a function of time during pressure treatment. Triangles, squares, and diamonds indicate the use of water and water-saccharose solutions with a mass fraction of 0, 0.25, and 0.5, respectively. For selected time instances, the distribution of the temperature increase in the autoclave is visualized in the insets underneath the diagram. (Rauh, 2008.)

While the cold container bottom yields a stable stratified configuration right from the start, the unstable density stratification at the cool top of the container may result in a Rayleigh-Bénard instability (Chen *et al.*, 2007). In the third and fourth insets of Figure 2.1, the instability is observed as cold liquid draining down from the top into the bulk.

Besides natural convection during pressure holding, forced convection may also arise during the pressure ramp. In systems where pressurizing is achieved by injection of pressure-transmitting liquid through a tube into the vessel, the incoming liquid is usually colder than that in the vessel (de Heij *et al.*, 2003). Because of higher contact area, the heat transfer between inflowing liquid and piping is more efficient than in the autoclave. Here, a free jet enters the vessel with high flow velocities and induced thermal heterogeneities can be large. Hartmann and Hartmann *et al.* showed the strong effect of the forced convection on the temperature and on the velocity field in the vessel (Hartmann, 2002; Kilimann *et al.*, 2005; Hartmann *et al.*, 2004). Recently, Song *et al.* (2009a, b) gave evidence of turbulent inflow and relaminarization in liquids with sufficiently strong pressure-dependence of the viscosity (Song *et al.*, 2009a, b).

Forced and natural convection enhance heat transfer through the container walls and a temperature decrease in the bulk. As shown in Figure 2.1, the intensity of the convection strongly depends on the viscosity of the pressure-transmitting liquid, which increases with the saccharose concentration. Rauh *et al.* (2009) showed that natural convection could be suppressed in pressure-transmitting liquids of sufficiently high viscosity (Rauh *et al.*, 2009). The effect of viscosity has been further studied in detail by Hartmann *et al.* (2004), Baars *et al.* (2007), and by Hartmann and Delgado (2002). Kulisiewicz and Delgado (2010) provided a review of available measurement methods for the determination of viscosity under high pressures.

To validate the numerical results, Delgado and co-workers also studied the temperature field experimentally. They showed that numerical and experimental results coincide within measurement uncertainty. Detailed information about the two-dimensional temperature and velocity fields inside the autoclave has been obtained by using liquid crystals as temperature indicators and applying high-pressure particle image velocimetry and thermography (Pehl and Delgado, 1999; Pehl *et al.*, 2000; Özmutlu *et al.*, 2006).

The aforementioned heterogeneous temperature field in the pressure autoclave and its temporal evolution may significantly affect the inactivation process of microorganisms and enzymes. For enzymes diluted in pressure-transmitting media, Rauh *et al.* (2009) have numerically studied inactivation and its heterogeneity in the presence of the spatiotemporal temperature field. For this purpose they modeled the inactivation kinetics of several enzymes in water as a first-order reaction

$$\frac{\partial A}{\partial t} + u \cdot \nabla A = -k_p(1 + k_T)A, \tag{2.6}$$

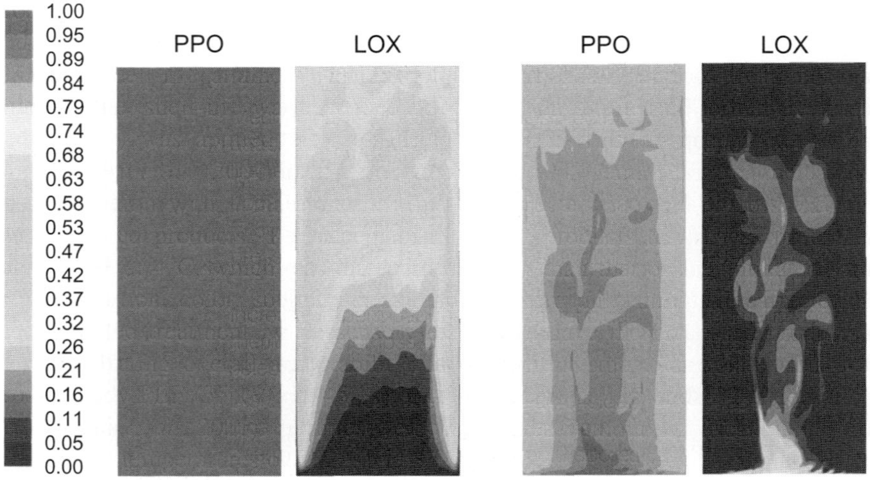

FIGURE 2.2

Activity of PPO and LOX at the end of the pressure-holding phase. Starting temperature on the left: 293.15 K, on the right: 323.15 K. (Hartmann and Delgado, 2005.)

where A and u indicate the enzyme activity and the local velocity of the liquid medium, respectively. As shown, they split the inactivation rate into a pressure-sensitive one for given starting temperature, k_p, and a temperature-sensitive one at high static pressure, k_T (Baars *et al.*, 2007).

Showcases of different outcomes for the enzyme activity at the end of the pressure-holding phase are shown in Figure 2.2. At a starting temperature of 293.15 K, the activity of polyphenoloxidase from avocado, modeled according to Weemaes *et al.* (1998), is rather pressure insensitive. Lipoxygenase (Denys *et al.*, 2000), on the other hand, reacts sensitively to the high-pressure treatment. Due to the temperature sensitivity, the inactivation is heterogeneous, roughly reflecting the footprint of the temperature stratification in the autoclave. In this particular case, the inactivation process is more efficient at lower temperatures.

At higher temperatures, the pressure sensitivities of the inactivation of both enzymes converge and thermal sensitivity rises (Rauh *et al.*, 2009; Weemaes *et al.*, 1998; Denys *et al.*, 2000). As shown for a starting temperature of 323.15 K in Figure 2.2, both enzymes qualitatively depict the same activity heterogeneities. Strong deviations between activity and temperature heterogeneities are apparent. Using numerical particle tracking, Kitsubun (2006) has shown that the discrepancies between activity and temperature heterogeneities are due to the convective transport of the enzymes during the pressure treatment: enzymes that are in a warm region at the end of the holding phase may have been in rather cold regions for most of the time, with different inactivation sensitivities. As a consequence, the enzymes experience varying mechanical and thermal histories, which originate different inactivation. Kitsubun found variations in the residual activity of *bacillus subtilis* α-amylase in the range of 0.59 to 0.72. This example underlines that the actual state depends on the entire treatment history of the process as outlined by Delgado *et al.* (2008).

Several authors have studied the role of heat loss through the container walls (de Heij *et al.*, 2002; Otero *et al.*, 2002; Rauh *et al.*, 2009; Hartmann *et al.*, 2004). Hartmann *et al.* (2004) showed that overall inactivation rates of *bacillus subtillis* α-amylase increase with thermal insulation. They point out that this is due to higher mean temperature in the vessel, which favors inactivation. Inhibiting heat losses through the container walls also reduces effectively thermal and thus inactivation inhomogeneities, as reported by Hartmann *et al.* (2004) and de Heij *et al.* (2002). Rauh *et al.* (2009) studied the impact of preheating the container walls to the temperature of the pressure-transmitting medium after compression and compared it to the effect of a higher starting temperature and of the suppression of convection by using a highly viscous liquid. With the preheated wall they found a much higher degree of uniformity for the temperature field than with the other methods. Preheating the walls also had a considerable impact on enzyme inactivation: for all the enzymes studied, i.e. polyphenoloxidase, lipoxygenase, β-glucanase, and *bacillus subtilis* α-amylase, they obtained the highest inactivation uniformity and the lowest overall enzyme activity.

The size of the autoclaves also may have a significant effect on the temperature distribution and hence on inactivation (Hartmann and Delgado, 2002; Kitsubun, 2006; Hartmann and Delgado, 2003). Hartmann and Delgado (2002, 2003) showed that overall inactivation rates of *bacillus subtillis* α-amylase increase with container size. They point out that this is due to the reduced effect of heat losses and higher mean temperature in the vessel, analogous to thermal insulation of the container walls. Kitsubun (2006) focused on the role of autoclave size on inactivation heterogeneity. He found that, in small vessels, effective heat transfer through the container walls reduces thermal and, thereby, activity heterogeneities. On the other extreme, in large chambers the heat transfer through the walls is inefficient for the large bulk volume, which results again in small heterogeneities even without turbulence mixing. Hence, maximum heterogeneities due to thermal effects occur in mid-size containers (Kitsubun, 2006).

The effect of packaging on thermal heterogeneities and its impact on inactivation has been studied by Hartmann and Delgado (2005), Hartmann (2002), and Hartmann and Delgado (2003). Besides serving as a mechanical barrier to prevent mixing and large-scale convection, they find, in line with the effect of the container walls, that the package material serves as a heat barrier. Thus, elevated temperature levels are maintained through a longer part of the pressure treatment in better-insulating packaging material, resulting in lower inactivation heterogeneity (Hartmann and Delgado, 2003, 2005). Nevertheless, Hartmann *et al.* reported variations in inactivation efficiency of *E. coli* suspended in milk in pouches depending on the relative position of the packaging inside the autoclave (Hartmann, 2002); see Figure 2.3.

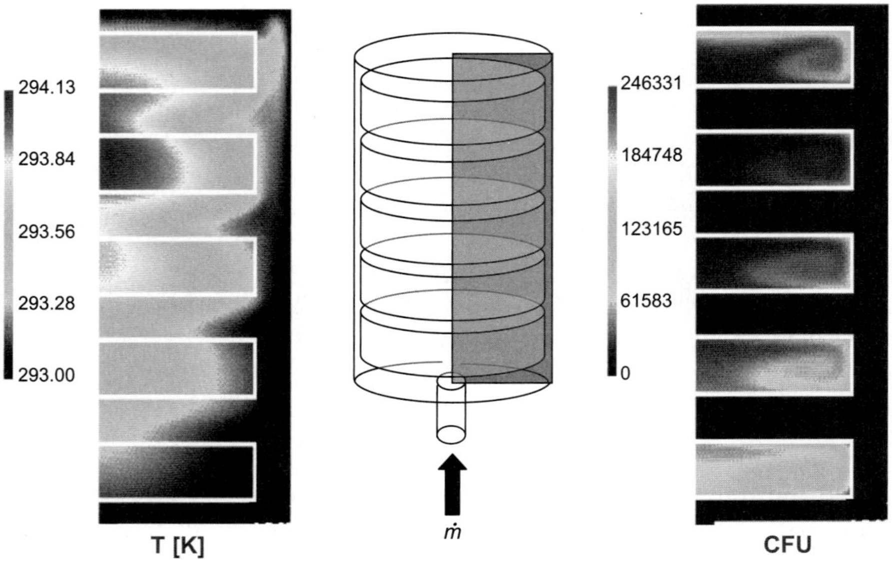

FIGURE 2.3

Temperature (left) and activity distribution of colony forming units of *E. coli* (right) in packaged milk. (Denys *et al.*, 2000.)

Inhomogeneous inactivation is not restricted to liquids. Due to different compressibility of the materials, the temperature rise during compression alters with the material distribution inside the vessel. Under these conditions, heat conduction inside solids or gels may also result in heterogeneous inactivation. Several authors have studied the temperature distribution in solid biomatter (Sanz *et al.*, 1997; Chen *et al.*, 2007; Denys *et al.*, 2000, 1999; Carroll *et al.*, 2003; Otero *et al.*, 2006). Denys *et al.* studied the inactivation of *bacillus subtilis* α-amylase immobilized in agar gel (Denys *et al.*, 1999). In the gel matrix, the heterogeneities are due to heat conduction. Similar results were obtained by the same group for heat conduction problems in other materials (Denys *et al.*, 2000). Taking into account the temperature distribution and convection in the pressure-transmitting medium, Otero *et al.* (2006) reported considerable thermal heterogeneities in solid foods (Carroll *et al.*, 2003). Besides heat conduction, solid food structure may also shelter its interior from the high pressure applied in the autoclave. Minerich and Labuza (2003) reported on indications of a pressure drop of about 9 MPa in ham during high-pressure processing, which could result in under-processing the interior of the product during inactivation (Otero *et al.*, 2006).

Process time-scale is another parameter that affects inactivation heterogeneity of microorganisms and enzymes. Quasi-isothermal processing and thus improved uniformity could be achieved in principle in processes that are either much slower or faster than the heat-transfer processes involved. Slower processes could be achieved, for instance, with slow compression rates (<0.1 MPa/s). However, in this case process time would exceed the time required for industrial throughput (Otero *et al.*, 2003; de Heij *et al.*, 2003). On the other extreme, with pressure ramps of the order of 60–600 MPa/s (Ardia *et al.*, 2004, Ardia, 2004, Valdramidis *et al.*, 2007), almost adiabatic conditions can be achieved with fast pressure ramp and release periods in short-time high-pressure processes (Rauh *et al.*, 2009). For these processes, Ardia *et al.* reported enhanced microbial inactivation and product quality compared to conventional high-pressure processes (Minerich and Labuza, 2003).

In this respect, information on the influence of time, temperature, and pressure on the microbial or enzymatic inactivation given in the form of a kinetic model is not sufficient for process-engineering purposes. In order to investigate the impact of treatment in a particular technical application, it is necessary to take into account the macroscopic aspects of the process, such as the heat and mass transport effects, size and form of the pressure vessel, product packaging, and the specific boundary and initial conditions. Inevitable temperature inhomogeneities and convective motion accompanying the pressurization result in an inhomogeneous treatment of particular molecules embedded in liquid or solid product. As a result, the global impact of pressure on the biochemical reactions depends on the entire treatment history of all individual molecules.

2.4 PULSED ELECTRIC FIELDS

Food treatment by means of PEF consists in application of high-intensity electric fields in pulsed mode. Most phenomena and mechanisms connected to PEF are elucidated by Barsotti *et al.* (1999), Barbosa-Cánovas and Altunakar (2006), Álvarez *et al.* (2006), Ravishankar *et al.* (2008), Jeyamkondan *et al.* (1999), Teopfl *et al.* (2006), and Toepfl *et al.* (2007). The field intensity is in the range 15–50 kV/cm and the pulse length is of the order of microseconds (Ravishankar *et al.*, 2008). The intensive electric field generates a potential difference across cell membranes high enough to cause the membranes to break down (Jeyamkondan *et al.*, 1999). Thus, as a result of cell permeabilization, the microorganisms are irreversibly inactivated (Teopfl *et al.*, 2006). PEF treatment is carried out in liquid media, usually in continuous mode, leading to an inherent coupling of fluid-dynamical effects with the microbial inactivation. Action of various fields on the fluid system inevitably leads to inhomogeneities and PEF treatment, being classified as non-thermal technology, can not be treated as an isothermal process.

The energy of the electrical pulses required for the permeabilization of the biological cells alters the level of the thermal energy in the treatment chamber due to ohmic heating. Thus, PEF can fulfill isothermal conditions only approximately for vanishing electrical energy and very high values of thermal diffusivity. However, permeabilization takes place only if a certain level of the electrical energy is exceeded. Zimmermann *et al.* (1974) prove the critical trans–membrane electrical potential required for the breakdown of the cell to be almost 1 V. Furthermore, isothermal processing leads to large treatment time and to a dissipation of energy to a large extent. Obviously, both consequences of isothermal conditions are in crucial contradiction to the goals of modern food technologies. Thus, under practical conditions, PEF always generates thermal fields that interfere with the electromagnetic field used for cell permeabilization and the flow field that serves as a basis of continuous transport of food capable of flowing. Hereby, any thermal heterogeneity produces density differences that can induce a natural convective flow in foods consisting of pure liquids or dispersed fluid systems. Natural convection occurs particularly in the presence of gradients perpendicular to the direction of gravity. In this sense, PEF does not act completely homogeneously, even when assuming materially homogeneous food.

The local interplay of the heating of the fluid due to the electric field and the heat transfer due to heat convection and conduction is of special importance in food treatment. In order to profit from the advantages of PEF as a non-thermal technology, the temperature at every location in the chamber should remain low enough not to damage the valuable nutritive and sensorial qualities. Different effects can arise: when the temperature locally increases too strongly due to field heterogeneities, the quality of the food (nutritional and functional properties) may suffer. If the temperature in the treatment region is low enough, the efficiency of the PEF process may still depend on temperature (Heinz *et al.*, 2001; Toepfl *et al.*, 2007). Hence, a detailed knowledge of temperature as well as field strength distributions in the chamber is necessary for an efficient application of PEF.

Analysis of inhomogeneous fields demands adequate experimental, theoretical and numerical methods. The experimental investigations presented in literature revealed that the electric field strength and treatment time are the major process parameters affecting the perforation efficiency as well as the inactivation rate of microorganisms and enzymes. The secondary influencing parameters are the characteristics of the PEF treatment chamber, the amplitude, frequency, and shape of the electric pulse, the material consistency, structural conformation, and viability of the biomatter, and the electrical as well thermo-fluid-dynamical properties of the liquid food (Barbosa-Cánovas and Altunakar, 2006; Jeyamkondan *et al.*, 1999). Owing to the complex interaction of the PEF process with the medium and spoiling agents, numerous experiments have been necessary to quantify the inactivation performance of PEF process for different foods, microorganisms and enzymes. Álvarez *et al.* (2006) and Mañas and Vercet (2006) summarized the inactivation measurements available in the literature for different microorganisms and enzymes, respectively.

The electrohydrodynamic phenomena occurring in the treatment chamber of PEF systems can, in general, be modeled by applying the conservation equations of mass, momentum, energy (see Equations 2.1–2.4), and additional conservation equations, charge (Gerlach *et al.*, 2008; Krauss *et al.*, 2010). The overlapping effects of momentum and thermal energy transport in the fluid flow, as well as the electrical field, makes pure mathematical analytical treatment almost impossible. However, numerical methods as used in computational fluid-dynamics (CFD) deliver the tools capable of simulating the mentioned fields involved in PEF treatment of food capable of flowing. Numerical methods represent also an adequate and, at the same time, powerful basis for optimizing as well as for scaling up PEF processes, as required in industrial applications. In this context the optimization of the processes in the strong interrelated electro-fluid-dynamical and thermal fields particularly requires the optimization of the initial and boundary conditions realized in the treatment chamber (Gerlach *et al.*, 2008).

The number of publications using numerical tools in this field has been rather limited until now. Toepfl *et al.* (2007), Qin *et al.* (1995), Misaki *et al.* (1982) and Góngora-Nieto *et al.* (2003) focus on the simulation of the electric field in the treatment chamber without considering the flow field and the temperature distribution. The objectives of such simulations are to improve the uniformity of the electric field in the treatment region in order to avoid over-processing or the possibility of dielectric breakdowns due to field intensity peaks. However, the electric field strength must exceed a critical value over the cross-sectional area of the chamber (Zimmermann *et al.*, 1974; Heinz *et al.*, 2001) to provide a sufficient treatment of all fluid volumes. Non-uniform distributions of the field strength can be caused by the geometry of the chamber composed by insulators and electrodes (Qin *et al.*, 1995; Misaki *et al.*, 1982) or by impurities in the treatment medium such as air bubbles or other dielectric materials like fat globules (Toepfl *et al.*, 2007; Góngora-Nieto *et al.*, 2003). A more complemented approach is a numerical investigation solving the coupled problem including the electric field, the fluid flow and the ohmic heating of the fluid (Gerlach *et al.*, 2008; Krauss *et al.*, 2010; Fiala *et al.*, 2001; Lindgren *et al.*, 2002).

Gerlach *et al.* (2008) provided a complete set of governing equations sufficient for modeling of PEF treatment in fluids treated as Newtonian media. He proposes using the set of Equations 2.1–2.4 simplified by considering the flow as incompressible, which allows the effect of pressure to be neglected. The correct formulation of energy conservation must take into consideration the effect of the electric field strength on the temperature distribution. Gerlach *et al.* (2008) expressed the transformation of the electrical energy to the thermal energy (joule heating) by taking into account the term Q in Equation 2.4, as

$$Q = \zeta f \sigma E^2, \tag{2.7}$$

where E is the electric field strength and σ the temperature-dependent electrical conductivity. Since PEF processes employ very short pulse widths ζ with the frequency f, so the unsteady joule heating can be modeled as a steady source term. Consequently ζ and f appear as coefficients in front of the joule heating term (7) (Gerlach *et al.*, 2008). In the literature, it is generally assumed that the PEF does not induce unsteady magnetic field, i.e. $\nabla \times E = 0$. Hence, the electrical field strength vector becomes the (negative) gradient of the electrical potential Ψ (Gerlach *et al.*, 2008):

$$E = -\nabla\Psi. \tag{2.8}$$

As a result of charge conservation, the electrical potential can be derived from

$$\nabla J = \nabla[\sigma\nabla\Psi] = 0, \tag{2.9}$$

where J represents the current density. Hence, modeling results in the system of non-linear partial differential Equations 2.1–2.4, 2.7–2.9 that describe the electro-fluid-dynamical as well as thermal fields taking place in a PEF chamber. It is worth mentioning, that these equations are strongly interrelated. In principle, the conservation equations for mass (Equation 2.1) and momentum (Equations 2.2, 2.3) provide a sufficient base for determining both the velocity field and the pressure. However, the temperature dependence of the material properties requires the simultaneous solution of the energy equation. In turn, the energy equation is linked to the model for the conservation of electrical charge (Equation 2.9) via Equation 2.7. An additional link to the other conservation laws represents the appearance of the temperature-dependent electrical conductivity σ in the charge balance (Equation 2.9).

In principle, dealing with Equations 2.4, 2.7, and 2.9 can contribute, at least to a certain level, to the design of treatment chambers that offer improved homogeneity of the electrical field. Particularly it requires that the minimum transmembrane potential is achieved in any point of the chamber for ensuring adequate homogeneous treatment. However, field intensity peaks must be avoided as they might cause over-treatment or even dielectric breakdowns in the medium (Toepfl *et al.*, 2007; Qin *et al.*, 1995; Góngora-Nieto *et al.*, 2003). This offers unique possibilities of exploiting the advantages of PEF as a non-thermal technology by monitoring and optimizing the whole "treatment history" (Rauh *et al.*, 2009) at every location in the treatment chamber.

Concerning the mathematical description of the PEF-induced microbial inactivation itself, two different modeling approaches are available in the literature. The most classical one treats inactivation of microorganisms or enzymes within the pure framework of kinetics. This corresponds to an inactivation purely dependent on the time for a given treatment procedure and food. In contrast to this, a novel modeling of inactivation (Delgado *et al.*, 2008; Delgado and Hartmann, 2002) which is becoming more and more accepted, balances the residual activity A or survival rate of the biotic system (e.g., microorganisms) as other conservative quantities discussed above. As shown by Delgado *et al.* (2008) for the first time, this is possible as A represents a conversion of mass which strictly obeys the partial continuum equation. This enables inactivation to be dealt with numerically in a similar manner to an arbitrary scalar quantity.

However, almost all investigations reported in the literature are based on the pure kinetic approach. The inactivation performance of the PEF process for a certain spoiling agent is measured after the specific food is exposed to field strength of E for a treatment time of $t^* = t\zeta f$ with t being the real time, ζ the pulse width and f the frequency. Hence, A can be calculated by

$$A = 100 \cdot \frac{A^*}{A_0^*}, \tag{2.10}$$

where A^* and A_0^* are the activities in the sample after and before the treatment, respectively. A is commonly modeled in two steps (Álvarez *et al.*, 2006). The primary model depicts the effect of the dominant process parameters, E and t and the secondary model formulates the influence of the characteristics of the medium and the enzyme. Based on the trends of the measurements, the following primary model of Hülsheger *et al.* (1981) found the most acceptance:

$$A = \left(\frac{t^*}{t_c^*}\right)^{\left(\frac{E-E_c}{a}\right)}, \tag{2.11}$$

where E_c and t_c^* represent the critical electric field strength and critical treatment time, respectively, above which the inactivation starts, and a is a constant factor characteristic for each microorganism or enzyme. Accordingly, it is clear from this equation that electric field strength is the most dominant factor in the inactivation process, followed by the treatment time. The secondary model formulates the relations for E_c, t_c^*, and a as functions of medium properties (e.g., temperature, pH, conductivity), pulse characteristics, and type of microorganism or enzyme (Álvarez *et al.* 2006).

It is worth mentioning that Equation 2.11 reflects only pure time dependency. However, as liquid food volumes with different values of A are transported in the chamber by forced and natural convection through regions of different electrical field strengths and temperatures, the novel transport model mentioned above must be employed for adequately describing the "treatment history". It reads

$$\frac{\partial A}{\partial t} + (u\nabla)A = \nabla(D\nabla A) + S, \tag{2.12}$$

with the diffusion coefficient, D. The balance equation (2.12) expresses that the temporal change of the residual activity depends on the convective and diffusive flow through the surface of the balance volume. However, additionally, a source term S is available which represents the case of an inactivation as a negative source term. Generally, it depends on the temperature T, the electrical field strength E and the treatment time t^*, and, if necessary, also on other influencing factors. The diffusion term on the right-hand side is neglected, because of its relatively small influence in comparison to the other terms present in the equation (Delgado *et al.*, 2008).

Different empirical models for the source term in the transport equation of A are suggested in the literature. For example, Giner *et al.* (2005) propose the following first-order fractional model of A based on the measurements conducted for pectinesterase:

$$A = A_0 \cdot \exp((4.2 \cdot 10^{-3} - 266 \cdot 10^{-6} E) \cdot t^*). \tag{2.13}$$

This model is a function of the treatment time t^* as well as the electric field strength E. Hence, the source term is simply the time derivative of the empirical A model (Equation 2.13):

$$\frac{dA}{dt} = S = A_0 \cdot (4.2 \cdot 10^{-3} - 266 \cdot 10^{-6} E) \cdot \zeta f \cdot A. \tag{2.14}$$

Equations 2.13 and 2.14 are empirical formulas based on experimental data provided at temperatures at the exit of a PEF chamber up to 313.15 K (Giner *et al.*, 2005). Thus, the validity of the equations is limited only to that temperature. The transport equation of A becomes

$$\frac{\partial A}{\partial t} + (u\nabla)A = S. \tag{2.15}$$

Hence, a model equation is proposed having the same form as the inactivation equation (2.6) developed by Hartmann *et al.* (Hartmann and Delgado, 2003; Hartmann *et al.*, 2003, Rauh *et al.*, 2009) for the HPP process. Equations 2.1, 2.2, 2.4–2.6, 2.14, and 2.15 form the coupled system of partial differential equations whose solutions with defined initial and boundary conditions and the constitutive parameters yield the description of the flow field, temperature field, electrical field, and the field of relative residual activity.

Based on the set of Equations 2.1, 2.2, 2.4–2.6, 2.14, and 2.15, Gerlach *et al.* (2008) and, more recently, Krauss *et al.* (2010), provided numerical solutions by means of the finite volume method. Thus, the complete electro-hydrodynamic fields could be studied in every location of the PEF chamber. Krauss *et al.* (2010) coupled the inactivation field of pectinesterase to the electro-hydrodynamic field and simulated the complete PEF process, according to Equation 2.15 based on the empirical model (Equation 2.13).

Various geometries of PEF treatment chambers were proposed including, for example, parallel plates and coaxial cylinders (Ravishankar *et al.*, 2008). Gerlach *et al.* (2008) and Krauss *et al.* (2010) studied a co-linear treatment chamber as used in Toepfl *et al.* (2007). A schematic sketch of this axisymmetric geometry is given in Figure 2.4. The chamber comprises a high-voltage electrode, a ground

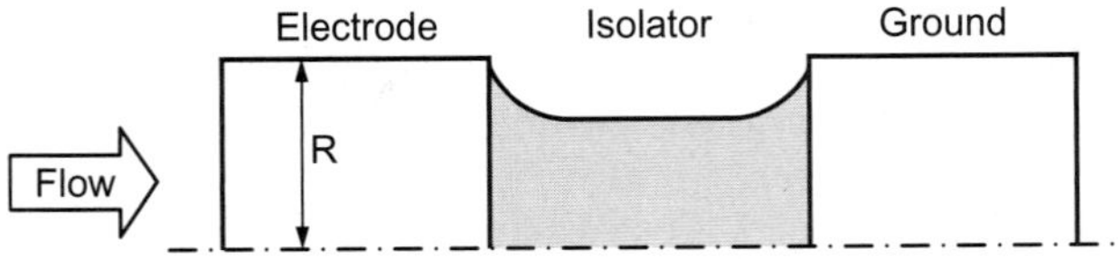

FIGURE 2.4

Half cross-section of a co-linear PEF chamber. (de Heij *et al.*, 2002; Delgado *et al.*, 2008.)

electrode, and an isolator. The flow enters into the chamber at the left side (inflow). The inner radius of the electrodes is R = 3 mm.

For making use of the benefits of continuous treatment technologies, food capable of flowing is forced through the treatment chamber. In general, this occurs by a pressure gradient generated by a flow machine such as a pump. As a consequence of this, overlapping electrical and forced convection fields, i.e. electro-fluid-dynamical effects appear.

Results of a simulation for the electro-hydrodynamic field are presented in Figure 2.5 (Krauss *et al.*, 2010). A long stable separation region behind the isolator is an important feature of the flow in the co-linear PEF chamber. This separation region is the main reason for the accumulation of heated fluid behind the isolator. Similar phenomena should be expected for any scalar variable like rest activity A. It is interesting to observe that the E field in Figure 2.5b is not symmetric on the isolator. This is mainly due to the temperature-dependent material properties, foremost the electrical conductivity σ.

Usually, the treatment time for a fluid particle exposed to the E field as it moves through the PEF chamber, is too short to provide sufficient microbial or

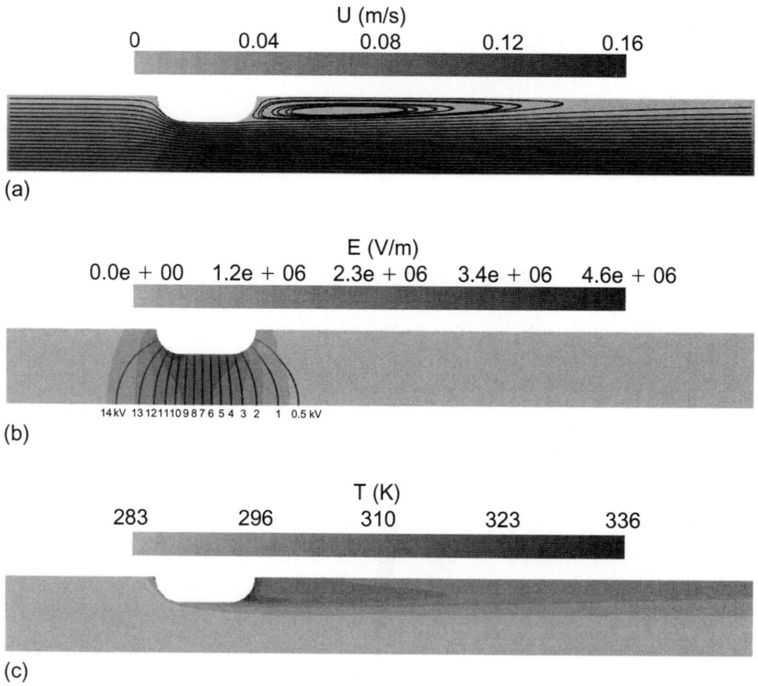

FIGURE 2.5

Electro-hydrodynamic fields in the co-linear PEF chamber for $f = 20$ Hz, $\zeta = 10\ \mu s$, $\Psi = 15$ kV and $\dot{m} = 5.2\ kg/h$: (a) velocity field and streamline, (b) electric field strength and electric potential, (c) temperature field. (Otero *et al.*, 2002.)

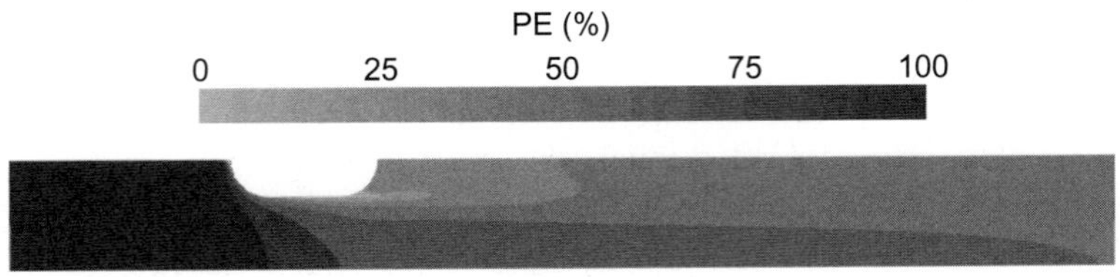

FIGURE 2.6

Residual activity field in the PEF chamber for $f = 20$ Hz, $\zeta = 10\ \mu s$, $\Psi = 15$ kV, $\dot{m} = 5.2\ kg/h$ and $n = 25$ (equivalent overall treatment time of 192 μs). (Delgado *et al.*, 2007.)

enzymatic inactivation effects. Therefore, a number of PEF chambers are serially connected to reach a reasonable reduction in activity A and the fluid is cooled to prevent excessive fluid temperatures between each PEF chamber (Bendicho *et al.*, 2003). Giner *et al.* (2005) readily utilized this method in the experiments. Obviously, the number of necessary treatment chambers can be reduced by increasing ζf, but with the cost of increasing detrimental effects of excessive temperature. Reduction of A along serially connected chambers can be approximated by modifying the computational treatment time to involve the number of chambers, i.e. the treatment time t^* is redefined as $t^* = t\zeta fn$, where n is the number of treatment chambers (Krauss *et al.*, 2010).

The simulated field of enzymatic activity A equivalent to 25 co-linear PEF chambers in series, corresponding to the electro-hydrodynamic fields in Figure 2.5 is shown in Figure 2.6. As can be seen from the E field (Fig. 2.5b) and A field (Fig. 2.6), the highest decay of residual activity occurs as the fluid approaches the isolator and is exposed to E for the first time.

Results of the numerical simulations (Gerlach *et al.*, 2008; Krauss *et al.*, 2010) prove that the effects of PEF treatment, such as the residual activity, average temperature in the isolator volume, and maximum temperature in whole PEF chamber, are strong functions of the chamber geometry and the process parameters, such as flow rate, frequency, pulse width, and electric potential. Those observations reveal the necessity of optimization of the geometry and process parameters in order to find the balance between the minimization of the residual activity A and minimization of the excessive temperatures. The considerations can be extended with additional activity models of pathogenic/spoilage microorganisms and enzymes, and useful microorganisms and enzymes, which would enable optimization of the PEF processing for multi-component mixtures existing in the liquid food. Implementation of temperature constraints in the optimization procedure allows elimination of designs generating excessive local temperatures which can be detrimental to valuable components of a specific food. Krauss *et al.* (2010) provide automatic numerical optimization algorithms of PEF chamber geometry and the PEF process having multiple objectives and constraints.

2.5 ULTRASOUND

The term "ultrasound" refers to the pressure waves at frequencies above 20 kHz. The power ultrasound applied in food treatment comprises sound waves with frequencies between 20 and 100 kHz and sound intensity of 10 to 1000 W/cm^2 (Feng *et al.*, 2008; Piyasena *et al.*, 2003). The use of ultrasound is often combined with other treatments at moderate intensities, e.g., temperature, pressure and chemical treatment (Raso *et al.*, 1998; Raso and Barbosa-Canovas, 2003).

However, the applications of ultrasound in the food industry involve not only microbial and enzyme inactivation, but also homogenization of multi-phase fluids, emulsification, heat and mass transfer enhancement, particularly in membrane technologies, crystallization enhancement, surface cleaning, etc. The interaction of ultrasound with biomatter relies on the micro-fluid-dynamical effects caused by the traveling acoustic wave with subsequent regions of compression and expansion (Sala *et al.*, 1995). In the local regions of lower pressure, transient gas bubbles appear, fed by the vapor and gases diluted in the liquid. The bubbles last for several acoustic cycles. At a certain critical stage, the ultrasonic wave pressure exceeds the void pressure, causing the bubbles to collapse violently. The implosion of gas bubbles creates both fluid-mechanical and chemical effects in the liquid. The fluid-mechanical effects include the formation of shock waves with local regions of high pressure and temperature, reaching up to 50 MPa and 5000 K (Feng *et al.*, 2008; Piyasena *et al.*, 2003) with high rates of temperature change up to 10^9 K/s (Feng *et al.*, 2008) and liquid micro-jets with local velocities exceeding 150 m/s. The chemical effects include the formation of free radicals and hydrogen peroxide, having bactericidal properties. Thus, the impact of power ultrasound on microorganisms and enzymes is a combination of sonochemical reactions and mechanical stresses caused by fluid-dynamical phenomena (Feng *et al.*, 2008).

The spatiotemporal effects of the ultrasound treatment on the fluid foods treated as a continuum can be analysed in terms of the basic set of governing equations discussed already in previous sections. In this case, the term Q in Equation 2.4 describing the heat generation in acoustic medium takes the form

$$Q = \nabla \cdot I,$$

where I is the acoustic intensity (Zeqiri *et al.*, 2007). Thus, it becomes evident that the acoustic treatment involves certain inhomogeneities in the distribution of the lethal acoustic pressure in combination with temperature. As shown by Gelat *et al.* (2005), by means of numerical simulation based on the finite element method, the acoustic pressure in a sonication chamber is strongly inhomogeneous. Furthermore, a strong dependence of acoustic pressure distribution on the temperature of the medium must be taken into account. Changes in fluid temperature as small as 2 K would cause the acoustic pressure distribution to change significantly, with regard to the form and location of intensity peaks (Hodnett and Zequiri, 2008). Additionally, the cavitation threshold (i.e. the minimum pressure oscillation necessary to produce cavitation) is a function of temperature, making the interrelation of temperature distribution and bio-inactivation intensity apparent.

It is clear that the inhomogeneities in the distribution of the lethal agent (in this case acoustic intensity) in the food-treatment vessel also cause inhomogeneities in the residual activity A of the biological system, such as microorganisms or enzymes. Analogically to the food-treatment technologies discussed above, the ultrasound treatment must be also considered with regard to transport of A in the chamber by forced and natural convection through regions of different acoustic intensities and temperatures. Thus, adequate description of the overall inactivation effect of ultrasound on biotic matter requires a transport model in the form of Equations 2.6 and 2.15.

However, a crucial difference must be taken into consideration: the micro-fluid-dynamical effects in connection to cavitation gas bubbles depend strongly on the availability of "nucleation centers", at which the bubbles are preferably generated. In a wide spectrum of food, a large number of possible nucleation centers such as non-dissolved particles, microbubbles, and partial solidification fronts are available. However, the nucleation centers must be expected to be stochastically distributed. Thus, this substantially increases the complexity of modeling and simulating the sonochemical and micro-fluid-mechanical effects. Consequently, all nucleation centers must be considered as boundary conditions which are stochastically distributed. No more than probabilistic-based statements on the local occurrence of cavitation bubbles which, subsequently, implode releasing a large amount of energy and momentum, are possible. From the literature, it must be concluded that there is a large gap in knowledge regarding the basic understanding on the effect of ultrasound on food capable of flowing. In future, substantial efforts must be undertaken to close this gap and, therefore, to create a suitable base for exploiting the obvious high potential of ultrasound treatment of food.

2.6 LIGHT PULSES

Pulsed-light systems used for food decontamination involve broadband spectrum light from ultraviolet to infrared applied in several pulses per second, each pulse lasting from 100 ns to 2 ms (Demirci and Panico, 2008). Despite the high intensity of light radiation, short pulses prevent the product from overheating and still ensure inactivation of microorganisms due to both photochemical and photothermal effects (Elmnasser *et al.*, 2007). However, the bactericidal effect of pulsed light is limited by its low degree of penetration. It is well suited for inactivation of microorganisms on solid surfaces, but applicability in fluid foods is rather limited to transparent liquids. Still, the penetration depth is rather small, depending on the optical properties of the medium (Demirci and Panico, 2008). The experimental investigations of Krishnamurthy *et al.* (2004) on inactivation of

Staphylococus aureus in milk matrix and of Hillegras and Demirci (2003) on inactivation of *Clostridium sporogenes* in honey proved that the overall decontamination effect of pulsed-light treatment decreases with the sample depth. The bactericidal effect declines with the sample depth, due to optical opaqueness of the matrix fluid (Demirci and Panico, 2008). Thus, it is evident that the pulsed-light treatment leads to an activity gradient in the fluid food, with the microbial activity increasing in the direction from the surface. It is clear that temperature gradients must also appear with warmer liquid close to the light source. As a result, diffusive and convective heat and mass transfer emerge, driven by the buoyancy effects or surface tension differences (Marangoni convection), leading to homogenization of the sample with respect to temperature and microbial activity. However, the typical treatment time by means of pulsed light is in the order of several seconds (2008), which must be considered too short time for providing sufficient homogeneity. Hence, the treatment of liquid foods with light pulses appears to be of rather limited potential, particularly if the aim is to up-scale. Still, large efforts must be made in order to overcome these difficulties. This is of crucial importance for getting a better estimation regarding the possibility of realising a homogeneous treatment of food as a fundamental requirement to be fulfilled, not only for ensuring the required consistency and aroma profile of food, but also its safety. Furthermore, to extend the investigation of light-related processes to the treatment with ultraviolet, mathematical modeling and numerical analyses have to be mentioned, carried out by Blatchley (1997), Bolton (2000) and Lyn and Blatchley (2005), amongst others.

2.7 OHMIC HEATING

Ohmic heating as a novel thermal food-processing technology refers to the resistive dissipation of electrical energy in the conducting food matter being in contact with electrodes. Usually, in the case of continuous electrical resistance heating, electrical fields with intensities not exceeding 1 kV/cm are used (Sastry, 2008). In the absence of hot walls, ohmic heating provides an effective and rapid method for thermal pasteurization, blanching, dehydration, etc. However, in contrast to PEF, the temperature increase in the liquid is the principal goal of the process and the microbial and enzyme inactivation is mainly (but not exclusively) attributed to thermal effects (Sastry, 2008). The electrically induced inactivation appears to play a secondary role at these values of electrical-field intensity. Ohmic heating technology can be applied, not only for liquids, but also for multiphase liquid–solid mixtures, particularly in the case of media that would be difficult to process using conventional heat exchangers (Sastry and Qiong, 1993).

The principle ideas concerning physical and particularly fluid-dynamical analysis of ohmic heating processes are identical to those presented in the section on PEF and do not require particular discussion. The set of governing Equations 2.1–2.4, 2.7–2.9, is also valid for studying effects of ohmic heating in fluid foods by the application of electric fields in continuous (not pulsed) mode (see also Sastry and Qiong (1993)). Obviously, for continuous application of electrical field, the term ζf in Equation 2.7 becomes equal to 1. This approach was used recently by Salih *et al.* (2009) and Jaeger *et al.* (2009), for the assessment of velocity and temperature patterns in a co-linear ohmic heater, by Ghnimi *et al.* (2009) in a fluid jet, and by Jun and Sastry (2005) for studying ohmic heating of fluid food in microgravity conditions. Fillaudeau *et al.* (2006) analyzed the temperature distribution and fouling effects in a parallel-plate chamber.

In comparison to the use of ultrasound and light pulses, the literature connected to ohmic heating must be estimated as providing a better base for treating fluid-mechanical effects. However, a large spectrum of questions remains unanswered. This is especially due to the fact that, in general, particular processes, but no systematic investigations, considering ohmic heating from a holistic point of view are presented in the literature. Nevertheless, evaluation and exploitation of the potentials of ohmic heating will depend basically of the systematic knowledge available.

2.8 CONCLUSION

The state of the art concerning fluid dynamics in novel thermal and non-thermal technologies has achieved unequal levels. For example, substantial progress has been made regarding high-pressure treatment and, partly, regarding PEF, dielectric and ohmic heating technologies, microwave and infrared heating. In contrast to this, fluid-mechanical effects regarding food treatment by means of pulsed light and ultrasound are far from being sufficiently understood. Overcoming the difficulties corresponding to the gaps in the available knowledge represents one of the most challenging tasks in literature.

References

Álvarez, I., Condón, S., Raso, J., 2006. Microbial inactivation by pulsed electric Fields. In: Raso, J., Heinz, V. (Eds.), Pulsed Electric Fields Technology for the Food Industry: Fundamentals and Applications. Springer, New York.

Ardia A., 2004. Process considerations on the application of high pressure treatment at elevated temperature levels for food preservation. Ph.D. Thesis, TU Berlin.

Ardia A., Heinz V., Knorr D., 2004. Very short treatment times for high pressure processing: a new concept. In: Proceedings of the International Congress on Engineering and Food (ICEF9), Montpellier, France, 1–5.

Baars, A., Rauh, C., Delgado, A., 2007. High pressure rheology and the impact on process homogeneity. High Press. Res. 27, 77–83.

Balny, C., 2004. Pressure effects on weak interactions in biological systems. J. Phys. Condens. Matter 16, S1245–S1253.

Barbosa-Cánovas, G.V., Altunakar, B., 2006. Pulsed electric fields processing of food: an overview. In: Raso, J., Heinz, V. (Eds.), Pulsed Electric Fields Technology for the Food Industry: Fundamentals and Applications. Springer, New York.

Barsotti, L., Merle, P., Cheftel, J.C., 1999. Food processing by pulsed electric fields. Physical aspects. Food Rev. Int. 15, 163–180.

Bendicho, S., Barrosa-Cánovas, G.V., Martín, O., 2003. Reduction of protease activity in simulated milk ultrafiltrate by continuous flow high intensity pulsed electric field treatments. J. Food Sci. 68, 952–957.

Blatchley III, E.R., 1997. Numerical modelling of UV intensity: application to collimated-beam reactors and continuous-flow systems. Water Res. 31, 2205–2218.

Bolton, J.R., 2000. Calculation of ultraviolet fluence rate distributions in an annular reactor: significance of refraction and reflection". Water Res. 34, 3315–3325.

Carroll, T., Chen, P., Fletcher, A., 2003. A method to characterise heat transfer during high pressure processing. J. Food Eng. 60, 131–135.

Cheftel, J.C., 1995. Review: high pressure microbial inactivation and food preservation. Food Sci. Technol. Int. 1, 75–90.

Chen, C.R., Zhu, S.M., Ramaswamy, H.S., Marcotte, M., Le Bail, A., 2007. Computer simulation of high pressure cooling of pork. J. Food Eng. 79, 401–409.

de Heij, W.B.C., van Schepdael, L.J.M.M., van den Berg, R., Bartels, P., 2002. Increasing preservation efficiency and product quality through control of temperature distributions in high pressure applications. High Press. Res. 22, 653–657.

de Heij, W.B.C., van Schepdael, L.J.M.M., Moezelaar, R., Hoogland, H., Matser, A.M., van den Berg, R.W., 2003. High-pressure sterilization: maximizing the benefits of adiabatic heating. Food Technol. 57, 37–41.

Delgado, A., Baars, A., Kowalczyk, W., Benning, R., Kitsubun, P., 2007. Towards adaptive strategies for high pressure bioprocesses. High Press. Res. 27, 7–16.

Delgado, A., Hartmann, C., 2002. Pressure treatment of food: instantaneous but not homogeneous effect. In: Winter, R. (Ed.), Advances in High Pressure Bioscience and Biotechnology II. Springer, Heidelberg, pp. 459–464.

Delgado, A., Rauh, C., Kowalczyk, W., Baars, A., 2008. Review of modelling and simulation of high pressure treatment of materials of biological origin. Trends in Food Sci. Technol. 19, 329–336.

Demirci, A., Panico, L., 2008. Pulsed ultraviolet light. Food Sci. Technol. Int. 14, 443–446.

Denys, S., Ludikhuyze, L., van Loey, A.M., Hendrickx, M.E., 1999. Modelling conductive heat transfer and process uniformity during batch high-pressure processing of foods. In: Ludwig, H. (Ed.), Trends in High Pressure Bioscience and biotechnology. Springer, Berlin, pp. 381–384.

Denys, S., van Loey, A.M., Hendrickx, M.E., 2000. Modeling conductive heat transfer during high pressure thawing processes: determination of latent heat as a function of pressure. Biotechnol. Prog. 16, 447–455.

Denys, S., van Loey, A.N., Hendrickx, M.E., Tobback, P.P., 1997. Modeling heat transfer during high-pressure freezing and thawing. Biotechnol. Prog. 13, 416–423.

Denys, S., van Loey, A.M., Hendrickx, M.E., 2000. A modelling approach for evaluation process uniformity during batch high hydrostatic pressure processing: combination of a numerical heat transfer model and enzyme inactivation kinetics. Innovative Food Sci. Emerg. Technol. 1, 5–19.

Elmnasser, N., Guillou, S., Leroi, F., Orange, N., Bakhrouf, A., Federighi, M., 2007. Pulsed-light system as a novel food decontamination technology: a review. Can. J. Microbiol. 53, 813–821.

Feng, H., Yang, W., Hielscher, T., 2008. Power ultrasound. Food Sci. Technol. Int. 14, 433–436.

Ferziger, J.H., Peric, M., 2002. Computational Methods for Fluid Dynamics. Springer, Berlin.

Fiala, A., Wouters, P., Van den Bosch, E., Creyghton, Y., 2001. Coupled electrical–fluid model of pulsed electric field treatment in a model food system. Innovative Food Sci. Emerg. Technol. 2 (4), 229–238.

Fillaudeau, L., Winterton, P., Leuliet, J.C., 2006. Heat treatment of whole milk by the direct joule effect—Experimental and numerical approaches to fouling mechanisms. J. Dairy Sci. 89, 4475–4489.

Gelat, P.N., Zeqiri, B., Hodnett, M., 2005. A finite-element model of the aperture method for determining the effective radiating area of physiotherapy treatment heads. Ultrasonics 43, 321–330.

Gerlach, D., Alleborn, N., Baars, A., Delgado, A., Moritz, J., Knorr, D., 2008. Numerical simulations of pulsed electric fields for food preservation: a review. Innovative Food Sci. Emerg. Technol. 9, 408–417.

Ghani, A.G.A., Farid, M.M., 2007. Numerical simulation of solid–liquid food mixture in a high pressure processing unit using computational fluid dynamics. J. Food Eng. 80, 1031–1042.

Ghnimi, S., Zaid, I., Maingonnat, J.F., Delaplace, G., 2009. Axial temperature profile of ohmically heated fluid jet: analytical model and experimental validation. Chem. Eng. Sci. 64, 3188–3196.

Giner, J., Grouberman, P., Gimeno, V., Martín, O., 2005. Reduction of pectinesterase activity in a commercial enzyme preparation by pulsed electric fields: comparison of inactivation kinetic models. J. Sci. Food Agric. 85, 1613–1621.

Góngora-Nieto, M.M., Pedrow, P.D., Swanson, B.G., Barbosa-Cánovas, G.V., 2003. Impact of air bubbles in a dielectric liquid when subject to high field strengths. Innovative Food Sci. Emerg. Technol. 4, 57–67.

Guerrero-Beltrán, J.A., Barbosa-Cánovas, G.V., Swanson, B.G., 2005. High hydrostatic pressure processing of fruit and vegetable products. Food Rev. Int. 21, 411–425.

Hartmann, C., Delgado, A., Szymczyk, J., 2003. Convective and diffusive transport effects in a high pressure induced inactivation process of packed food. J. Food Eng. 59, 33–44.

Hartmann, C., 2002. Numerical simulation of thermodynamic and fluid-dynamic processes during the high-pressure treatment of liquid food systems. Innovative Food Sci. Emerg. Technol. 3, 11–18.

Hartmann, C., Delgado, A., 2002. Numerical simulation of convective and diffusive transport effects on a high-pressure-induced inactivation process. Biotechnol. Bioeng. 79, 94–104.

Hartmann, C., Delgado, A., 2003. The Influence of Transport Phenomena During High-Pressure Processing of Packed Food on the Uniformity of Enzyme Inactivation. Biotechnol. Bioeng. 82, 725–735.

Hartmann, C., Delgado, A., 2005. Numerical simulation of thermal and fluiddynamical transport effects on a high pressure induced inactivation. Simul. Model. Pract. Theory 13, 109–118.

Hartmann, C., Schuhholz, J.-P., Kitsubun, P., Chapleau, N., Le Bail, A., Delgado, A., 2004. Experimental and numerical analysis of the thermofluiddynamics in a high-pressure autoclave. Innovative Food Sci. Emerg. Technol. 5, 399–411.

Heinz, V., Álvarez, I., Angersbach, A., Knorr, D., 2001. Preservation of liquid foods by high intensity pulsed electric fields—Basic concepts for process design. Trends in Food Sci. Technol. 12, 103–111.

Hillegras, S.L., Demirci, A., 2003. Inactivation of Clostridium sporogenes in clover honey by pulsed UV-light treatment. CIGR J. AE Sci. Res. and Dev. V 03–009.

Hodnett, M., Zequiri, B., 2008. Toward a reference ultrasonic cavitation vessel. IEEE Trans. Ultraso. Ferrolelectrics Freq. Contr. 55, 1809–1822.

Hülsheger, H., Potel, J., Niemann, E.-G., 1981. Killing of bacteria with electric pulses of high field strength. Radiat. Environ. Biophys. 20, 53–65.

Jaeger, H., Meneses, N., Knorr, D., 2009. Impact of PEF treatment inhomogeneity such as electric field distribution, flow characteristics and temperature effects on the inactivation of E. coli and milk alkaline phosphatase. Innovative Food Sci. Emerg. Technol. 10, 470–480.

Jeyamkondan, S., Jayas, D.S., Holley, R.A., 1999. Pulsed electric field processing of foods: A review. J. Food Prot. 62 (9), 1088–1096.

Jun, S., Sastry, S., 2005. Modeling and optimization of ohmic heating of foods inside a flexible package. J. Food Process Eng. 28, 417–436.

Khurana, M., Karwe, M.V., 2009. Numerical prediction of temperature distribution and measurement of temperature in a high hydrostatic pressure food processor. Food Bioprocess Technol. 2, 279–290.

Kilimann, K.V., Kitsubun, P., Delgado, A., Gänzle, M.G., Chapleau, N., Le Bail, A., et al., 2005. Experimental and numerical study of heterogeneous pressure–temperature-induced lethal and sublethal injury of lactococcus lactis in a medium scale high-pressure autoclave. Biotechnol. Bioeng. 94, 655–666.

Kitsubun P., 2006. Numerical Investigation of Thermofluiddynamical Heterogeneities during High Pressure Treatment of Biotechnological Substances. Ph.D. Thesis, TU Munich

Knoerzer, K, Juliano, P, Gladman, S, Versteeg, C., Fryer, P.J., 2007. A computational model or temperature and sterility distributions on a pilot-scale high-pressure high-temperature process. AIChE J. 53, 2996–3010.

Knorr, D., Mathys, A., 2008. Ultrahochdrucktechnik für innovative Behandlungsverfahren von Lebensmitteln. Chemie Ing. Tech. 80, 1069–1080.

Kowalczyk, W., Delgado, A., 2007. Dimensional analysis of thermo-fluid-dynamics of high hydrostatic pressure processes with phase transition. Int. J. Heat and Mass Transf. 50, 3007–3018.

Kowalczyk, W., Hartmann, C., Delgado, A., 2004. Modelling and numerical simulation of convection driven high pressure induced phase changes. Int. J. Heat and Mass Transf. 47, 1079–1089.

Krauss, J., Ertunç, Ö, Rauh, C., Delgado, A., 2010. Novel, multi-objective optimization of pulsed electric field (PEF) treatment for liquid food treatment. In: Knoerzer, K. (Ed.), Multiphysics Simulation of Emerging Food Processing Technologies.

Krishnamurthy, K., Demirci, A., Irudayaraj, J., 2004. Inactivation of Staphylococcus aureus by pulsed UV-light treatment. J. Food Prot. 67, 1027–1030.

Kulisiewicz, L., Delgado, A., 2010. High-pressure rheological measurement techniques: a review. Appl. Rheol. 20, 13018.

Leistner, L., 1992. Food preservation by combined methods. Food Res. Int. 25, 151–158.

Lindgren, M., Aronsson, K., Galt, S., Ohlsson, T., 2002. Simulation of the temperature increase in pulsed electric field (PEF) continuous flow treatment chambers. Innovative Food Sci. Emerg. Technol. 3 (3), 233–245.

Ludikhuyze, L., Van Loey, A., Indrawati, Smout, C., Hendrickx, M., 2003. Effects of combined pressure and temperature on enzymes related to quality of fruits and vegetables: from kinetic information to process engineering aspects. Crit. Rev. Food Sci. Nutr. 43, 527–586.

Lyn, D.A., Blatchley III, E.R., 2005. Numerical computational fluid dynamics-based models of ultraviolet disinfection channels. J. Environ. Eng 131, 838–850.

Mañas, P., Vercet, A., 2006. Effect of pulsed electric fields on enzymes and food constituents. In: Raso, J., Heinz, V. (Eds.), Pulsed Electric Fields Technology for the Food Industry: Fundamentals and Applications. Springer, New York.

Minerich, P.L., Labuza, T.P., 2003. Development of a pressure indicator for high hydrostatic pressure processing of foods. Innovative Food Sci. Emerg. Technol. 4, 235–243.

Misaki, T., Tsuboi, H., Itaka, K., Hara, T., 1982. Computation of three-dimensional electric field problems by a surface charge method and its application to optimum insulator design. IEEE Trans. Power Ap. Syst. 101, 627–634.

Northrop, D.B., 2002. Effects of high pressure an enzymatic activity. Biochim. Biophys. Acta 1595, 71–79.

Otero, L., Molina-García, A.D., Sanz, P.D., 2000. Thermal effect in foods during quasi-adiabatic pressure treatments. Innovative Food Sci. Emerg. Technol. 1, 119–126.

Otero, L., Molina-García, A.D., Ramos, A.M., Sanz, P.D., 2002. A model for real thermal control in high-pressure treatment of foods. Biotechnol. Prog. 18, 904–908.

Otero, L., Ramos, A.M., de Elvira, C., Sanz, P.D., 2006. A model to design high-pressure processes towards an uniform temperature distribution. J. Food Eng. 78, 1463–1470.

Otero, L., Sanz, P.D., 2003. Modelling heat transfer in high pressure food processing: a review. Innovative Food Sci. Emerg. Technol. 4, 121–134.

Özmutlu, Ö., Hartmann, C., Delgado, A., 2006. Momentum and energy transfer during phase change of water under high hydrostatic pressure. Innovative Food Sci. Emerg. Technol. 7, 161–168.

Pehl, M., Delgado, A., 1999. An in-situ technique to visualize temperature and velocity fields in liquid biotechnical substances at high pressure. In: Ludwig, H. (Ed.), Advances in High Pressure Bioscience and Biotechnology. Springer, Heidelberg, pp. 519–522.

Pehl, M., Werner, F., Delgado, A., 2000. First visualization of temperature fields in liquids at high pressure using thermochromic liquid crystals. Exp. Fluids 29, 302–304.

Piyasena, P., Mohareb, E., McKellar, R.C., 2003. Inactivation of microbes using ultrasound: a review. Int. J. Food Microbiol. 87, 207–216.

Qin, B., Zhang, Q., Barbosa-Cánovas, G.V., Swanson, B.G., Pedrow, P.D., 1995. Pulsed electric field treatment chamber design for liquid food pasteurization using a finite element method. Trans. ASAE 38, 557–565.

Raso, J., Barbosa-Canovas, G.V., 2003. Nonthermal preservation of foods using combined processing techniques. Crit. Rev. Food Sci. Nutr. 43, 265–285.

Raso, J., Palop, A., Pagan, R., Condon, S., 1998. Inactivation of Bacillus subtilis spores by combining ultrasonic waves under pressure and mild heat treatment. J. Appl. Microbiol. 86, 849–854.

Rauh C., 2008. Modellierung und Simulation von Kurzzeit-Ultra-Hochdruckprozessen. PhD Thesis, U Erlangen-Nuremberg.

Rauh C., Baars A., Delgado A., 2006. Analysis of inhomogeneous thermofluiddynamical processes in short time high pressure treatment of liquid foods. In: Proceedings of the 4th International Conference on High Pressure Bioscience and Biotechnology, Tsukuba, Japan, 186–191.

Rauh, C., Baars, A., Delgado, A., 2009. Uniformity of enzyme inactivation in a short-time high-pressure process. J. Food Eng. 91, 154–163.

Ravishankar, S., Zhang, H., Kempkes, M.L., 2008. Pulsed Electric Fields. Food Sci. Technol. Int. 14, 429–432.

Ross, A., Griffiths, M., Mittal, G., Deeth, H., 2003. Combining nonthermal technologies to control foodborne microorganisms. Int. J. Food Microbiol. 89, 125–138.

Sala, F.J., Burgos, J., Condon, S., Lopez, P., Raso, J., 1995. Effect of heat and ultrasound on microorganisms and enzymes. In: Gould, G.W. (Ed.), New Methods of Food Preservation. Blackie Academic & Professional, London.

Salih, E.A., Choong, T.S.Y., Sergie, S.Y., Chin, N.L., Ibrahim, O.M., 2009. The effect of buoyancy force in computational fluid dynamics simulation of a two-dimensional continuous ohmic heating process. Am. J. Appl. Sci. 6, 1902–1908.

Sanz, P.D., Otero, L., de Elvira, C., Carrasco, J.A., 1997. Freezing process in high-pressure domains. Int. J. Refrig. 20, 301–307.

Sastry, S., 2008. Ohmic heating and moderate electric field processing. Food Sci. Technol. Int. 14, 419–422.

Sastry, S., Qiong, L., 1993. Models for ohmic heating of solid–liquid mixtures. ASME HTD 254, 25–33.

Smeller, L., 2002. Pressure – temperature phase diagrams of biomolecules. Biochim. Biophys. Acta 1995, 11–29.

Song, K., Al-Salaymeha, A., Jovanovic, J., Rauh, C., Delgado, A., 2009b. Experimental in-situ investigations of turbulence under high pressure. Ann. N. Y. Acad. Sci. in press.

Song, K., Regulski, W., Jovanovic, J., Rauh, C., Delgado, A., 2009a. In-situ Investigation of the Turbulent-Laminar Transition of Temperature Fluctuations during the Pressure Building up to 300 MPa. High Press. Res. in press.

Teopfl, S., Mathys, A., Heinz, V., Knoor, D., 2006. Review: Potential of high hydrostatic pressure and pulsed electric fields for energy efficient and environmentally friendly food processing. Food Rev. Int. 22, 405–423.

Ting, E., Balasubramaniam, V.M., Raghubeer, E., 2002. Determining thermal effects in high-pressure processing. Food Technol. 56, 31–35.

Toepfl, S., Heinz, V., Knorr, D., 2007. High intensity pulsed electric fields applied for food preservation. Chem. Eng. Process. Process Intensification 46 (6), 537–546.

Valdramidis, V.P., Geeraerd, A.H., Poschet, F., Ly-Nguyen, B., van Opstal, I., van Loey, A.M., et al., 2007. Model based process design of the combined high pressure and mild heat treatment ensuring safety and quality of a carrot simulant system. J. Food Eng. 78, 1010–1021.

Weemaes, C.A., Ludikhuyze, L.R., Van Den Broeck, I., Hendrickx, M.E., 1998. Effect of pH on Pressure and Thermal Inactivation of Avocado Polyphenol Oxidase: A Kinetic Study. J. Agric. Food Chem. 46, 2785–2792.

Zeqiri, B., Gelat, P.N., Barrie, J., Bickley, C.J., 2007. A novel pyroelectric method of determining ultrasonic transducer output power: device concept, modelling and preliminary studies. IEEE Trans. Ultrason. Ferrolelectrics Freq. Contr. 54, 2318–2330.

Zimmermann, U., Pilwat, G., Riemann, F., 1974. Dielectric breakdown of cell membranes. Biophys. J. 14, 881–899.

CHAPTER

3 Fluid Rheology in Novel Thermal and Non-Thermal Processes

P.J. Cullen

School of Food Science and Environmental Health, Dublin Institute of Technology, Dublin, Ireland

3.1 INTRODUCTION

Processed fluid foods are typically complex in nature, with infinite mixtures of ingredients combined to meet desired nutritional and sensory characteristics. Such multiphase and structured fluids may display complex rheological behavior offering pleasing textural and mouthfeel profiles to consumers. Due to their inherent nature and rheological behavior, processing such fluids to ensure safety and to extend shelf-life is challenging. For a treatment process to be accepted, treatment homogeneity must be validated. Ideally the process will have:

1. A uniform treatment field. This will depend on the inherent nature of the treatment (pulsed electric field (PEF), high-pressure processing (HPP), microwave, etc). Distortions may arise due to the nature of the treated fluid, such as the influence of suspended particulates.
2. A homogenous fluid. Difficulties arise due to the variable nature of fluid properties (velocity gradients, viscosity gradients, wall effects, etc) and the nature of multiphase fluids. Such challenges are magnified when dealing with non-Newtonian fluids where the mixing effects of natural convection may be suppressed due to high viscosities or yield stresses.

Consequently, rheology is typically found to govern the efficacy of many thermal and non-thermal food-treatment processes. Treatment systems will be designed to deal with specific, often model, fluid properties with an objective of treatment homogeneity. As food displays a vast array of rheological behavior, such as shear thinning, plasticity, and viscoelasticity, technologies may be suitable only for a limited range of fluid properties. However there are numerous approaches available to the food technologist to deal with fluid foods as outlined

Novel Thermal and Non-Thermal Technologies for Fluid Foods. DOI: 10.1016/B978-0-12-381470-8.00003-7

in this book. Some approaches may operate under batch conditions where fluid flow may be limited to natural convection or indeed absent while others operate under continuous conditions where shear flow exists.

Although rheology may govern the efficacy of these processes, few studies have validated these technologies for various rheological properties. Treatment of multiphase fluids is challenging due to the different properties of the phases. On top of this, food dispersions (suspensions, emulsions, gas–liquid) are often heterogeneous in nature both in terms of spatial distribution and particle size/shape distributions. For non-Newtonian fluids the viscosity will be dependent upon the shear rate which varies with treatment geometry, flow rate, and temperature. For dispersions, boundary effects due to the confines of the treatment geometry may be evident, leading to slip layers. Even the treatment process itself may modify the fluids properties. Consequently it is difficult to validate such systems for complex rheologies. However, if treatment homogeneity can be validated for fluid foods displaying more complex behavior, products which meet safety requirements yet retain high nutritional or sensory profiles could be processed. Apart from rheology influencing the efficacy of the process, the process itself may in turn significantly influence the final product rheology and, ultimately, consumer acceptability. This chapter provides an overview of food rheology with particular reference to novel thermal and non-thermal technologies.

3.2 FLUID RHEOLOGY

Rheology is defined as the science of deformation and flow, it provides insight into material phenomena, which are governed by the microscopic scale at a macroscopic level. Rheology entails the study of materials with properties described by relationships between force and deformation. As rheology describes how matter responds to applied stress or strain, it finds application in diverse areas such as product development, process engineering calculations, quality control, stability studies, and correlations to sensory data. Many processed foods are formulated to display desired rheological behavior under specific stress conditions such as gravity, pouring, mouthfeel, etc.

Shear flows occur when force is applied parallel to a face of a fluid element (shear stress, σ) and results in a shear deformation where there is an angular displacement of parallel surfaces in the fluid element (shear strain, γ). The rate at which the shear deformation occurs is the shear rate ($\dot{\gamma}$). Shear flow is the dominant flow type found in most processes, where flow is initiated. The typical rheological behaviors of fluid foods under steady shear conditions are depicted in the rheogram in Figure 3.1. The most basic behavior is the case in which the shear stress is proportional to the shear rate as seen in curve 4, where the slope is the Newtonian viscosity (μ) as was originally defined by Newton's viscosity law.

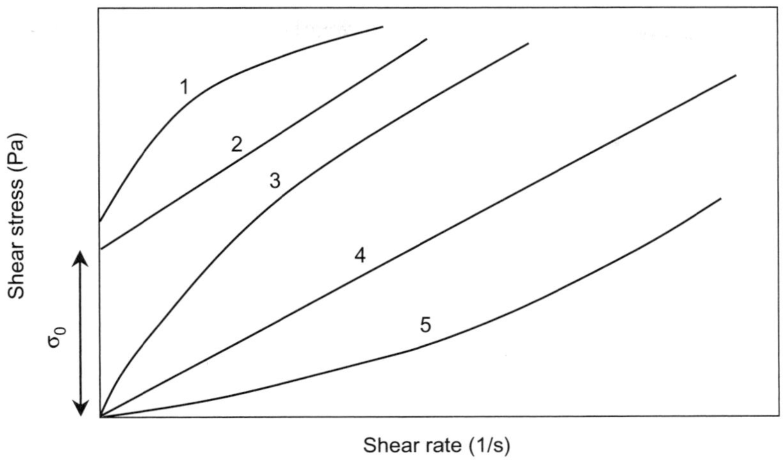

FIGURE 3.1

Flow curves for typical time-independent fluids. 1, viscoplastic fluid; 2, bingham fluid; 3, pseudoplastic fluid; 4, Newtonian fluid; 5, dilatant fluid.

Curves 3 and 5 describe shear thinning (or pseudoplastic) and shear thickening (or dilatent) behavior, respectively, and are both described by the power law model. Curves 1 and 2 describe fluids that have a dynamic yield stress (σ_y) that must be overcome before fluid flow commences. At stress levels greater than the yield stress, Bingham fluids (curve 2) exhibit shear stress that is proportional to shear rate with the slope described as the plastic viscosity (μ_{pl}), while fluids exhibiting continued shear thinning behavior are described as viscoplastic (curve 1). This range of behaviors can be generally described using the Herschel–Bulkley model (Steffe, 1996):

$$\sigma = \sigma_y + K\dot{\gamma}^n, \tag{3.1}$$

where K is known as the consistency coefficient and n is known as the flow behavior index. It directly describes the behavior seen in curve 1 of Figure 3.1, as well as that in curve 2 when $n = 1$ to give the Bingham model where $K = \mu_{pl}$. When the yield stress is zero ($\sigma_y = 0$), Equation 3.1 takes the form of the power law model with the pseudoplastic behavior of curve 3 modeled when $n < 1$ and dilatent behavior of curve 5 modeled when $n > 1$. In the case where both $\sigma_y = 0$ and $n = 1$, Equation 3.1 reduces to the Newtonian case of curve 4 where K becomes the Newtonian viscosity (μ).

Shear thinning behavior is observed with structured foods, where viscosity decreases with applied shear. Consequently, the viscosity at any given shear rate must be reported as apparent viscosity ($\eta_a = \sigma/\dot{\gamma}$). The fluid's microstructure, which may be due to either macromolecule entanglement or particle–particle interaction, is broken down under shear. Rheograms of such complex fluids typically display two regions of near-constant viscosity connected by a shear thinning region. The zero shear viscosity (η_0) corresponds to a region where the fluid is completely structured. The intermediate shear thinning region, typically characterized by the power law, corresponds to structural breakdown. The second Newtonian infinite shear viscosity (η_∞) region corresponds to complete structural breakdown. The shear rate will be governed by the system design and operating parameters along with the fluid's properties. However, for continuous systems and batch systems which initiate shear flow, the rheological behavior of these structured fluid materials are generally found to fall entirely in the shear thinning region.

Where a fluid structure can recover, the shear thinning behavior may be independent of time, since at a constant shear rate there is an equilibrium between the rate of structural breakdown and recovery. It is also possible for the viscosity of fluids to increase as a function of shear. Shear thickening (or dilatent) fluids are frequently solid suspensions, such as concentrated raw corn starch in water (Steffe, 1996). As observed with shear thinning, the shear thickening phenomenon occurs due to the rearrangement of the fluid's microstructure. For example, in the case of concentrated raw corn starch in water, at low shear rates the water has a lubricating effect such that the starch particles can slide past each other easily, thus exhibiting a low apparent viscosity. However, as the shear rate increases, increased resistance caused by particle to particle interactions causes the apparent viscosity to increase substantially (Steffe, 1996).

Materials with a yield stress (plastic) do not flow until a minimum yield stress (σ_y) is exceeded. Below this minimum stress, the materials structure is sufficiently rigid to prevent flow and instead stores the applied energy (Fig. 3.1). Upon breach of the yield stress value, the structure disintegrates and flow is initiated. Although the existence of a yield stress may be argued (Barnes and Walters, 1985), based upon the theory that everything flows given sufficient time or sufficiently sensitive measurement equipment, it is an engineering reality influencing process design, product stability, and sensory assessment. The ability to build in a yield stress within food systems is one of the most valuable strategies available in food product development. The use of structure-modifying ingredients such as biopolymers allows the design and control of desired rheological behavior at varying applied stress conditions within food systems.

The rebuilding of fluid structure may also be dependent upon time. Thixotropy is defined as the progressive decrease in viscosity with time for a constant applied shear stress, followed by a gradual recovery when the stress is removed. The mechanical properties of thixotropy systems result from the competition between the spontaneous restructuring of the microstructure at rest and its destruction under shear. Such materials exhibit a complex rheological behavior comprising of shear-thinning, time-dependent viscosity, and a yield stress.

3.3 INFLUENCE OF RHEOLOGY ON PROCESS EFFICACY

Thermal processing of fluid foods which display complex rheology, such as purees, pastes, soups, and jams, has always been a challenge for the food industry. To improve treatment homogeneity, various forced convection approaches such as impellers, static mixers, and scraped surface heat exchangers are typically employed. Nevertheless, over-processing is common to ensure safety requirements are met, leading to avoidable nutritional and organoleptic losses. Although certain novel thermal and non-thermal technologies may overcome some of the difficulties arising with processing such fluids, treatment homogeneity still remains a significant challenge. Knowledge of the spatial treatment inside any treatment chamber is key to ensuring treatment homogeneity and process acceptability for any given technology.

Viscous or plastic fluids may suppress convection currents, for example with microwave heating. Shear stresses experienced by fluid elements within the treatment geometry may be non-uniform. Shear thinning fluids which exhibit yield stresses may result in the formation of regions of flow and stagnant fluid elsewhere within the confines of the treatment geometry or fluid container which may lead to under-processing. Inducing macro mixing of viscous fluids may lead to increased treatment homogeneity. With continuous systems, multiple treatment zones and/or static mixers have been proposed to mitigate under-processing for ohmic heating (Zhang *et al.*, 1995), PEF (Schrive *et al.*, 2006), Dense-Phase Carbon Dioxide (DPCD) (Ferrentino *et al.*, 2009), microwaves (Kumar *et al.*, 2008), and UV irradiation (Altic *et al.*, 2007).

For a given treatment geometry and set of boundary conditions, the flow regime may range from laminar to turbulent as quantified by the Reynolds number (R_e). For a Newtonian fluid flow in a simple circular pipe R_e is given by:

$$R_e = \frac{\rho VD}{\mu}, \tag{3.2}$$

where V, D, σ, and μ are fluid velocity, pipe diameter, fluid density, and fluid viscosity, respectively.

Laminar flow: $R_e < 2000$
Transitional flow: $2000 < R_e < 4000$
Turbulent: $R_e > 4000$

At high Reynolds numbers, inertial forces dominate, changes in viscosity have no effect, and infinitesimal disturbances grow into complex interacting structures (turbulence) leading to effective fluid mixing and perhaps a more uniform treatment. For example turbulent flow within PEF treatment chambers have been reported to provide a more uniform treatment. The presence of particulate matter, walls, baffles, impeller blades, static mixer elements, etc., may result in the formation of highly distorted eddies, further enhancing the dynamic lifecycle of turbulence. Conversely, steady-state laminar flow conditions are found for low-velocity and high-viscosity fluids. Such conditions result in poor fluid mixing, as fluid motion is dominated by linear, viscous forces instead of nonlinear inertial forces. Essentially, viscosity dampens out any infinitesimal disturbances. By inserting a series of flow orientation elements along the axis of a straight pipe, the flow can be periodically split and remixed at the junctions along with stretching and folding within the elements. Static mixers may function in either the laminar, transitional, or turbulent flow regimes.

Flow regime predictions are only valid for constant flow and become more difficult where flow is pulsatile or radial and axial geometries are inconstant. The situation is further complicated if the contained fluid is non-Newtonian. In general, the pattern of the flow varies with velocity, physical properties of the fluid and the geometry.

3.3.1 Continuous Flow-Through Systems

Many flow-through systems such as ohmic heating and PEF have treatment chambers analogous to pipe flow. For non-Newtonian fluids the Ostwald de Waele power law may be employed to describe flow:

$$\frac{1}{r}\frac{\partial}{\partial r}\left(rK\left(\frac{\partial v}{\partial r}\right)^{n}\right)-\frac{\partial \rho}{\partial z}-g\rho(1-\beta(T-\langle T\rangle))=\rho\bar{v}\cdot\nabla v, \tag{3.3}$$

where v is the axial velocity, K the consistency index, and n the flow behavior index. A natural convection term, due to temperature differences between the wall and the centre of the tube, is included. Viscous dissipation due to radial velocities may be neglected as these velocities will be orders of magnitude lower than the axial velocities. The radial velocities can be determined from the continuity equation:

$$\frac{\partial v}{\partial z}+\frac{1}{r}\frac{\partial}{\partial r}\left(\frac{\partial (ru)}{\partial r}\right)=0, \tag{3.4}$$

where the boundary conditions are (1) axial symmetry, (2) zero velocity at the wall, and (3) fully developed flow.

3.3.1.1 Ohmic Heating

For food-heating purposes a flat temperature distribution is important (Goullieux and Pain, 2005). For ohmic heaters, the temperature profile causes the flow patterns to flatten due to natural convection and a viscosity gradient between the wall and the centre. Natural convection improves heater performance for low viscosities (Goullieux and Pain, 2005). However influences due to non-Newtonian behavior become apparent when the flow behavior index (n) is less than 0.6. Muller *et al.* (1994) employed such numerical analysis to show that the axial velocity distribution oscillates due to natural convection, with the centerline velocity oscillating around the mean velocity, decaying with amplitude and frequency.

Salih *et al.* (2009) developed a two-dimensional model describing the flow, temperature and electric field distribution in a cylindrical continuous ohmic heating cell for a non-Newtonian power law guava juice fluid. A buoyancy driving force is caused by density differences within the liquid due to spatial temperature variation between the wall and the center. It distorts the velocity profiles and increases to a maximum value of the velocity from the center to near walls of the heating cell. Because the density and the viscosity of the fluids decreased with increasing temperature near the region of the walls, there is generally an increase in velocity. The authors reported that this provides better mixing in the cell which resulted in a more uniform temperature distribution compared to simulations which ignore the buoyancy effect.

Ayadi *et al.* (2008) evaluated the capability of a rectangular ohmic unit to provide a homogenous heat treatment of a complex dairy fluid as a function of fluid rheology, flow rate, and fouling presence. Under non-fouling conditions, the temperature gradient depended on the flow regime and the fluid's rheological behavior. Under fouling conditions, the temperature gradient obtained at different Reynolds numbers exhibited different trends. These differences were explained by the effect of differential electrical conductivities between the bulk and the deposit, and the balance between heat generation by electrical power dissipation and thermal loss by convection (with the fluid) and conduction (with the electrode surfaces).

3.3.1.2 Microwave Heating

As for ohmic heating, non-uniform temperature distribution is a major hurdle in the implementation of continuous-flow microwave heating. Gentry and Roberts (2005) evaluated a continuous-flow microwave pasteurization system which consisted of helical coils within a large cavity oven. Process lethality was verified based on inoculation of *Escherichia coli* 25922 in apple cider, in which the pasteurization process resulted in a 5-log calculated R_e number showing laminar flow, which would indicate a parabolic profile for the fluids being pumped through a straight tube. However, the use of helical coils creates secondary flow perpendicular to the main direction of flow due to the centrifugal force exerted by the curvature of the coil, and this secondary flow results in the maximum velocity being pushed outward from the centre. The Dean (D_e) quantifies this secondary flow, which is the product of the Reynolds number and the square-root of the curvature ratio:reduction.

$$D_e = \frac{\rho VD}{\mu}\left(\frac{D}{2r}\right)^{1/2}. \tag{3.5}$$

The reported Dean numbers were all above 100, which are considered high. Prediction of velocity profiles of laminar flow in helically coiled pipes having such Dean numbers showed profiles resembling plug flow. Such a velocity profile is suitable for treating fluids where the fluid particles with maximum velocity span across most of the tube.

For higher-viscosity fluids an alternative approach is required. Kumar *et al.* (2008) studied the efficiency of static mixers as a means of temperature equalization for processing green pea and carrot purees in a continuous-flow microwave system. It can be seen from Figure 3.2a that there is a substantial temperature difference without static mixers at the exit of the first applicator. It can be seen from Figure 3.2b that the temperature difference is significantly reduced by the installation of static mixers. The results illustrate the effectiveness of temperature equalization in eliminating temperature differences in the cross-sectional area. The static mixers were installed at the exit of each of the microwave applicators. This approach will work for heating applications assuming there is little time lapse

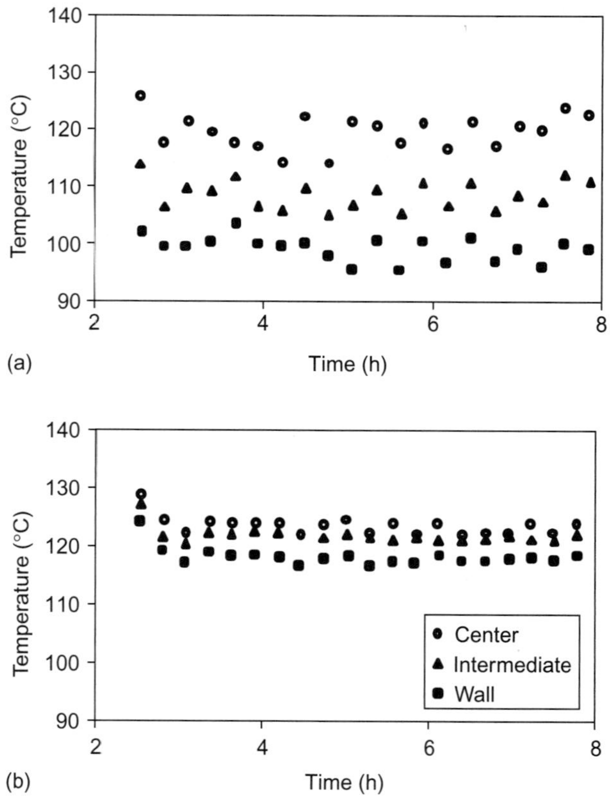

FIGURE 3.2

Temperature profile during processing of green pea puree in the 60-kW microwave system. *[Reproduced from Kumar* et al. *(2008).]*

between the treatment and mixing. However for non-thermal approaches the mixing must take place within the treatment chamber itself as there will be no residual effect once the fluid exits.

3.3.1.3 Pulsed Electric Field

Schrive *et al.* (2006) studied PEFs to evaluate the synergy between electrical parameters and hydraulic parameters with regard to cell death. To investigate the effect of hydrodynamics, a modified treatment chamber with "turbulence promoters" made of thin vibrating plastic elements fastened at the inlet and outlet provided drastically different R_e numbers. The R_e number was estimated to be about 100-times higher than without the turbulence promoters. The investigators employed a modified Sherwood number (Sh_m) where diffusivity is dependent on the electric field. Global mass transfer arising from diffusive and convective phenomena requires a dimensionless number approach using the Schmidt number ($Sc = \nu/D$) and the Reynolds number (R_e). Globally, the analogy between mass transfer and death rate is transcribed in a model where, the survival rate is related to the modified Sherwood number (Equation 3.6) taking into account the electromotive driving force and all types of process parameters: biological (mainly through cell radius), electrical (through electric field and electric resistance), and hydraulic (through the Reynolds number and the number of recyclings).

$$Sh_m = (Re \times n_r)^{1/2} \times \left(\frac{a^4 \times n_{total}^4 \times [(E - E_c)/2]^2 \times \eta}{\mu \times D_0} \right)^{1/3}, \tag{3.6}$$

where E is the electric field, E_c is the critical electric field leading to cell death, n_{total} is the total number of pulses, a is the the bacterial cell radius, η is the magnetic permeability, μ the Einstein viscosity, D_0 the temperature-dependent diffusion coefficient.

The authors reported that the PEF effect was drastically improved for a modified Sherwood number above 100, while pasteurization can be achieved for $Sh_m > 1500$ (Fig. 3.3). Survival rates obtained show that the greater the turbulence, the lower the survival rate.

3.3.2 Batch Systems

For batch systems, natural convection may be present if there is a thermal gradient or suppressed with a sufficient yield stress. Forced convection may be induced with the use of impellers, gas flow, etc. In batch systems where no flow is induced, parts of the food may be under-processed, noticeable as tailing effects in the microbial inactivation kinetics.

As discussed in Chapter 2, even for systems such as high pressure where the principle control parameter of pressure acts mostly instantaneously and uniformly, the additional parameter of compression heat may be non-uniform. Depending on the temperature sensitivity of the kinetics of the change in target attributes

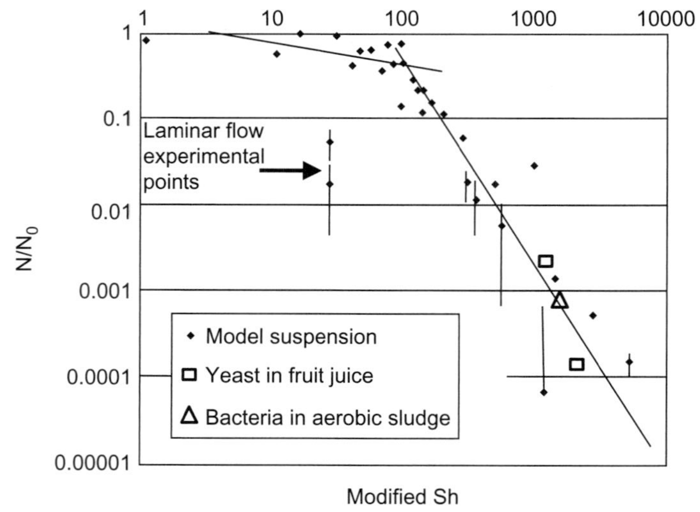

FIGURE 3.3

Survival rate versus modified sherwood number.

[Reproduced from Schrive et al. *(2006).]*

(e.g., safety, quality), the temperature non-uniformity can result in process impact non-uniformity (Van der Plancken *et al.*, 2008). When, for example, the compression heat of the high-pressure vessel wall, of the pressure transmitting medium, of the food packaging, of the food product, or even different constituents of a food product differs, a gradient is established, which causes heat transfer to occur. As pressure increases the temperature will depend upon thermodynamic and thermophysical parameters of the food treated. Density differences arising from temperature gradients may lead to natural convection (Rauh *et al.*, 2009, Khurana and Karwe, 2009) and consequently the rheology of such systems may influence uniformity. However, scant data are reported on the influence of complex fluid phenomena such as yield stress on treatment uniformity.

For ultrasound processing, the effectiveness of cavitation is influenced by the viscosity of the fluid. In viscous systems, ultrasound diffusion is disrupted and this reduces the degree to which cavitation occurs. Low-frequency, high-power ultrasound is better at penetrating viscous products than higher-frequency, high-power ultrasound, which is more easily dispersed within a viscous liquid. However, again, the reported influence of complex rheology on process efficacy is scant.

3.3.3 Solid–Liquid Mixtures

Numerous studies have investigated the efficacy of novel thermal and non-thermal treatment approaches for solid–liquid mixtures. Distortions in the treatment homogeneity due to solid particles have been reported for ohmic heating, infrared heating, PEF, etc. Particle size, orientation, and/or concentration may influence the treatment, however particles will also influence the flow regime. Studies to date have not typically focused on the hydrodynamics of solid–liquid mixtures within novel thermal or non-thermal treatment chambers.

Shim *et al.* (2010) used computational fluid dynamics (CFD) simulations of the thermodynamic performances of a multi-component food mixture during ohmic heating. Simulations identified over-processed zones at the top and bottom sides of particles (carrot, meat, potato) where higher electron density would be anticipated (Fig. 3.4). The authors suggest that these non-uniform temperature distributions could be resolved by forced convection flow into the system. They also propose that future modeling of ohmic heating patterns for multiphase food products in a three-dimensional environment is needed as a future study because the flow regime with moving particle orientation inside the field strength distribution is crucial. Also, solid particles in high concentrations will lead to heating in a different way because a greater proportion of the total current can flow through a number of particles resulting in higher energy generation even though they are less conductive. A new model should be capable of addressing the particle concentrations and orientations in a continuous-flow system.

Abdul Ghani and Farid (2007) studied the effects of forced and natural convection currents at the early stages of compression during HPP of a non-homogenous food. Velocity profile, temperature distribution and the shape of the hottest zone (HZ) during the compression were examined. The HZ in the cylinder (the location of the highest temperature at a given time) was found in the middle of the solid food due to the higher heat of compression of beef fat and due to the fact that heat is transferred by conduction only. Due to the higher temperature of the solid, heat is transferred from the solid to the water by free and forced convection heat transfer.

3.4 EFFECTS OF TREATMENT ON FLUID RHEOLOGY

The effects of novel thermal or non-thermal treatments on the rheological properties of fluid fluids may be insignificant, resulting in similar flow and textural behavior as non-processed produce; conversely, significant changes may be induced which may be desirable or undesirable in nature. These changes are typically found to be both product and process specific. A recent book entitled "*Novel Food Processing: Effects on Rheological and Functional Properties*" (Ahmed *et al.*, 2009) deals in detail with changes induced in the micro-structural and rheological properties of foods due to such processes. Some of the rheological changes induced for fluid foods are summarized below.

3.4.1 Pulsed Electric Fields

PEF treatment is based on the application of high-intensity electric fields to a food as it flows through the confines of a treatment gap between two electrodes.

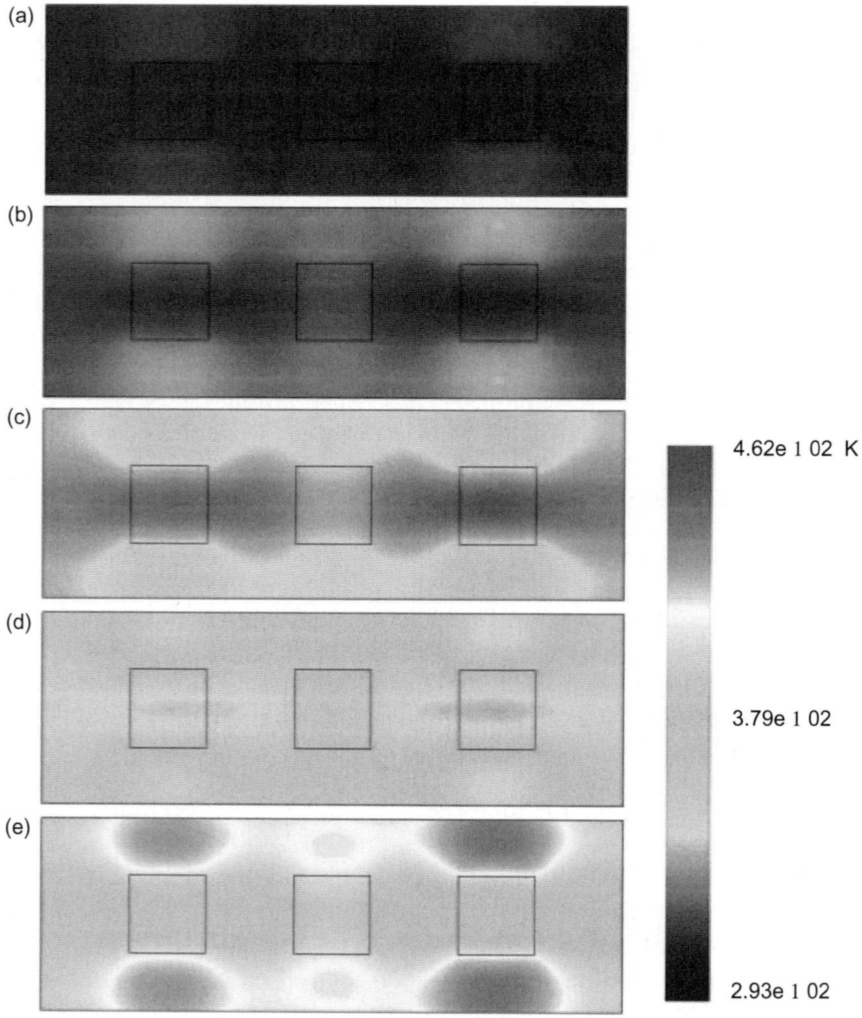

FIGURE 3.4

Simulated thermal patterns of solid (1, carrot; 2, meat; 3, potato) and liquid foods inside the ohmic cell after heating of (a) 50 s, (b) 100 s, (c) 150 s, (d) 200 s, and (e) 250 s. *[Reproduced from Shim* et al. *(2010).]*

Such electric fields induce transmembrane potential on cell membranes resulting in electroporation and electrofusion leading to bacterial inactivation. The field may, in turn, induce changes in the fluid structure. Garcia-Amezquita *et al.* (2009) examined the effect of PEF on the native size distribution of fat globules in bovine cheese-making milk. Although PEF processing did not modify the true mean diameter of milk-fat globules, it induced small globules to clump together, causing an apparent increment in the population of larger milk-fat globules. Sampedro *et al.* (2006) reviewed the application of PEF in egg and egg derivatives and concluded that treatment did not cause notable changes in proteins in a solution of ovalbumin and dialyzed fresh egg white. However, some structural changes and functional modifications were observed in fresh egg white as a result of PEF treatment. The texture and microstructure of gels were affected by the application of PEF, and therefore PEF treatment conditions in egg white must be optimized to minimize possible modifications. Aguilo-Aguayo *et al.* (2009) investigated the effects of pulse frequency (50–250 Hz), pulse width (1.0–7.0 mus), and polarity (monopolar or bipolar) of high-intensity PEF treatments (35 kVcm − 1 and 1000 mus) on the viscosity and the pectin methylesterase (PME) and polygalacturonase (PG) activities of tomato and strawberry juices. Apparent viscosity of strawberry juices increased slightly when frequencies higher than 100 Hz and 1 mus monopolar pulses were applied to the juice. Tomato juice apparent viscosity increased within the range of the assayed conditions, achieving the highest values at 250 Hz and 7.0 mus in bipolar mode. At the same conditions the lowest residual PME (RAPME = 10%) and PG (RAPG = 45%) activities were observed in the juice. Treatments causing the greatest increase in strawberry juice apparent viscosity also led to the lowest RAPME (10%) and RAPG (75%) values. In contrast, viscosity loss was promoted under the rest of the assayed PEF conditions despite the low RAPME values (<20%) achieved. Aguilo-Aguayo *et al.* (2010) also reported that PEF-treated watermelon juices exhibit better physical properties such as color and viscosity when compared with thermally treated juices throughout storage.

3.4.2 High-Pressure Processing

HPP of food is mainly applied for pasteurization purposes. By avoiding the adverse quality effects of thermal treatments for long times, HPP may provide distinct product-quality merits over conventional heat processing (Krebbers *et al.*, 2003). Other applications include pressure shift freezing, thermal assisted sterilization, enzyme inactivation, and optimization of functionality. From a rheology perspective, HPP may not only result in products which are similar to untreated foods, but may also induce rheological changes which result in desired functional properties.

For some fluids such as juices, HPP is found not to significantly change the viscosity relative to untreated or thermally treated produce. At higher viscosities, effects may become apparent. Ahmed *et al.* (2005) reported improved rheology for HPP treated mango pulp relative to control samples. Krebbers *et al.* (2003) examined the combined effects of high-pressure and thermal treatments for processing of tomato puree, reporting that pressure treatment with or without heat induced a reduction in viscosity compared to conventional sterilized samples. However the water-binding capacity was improved.

For foods with protein, denaturation can occur during HPP, which may or may not be reversible depending upon protein type and treatment conditions. Significant amounts of research have examined the effects of HPP on milk functionality. Milk protein structural changes may be induced, influencing subsequent coagulation and the final rheological behavior of dairy products. The approach has been adopted in particular to improve the stringiness and meltability of mozzarella cheese. Molina *et al.* (2000) reported increased yield for reduced-fat cheese through enhanced protein and moisture retention after HPP of skim milk.

Huppertz *et al.* (2004) reported that HPP treatment of milk resulted in whey protein denaturation inducing interactions between proteins and the Milk Fat Globule Membrane (MFGM). Creaming was also reduced by up to 70%. Gervilla *et al.* (2001) also reported that HPP did not increase spontaneous lipolysis, indicating that the changes to the average size of milk fat globules in HPP-treated milk are not caused by disruption of the MFGM. Garcia-Amezquita *et al.* (2009) evaluated the effects of HPP and conventional thermal treatments on fat-globule size distribution and ζ-potential size distribution of fat globules in bovine cheese-making milk. Milk was processed by HPP at 400 and 500 MPa for 0–20 min. HPP treatments induced fat globule flocculation, increasing their mass moment mean diameter.

Hemar *et al.* (2010) studied the flow behavior of aqueous solutions of gelatin and skim milk–gelatin mixtures treated by HPP at 5°C for 15 min, at pressures between 150 and 600 MPa. HPP treatment did not affect the flow behavior of gelatin alone, nor that of the skim milk–gelatin mixtures made with 0.4 wt% gelatin. However, at gelatin concentration of 0.4 wt%, the mixtures treated with 300 and 450 MPa exhibited a peculiar flow behavior, where at intermediate shear rates the viscosity was higher than that of the non-treated mixture or the mixtures treated at 150 and 600 MPa. Particle size measurements showed that, for gelled mixtures, 300-MPa HPP treatment resulted in an increase in the particle size, while at all other pressure treatments (150 MPa), a shift in particle size distribution to lower sizes was observed. Microstructural analysis showed that these skim milk–gelatin mixtures were phase-separated with a gelatin continuous phase, this was confirmed by dynamic rheological measurements which showed that qualitatively the viscoelastic properties of the mixtures were the same.

As for the benefits found for cheese manufacture, numerous studies have investigated the effects on the rheological properties of yoghurts made from HPP milk. Walker *et al.* (2006) reported that HPP-treated fruit yoghurt had a smoother and thicker appearance relative to control samples. However, only limited benefits within narrow processing windows were obtained when yoghurts made from HPP-treated milks were compared to yoghurts made with heat-treated milk (Needs *et al.*, 2000; Capellas and Needs, 2003; Harte *et al.*, 2003; Penna *et al.*, 2007). Penna *et al.* (2007) reported that the microstructure of heat-treated milk yoghurt had fewer interconnected chains of irregularly shaped casein micelles, forming a network that enclosed the void spaces, in comparison to the microstructure of HPP yoghurt which had more interconnected clusters of densely aggregated protein of reduced particle size. This resulted in a smoother, more regular surface, and more uniform size distribution. The combined HPP and heat milk treatments led to compact yoghurt gels with increasingly larger casein micelle clusters interspaced by void spaces, and exhibiting a high degree of cross-linking. The rounded micelles tended to fuse and form small irregular aggregates in association with clumps of dense amorphous material, which resulted in improved gel texture and viscosity. A schematic diagram of the effects of heat treatment and HPP of milk is shown in Figure 3.5. Udabage *et al.* (2010) investigated the effects of HPP treatment with and without simultaneous heating of yoghurt milk on the physical properties of stirred yoghurt. They reported that HPP of skim milk at 25°C with a prior or later heat treatment resulted in stirred yoghurts of similar viscosities to those made with traditional heat treatment, but lower viscosities were obtained when yoghurts were made from milk HPP treated at elevated temperatures. They concluded that from a rheological perspective that the use of HPP-processed milk may not be of significant commercial interest, unless the substitution of the standard heat treatment given to yoghurt milk by HPP treatment imparts significantly improved functional properties to the final yoghurt. Stating that as HPP promotes whey protein denaturation, there is a possibility to reduce the severity of the standard heat treatment given traditionally to yoghurt milk by incorporating HPP step facilitating strategies to incorporate heat-sensitive ingredients in the yoghurt formulation. Furthermore, they suggest a potential for producing high-protein drinking yoghurts of low viscosity, which may be of particular interest to elderly and other groups requiring convenient/nutritious fermented beverages.

Anema (2010) demonstrated that the initial pH of the milk at pressure treatment markedly influenced the rheological properties of acid gels prepared from the pressure-treated milks. This pressure treatment resulted in a marked disruption of the casein micelles in the milk; however, these changes in the casein micelles could not account for the effect of pH and pressure on the properties of the acid milk gels. Although whey protein denaturation in pressure-treated milks displayed a strong dependence on pH, it was evident that the denaturation level alone could not account for the changes in the final shear modulus, G', and yield stresses of the acid gels. The authors proposed that the pH and the magnitude of pressure affects the interactions of the denatured β-Lactoglobulin (β-LG) with the casein proteins during a pressure-treatment cycle. These changes in interactions affect the ability of the denatured β-LG to participate in the acid gel structures so that, at low pressures and/or initial pHs, the denatured β-LG acts predominantly as an inert filler in the acid gel structure whereas, at higher pressures and/or initial pHs, the denatured β-LG actively participates in the formation of the gel network during acidification.

Johnston and Gray (2008) examined the individual contributions of heat and HPP on the rheology of waxy maize starch dispersions in skim milk. Pressure processing at 200, 400, or 600 MPa for 20 min provided a greater increase in the shear modulus, G' than heat treatment at 85°C for 30 min. Measurements of relative apparent viscosity showed that pressure-treated samples were more susceptible to structure loss on shearing.

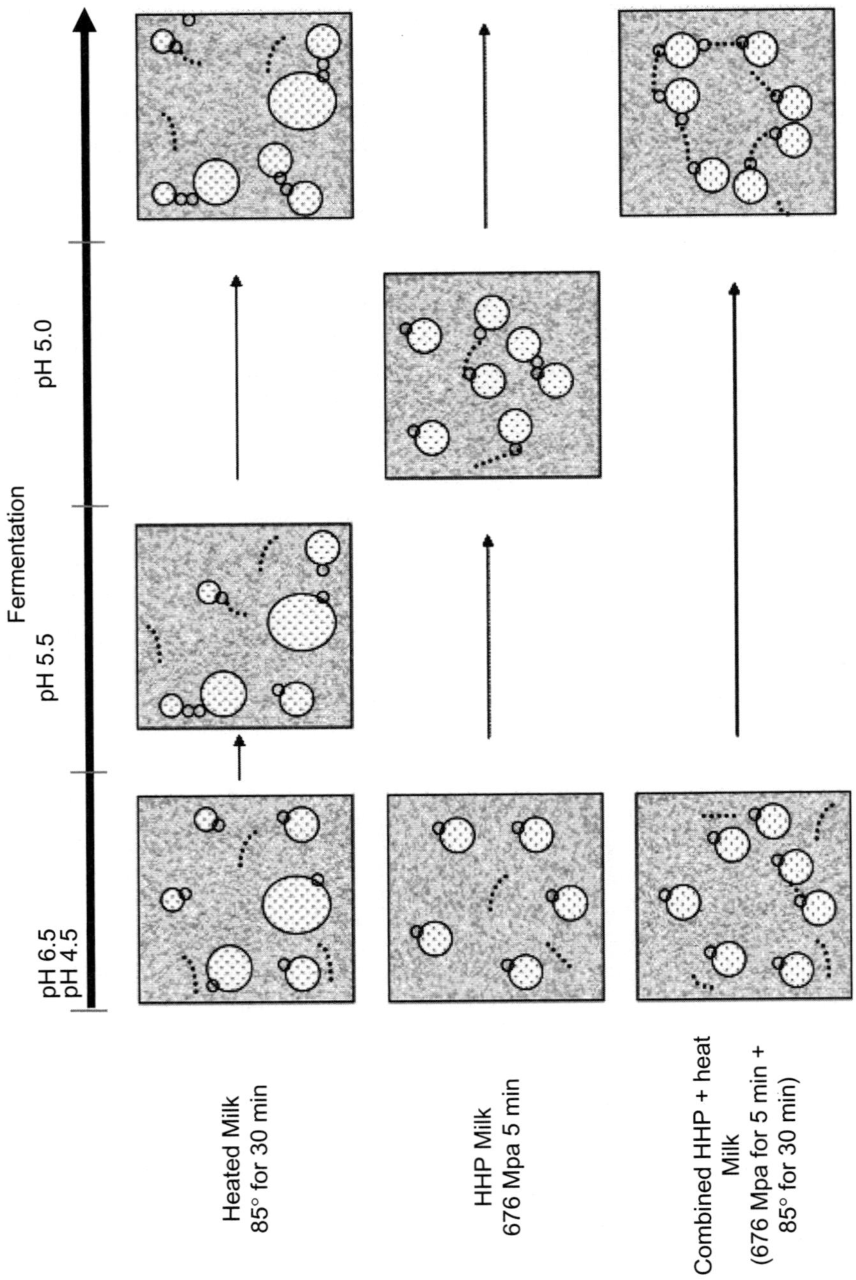

FIGURE 3.5

Schematic diagram of the effect of heat, HPP, and combined heat + HHP on casein micelle microstructure.

[Reproduced from Penna et al. *(2007).]*

Ahmed *et al.* (2003) studied the effect of HPP on thixotropic whole liquid egg, albumen, and yolk, reporting protein structure breakdown. However the protein coagulation was found to be reversible.

3.4.3 Power Ultrasound

The effect of ultrasound on food rheology is principally due to cavitation; the formation, growth, and, in some cases, implosion of bubbles within liquids. Two different types of cavitation phenomena can be generated by acoustic waves, namely inertial and noninertial cavitation. Inertial cavitation involves large-scale variations in bubble size (relative to the equilibrium size) over a timescale of a few acoustic cycles, where the rapid growth terminates in bubble collapse with varying degrees of intensity. Noninertial cavitation (stable) involves small-amplitude oscillations (compared to bubble radius). The thermal, mechanical, and chemical effects of high-intensity ultrasound have been attributed to the rapid formation and collapse of cavitational bubbles, generating intense normal and shear stresses. Implosion of cavitation bubbles leads to energy accumulation in hot spots where temperatures of 5000°C and pressures of 100 MPa have been measured (Suslick, 1988). As a result of these conditions, water molecules can be broken, generating highly reactive free radicals that can react with other molecules. Cavitational thermolysis may produce hydroxyl radicals and hydrogen atoms that can be followed by formation of hydrogen peroxide and, in the absence of oxygen, hydroperoxyl radicals (Ashokkumar and Grieser, 1999).

These transient reactive species can subsequently react with carbohydrates. In addition, hydrolysis and cleavage due to the strong mechanical forces have been reported for a variety of polysaccharides (Kardos and Luche, 2001), reducing molecular weight and changing the rheological properties of food. Mechanical stress, generated by shock waves derived from bubble implosion or from microstreaming derived from bubble size oscillation, may also break large macromolecules or particles. Because of the nature of these ultrasound effects on molecules or particles dissolved or suspended in liquids, it is expected that the large multimolecular structures such as casein micelles and fat globules in milk will be affected by ultrasound.

Viscosity of tomato juice/puree is highly dependent on pectic substances that form an entanglement where other particles are physically entrapped (Rao and Cooley, 1992). Tomato juices are pseudogels, whose flow properties depend on the interaction or entanglement of cell particles (mostly cell walls), soluble pectin concentration, and the chemical properties of the latter. In tomato juices, the combined treatments of heat and ultrasound under pressure: Mano-Thermo-Sonication (MTS) results in a thicker consistency and initial apparent viscosities compared with unprocessed juice (Vercet *et al.*, 2002). MTS treatment of pure pectin solutions yielded molecules with lower apparent viscosities due to size reduction (Mason *et al.*, 2005). A similar study conducted by Tiwari *et al.* (2010) observed a significant reduction in apparent viscosity for 2% w/v pectin dispersion.

Seshadri *et al.* (2003) also reported a similar influence of sonication on the gel strength properties of pectin. It is difficult to predict what might be expected from a modification of pectin properties in gels or pseudogels derived from pectin. Longer molecules show a higher resistance to flow; however, shorter ones can interact in a different way with suspended particles also leading to an increased resistance to flow.

Vercet *et al.* (2002) reported that the rheological properties of tomato paste are improved by MTS. In addition, they observed that the thixotropic and pseudoplastic behavior of tomato paste was not affected by MTS. These are important properties to maintain the characteristic mouthfeel of the product. Particle interactions and particle size both play a role in determining tomato consistency. An increase in the number and/or intensity of interactions leads to increased consistency. These interactions can be chemical and/or physical. Pectin molecules are the main agents of physical interactions (Voragen *et al.*, 1995), whereas chemical interactions are dependent on many parameters (Tsai and Zammouri, 1988). Consistency is dependent on particle size in a complex manner. Reducing particle size leads to a decrease in viscosity. However, there is a point where viscosity begins to increase when particle size decreases (Beresovsky *et al.*, 1995). This has been ascribed to an increase in the number of interactions between particles or to the fact that smaller particles fit better within the pectin network. The mechanism by which MTS improves the rheological properties of tomato pastes is related to cavitation. This phenomenon results in the breakage of molecules or particles. Particle size reduction and molecule breakage induced by ultrasound has been described in the literature (Price, 1990). Application of MTS to break pectin molecules in a purified pectin solution was reported by Mason *et al.* (2005). It is also possible that ultrasound promotes protein denaturation (Villamiel and de Jong, 2000). Denatured proteins can adhere nonspecifically to tomato particles and facilitate better interaction between particles.

Krešić *et al.* (2008) reported that ultrasound treatment of whey proteins caused a significant increase in apparent viscosity. The changes in flowing behavior upon ultrasound treatment are a consequence of changes in binding capacity for water, where altered protein structure after ultrasound treatment results in hydrophilic parts of amino acids opening toward water surroundings leading to higher binding of water molecules.

3.4.4 Irradiation

Food irradiation has been studied extensively for its efficacy as a food-preservation tool. However, with regard to possible structural changes in foods, most of the studies to date have focused on the textural changes induced in solid foods, in particular meats and vegetables. Those studies which examine fluids have principally examined the depolymerization of biopolymers.

The effects of irradiation on food rheology are found to depend on the dose and food treated. Matsuda and Sabato (2004) reported that gamma irradiation did not impair the viscosity of honey at doses of 5 and 10 kGy. A number of studies have examined the effects of irradiation on the rheology of liquid eggs. Min *et al.* (2005) reported a dramatic decrease in the viscosity of egg white with irradiation, which may facilitate the separation of egg white and yolk. Studies on the effect of irradiation dose on the gelation properties of liquid eggs suggest that doses below 3.5 kGy does not affect gelation significantly. Low doses (<2 kGy) do not appear to alter the thermal characteristics of egg white proteins, as sensory analysis of hard cooked egg whites could not differentiate against control samples.

Dogan *et al.* (2007) determined that the viscosity and consistency of selected hydrocolloid solutions decreased with increasing radiation dose (Fig. 3.6). Guar gum was particularly sensitive to irradiation, conversely the rheological characteristics of pectin and salep solutions did not change much with irradiation. The authors conclude that these hydrocolloids could be irradiated for preservation purposes.

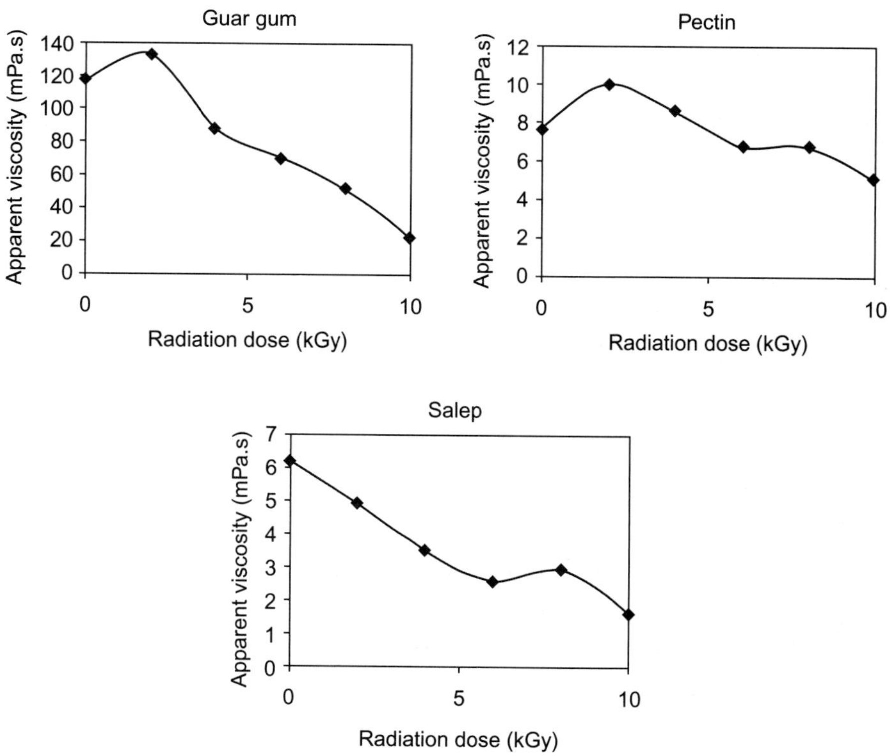

FIGURE 3.6

Apparent viscosity variation of samples with increasing irradiation dose.
[Reproduced from Dogan et al. *(2007).]*

3.4.5 Ultraviolet and Pulsed White Light

Due to the low penetration capacity of UV light into low-UV transmittance (UVT) or almost opaque substances such as liquid foods, it was initially intended to be used only for disinfection of food surfaces or packaging. Its use in fluid foods was originally limited to highly transparent liquids; however, development of new reactors with optimized hydraulics and thin-film design now allows delivery and exposure of target microorganisms to UV light in low UVT fluids (Chapter 8). Bhat and Karim (2009) studied the effects of UV irradiation on the rheological properties of fish gelatin. Treated samples exhibited significant improvement in the gel strength, marked reduction in viscosity, with significant changes in the melting enthalpy (Fig. 3.7). The authors reported two opposing effects of UV: a marked decrease in viscosity, which suggests a chain fragmentation, and a significant increase in the gel strength, which suggests a crosslinking that forms a stronger gel network. It was suggested that in the flow measurement, the polymer chain would align in the direction of shear flow; therefore, it is envisaged that the shorter polypeptide chain in the irradiated sample would align easily, hence manifest lower viscosity. However, upon cooling below the gelling point, extensive inter-molecular binding between the shorter cross-linked polypeptide chains in the irradiated fish gelatin would develop during the maturation period, thus giving rise to a stronger gel than that of the control sample. The cross-linking appears to dominate over the effect of chain degradation as manifested by enhancement in gel strength of UV-irradiated fish gelatin.

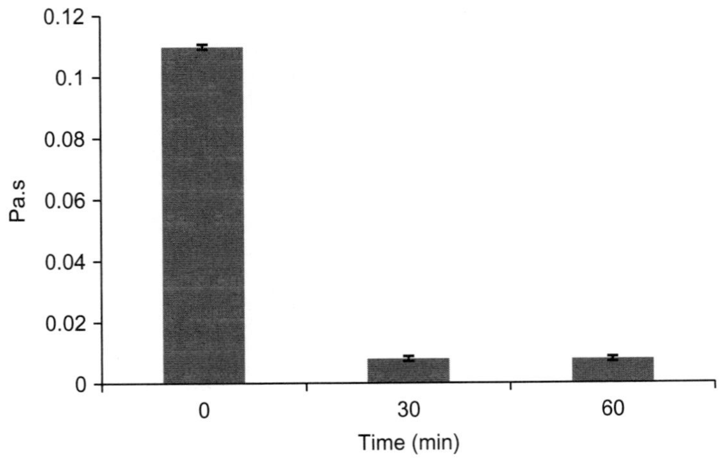

FIGURE 3.7

Changes in the viscosity (flow curve according to Newtonian model) of fish gelatin granules exposed to UV treatments for 30 and 60 min (n = 3, ±SD).

[Reproduced from Bhat and Karim (2008).]

Yau-Hoong *et al.* (2009) examined the effects of UV irradiation on the physicochemical and functional properties of gum Arabic. They showed that UV-irradiated samples exhibited better emulsification properties than unirradiated samples. No significant changes occurred to the molecular structure. Free amino group analysis indicated that mild UV irradiation (30 min) could induce cross-linking of gum arabic; however, viscosity breakdown was observed for samples exposed to UV irradiation for longer times (90 and 120 min).

3.4.6 Ozone

Ozonation has been reported to decrease molecular weight and viscosity of food hydrocolloids, which may be desirable or undesirable depending on the end use. For example, Yue *et al.* (2008) reported an application of ozone in reducing the molecular weight of chitosan. This innovative process involves preparation of low-molecular-weight water-soluble chitosan. Chitosan, a natural polysaccharide obtained by the deacetylation of chitin from crustacean shells, is generally used as binding, gelling, thickening, and stabilizing agents. Seo *et al.* (2007) investigated the potential of ozone in depolymerization of this polysaccharide, which is generally done by using enzymatic or chemical processes. They reported a decrease in viscosity and molecular weight of chitosan. The molecular weight of ozone-treated chitosan in acidic conditions, i.e. acetic acid solution, caused a significant decrease with increase in ozone treatment compared to ozone treatment in water. A decrease of 92% (104 kDa) was found after a treatment time of 20 min compared to the untreated chitosan (1333 kDa). Since the molecular weight of a polymer is associated with viscosity, a similar decrease in viscosity of chitosan solution with ozone treatment is expected. No *et al.* (1999) reported a decrease of 63% (206 mPa · s) during ozone treatment of 10 min with ozone concentration of 0.5 ppm compared to untreated chitosan solution (556 mPa · s). Tiwari *et al.* (2008) studied the effect of ozone treatment on aqueous dispersions of pectin, guar gum, and carboxyl methyl cellulose. A decrease of 95.5%, 81.6%, and 31.7% in apparent viscosity compared to control was reported for guar, Carboxy Methyl Cellulose (CMC), and pectin dispersions, respectively, at an ozone concentration of 7.8% w/w for 10 min.

Degradation of the chitosan polymer is mainly due to the strong oxidative properties of ozone, which selectively degrade β–D-glucoside bonds between units by the electrophilic attack on the C(1)–H bond by ozone molecules (Kabalnova *et al.* 2001). Degradation of polymers may be due to the formation of other highly reactive species, such as $\cdot OH$, $HO_2\cdot$, $\cdot O_2^-$, and $\cdot O_3^-$, which facilitates degradation.

Degradation of organic polymers and subsequent reduction in viscosity in the presence of ozone could be due to either direct reaction with ozone or indirect reaction because of secondary oxidators. Direct reaction is described by the Criegee mechanism (Criegee, 1975) where ozone molecules undergo 1–3 dipolar cycloadditions with the double bonds present, leading to the formation of ozonides (1,2,4-trioxolanes) from alkenes and ozone with aldehyde or ketone oxides as decisive intermediates, all of which have finite lifetimes (Criegee, 1975). This leads to the oxidative disintegration of ozonide and formation of carbonyl compounds, while oxidative work-up leads to carboxylic acids or ketones. Ozone attacks OH radicals, preferentially to the double bonds in organic compounds. Degradation of organic compounds is also reported to be due to various intermediate radical formations leading to electrophilic and nucleophilic reactions occurring with aromatic compounds that are substituted with an electron donor (e.g., OH^-) having high electron density on the carbon compounds in ortho- and para- positions. Chemical reactions involve breakage of old bonds and formation of new bonds and according to bond dissociation energy theory, the lower the bond dissociation energy the more active the bond.

Sadowska *et al.* (2008) reported an increase in viscosity and showed that viscosity is a function of molecules' dimensions and orientations. The decrease in the degree of unsaturation due to ozonation and the increase in molar mass both contribute to the increase in viscosity of the ozonated oils. Oil viscosity decreases with temperature and increasing degree of unsaturation of triglyceride fatty chains, which is due to reduced intermolecular forces and increased thermal movement among molecules. Ozonolysis of fatty acids, olive oil, and soybean oil was suggested to follow the Criegee mechanism. The reaction of ozone with vegetable oils occurs almost exclusively with carbon–carbon double bonds present in unsaturated fatty chains. The disappearance of unsaturation and the formation of ozonide were almost equal. The ozonide to aldehyde ratio was always above 90%, which indicates that the major product in the early stage of the reaction was ozonide. The ozonation time for complete consumption of the double bonds was 10 times longer for the oils than for the pure fatty acids (Sadowska *et al.*, 2008). Conversely, Diaz *et al.* (2005) reported a decrease in viscosity of ozonized coconut oil with ethanol compared to ozonized coconut oil with water. This behavior may be due to the solubility of oil with ethanol. When water is used in the systems, the viscosity increased due to the poor solubility of oil with water. The reaction of ozone with oil occurs exclusively with carbon–carbon double bonds present in unsaturated fatty chains. The disappearance of unsaturation (double bond) leads to the formation of ozonide.

3.4.7 Dense-Phase Carbon Dioxide

There are a limited number of studies available in the literature regarding the effect of DPCD on product quality and its influence on the rheological properties of foods. DPCD as a nonthermal pasteurization technique for beer was found to preserve aroma, flavor, foam capacity, and stability coupled with reducing beer haze and extending shelf-life (Dagan and Balaban, 2006). Parton *et al.* (2007) tested a continuous DPCD system for liquid foods such as grape must, orange juice, and tomato paste. No qualitative physical or chemical changes after high-pressure CO_2 treatment were reported with respect to control samples. DPCD treatment of orange juice improves some physical and quality attributes such as cloud formation and stability (Arreola *et al.*, 1991). Moreover, color and cloudiness of DPCD-treated juice was preferred over untreated juice. Dagan and Balaban (2006) employed a continuous DPCD system for pasteurization of beer and reported a reduction in haze by DPCD processing from 146 NTU to 95 NTU compared to fresh beer at processing conditions of 26.5 MPa, 21°C, 9.6% CO_2, and 4.77 min residence time. Haze is an important rheological property of beer that can be defined as the formation of a colloidal suspension that scatters light and makes a beverage appear cloudy. However, they reported a slight decrease in foaming capacity and stability by DPCD. Changes reported for foam characteristics due to DPCD processing may have been caused by the extraction of yeast cell membrane or cell wall parts that may have changed the amount of hydrophobic compounds in the beer, therefore affecting foaming.

DPCD can inactivate pathogenic and spoilage microorganisms and enzymes pertinent to milk and milk products (Hong and Pyun, 2001). A major disadvantage for potential applications of DPCD in dairy products is its negative influence on the rheological properties of milk due to changes in milk protein (casein). As CO_2 dissolves in the aqueous portion of food it undergoes reaction of CO_2 with water to form carbonic acid, thus lowering pH. This further dissociates to yield bicarbonates, carbonates, and H^+ ions, lowering extracellular pH and aiding inactivation on microorganisms. This lowering of pH in milk causes precipitation of casein, which has an isoelectric point of pH 4.6. This pH-lowering effect of DPCD can be employed in casein production (Tomasula *et al.*,1997; Hofland et al., 1999).

Precipitation of casein has a negative influence on consumer sensory perception. Dairy products prepared from DPCD-processed milk, such as cheese, have resulted in improved rheological parameters of cheese making such as reduced clotting time, augmentation of curd hardness, and whey losses along with a slight increase of cheese yield (Ruas-Madiedo *et al.*, 2002). DPCD treatment of milk is reported to reduce particle size distribution compared to controls. The change in the particle size distribution indicates a modification to the fat when milk was treated with CO_2 and pressure.

3.4.8 Ohmic Heating

The changes of rheological properties of liquid foods during ohmic heating are discussed in Section 11.5.7.

3.5 CONCLUSIONS

There are significant opportunities and challenges to the adoption of novel thermal and non-thermal technologies for foods displaying complex rheologies. There are few reported studies on the effects of complex flows on the process efficacy for food-treatment systems. If process validation can be achieved, foods with desirable organoleptic properties can be produced. The development of global models describing the influence of both the control parameters and the intrinsic fluid properties are required to ensure both treatment homogeneity and the prevention of overtreatment.

References

Abdul Ghani, A.G., Farid, M.M., 2007. Numerical simulation of solid–liquid food mixture in a high pressure processing unit using computational fluid dynamics. J. Food Eng. 80 (4), 1031–1042.

Aguilo-Aguayo, I., Soliva-Fortuny, R., Martin-Belloso, O., 2009. Changes in viscosity and pectolytic enzymes of tomato and strawberry juices processed by high-intensity pulsed electric fields. Int. J. Food Sci. Technol. 44 (11), 2268–2277.

Aguilo-Aguayo, I., Soliva-Fortuny, R., Martin-Belloso, O., 2010. Color and viscosity of watermelon juice treated by high-intensity pulsed electric fields or heat. Innovative Food Science and Emerging Technologies. 11 (2), 299–305.

Ahmed, J., Ramaswamy, H.S., Alli, I., Ngadi, M., 2003. Effect of high-pressure on rheological characteristics of liquid egg. Lebensmittel-wissenschaft und-technologie 36, 517–524.

Ahmed, J., Ramaswamy, H.S., Hiremath, N., 2005. The effect of high pressure treatment on rheological characteristics and colour of mango pulp. Int. J. Food Sci. Technol. 40, 885–895.

Ahmed, J., Ramaswamy, H.S., Kasapis, S., Boye, J.I., 2009. Novel Food Processing: Effects on Rheological and Functional Properties. CRC Press.

Altic, L., Rowe, M.T., Grant, I.R., 2007. UV Light Inactivation of Mycobacterium avium subsp. paratuberculosis in Milk as Assessed by FASTPlaqueTB Phage Assay and Culture. Appl. Environ. Microbiol. 73 (11), 3728–3733.

Anema, S.G., 2010. Effect of pH at pressure treatment on the acid gelation of skim milk. Innovative Food Science and Technology 40, 885–895.

Arreola, A.G., Balaban, M.O., Marshall, M.R., Peplow, A.J., Wei, C.I., Cornell, J.A., 1991. Supercritical CO2 effects on some quality attributes of single strength orange juice. J. Food Sci. 56 (4), 1030–1033.

Ashokkumar, M., Grieser, F., 1999. Ultrasound assisted chemical process. Rev. Chem. Eng. 15 (1), 41–83.

Ayadi, M.A., Benezech, T., Chopard, F., Berthou, M., 2008. Thermal performance of a flat ohmic cell under non-fouling and whey protein fouling conditions. LWT 41, 1073–1081.

Barnes, H.A, Walters, K., 1985. The yield stress myth? Rheol. Acta 24 (4), 323.

Beresovsky, N., Kopelman, I.J., Mizrahi, S., 1995. The role of pulp interactions in determining tomato juice viscosity. J. Food Process. Preserv. 19, 133–146.

Bhat, R., Karim, A.A., 2009. Ultraviolet irradiation improves gel strength of fish gelatine. Food Chem. 113, 1160–1164.

Capellas, Needs, 2003. Physical properties of yogurt prepared from pressure-treated concentrated or fortified milks. Milchwissenschaft 58 (2003), 46–48.

Criegee, 1975. Mechanism of Ozonolysis. doi:10.1002/anie197507451.

Dagan, G.F., Balaban, M.O., 2006. Pasteurization of beer by a continuous dense-phase CO2 system. J. Food Sci. 71 (3), 715–719.

Diaz, M.F., Nunez, N., Quincose, D., Diaz, W., Hernandez, F., 2005. Study of Three Systems of Ozonized Coconut Oil. Ozone: Sci. Eng. 27 (2), 153–157.

Dogan, M., Kayacier, A., Ic, E., 2007. Rheological characteristics of some food hydrocolloids processed with gamma irradiation. Food Hydrocolloids 21 (3), 392–396.

Ferrentino, G., Plaza, M.L., Ramirez-Rodrigues, M., Ferrari, G., Balaban, M.O., 2009. Effects of dense phase carbon dioxide pasteurization on the physical and quality attributes of a red grapefruit juice. J. Food. Sci. 74 (6), 333–341.

Garcia-Amezquita, L.E., Primo-Mora, A.R., Barbosa-Cánovas, G.V., Sepulveda, D.R., 2009. Effect of nonthermal technologies on the native size distribution of fat globules in bovine cheese-making milk. Innovative Food Sci. Emerg. Technol. 10, 491–494.

Gentry, T.S., Roberts, J.S., 2005. Design and evaluation of a continuous flow microwave pasteurization system for apple cider. LWT 38, 227–238.

Gervilla, R., Ferragut, V., Guamis, B., 2001. High hydrostatic pressure effects on colour and milk-fat globule of ewe's milk. J. Food Sci. 66 (6), 880–885.

Goullieux, A., Pain, J.P., 2005. Ohmic heating, in Emerging Technologies for Food Processing. Edited by: Da-Wen Sun, Elsevier.

Harte, F., Luedecke, L., Swanson, B.G., Barbosa-Cánovas, G.V., 2003. Low fat set yogurt made from milk subjected to combinations of high hydrostatic pressure and thermal processing. J. Dairy Sci. 86 (4), 1074–1082.

Hemar, Y., Hui Liu, L., Meunier, N., Woonton, B.W., 2010. The effect of high hydrostatic pressure on the flow behavior of skim milk–gelatin mixtures. Innovative Food Sci. Emerg. Technol. 11 (3), 432–440.

Hofland, G.W., Van Es, M., Luuk, A.M., Van der, W., Witkamp, G.J., 1999. Isoelectric precipitation of casein using high pressure CO2. Ind. Eng. Chem. Res. 38 (12), 4919–4927.

Hong, S.I., Pyun, Y.R., 2001. Membrane damage and enzyme inactivation of L. plantarum by high pressure CO2 treatment. Int. J. Food Microbiol. 63, 19–28.

Huppertz, T., Fox, P.F., Kelly, A.L., 2004. High pressure treatment of bovine milk: Effects on casein micelles and whey proteins. Journal of Dairy Research, 71.

Johnston, D.E., Gray, R.H., 2008. The effects of operation sequence on the rheology of waxy maize starch dispersions in skim milk subjected to heat and high pressure treatments. Milchwissenschaft 63 (1), 15–18.

Kabalnova, N.N., Murinov, K.Y., Mullagaliev, R., Krasnogorskaya, N.N., Shereshovets, V.V., Monakov, Y.B., et al., 2001. Oxidative destruction of chitosan under effect of ozone and hydrogen peroxide. J. Appl. Polym. Sci. 81, 875–881.

Kardos, N., Luche, J.L., 2001. Sonochemistry of carbohydrate compounds. Carbohydr. Res. 332, 115–131.

Khurana, M., Karwe, M.V., 2009. Numerical prediction of temperature distribution and measurement of temperature in a high hydrostatic pressure food processor. Food Bioprocess Technol. 2, 279–290.

Krebbers, B., Matser, A.M., Hoogerwerf, S.W., Moezelaar, R., Tomassen, M.M., van den Berg, R.W., 2003. Combined high-pressure and thermal treatments for processing of tomato puree: evaluation of microbial inactivation and quality parameters. Innovative Food Sci. Emerg. Technol. 4, 377–385.

Krešić, G., Lelas, V., Jambrak, A.R., Herceg, Z., Brnčić., S., 2008. Influence of novel food processing technologies on the rheological and thermophysical properties of whey proteins. J. Food Eng. 87 (1), 64–73.

Kumar, P., Coronel, P., Truong, V.D., Simunovic, J., Swartzel, K.R., Sandeep, K.P., et al., 2008. Overcoming issues associated with the scale-up of a continuous flow microwave system for aseptic processing of vegetable purees. Food Res. Int. 41, 454–461.

Mason, T.J., Riera, E., Vercet, A., Lopez-Bueza, P., 2005. Application of ultrasound. In: Sun, D.-W. (Ed.), Emerging Technologies for Food Processing. Elsevier, Cambridge, MA, pp. 323–351.

Matsuda, A.H., Sabato, S.F., 2004. Effect of irradiation on Brazilian honeys' consistency and their acceptability. Radiat. Phys. Chem. 71 (2004), 107–110.

Min, B.R., Nam, K.C., Lee, E.J., Ko, G.Y., Trampel, D.W., Ahn, D.U., 2005. Effect of irradiating shell eggs on quality attributes and functional properties of yolk and white. Poultry Sci. 84 (11), 1791–1796.

Molina, E., Álvarez, M, Ramos, M., Olano, A, López-Fandiño, R., 2000. Use of high-pressure-treated milk for the production of reduced-fat cheese. Int. Dairy J. 10, 467–475.

Muller, F.L., Pain, J.P., Villion, P., 1994. On the behavior of non-Newtonian liquids in collinear ohmic heaters. In proceedings of 10th international Heat Transfer Conference, Brighton, pp. 285–290.

Needs, E.C., Capellas, M., Bland, A.P., Manoj, P., MacDougal, D., Paul, G., 2000. Comparison of heat and pressure treatments of skim milk, fortified with whey protein concentrate, for set yoghurt preparation: Effects on milk proteins and gel structure. J. Dairy Res. 67, 329–348.

No, H.K., Kim, S.D., Kim, D.S., Kim, S.K., Meyers, S.P., 1999. Effect of physical and chemical treatment on chitosan viscosity. J. Chitin Chitosan 4 (4), 177–183.

Parton, T., Bertucco, A., Elvassore, N., Grimolizzi, L., 2007. A continuous plant for food preservation by high pressure CO2. J. Food Eng. 79, 1410–1417.

Penna, A.L.B., Subbarao-Gurram, Barbosa-Cánovas, G.V., 2007. High hydrostatic pressure processing on microstructure of probiotic low-fat yogurt. Food Res. Int. 40, 510–519.

Price, G.J., 1990. The use of ultrasound for the controlled degradation of polymer solutions. Advances in Sonochemistry 1, 231–287.

Rao, M.A., Cooley, H.J., 1992. Rheology of tomato pastes in steady dynamic shear. J. Texture Stud. 12, 521–538.

Rauh, C., Baars, A., Delgado, A., 2009. Uniformity of enzyme inactivation in a short-time high-pressure process. J. Food Eng. 91, 154–163.

Ruas-Madiedo, P., Alonso, L., Delgado, T., Bada-Gancedo, J.C., de Los Reyes-Gavilán, C.G., 2002. Manufacture of Spanish hard cheeses from CO2-treated milk. Food Res. Int. 35 (7), 681–690.

Sadowska, J., Johansson, B., Johannessen, E., Friman, R., Broniarz-Press, L., Rosenholm, J.B., 2008. Characterization of ozonated vegetable oils by spectroscopic and chromatographic methods. Chem. Phys. Lipids 151, 85–91.

Salih, E.A., Choong, T., Sergie, S.Y., Chin, N.L., Ibrahim, O.M., 2009. The Effect of Buoyancy Force in Computational Fluid Dynamics Simulation of a Two-Dimensional Continuous Ohmic Heating Process. American J. Appl. Sci. 6 (11), 1902–1908.

Sampedro, F., Rodrigo, D., Martínez, A., Barbosa-Cánovas, G.V., Rodrigo, M., 2006. Application of Pulsed Electric Fields in Egg and Egg Derivatives. Food Sci. Technol. Int. 12 (5), 397–405.

Schrive, L., Grasmick, A., Moussiere, S., Sarrade., S., 2006. Pulsed electric field treatment of Saccharomyces cerevisiae suspensions: A mechanistic approach coupling energy transfer, mass transfer and hydrodynamics. Biochem. Eng. J. 27, 212–224.

Seo, S., King, J.M., Prinyawiwatkul, W., 2007. Simultaneous depolymerization and decolorization of chitosan by ozone treatment. J. Food Sci. 72 (9), C522–C526.

Seshadri, R., Weiss, J., Hulbert, G.J., Mount, J., 2003. Ultrasonic processing influences rheological and optical properties of high-methoxyl pectin dispersions. Food Hydrocolloids 17, 191–197.

Shim, J.Y., Lee, S.H., Jun, S., 2010. Modeling of ohmic heating patterns of multiphase food products using computational fluid dynamics codes. J. Food Eng. 99 (2), 136–141.

Steffe, J.F., 1996. Rheological Methods in Food Process Engineering. Second ed. Freeman Press, Michigan.

Suslick, K., 1988. Ultrasound: Its Chemical, Physical and Biological Effects. VCH, New York.

Tiwari, B.K., Muthukumarappan, K., O'Donnell, C.P., Cullen, P.J., 2008. Modeling colour degradation of orange juice by ozone treatment using response surface methodology. J. Food Eng. 88, 553–560.

Tiwari, B., Muthukumarappan, K., O' Donnell, C.P., Cullen, P.J., 2010. Rheological properties of sonicated guar, xanthan and pectin dispersions. B. K. Int. J. Food Prop. 13 (2), 223–233.

Tomasula, P.M., Craig, J.C., Boswell, R.T., 1997. A continuous process for casein production using high pressure CO2. J. Food Eng. 33 (3/4), 405–419.

Tsai, S.C., Zammouri, K., 1988. Role of interparticular van der Waals forces in rheology of concentrated suspensions. J. Rheol. 32, 737–750.

Udabage, P., Augustin, M., Versteeg, C., Puvanenthiran, A., Yoo, A., Narissara, A., et al., 2010. Properties of low-fat stirred yoghurtsmade from high-pressure-processed skimmilk. Innovative Food Sci. Emerg. Technol. 11, 32–38.

Van der Plancken, I., Grauwet, T., Oey, I., Van Loey, A., Hendrickx, M., 2008. Impact evaluation of high pressure treatment on foods: considerations on the development of pressure–temperature–time integrators (pTTIs). Trends Food Sci. Technol. 19, 337–348.

Vercet, A., Sanchez, C., Burgos, J., Montanes, L., Lopez-Buesa, P., 2002. The effects of manothermosonication on tomato pectic enzymes and tomato paste rheological properties. J. Food Eng. 53, 273–278.

Villamiel, M., de Jong, P., 2000. Influence of high-intensity ultrasound and heat treatment in continuous flow on fat, proteins and native enzymes of milk. J. Agric. Food Chem. 48, 472–478, Volume 11, Issue 2, April 2010, Pages 265–273.

Voragen, A.G.J., Pilnik, W., Thibault, J.F., Axelos, M.A.V., Renard, C.M.G.C., 1995. Pectins. In: Stephen, A.M. (Ed.), Food Polysaccharides and Their Applications. Marcel Dekker, New York, pp. 287–340.

Walker, M.K., Farkas, D.F., Loveridge, V., Meunier-Goddik, L., 2006. Fruit Yogurt processed with High Pressure. Intl. J. Food Sci. Tech. 41, 464–467.

Yau-Hoong Kuan, Rajeev Bhat, Chandra Senan, Williams, P.A., Karim, A.A., 2009. Effects of ultraviolet irradiation on the physicochemical and functional properties of gum arabic. J. Agric. Food Chem. 57 (19), 9154–9159.

Yue, W., Yao, P., Wei, Y., Li, S., Lai, F, Liu, X., 2008. An innovative method for preparation of acid-free-water-soluble low-molecular-weight chitosan (AFWSLMWC). Food Chem. 108 (3), 1082–1087.

Zhang, Q.H., Qin, B.L., Barbosa-Cànovas, G.V., Swanson, B.G., 1995. Inactivation of E. coli for food pasteurization by high-strength pulsed electric fields. J. Food Process. Preserv. 19 (2), 103–118.

CHAPTER

4 Pulsed Electric Field Processing of Fluid Foods

Pedro Elez-Martínez, Ángel Sobrino-López, Robert Soliva-Fortuny and Olga Martín-Belloso

Department of Food Technology, University of Lleida, Spain

4.1 INTRODUCTION

Consumers are currently demanding fresh-like appearance and taste of foods with minimal changes in their nutritional content. As a result, the food industry is focusing on the development of new preservation techniques in order to meet consumer expectations. Pulsed electric fields (PEF) is a non-thermal food processing technology that has been studied in recent years as an alternative or as a complement to heat treatments. Inactivation of pathogenic and spoilage microorganisms can be attained together with a minimal impact on food quality and nutritional properties, which suggests that it is possible to obtain safe high-quality PEF-treated fluid foods with an acceptable shelf-life.

Despite the specific uses of electric fields, research regarding PEF application and its effects is addressed towards liquid products. In fact, potential implementation throughout continuous-flow processes and versatility favor the application of PEF on a great variety of fluid foods, such as juices (Espachs-Barroso *et al.*, 2003; Elez-Martínez *et al.*, 2009), beverages (Evrendilek *et al.*, 2004a; Garde-Cerdán *et al.*, 2008; Puértolas *et al.*, 2009; Zhao *et al.*, 2009), milk (Sampedro *et al.*, 2005; Sobrino-López and Martín-Belloso, 2009), dairy products (Sampedro *et al.*, 2005), and liquid egg (Sampedro *et al.*, 2006a). Most of the scientific information about the effects of PEF on fluid foods is focused on microbial and enzyme inactivation. However, despite being a recent matter of interest, several studies exist about the effects of PEF on quality, sensory characteristics, and bioactive compounds of fluid foods.

This chapter presents the most outstanding results obtained so far concerning the implementation of PEF for fluid food processing. The effects of PEF on those foods in relation to microorganisms, enzymes, quality and bioactive components are described here.

Novel Thermal and Non-Thermal Technologies for Fluid Foods. DOI: 10.1016/B978-0-12-381470-8.00004-9

4.2 PRINCIPLES OF OPERATION

PEF processing involves the application of short pulses (μs) of high voltage (kV/cm) to foods placed between two electrodes. Application of PEF is restricted to food products that can withstand high electric fields, have low electrical conductivity, and do not contain or form bubbles. The particle size of the food may also be a limitation (IFT, 2001).

In general, PEF treatment systems are composed of PEF treatment chambers, a pulse generator, a fluid-handling system, and monitoring systems (Min *et al.*, 2007). The treatment chamber is used to house electrodes and deliver a high voltage to the food material. It is generally composed of two electrodes held in position by insulating material, thus forming an enclosure containing the food material. Therefore, the proper design of the treatment chamber is an essential component for the efficiency of the PEF technology (Alkhafaji and Farid, 2007). Optimization of the treatment chamber design is of great importance to properly apply the electrical field but also to avoid unexpected phenomena during the treatment, such as heat accumulation, electric field perturbations, deposition of particles, gas-bubble formation, and arching (Huang and Wang, 2009). With regards to PEF treatment-chamber configuration, parallel, coaxial and colinear configurations are used for processing foods by PEF (Fig. 4.1). The parallel-plate chambers are generally separated by a gap that is considerably smaller than the electrode's surface (Dunn and Pearlman, 1987; Zhang *et al.*, 1995a). Treatment chambers with parallel plate-electrodes provide a uniform electrical-field distribution along the gap axes and electrode surfaces, but create a field enhancement problem at the edge of the electrodes (Huang and Wang, 2009). On the other hand, coaxial and cofield chambers are widely used due to their simplicity in structure (Zhang *et al.*, 1995a). Electrical current flows perpendicularly to food flow in coaxial PEF treatment chambers and in parallel to food flow in cofield flow PEF treatment chambers (Dunn, 2001). The cofield chamber configuration has advantageous fluid dynamics, highly desirable for food processing and convenient for cleaning in place (Góngora-Nieto *et al.*, 2002).

Treatment chamber design has evolved from static to continuous treatment chambers. Static chambers are mainly suitable for laboratory use in order to evaluate the influence of any relevant parameter critical to the process efficiency, while for large-scale operations, continuous chambers seem to perform much better. The main limitation of parallel chambers is the batch nature of the process. However, while cofield and coaxial chambers allow continuous treatment, the electric fields are not as homogenous as in the case of parallel chambers (Góngora-Nieto *et al.*, 2002).

Chamber electrodes, as well as insulating components, must be made of food-grade, chemically inert materials in order to prevent food contamination. In addition, PEF treatment chamber materials should be washable and autoclavable (Huang and Wang, 2009).

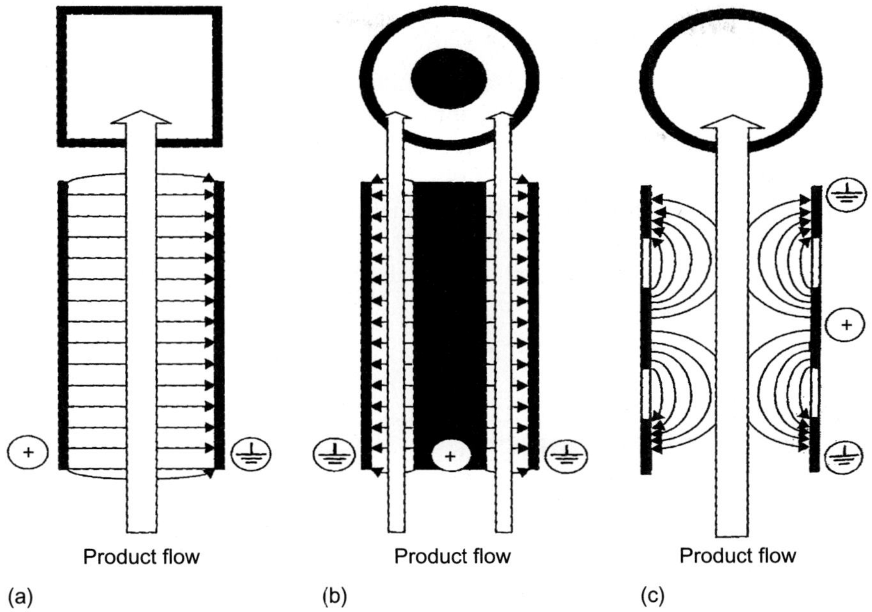

FIGURE 4.1

Configurations of treatment chambers for continuous PEF treatment: (a) parallel plate, (b) coaxial, and (c) co-linear configuration.

[Reproduced with permission from Toepfl et al. *(2005).]*

With regards to continuous PEF treatments, two different fluid handling systems can be used to allow the treatment of liquid foods (Evrendilek and Zhang, 2005). On the one hand, the fluid can be repeatedly pumped though the system, as many times as needed, to apply the desired treatment time in a stepwise circulation mode. On the other hand, a recirculation mode is also possible. In this latter case, liquid is pumped to the feeding container without interruption until the whole treatment is applied.

Pulse generators convert low voltage into high voltage and provide the latter to PEF chambers (Qin *et al.*, 1995a). Generation of PEF requires a fast discharge of electrical energy within a short period of time. Square, exponential decay, or oscillatory pulses are generally used for PEF treatment of fluid foods (Bendicho *et al.*, 2002a; Espachs-Barroso *et al.*, 2003; Sampedro *et al.*, 2006b). Because discharge time is extremely short, heating of the foods is minimized (Zhang *et al.*, 1995a).

Temperature- and pulse-monitoring systems are the main equipment used to supervise PEF processing. Temperature is monitored with thermocouples, while pulses are monitored with high-voltage probes, current monitors, and oscilloscopes (Min *et al.*, 2003a).

4.3 PROCESS CONTROL PARAMETERS

The feasibility of processing fluid foods by PEF is related to the potential of this technology for inactivating microorganisms and quality-related enzymes, while maintaining sensory and nutritional properties. The critical factors determining the efficiency of processing foods by PEF can be classified as (i) treatment parameters, (ii) microorganism, enzyme, and quality-related compounds characteristics, and (iii) product parameters. Tables 4.1, 4.2, and 4.3 summarize the effects of process control parameters on microbial survival, enzyme inactivation and health related compounds.

4.4 TREATMENT PARAMETERS

Processing factors such as electric field strength, treatment time, pulse shape, pulse width, pulse frequency, pulse polarity, and temperature, as well as the treatment in batch or continuous flow mode have been reported to have influential effects on the efficiency of PEF technology for processing fluid foods.

Electric field strength and total treatment time are the most studied PEF treatment parameters. An increase in these factors generally tends to increase the inhibition of microorganisms (Góngora-Nieto *et al.*, 2002; Min *et al.*, 2007; Mosqueda-Melgar *et al.*, 2008c). However, the effects of pulse frequency and pulse width on microbial inactivation have not been yet completely understood. For treatments with similar field strength and number of pulses, inactivation of microorganisms increased with the increase in pulse width (Martín *et al.*, 1997; Martín-Belloso *et al.*, 1997; Aronsson *et al.*, 2001; Mabrook and Petty, 2003). Nonetheless, when microorganisms were subjected to the same treatment time and field strength, microbial inactivation decreased with pulse width (Elez-Martínez *et al.*, 2004, 2005; Sampedro *et al.*, 2006c). While some results have shown that an increase of the pulse frequency improves microbial inactivation (Aronsson *et al.* 2001; Otunola *et al.*, 2008), others have shown no significant effects (Mosqueda-Melgar *et al.* 2007; 2008a).

PEF may be applied in the form of exponentially decaying and square-wave pulses. Studies comparing the efficacy of different pulse wave forms have indicated that both are effective for microbial inactivation; nonetheless square wave pulses save more energy and require less cooling effort (Góngora-Nieto *et al.*, 2002; De Haan and Willcock, 2002; Kotnik *et al.*, 2003). In terms of pulse polarization, bipolar pulses are more efficient than monopolar pulses regarding the inhibition of microorganisms (Qin *et al.*, 1994; Ho *et al.*, 1995; Elez-Martínez *et al.*, 2004, 2005).

PEF processing of foods may be carried out in batch mode or in a continuous process. In general, continuous processes reached higher microbial inactivation rates than batch treatments (Martín *et al.*, 1997). PEF treatments at moderate temperature (50–60°C) have exhibited a synergistic effect on the inactivation of microorganisms (Rastogi, 2003). An increase in PEF treatment temperature leads to higher microbial inactivation levels (Shin and Pyun, 1999; Aronsson and Röner 2001; Lindgren *et al.*, 2002).

Processing factors such as electric field strength, treatment time, pulse shape, pulse width, pulse frequency, pulse polarity, electrical energy density, and temperature, as well as the flow nature of the treatment have been reported to have a determinant influence on the inactivation of quality-related enzymes. Among them, electric field strength and treatment time have been the most studied parameters. Hence, the effectiveness of enzyme inactivation by PEF is higher when electric-field strength and treatment time increased (Yeom *et al.*, 2002; Bendicho *et al.*, 2002b, 2003; Min *et al.*, 2003b; Rodrigo *et al.*, 2003a; Cortés *et al.*, 2005; Elez-Martínez *et al.*, 2006a, 2007; Marsellés-Fontanet and Martín-Belloso, 2007; Aguiló-Aguayo *et al.*, 2008b; Giner *et al.*, 2009). Enzyme inactivation caused by PEF increases with a rise in pulse frequency and width (Elez-Martínez *et al.*, 2006a, 2007; Bendicho *et al.*, 2003, Li *et al.*, 2008; Aguiló-Aguayo *et al.*, 2009a).

Controversial results regarding pulse polarity have been obtained depending on the enzyme and the food under study. Bipolar pulses led to greater pectin methyl esterase (PME) inactivation in orange juice than monopolar pulses (Elez-Martínez *et al.*, 2007). However, PME activity in strawberry juice was not affected when a PEF treatment was applied in mono- or bipolar mode (Aguiló-Aguayo *et al.*, 2009a). Aguiló-Aguayo *et al.* (2008b) concluded that the greatest peroxidase (POD) inactivation in tomato juice was obtained after PEF processing with bipolar pulses, while the maximum inactivation of orange juice POD was achieved in monopolar mode (Elez-Martínez *et al.*, 2006a).

PEF processing can be carried out using batch or continuous flow equipments. Bendicho *et al.* (2005) observed that the inhibition of the activity of a protease from *Bacillus subtilis* greatly depended on the flow mode of the PEF-processing devices. Moreover, temperature during processing plays an important role on the enzyme inactivation by PEF. The majority of studies conclude that enzyme activity is diminished by increasing the temperature during PEF processing (Van Loey *et al.*, 2002; Yeom *et al*, 2002; Min *et al.*, 2003b; Rodrigo *et al.*, 2003a; Yang *et al.*, 2004a; Sampedro *et al.*, 2009a).

Nutritional properties of fluid foods processed by PEF depend also on the treatment conditions. According to Sánchez-Moreno *et al.* (2009), the retention of vitamin C after PEF processing is influenced by the different treatment conditions applied to the product as well as by the different PEF equipments used. In general, the lower the electric field strength and the treatment time, the higher the vitamin retention of PEF-treated foods (Bendicho *et al.*, 2002c; Torregrosa *et al.*, 2006; Elez-Martínez and Martín-Belloso, 2007; Oms-Oliu *et al.*, 2009). Bipolar pulses have been shown to lead to higher vitamin C contents, whereas the lower the pulse frequency and the pulse width, the higher the vitamin C retention

(Elez-Martínez and Martín-Belloso, 2007; Odriozola-Serrano *et al.*, 2009a; Oms-Oliu *et al.*, 2009). PEF processing parameters affected the concentration of carotenoids of fluid foods. Some authors have reported that an increase in the value of treatment parameters, meaning an increase in the treatment intensity, led to a greater content of carotenoids in PEF-treated fluid foods (Torregrosa *et al.*, 2005; Odriozola-Serrano *et al.*, 2008b; Oms-Oliu *et al.*, 2009). Phenolic content of PEF-processed foods is influenced by treatment conditions. In this sense, both pulse width and frequency, as well as pulse polarity determine the profile and total phenolic content of fluid foods (Odriozola-Serrano *et al.*, 2009a). As a result of the effects of PEF treatment parameters on health-related compounds, these processing parameters also impact on the antioxidant capacity of foods (Elez-Martínez *et al.*, 2007; Odriozola-Serrano *et al.*, 2009a; Oms-Oliu *et al.*, 2009).

In general, PEF processing shows little effect on the quality parameters of foods. While color, pH, total acidity, soluble solids, or formol index are only slightly affected by changing PEF treatment parameters (Cortés *et al.*, 2005; Akin and Evrendilek, 2009), viscosity may vary significantly with pulse frequency, width, and polarity (Aguiló-Aguayo *et al.*, 2009a).

4.5 MICROORGANISM, ENZYME AND QUALITY- AND HEALTH-RELATED COMPOUND CHARACTERISTICS

The effectiveness of PEF in inactivating microorganisms changes with the type and the growth stage of microorganisms as well as the initial concentration of viable cells. Many publications have demonstrated that yeasts are more sensitive to PEF treatment than vegetative bacteria, and within bacteria, Gram-negative bacteria are more susceptible to treatments (Zhang *et al.*, 1994a; Wouters and Smelt, 1997; Qin *et al.*, 1998; Aronsson *et al.*, 2001; García *et al.*, 2007; Mosqueda-Melgar *et al.*, 2008c). In general, bacterial spores are resistant to PEF treatments, but after germination they become PEF-sensitive (Marquez *et al.*, 1997; Barbosa-Cánovas *et al.*, 1998; Barsotti and Cheftel, 1999; Shin *et al.*, 2010). Moreover, different strains of the same microorganisms showed dissimilar resistance to PEF processing (Somolinos *et al.*, 2008a; Saldaña *et al.*, 2009). Cells in the logarithmic phase were more sensitive to PEF treatment than those in the stationary phase, as was found by Gásková *et al.* (1996), Pothakamury *et al.* (1996), Álvarez *et al.* (2000), Rodrigo *et al.* (2003b), Somolinos *et al.* (2008b), and Sagarzazu *et al.* (2010). However, microorganisms inoculated in juices are more sensitive to PEF in the stationary phase (Molinari *et al.*, 2004) The effects of the initial microbial concentration on the inactivation rate achieved by PEF is not clear (Donsi *et al.*, 2007). Several authors reported that microbial inactivation is not a function of the initial concentration (Zhang *et al.*, 1995b; Álvarez *et al.*, 2000; Selma *et al.*, 2003). However, Zhang *et al.* (1994b, 1994c) and

Molinari *et al.* (2004) reported that initial concentration of the microorganism was inversely correlated with its survival fraction after a PEF treatment.

The type, source, and concentration of the enzyme have a significant effect on the reached inactivation level (Castro *et al.*, 2001; Marsellés-Fontanet and Martín-Belloso, 2007; Aguiló-Aguayo *et al.*, 2008a, 2009b, 2010a; Riener *et al.*, 2008a). For instance, polyphenoloxidase (PPO) from pear required stronger PEF treatment conditions than apple PPO to reach the same degree of inactivation (Giner *et al.*, 2001). However, Rodrigo *et al.* (2003a) reported greater PME inactivation for higher initial PME concentrations.

The impact of PEF processing on quality-related properties and nutritional-related compounds is influenced by the nature of these components. Color, viscosity, hydroxymethyl furfural, soluble solids, pH, and total acidity are affected with different intensity by similar PEF treatments (Elez-Martínez *et al.*, 2006b; Rivas *et al.*, 2006b; Charles-Rodríguez *et al.*, 2007; Akin and Evrendilek, 2009; Aguiló-Aguayo *et al.*, 2008c, 2009a, c, 2010b). The effects of PEF processing on the flavor profile of foods rely on the different impacts of PEF treatments on individual volatile compounds (Yeom *et al.*, 2000a; Min *et al.*, 2003a; Aguiló-Aguayo *et al.*, 2009d; Li *et al.*, 2009; Sampedro *et al.*, 2009a; Valappil *et al.*, 2009). Health-related compound families, such as vitamins, phenolics, carotenoids, amino acids, and fatty acids, are affected in a different way by PEF processing (Soliva-Fortuny *et al.*, 2009). Morever, impact of PEF treatment is different depending on the individual health-related compound belonging to the same family studied (Bendicho *et al.*, 2002c; Sánchez-Moreno *et al.*, 2005a; Torregrosa *et al.*, 2005; Garde-Cerdán *et al.*, 2007; Zulueta *et al.*, 2007; Odriozola-Serrano *et al.*, 2008cd; Riener *et al.*, 2008a).

4.6 PRODUCT PARAMETERS

The effectiveness of PEF processing is directly related to both physical and chemical properties of products as well as to their composition. Critical product parameters include electric conductivity, density, viscosity, pH, and water activity.

In general, PEF treatment is more effective for the microbial inhibition in foods with low electrical conductivity (Min *et al.*, 2007). However, Álvarez *et al.* (2000) reported that conductivity did not influence the inactivation of *Salmonella senftenberg* and Gásková *et al.* (1996) found that the inactivation of *Saccharomyces cerevisiae* was inversely related to the medium conductivity. Therefore, it can be concluded that the effect of conductivity on microbial destruction depends on the characteristics of microorganisms (Mosqueda-Melgar *et al.*, 2007). Likewise, the effect of pH on the inhibition of microorganisms by PEF depends on the characteristics of the microorganisms studied (Mosqueda-Melgar *et al.*, 2007; Sagarzazu *et al.*, 2010; Saldaña *et al.*, 2009, 2010). Although the role of water activity in the survival of microorganisms subjected to PEF treatment has not been widely studied, it was found that low water activities appear to protect microorganisms from PEF (Aronsson and Rönner, 2001; Min *et al.*, 2002; Arroyo *et al.*, 2010). The influence of the composition of the medium on microbial inactivation by PEF is not well known. While some investigators have reported a protective effect of several food constituents such as xanthan gum (Ho *et al.*, 1995), proteins (Martín *et al.*, 1997; Jaeger *et al.*, 2009) or fats (Grahl and Märkl, 1996), others did not observe differences between microbial inactivation in buffers with different components and in a complex medium such as milk (Reina *et al.*, 1998; Dutrueux *et al.*, 2000; Mañas *et al.*, 2001).

The inactivation effect of PEF on enzymes is affected by the electrical conductivity of the media. Hence, the higher the electrical conductivity of the medium, the more effective the PEF inactivation of pepsin and PPO (Yang *et al.*, 2004ab). However, medium conductivity did not influence the PPO and POD stability towards PEF processing (Van Loey *et al.*, 2002). The effect of electrical conductivity of foods on the PEF-resistance of enzymes is directly related to the PEF processing parameters (Aguiló-Aguayo *et al.*, 2009a). The inactivation effect of PEF on enzymes is affected by the pH of the media (Yang *et al.*, 2004b). Pepsin is more stable to PEF at its optimum pH 2.0 (Yang *et al.*, 2004a). Van Loey *et al.* (2002) observed that lowering the pH to 4 decreased the stability of POD treated by PEF, while changes in pH did not influence the stability of PPO to PEF treatment. Aguiló-Aguayo *et al.*, (2009a) suggested an interactive effect between PEF treatment parameters and pH of food regarding the inactivation of pectolytic enzymes. The influence of the medium composition on the effectiveness of PEF to inactivate enzymes is still not clear. Some authors have reported that the fat content of the medium affects the inactivation of enzymes by PEF (Castro *et al.*, 2001; Bendicho *et al.*, 2003). However, Grahl and Märkl (1996), Ho *et al.* (1997), and Van Loey *et al.* (2002) did not observe any significant change in the activity of alkaline phosphatase either in milk or in aqueous solution. Controversial results have been found regarding the effect of protein content of the medium on enzyme inactivation by PEF. Casein has exhibited a protective effect on protease against PEF processing (Vega-Mercado *et al.*, 2001). However, the protein content had no effect on the inactivation of PPO, POD and lipoxygenase (LOX) by PEF (Van Loey *et al.*, 2002). Aguiló-Aguayo *et al.* (2009a) reported that an increase in the soluble solids concentration of foods led to higher levels of inactivation of PEF-treated pectolytic enzymes.

The impact of PEF on quality properties such as color, viscosity, pH, acidity, soluble solids, and flavor has been shown to depend upon the food matrix that is considered (Min *et al.*, 2003a; Elez-Martínez *et al.*, 2006b; Charles-Rodríguez *et al.*, 2007; Aguiló-Aguayo *et al.*, 2008c, 2009cd; Zhao *et al.*, 2008; Akin and Evrendilek, 2004a, 2009; Valappil *et al.*, 2009; Morales-de la Peña *et al.*, 2010). In this sense, similar conclusions are obtained when looking at the effects of food parameters on the nutritional composition of PEF-processed foods (Sánchez-Moreno *et al.*, 2009; Soliva-Fortuny *et al.*, 2009).

4.7 MICROBIAL INACTIVATION MECHANISM

4.7.1 Pulsed Electric Field Inactivation Mechanism

Many studies have focused on the causes and reasons for cell death due to electric-field process. Exposure of microorganisms to a pulsed electric field directly affects the integrity of the cell by a mechanism known as electroporation. It is generally accepted that cell electroporation is caused by mechanical breakdown of the cell membrane due to the compression exerted by the accumulation of free charges at both sides of the membrane (Zimmermann, 1986). Such instability leads to appreciable morphological changes in the cell envelope and/or cytoplasm disorganization and, after that, it is followed by electropermeabilization, consisting of shrinkage and leakage of intracellular content, such as cytoplasmic and nuclear material (Picart *et al.*, 2002; Elez-Martínez *et al.*, 2005; Choi *et al.*, 2008), and pore-formation in the cell wall and/or cell membrane (Rowan *et al.*, 2000; García *et al.*, 2007). Nevertheless, the injuries caused by the electric field seem to obey a gradation ranging from an all-or-nothing activity to the occurrence of sublethal injuries. The degree of the damages infringed seem to be dependent on (i) the type of microorganism, (ii) the pH of the treatment, and (iii) the harshness of the PEF treatment in terms of treatment time and field strength (García *et al.*, 2003, 2005, 2007). In general, yeasts are less resistant than bacteria and, among them, the different cell structure between Gram-positive and Gram-negative may justify their distinct behavior. While Gram-positive bacteria are sublethally injured at neutral pH, major damages on Gram-negative cells occur at acidic pH (García *et al.*, 2005). However, other factors may interfere on the PEF mechanism against cell envelopes according to the contradictory pattern followed by Gram-negative and Gram-positive bacteria.

With respect to the repairing mechanism, injuries mended immediately after ceasing the electric processing may involve reorganization of the cytoplasmic membrane (Garcia *et al.*, 2006; Somomolinos *et al.*, 2008b), while long-lasting damages may require more complex metabolic reactions, such as energy and lipid synthesis (Garcia *et al.*, 2006) or protein synthesis (Somomolinos *et al.*, 2008b). The study of the PEF mechanism on the cell membrane may lead to the development of new techniques and preservation strategies in the sense of enhancing the electric field activity over the target structures or prolonging their effect.

4.7.2 Effects of Pulsed Electric Fields on Microorganisms

The effect of PEF on pathogenic and spoilage microorganisms has been studied in different fluid foods. In general, the PEF process is effective in destroying microorganisms (Dutrueux *et al.*, 2000; Evrendilek *et al.*, 2000; Elez-Martínez *et al.*, 2004; Bazhal *et al.*, 2006; Sobrino-López *et al.*, 2006; Mosqueda-Melgar *et al.*, 2007, 2008a, b; Monfort *et al.*, 2010; Morales-de la Peña *et al.*, 2010). It is generally recognized that PEF treatments are as effective as traditional pasteurization heat treatments with regard to microbial destruction (Yeom *et al.*, 2000a).

The destruction of microorganisms by PEF has been demonstrated in different fluid foods, as shown in Table 4.1. A great number of studies have been carried out regarding the impact of PEF on microorganisms inoculated in juices. However, less information has been found in the scientific literature related to animal food products, such as milk, dairy products, or liquid egg.

Table 4.1 Effects of Pulsed Electric Fields on Microorganisms in Fluid Foods

Microorganism	Food	Treatment Conditions	Microbial Reductions	Reference
Escherichia coli	Apple juice	E = 30 kV/cm, n = 43, τ = 4 μs, T = 25°C	5	Evrendilek *et al.* (2000)
	Apple juice	E = 31 kV/cm, n = 50.5, τ = 4 μs, T = 10°C	2.6	Evrendilek and Zhang (2005)
	Watermelon juice	E = 35 kV/cm, n = 500, τ = 4 μs, T = 40°C	4	Mosqueda-Melgar *et al.* (2007)
	Orange juice	E = 40 kV/cm, n = nd, τ = nd, T = 56°C	6.3	McNamee *et al.* (2010)
	Pea soup	E = 33 kV/cm, n = 30, τ = 2 μs, T = 55°C	6.5	Vega-Mercado *et al.* (1996)
	Liquid whole egg	E = 9–15 kV/cm, n = 138, τ = 2 μs, T = 50–60°C	4	Bazhal *et al.* (2006)
	Skim milk	E = 24 kV/cm, n = 50.4, τ = 2.8 μs, T = 10°C	1.96	Evrendilek and Zhang (2005)
	Carrot juice-based beverage	E = 13–27 kV/cm, n = nd, τ = nd, T = 12°C	3.5	Akin and Evrendilek (2009)
	Green tea beverage	E = 38 kV/cm, n = nd, τ = nd, T = 20°C	5.6	Zhao *et al.* (2008)
	Tropical fruit smoothie	E = 34 kV/cm, n = nd, τ = nd T = 55°C	5.2	Walkling-Ribeiro *et al.* (2008)

(*Continued*)

Table 4.1 (Continued)

Microorganism	Food	Treatment Conditions	Microbial Reductions	Reference
Listeria innocua	Skim milk	E = 30–40 kV/cm, *n* = nd, $\tau = 3\ \mu s$, T = 37°C	2.5	Férnandez-Molina *et al.* (2006)
	Skim milk	E = 41 kV/cm, *n* = 63 pulses, $\tau = 2.5\ \mu s$, T = 37°C	3.9	Dutrueux *et al.* (2000)
	Orange juice	E = 40 kV/cm, *n* = nd , τ = nd, T = 56°C	3.7	McNamee *et al.* (2010)
	Fruit juice–soymilk beverage	E = 35 kV/cm, *n* = 350, $\tau = 4\ \mu s$, T = 32°C	5	Morales-de la Peña *et al.* (2010)
Listeria monocytogenes	Melon juice	E = 35 kV/cm, *n* = 500, $\tau = 4\ \mu s$, T = 40°C	4.3	Mosqueda-Melgar *et al.* (2007)
	Watermelon juice	E = 35 kV/cm, *n* = 500, $\tau = 4\ \mu s$, T = 40°C	3.8	Mosqueda-Melgar *et al.* (2007)
Salmonella enteritidis	Melon juice	E = 35 kV/cm, *n* = 312.15, $\tau = 4$ s, T = 40°C	3.8	Mosqueda-Melgar *et al.* (2007)
	Watermelon juice	E = 35 kV/cm, *n* = 500, $\tau = 4\ \mu s$, T = 40°C	4.3	Mosqueda-Melgar *et al.* (2007)
	Apple juice	E = 35 kV/cm, *n* = nd, $\tau = 4\ \mu s$, T = 40°C	4.5	Martinez-Viedma *et al.* (2008)
	Pear juice	E = 35 kV/cm, *n* = 400, $\tau = 4\ \mu s$, T = 40°C	4.8	Mosqueda-Melgar *et al.* (2008a)
	Orange juice	E = 35 kV/cm, *n* = 425, $\tau = 4\ \mu s$, T = 40°C	5.2	Mosqueda-Melgar *et al.* (2008a)
	Strawberry juice	E = 35 kV/cm, *n* = 425, $\tau = 4\ \mu s$, T = 40°C	4.4	Mosqueda-Melgar *et al.* (2008a)
	Tomato juice	E = 35 kV/cm, *n* = 50–200, $\tau = 4\ \mu s$, T = 38°C	4	Mosqueda-Melgar *et al.* (2008b)

(Continued)

Table 4.1 (Continued)

Microorganism	Food	Treatment Conditions	Microbial Reductions	Reference
Salmonella typhimurium	Liquid whole egg	E = 45 kV/cm, *n* = 10, τ = 3 μs, T = 20°C	4	Monfort *et al.* (2010)
Staphylococcus aureus	Milk	E = 35 kV/cm, *n* = 150, τ = 8 μs, T = nd	4.5	Sobrino-López *et al.* (2006)
	Liquid whole egg	E = 40 kV/cm, *n* = 6, τ = 3 μs, T = 20°C	3	Monfort *et al.* (2010)
	Green tea beverage	E = 38 kV/cm, *n* = nd , τ = nd, T = 20°C	4.9	Zhao *et al.* (2008)
Pseudomonas fluorescens	Skim milk	E = 30–40 kV/cm, *n* = nd, τ = 3 μs, T = 37°C	2.5	Férnandez-Molina *et al.* (2006)
Lactobacillus brevis	Orange juice	E = 35 kV/cm, *n* = 250, τ = 4 μs, T = 32°C	5.8	Elez-Martínez *et al.* (2005)
	Fruit juice-soymilk beverage	E = 35 kV/cm, *n* = 200, τ = 4 μs, T = 32°C	5	Morales-de la Peña *et al.* (2010)
Lactobacillus plantarum	Orange juice–milk beverage	E = 15–40 kV/cm, *n* = 0–350, τ = 2 μs, T = 55°C	5	Sampedro *et al.* (2006)
Saccharomyces cerevisiae	Apple juice	E = 20 kV/cm, *n* = 10.4, τ = 2 μs, T = 30°C	4	Cserhalmi *et al.* (2002)
	Orange juice	E = 35 kV/cm, *n* = 250, τ = 4 μs, T = 39°C	5.1	Elez-Martínez *et al.* (2004)

E = Field strength, n *= number of pulses, τ = pulse width, T = process temperature, nd = no data.*

It is widely accepted that microbial inactivation depends not only on the process variables, but also on the target microorganisms and food media. As reported in several studies, microbial death increases as long as process variables, such as field strength, treatment time, number of pulses and pulse width, are set at higher values. In accordance with that, Sobrino-López *et al.* (2006) obtained a maximal inactivation of 4.5 log cycles when 150 bipolar pulses of 8 μs at 35 kV/cm were applied to skim milk inoculated with *S. aureus*. Other authors, such as Elez-Martínez *et al.* (2005), reported up to 5.8 log cycles in reducing *L. brevis* when orange juice was processed by 250 bipolar pulses of 4 μs at 35 kV/cm. Zhao *et al.* (2008) inoculated *E. coli* or *S. aureus* in green tea extract before PEF processing. In their study, the highest destruction of *E. coli* (5.6 log units) and *S. aureus* (4.9 log units) populations were obtained when applying treatments of 38.4 kV/cm for 160 and 200 μs, respectively. The disparity between the maximal inactivation reported at each case and the respective PEF treatment applied may be attributed to factors depending on the fluid media, the microorganisms inoculated, and the influence of other environmental or uncontrolled variables or, particularly, the incidence of the mutual effect of two or more variables. Nonetheless, technical limitations of the equipment, the experimental designs proposed, the few variables considered simultaneously or the considered ranges limited a deeper study of the variables involved. For instance, research on the mutual influence of process parameters showed that a simultaneous increment on treatment time and field intensity led to lower counts of cell population than the single increase of any of the two variables (Sobrino *et al.*, 2006). Zhao *et al.* (2008) registered an extra reduction of more than 4 log units in *E. coli* populations by increasing both treatment time from 160 to 200 μs and field strength from 27.4 to 38.4 kV/cm than any of them individually.

The effects of other variables, such as pulse width, are not completely clear. On the one hand, Rivas *et al.* (2006a) verified that lengthening pulses produced no variation in microbial enumeration of *E. coli* in a milk beverage and orange juice. The same pattern was observed by Elez-Martínez *et al.* (2004, 2005) in orange juice inoculated with *S. cerevisiae* and *L. brevis*. Similarly, decreasing the pulse length from 2 μs to 300 ns did not modify the number of *L. plantarum* cells at constant energy input (Fox *et al.*, 2008). The latter results are in disagreement with the fact that a rise in pulse width should cause a drop in microbial population due to the high level of energy applied (Martín *et al.*, 1997; Evrendilek *et al.*, 2004b). However, the pulse width effect may be directed or influenced by other factors. In fact, the effect of pulse duration can be intensified by increasing the number of pulses and/or the strength of the electric field, thus enabling the possibility of a greater lethality (Aronsson *et al.*, 2001; Elez-Martínez *et al.*, 2004; Sobrino *et al.*, 2006). However, after a certain value of pulse width, an increase in pulse duration did not produce any further increase in the lethality of the treatment (Sobrino *et al.*, 2006). The role of other electrical variables, such as pulse frequency, on cell destruction has received very little attention. Elez-Martínez *et al.* (2004, 2005) observed an increase of *S. cerevisiae* and *L. brevis* inactivation in orange juice when applying treatments of low frequency. In whole milk, an increase in pulse frequency markedly enhanced *L. innocua* inactivation, although the final reduction was strongly dependent on fat content (Picart *et al.*, 2002).

Other authors have reported enhancements of the PEF effect led by temperature of the food media. In particular, heating skim milk from 13 to 33°C accelerated the inactivation of *P. fluorescens* and *L. innocua* as electric field strength, treatment time, and/or energy input increased (Fernández-Molina *et al.*, 2006). These results may be explained by the alteration of the physical stability and fluidity of the cell membrane due to thermal injuries (Sampedro *et al.*, 2006c). In the opposite extreme, Zhao *et al.* (2008) found that storage at a refrigeration temperature of 4°C of PEF-treated green tea acted synergistically in a bacteriostatic and even bactericidal way against survivors.

With respect to factors dealing with resistance of the microorganisms to PEF, bacterial spores are generally the most resistant structures and, beneath them, bacteria and yeasts. In particular, Gram-negative bacteria are less resistant than Gram-positive bacteria owing to their dissimilar membrane structure while great sensitivity of yeasts is attributed to their larger size. The critical field strength reported to inactivate spores contrasts with that of vegetative cells of bacteria. For instance, up to 60 and 100 kV/cm of critical field intensity were registered to achieve inhibition of *B. coagulans* and *B. subtilis* spores, respectively (Choi *et al.*, 2008). Evrendilek *et al.* (2004a) found that lethal PEF treatment conditions (216 μs of treatment time, 22 kV/cm) against *S. uvarum* and *R. rubra* inoculated in beer was less severe than the conditions needed to satisfactorily inactivate a bacterial culture of *L. plantarum*, *P. damnosus* and *B. subtilis* (175 μs of treatment time, 41 kV/cm). Moreover, cell size and shape of the microorganisms considered may also justify the different degrees of inactivation. Since larger cells are more susceptible to electrical breakdown, the critical field intensity required to induce electroporation of yeasts is lower than that of bacteria. Toepfl *et al.* (2007) estimated a critical value of 2 kV/cm of field strength to hit *S. cerevisae* in comparison to the 15 kV/cm required to inactivate the smaller cells of *L. innocua*. However, orientation of cells with respect to the external electric field played an important role on the number of target cells affected. In the case of *S. cerevisae*, a critical field strength of 4 kV/cm was required to inactivate cells having unfavorable and favorable orientations, because the longer axis is directed in parallel to the electric field (Toepfl *et al.*, 2007). Recently, the effect of growth temperature on PEF inactivation of different microbial cultures at their stationary phase was also studied. Cells of *E. coli* grown at the lowest temperature of 10°C and inoculated in buffer media showed the highest resistance to a high-intensity pulsed electric field (HIPEF) treatment of 200 μs at 22 kV/cm and those incubated at 42°C were the more sensitive (Cebrián *et al.*, 2008). The unaltered fluidity of the membranes of those cells grown at 10°C may facilitate re-orientation of membrane components affected by the electric field and would allow resealing of many injured cells after ceasing PEF.

With reference to the weight of food media properties on final inactivation, electrical conductivity, chemical composition, structure and physical features, such as pH, were highlighted for their importance. The use of dilute solutions, such as buffers, model food media or juices, is useful in isolating the effects and behavior of one or more factors. However, results obtained from the latter media can not be extrapolated to complex liquid foods since microbial inactivations on model food solutions are habitually higher than those reached in complex foods. A good example can be found when comparing the effect of PEF treatments on a simulated milk ultra filtrate (SMUF) and milk (Bendicho *et al.*, 2002a). Also, electrical properties of liquid foods can be modified by changes on their physicochemical variables (Mabrook and Petty, 2003). In milk media, the electrical charges of milk, which contribute to the electrical conductivity, can be altered by the fat content, pH, and temperature. The immediate consequence is the alteration in the electric field distribution and shortening of pulse duration (Fernández-Molina *et al.*, 2006), which may explain some described contradictory results. While Sobrino-López *et al.* (2006) and Michalac *et al.* (2003) found that the fat content of milk produced no differences in lethality, Picart *et al.* (2002) reported a protective effect of that variable when processing dairy cream. With respect to acidification of liquid media, a decrease in pH of milk did not affect the survival fraction of *S. aureus* (Sobrino-López *et al.*, 2006) or *P. fluorescens* (Fernández-Molina *et al.*, 2005a). In contrast, Aronsson and Rönner (2001) observed that the effectiveness of the PEF process significantly enhanced the inactivation of *E. coli* when pH fell from 7.0 to 4.0.

4.8 PULSED ELECTRIC FIELD IN COMBINATION WITH OTHER TECHNOLOGIES

From a hurdle concept, the combination of two or more preservation methods may promote new intervention processes in which microbial stress and cell death may be enhanced by both the different nature and inactivation mechanism of each treatment. Hence, the simultaneous combination of various techniques may allow the achievement of greater reductions in cell populations in the treated food or lowering the intensity of both preservation methods without changing the microbial goal set up for commercialization. The second strategy is of relevant interest to the study of PEF applied simultaneously with heat since the detrimental effect of thermal processing could be reduced and high microbial and storage stability could be attained (Odriozola-Serrano *et al.*, 2006). Spores are known to be resistant to thermal processing. However, a thermal shock during PEF treatment produced great inactivation of those microbial forms attributable to additional structural fatigue (Choi *et al.*, 2008). Despite the high intensity of the electrical parameters, PEF (500 pulses at 110 kV/cm) in combination with a thermal treatment (115°C for 5 s) only caused an extra reduction of the spore counts of *B. subtilis* in a model food of 3.5 log units in comparison with counts observed by the singly applied thermal processing (Choi *et al.*, 2008). The advantage of using heat in parallel to PEF is especially attractive in the preservation of dairy products since shelf-life, microbial acceptance, and sensory properties are of key importance for their commercialization. Fernández-Molina *et al.* (2005b) processed skim milk by heat (80°C, 6 s) before applying PEF (30 kV/cm, 60 μs) and could prolong its shelf-life by up to 30 days, which was equivalent to doubling the shelf-life attained with any of the individually applied treatments. In another case, shelf-life was extended beyond 60 days when a thermal treatment (72°C, 15 s) was applied prior to PEF (35 kV/cm, 11.5 μs) (Sepulveda *et al.*, 2005). Yeom *et al.* (2004) used a PEF treatment (30 kV/cm, 32 μs treatment time) in combination with a mild thermal treatment (60°C, 30 s) to process flavored yoghurt-based products. The acceptability of the products tested was prolonged for 90 days at 4°C, which was three times longer than that of the untreated sample. In addition, color, soluble solids, and pH were only minimally affected by the combined treatment, while sensory evaluation gave the processed products a rating indicating overall acceptability. In a different study, chocolate milk was processed by PEF (35 kV/cm, 45 μs) prior to heating at 105 and 112°C for 31.5 s (Evrendilek *et al.*, 2001). Samples were then stored without significant decay for 119 days at temperatures of 4, 22, and 37°C.

In parallel to mild thermal treatments, the joint effects of PEF and other non-thermal treatments, such as the addition of natural preservatives, pulsed ultraviolet (UV) light or ultrasonications (US), may also exert a synergistic activity against microorganisms and may represent an alternative to classical thermal methods. PEF-stressed cells showed more sensitivity to the presence of certain antimicrobial compounds. A concentration of 2,000 ppm of sorbic acid in buffer solution acted synergistically with PEF, inactivating up to 5 log reductions of *D. bruxellensis* and *S. cerevisiae*, thus representing 2 extra log units more than those reached with a treatment of 50 pulses at 12 kV/cm (Somomolinos *et al.*, 2007). Smith *et al.* (2002) found that the addition of 4,250 IU/ml of lysozyme and a PEF treatment of 50 pulses at 80 kV/cm produced a great reduction in the natural flora of milk. The use of naturally occurring bacteriocins has received special attention among preservatives. More than 6.0 log reductions of *S. aureus* were obtained by adding 20 IU/ml of nisin to milk prior to treating by PEF (2,400 μs of treatment time at 35 kV/cm) (Sobrino-López *et al.*, 2006). Moreover, treating milk by a combination of PEF (1,200 μs of treatment time at 35 kV/cm), added nisin (1 IU/ml) and lysozyme (300 IU/ml) exhibited great synergism (6.2 log reductions) against *S. aureus* (Sobrino-López and Martín-Belloso, 2008). Mosqueda-Melgar *et al.* (2008a) studied the inactivation of *Salmonella enteritidis* in apple and orange juices processed by PEF with the addition of antimicrobials. Combining a treatment of 35 kV/cm for 1000 μs with 0.1% cinnamon oil resulted in a 6 log decrease. Similarly, Mosqueda-Melgar *et al.* (2008b) reported more than 6 log reductions in the counts of *S. enteritidis* inoculated in tomato juice when applying similar treatment conditions (Fig. 4.2).

More recently, applying PEF sequentially with UV light resulted in an additive effect. Up to 5.3 log unit reductions in *E. coli* populations were registered after processing apple juice with 11.3 pulses of 3.5 μs width at 60 kV/cm prior to UV exposure (Gachovska *et al.*, 2008). In the combined treatment of PEF and US, the survival fraction of *S. enteriditis* descended additively 2.5 log units when liquid

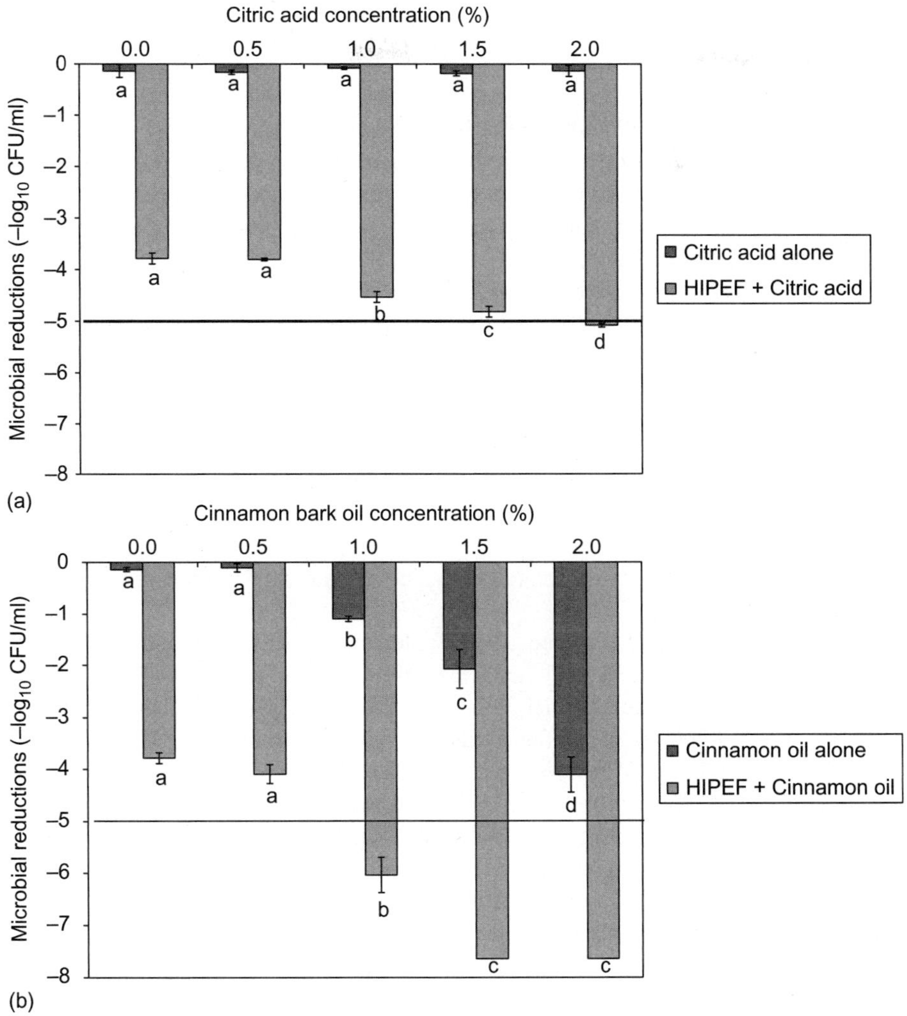

FIGURE 4.2

Inactivation of *Salmonella enterica* ser. Enteritidis in tomato juice by combining PEF with citric acid (a) or cinnamon bark oil (b). Treatment conditions: 35 kV/cm for 1000 μs at 100 Hz, 4 μs pulse width and 38.8 ± 1.8°C.

[Adapted with permission from Mosqueda-Melgar et al. *(2008b).]*

whole egg was treated with 50 pulses at 56.7 kV/cm and the US conditions were 50 W at 55°C for 5 min with 25 ml volume (Huang *et al.*, 2006). Greater inactivation of *L. innocua* (up to 6.7 log units) was registered when milk pre-heated at 55°C for 60 s followed by US at 400 W for 80 s was electrically processed at 50 kV/cm with an energy density of 576 kJ/l (Noci *et al.*, 2009).

4.9 EFFECTS ON FLUID FOOD NUTRITIONAL AND QUALITY PARAMETERS

4.9.1 Effects on Food-Quality Related Enzymes

The mechanisms involved in the inactivation of enzymes by PEF are not fully understood. It is believed that PEF may cause denaturation of enzymes, probably by changing their conformational state (Yang *et al.*, 2004a; Zhong *et al.*, 2005, 2007; Zhao and Yang, 2008).

Inactivation studies on enzymes show that, depending on the enzyme itself, the media where it is suspended, and the PEF treatment conditions, some enzymes are almost completely inactivated, while others are not affected or even exhibit an enhancement of their initial activity. In general, more severe PEF treatments are required for enzymes than for microorganisms to obtain significant reduction levels (Ho *et al.*, 1997). In addition, the sensitivities of different enzymes to PEF may substantially differ from one to another. Hence, as some enzymes are interesting for the food industry, some foods could be treated by PEF to destroy microorganisms while maintaining the activity of useful enzymes.

The inactivation of quality-related enzymes in fluid foods processed by PEF is shown in Table 4.2.

Yeom *et al.* (2000b) reported an 88% depletion of PME activity in orange juice after a PEF treatment of 59 μs at 35 kV/cm. About a 80% of PME activity was inactivated when orange juice was treated during 1500 μs (4-μs pulse width) at 35 kV/cm and 200 Hz in bipolar mode without exceeding 45°C (Elez-Martínez *et al.*, 2007). However, Van Loey *et al.* (2002) could not achieve more than 10% inactivation of PME in orange juice with a treatment of 1 ms at 35 kV/cm. Some authors have reported synergistic effects between PEF and thermal treatment on the inactivation of PME in fruit juices (Yeom *et al.*, 2002; Rodrigo *et al.*, 2003a). An increase in electric field strength caused greater levels of PME inactivation with an increase in temperature during PEF treatment. Aguiló-Aguayo *et al.* (2009a) observed that great PME activity depletion in tomato juice was achieved by increasing pulse frequency and/or pulse width, irrespective of the treatment polarity. PME activity in strawberry juice was not affected when a PEF treatment was applied in mono- or bipolar mode (Aguiló-Aguayo *et al.*, 2009a).

PEF treatments significantly reduced the polygalacturonase (PG) activity of a commercial enzyme formulation. PG initial activity was reduced by 98% after a PEF treatment of 32 ms HIPEF at 10 kV/cm. The temperature of the samples under PEF treatments was lower than 25°C (Giner *et al.*, 2003). Aguiló-Aguayo *et al.* (2009a) reported that PG activity in tomato juice was depleted by means of increasing pulse frequency and width.

Marsellés-Fontanet and Martín-Belloso (2007) observed that PPO activity in grape juice treated by PEF was lessened as electric field strength and treatment time increased. Consistently, Schilling *et al.* (2008) observed a similar trend in the inactivation of PPO of PEF-treated apple juice.

Table 4.2 Effects of Pulsed Electric Fields on Food Quality-Related Enzymes in Fluid Foods

Enzyme	Food	Treatment Conditions	Inactivation (%)	Reference
PME	Orange juice	E = 35 kV/cm, *t* = 59 μs, τ = nd, T = 60.1°C	88	Yeom *et al.* (2000b)
		E = 35 kV/cm, *t* = 1500 μs, τ = 4 μs, T = 35°C	80	Elez-Martínez *et al.* (2007a)
	Strawberry juice	E = 35 kV/cm, *t* = 1000 μs, τ = 7 μs, T = 35°C	92	Aguiló-Aguayo *et al.* (2009a)
	Tomato juice	E = 35 kV/cm, *t* = 1000 μs, τ = 7 μs, T = 35°C	92	Aguiló-Aguayo *et al.* (2009a)
	Orange–carrot juice	E = 35 kV/cm, *t* = 340 μs, τ = nd, T = 35°C	81.4	Rodrigo *et al.* (2003a)
	Gazpacho	E = 35 kV/cm, *t* = 1500 μs, τ = 4 μs, T = 40°C	87	Giner-Seguí *et al.* (2009)
	Orange juice–milk beverage	E = 30 kV/cm, *t* = 50 μs, τ = 2.5 μs, T = 80°C	90	Sampedro *et al.* (2009b)
PG	Strawberry juice	E = 35kV/cm, *t* = 1000 μs, τ = 1 μs, T = 35°C	28	Aguiló-Aguayo *et al.* (2009a)
	Tomato juice	E = 35 kV/cm, *t* = 1000 μs, τ = 7 μs, T = 35°C	62	Aguiló-Aguayo *et al.* (2009a)
PPO	Apple juice	E = 35 kV/cm, *t* = nd, τ = nd, T = 74°C	93	Schilling *et al.* (2008)
	Apple juice	E = 31 kV/cm, *t* = 40,000 μs, τ = 40 μs, T = 60°C	32	Van Loey *et al.* (2002)
	Grape juice	E = 30 kV/cm, *t* = 5000 μs, τ = 4 μs, T = 35–40°C	100	Marsellés-Fontanet and Martín-Belloso (2007)

(*Continued*)

Table 4.2 (Continued)

Enzyme	Food	Treatment Conditions	Inactivation (%)	Reference
POD	Orange juice	E = 35 kV/cm, t = 1500 μs, τ = 4 μs, T = 35°C	100	Elez-Martínez *et al.* (2006a)
	Grape juice	E = 30 kV/cm, t = 5000 μs, τ = 4 μs, T = 35–40°C	50	Marsellés-Fontanet and Martín-Belloso (2007)
	Apple juice	E = 35 kV/cm, t = nd, τ = nd, T = 74°C	100	Schilling *et al.* (2008)
	Watermelon	E = 35 kV/cm, t = 1000 μs, τ = 7 μs, T = 35°C	99	Aguiló-Aguayo *et al.* (2010a)
	Tomato juice	E = 35 kV/cm, t = 2000 μs, τ = 7 μs, T = 35°C	100	Aguiló-Aguayo *et al.* (2008b)
	Milk	E = 35 kV/cm, t = 75 μs, τ = 1 μs, T = nd	0	Riener *et al.* (2008a)
	Milk	E = 31 kV/cm, t = 40,000 μs, τ = 4 μs, T = 60°C	0	Van Loey *et al.* (2002)
	Horchata	E = 25 kV/cm, t = 300 μs, τ = nd, T = 35°C	72	Cortés *et al.* (2005)
LOX	Tomato juice	E = 35 kV/cm, t = 50 μs, τ = nd , T = 30°C	80	Min *et al.* (2003c)
		E = 35 kV/cm, t = 1000 μs, τ = 1 μs, T = 35°C	20	Aguiló-Aguayo *et al.* (2009b)
	Strawberry juice	E = 35 kV/cm, t = 1000 μs, τ = 1 μs, T = 40°C	35	Aguiló-Aguayo *et al.* (2008a)
	Watermelon juice	E = 35 kV/cm, t = 1000 μs, τ = 1 μs, T = 35°C	52	Aguiló-Aguayo *et al.* (2010a)
	Green pea juice	E = 20 kV/cm, t = 400 μs, τ = 1 μs, T = nd	0	Van Loey *et al.* (2002)
	Soymilk	E = 42 kV/cm, t = 1036 μs, τ = 2 μs, T = 25°C	88	Li *et al.* (2008)

(*Continued*)

Table 4.2 (Continued)

Enzyme	Food	Treatment Conditions	Inactivation (%)	Reference
ALP	Raw milk	E = 6.7–20 kV/cm, *t* = 800 μs, τ = 4 μs, T = 70°C	74	Van Loey *et al.* (2002)
	Whole milk	E = 18.8 kV/cm, *t* = 28,000 μs, τ = 70 μs, T = nd	59	Castro *et al.* (2001)
	Milk	E = 35 kV/cm, *t* = 75 μs, τ = 1 μs, T = nd	29	Riener *et al.* (2008a)
Protease	Skim milk	E = 14–15 kV/cm, T = nd, τ = nd, T = nd	60	Vega-Mercado *et al.* (2001)
		E = 35.5 kV/cm, *t* = 866 μs, τ = nd, T = 46°C	81.1	Bendicho *et al.* (2003)
	Whole milk	E = 35.5 kV/cm, *t* = 866 μs, τ = nd, T = 46°C	57.1	Bendicho *et al.* (2003b)
	Milk	E = 35 kV/cm, *t* = 75 μs, τ = 1 μs, T = nd	37	Riener *et al.* (2008a)
Lipase	Raw milk	E = 21.5 kV/cm, T = 20°C	60	Grahl and Märkl (1996)
	Milk	E = 35 kV/cm, *t* = 75 μs, τ = 1 μs, T = nd	14	Riener *et al.* (2008)a
β-glucosidase	Strawberry juice	E = 35 kV/cm, *t* = 1000 μs, τ = 7 μs, T = 40°C	27	Aguiló-Aguayo *et al.* (2008a)
HPL	Tomato juice	E = 35 kV/cm, T = 1000 μs, τ = 7 μs, T = 35°C	93	Aguiló-Aguayo *et al.* (2009b)

E = field strength, t *= treatment time, τ = pulse width, T = process temperature, nd = no data.*

A 100% POD inactivation was achieved in orange juice processed at 35 kV/cm for 1500 μs (4-μs pulse width) and 200 Hz in bipolar mode without exceeding 35°C. Residual POD activity was reduced to 6.9% when orange juice was treated at 45 Hz (35 kV/cm, 600 μs, bipolar pulses of 4 μs) and it was observed that higher inactivations were achieved when pulse frequency was increased

(Elez-Martínez *et al.*, 2006a). A study carried out by Aguiló-Aguayo *et al.* (2008b) focused on POD inactivation in tomato juice treated by PEF indicated that enzyme inactivation was directly proportional to the treatment time. Moreover, they concluded that the greatest POD inactivation in tomato juice was obtained after applying a bipolar treatment. POD activities in grape juice treated by PEF were depleted as electric field strength and treatment time increased (Marsellés-Fontanet and Martín-Belloso, 2007). Cortés *et al.* (2005) studied the effect of PEF on POD activity of "horchata", a Spanish vegetable beverage, reporting a 72.4% enzyme inactivation when treating at 25 kV/cm for 300 μs. However, Van Loey *et al.* (2002) did not find any effect on milk POD after exposing milk samples to PEF treatments of 500 μs at 19 kV/cm. The same fact was observed by Riener *et al.* (2008a) when processing fresh raw milk at 35 kV/cm during 75 μs.

An 80% reduction in LOX activity was observed when tomato juice was exposed to PEF at 35 kV/cm for 50 or 60 μs (3-μs pulse width) at 30°C (Min *et al.*, 2003b). Li *et al.* (2008) evaluated the effects of pulse strength, treatment time, and pulse width on LOX inactivation in soymilk. Their results indicate that residual LOX activity decreased with the increase in the processing parameters. LOX inactivation in strawberry juice was achieved by PEF treatments and different pulse frequencies did not have any effect on the enzyme inactivation (Aguiló-Aguayo *et al.* 2008a). On the contrary, LOX was not inactivated in pea juice after a PEF treatment of 400 pulses of 1 μs at field strength of 20 kV/cm and frequency of 1 Hz (Van Loey *et al.*, 2002).

Castro *et al.* (2001) studied the PEF inactivation of alkaline phosphatasse (ALP) in non-fat milk, 2% fat milk, and whole milk. PEF treatment was able to reduce up to 65% of ALP activity after 70 pulses of 740 μs at 18.8 kV/cm in skim milk. When 2% milk and whole milk were treated, ALP activity was reduced up to 59% after a 70-pulse treatment at 18.8 kV/cm with pulses of 400 μs. A treatment of 35 kV/cm for 75 μs inhibited ALP activity in milk a 29% (Riener *et al.*, 2008a). Nonetheless, Van Loey *et al.* (2002) did not observe any significant ALP reduction in PEF-processed milk after applying treatments of 200 pulses of 2 μs at 20 kV/cm.

Samples of milk with a commercial protease from *Bacillus subtilis* were processed at 67 Hz (35.5 kV/cm for 866 μs) with monopolar pulses of 7 μs at temperatures below 46°C. In skim milk, inactivation levels varied from 37.9% to 81.1%, depending on whether treatments of 35.5 kV/cm for 866 μs were performed at 67 or 111 Hz. In whole milk, both treatments led to 37.9 and 57.1% inactivation, respectively (Bendicho *et al.*, 2003). When fresh raw milk was processed at 35 kV/cm for 75 μs, protease activity was decreased to 37% (Riener *et al.*, 2008a).

Grahl and Märkl (1996) observed a reduction of about a 60% in lipase activity in raw milk after subjecting milk samples at 20 pulses of 21.5 kV/cm. Another lipase whose behavior has been studied after exposure to different PEF devices is a lipase from *Pseudomonas fluorescens* suspended in simulated milk ultrafiltrate.

Using batch-mode PEF equipment, a 62.1% maximum activity depletion was achieved after 80 pulses at 27.4 kV/cm at 34°C. However, when PEF treatments were applied in a continuous flow mode, an inactivation rate of just 13% was achieved, after applying 80 pulses at 37.3 kV/cm and 3.5 Hz at 35°C (Bendicho *et al.*, 2002b). Riener *et al.* (2008a) reported an inhibition of 14% in lipase of fresh raw milk when samples were treated for 75 μs at 35 kV/cm.

The impact of PEF on other quality-related enzymes of foods, such as hydroperoxide lyase (HPL) or β-glucosidase (β-gluc), was studied. Tomato juice HPL was inactivated by approximaely 93% when it was processed at 35 kV/cm for 1000 μs with a 7-μs pulse width and 250 Hz (Aguiló-Aguayo *et al.*, 2009b). While β-gluc of strawberry juice was activated by 112% when samples were treated at 50 Hz and 1-μs pulse width (35 kV/cm for 1000 μs), it was inactivated by 27% after processing juice at 250 Hz and 7-μs pulse width (35 kV/cm for 1000 μs) in bipolar mode (Aguiló-Aguayo *et al.*, 2008a).

Because of the resistance of several enzymes to PEF, treatments combining PEF with other hurdles such as mild heat or the use of some additives were studied on different liquid foods with promising results (Hodgins *et al.*, 2002; Riener *et al.*, 2008b, 2009).

4.9.2 Effects on Quality Parameters

It has been reported that PEF processing of different fruit juices did not significantly affect their quality characteristics, such as color, viscosity, acidity, soluble solids, or flavor, among others. Thus, PEF technology is able to preserve products maintaining their fresh-like characteristics.

Yeom *et al.* (2000a) studied aroma loss, browning index, color, and variation of soluble solids and pH in freshly squeezed single-strength orange juice. They applied treatments with PEF (35 kV/cm, 59 μs) and heat (94.6°C, 30 s). Regarding the retention of volatile components (α-pinene, myrcene, octanal, d-limonene, and decanal) they observed that PEF-treated juices better retained volatile compounds at 4°C than heat-treated juices. However, PEF-treated juices exhibited less browning and minimal modification of pH and soluble solids content. Cserhalmi *et al.* (2006) analyzed the effects of PEF-processing on citrus juice made from grapefruit, lemon, orange, and tangerine, reporting no differences on the physical and chemical properties between the treated and the untreated samples. Namely, factors such as electrical conductivity, pH, soluble solids, viscosity, color, nonenzymatic browning index, and hydroxymethylfurfural content did not vary after 100 μs of treatment time at 28 kV/cm. Interestingly, volatile flavor compounds of PEF-treated juice were essentially equal to those present in unprocessed juice. Elez-Martínez *et al.* (2006b) studied the effects of PEF processing (35 kV/cm, 1000 μs, bipolar 4-μs at 200 Hz) on quality-related parameters of orange juice during storage at 4 and 22°C. PEF-treated juice retained better colour than heat-pasteurized juice but no differences in pH, acidity, and °Brix were found between treatments. It is interesting to remark that sensory evaluation of texture, flavor, and overall acceptability were ranked highest for control orange juice, followed by PEF-processed juice and by thermally processed juice (Min *et al.*, 2003a).

A study on fresh apple juice showed that soluble solids, pH, and acidity of PEF-treated juice did not differ from values of untreated samples (Zárate-Rodríguez *et al.*, 2000). In another work, Evrendilek *et al.* (2000) studied the shelf-life (4, 22 and 37°C for 70 days) and changes in colour, pH, soluble solids, density, viscosity, conductivity, and sensory evaluation of fresh apple cider treated by PEF and PEF + heat (60°C, 30 s) and reconstituted apple juice treated by PEF. The authors did not find any differences in color in the apple juice or cider. In the sensory study, slight differences were observed, 37% of the participants preferred the cider treated with PEF, whereas 39% preferred the control sample, and 24% did not show a preference for any particular sample. Aguilar-Rosas *et al.* (2007) observed that losses in volatile compounds of apple juice processed by PEF were lower than those observed for a heat treatment.

Physico-chemical and sensory characteristics (color, pH, acidity, soluble solids, viscosity, aromas) were much better maintained in a PEF-treated tomato juice than in a heat-treated juice (Min and Zhang, 2003; Min *et al.*, 2003c; Aguiló-Aguayo *et al.*, 2008c). Moreover, flavor and overall acceptability of PEF-processed tomato juice were preferred to those of thermally processed juice (Min *et al.*, 2003c).

Aguiló-Aguayo *et al.* (2009a) demonstrated that PEF processing parameters such as pulse frequency, width and polarity affect the viscosity of a strawberry juice. In addition, they found that PEF-processed watermelon juices retained better color characteristics than those that were thermally treated (Aguiló-Aguayo *et al.* 2009b). Aguiló-Aguayo *et al.* (2009d) have also observed that, after PEF processing of strawberry juices, some volatile compounds such as 2,5-dimethyl-4-hydroxy-3(2h)-furanone, ethyl butanoate, and 1-butanol were better maintained than in heat-treated samples, resulting in a greater flavor stability.

Acidity and pH of milk was not affected by PEF treatment (Odriozola-Serrano *et al.*, 2006). Evrendilek *et al.* (2001) studied the changes during storage in color, pH, soluble solids, and conductivity in milk with chocolate using treatment with PEF and PEF + heat and compared the results with an untreated sample. Measurement of the a, b, and L parameters at 4°C revealed that the treatments of PEF + 105°C and PEF + 112°C did not cause changes in color.

Noticeable changes on the odor, color, taste, sourness, and overall attributes of PEF-processed melon and watermelon juices were not detected by panelists compared to untreated juices, with the exception of taste in melon juice. However, significant changes in those attributes were observed for both juices after a thermal treatment (Mosqueda-Melgar *et al.*, 2008d).

Morales-de la Peña *et al.* (2010) evaluated the effects of PEF on different quality parameters (pH, soluble solid content, acidity, and viscosity) of a fruit juice–soymilk beverage and reported that most of the quality attributes of the beverage were not significantly affected by PEF processing. The average loss in the concentration of volatile compounds in an orange juice–milk-based beverage ranged between 16 and 43% after thermal treatment; while the average loss after PEF treatment ranged between 8.3 and 13.7% (Sampedro *et al.*, 2009a).

Evrendilek *et al.* (2004a) did not find significant differences in foam condition, color and overall acceptability of untreated and PEF-treated beer samples, except for flavor and mouthfeeling.

4.9.3 Effects on Nutritional Parameters

PEF treatments have been reported to lightly affect those compounds with biological activity. During recent years, research has been conducted studying the feasibility of PEF technology in obtaining liquid products with increased concentrations of health-related compounds.

Several authors have studied the effects of PEF on vitamins. Vitamin C attracts most of the attention because of its heat-lability, especially when exposed to oxidative environments. Results by different authors indicate that differences between the vitamin C contents of PEF-pasteurized beverages and those of the just-made products may range from negligible to roughly 10%, depending on the substrate and treatment conditions. In turn, these losses never exceeded the decreases caused by heat-pasteurization, in some cases even representing less than 50% of the heat-treatment-induced depletion in beverages such as orange juice (Qiu *et al.*, 1998; Min *et al.*, 2003a; Sánchez-Moreno *et al.*, 2005a; Elez-Martínez *et al.*, 2006b; Elez-Martínez and Martín-Belloso, 2007), apple juice (Evrendilek *et al.*, 2000), strawberry juice (Odriozola-Serrano *et al.*, 2008e), tomato juice (Min *et al.*, 2003c, Odriozola-Serrano *et al.*, 2008a), orange–carrot juice (Torregrossa *et al.*, 2006), a vegetable based cold soup (Elez-Martínez and Martín-Belloso, 2007), and a fruit juices–soy milk blend (Morales-de la Peña *et al.*, 2010). Some differences between studies could be related to the different equipment and raw materials used, as well as to the differences in the applied treatment conditions. In a similar way, no significant changes or very little variations in the contents of other vitamins have been observed when applying PEF treatments. Changes in the vitamin A content of PEF-pasteurized orange and orange–carrot juices have been reported to be less important than in heat-treated juices (Cortés *et al.*, 2006a, 2006b; Torregrosa *et al.*, 2006). However, Bendicho *et al.* (2002c) evaluated the effect of PEF-treatments on water-soluble (riboflavin, thiamine, and ascorbic acid) and fat-soluble vitamins (cholecalciferol and tocopherol) in milk and observed no changes in the vitamin content, with the exception of ascorbic acid, which decreased slightly but less than when an equivalent thermal treatment was applied.

Major changes have not been described regarding the structure or bioactivity of isoprene compounds in PEF-processed products. Cortés *et al.* (2006b) did not find relevant changes in the amounts of individual carotenoids with antioxidant activity (β-carotene, β-cryptoxanthin, zeaxanthin and lutein) in PEF-treated orange juice, thus improving the content of these compounds with respect to a heat-treated juice as well as their stability throughout cold storage (Cortés *et al.*, 2006a). Torregrosa *et al.* (2005) reported an increase in the contents of individual carotenoids (neoxanthin + 9-*cis*-violaxanthin, antheraxanthin, mutatoxanthin, lutein, zeaxanthin, α-cryptoxanthin, β-cryptoxanthin, *cis*-β-cryptoxanthin 9-*cis*-α-carotene, α-carotene, phytoene + phytofluene, β-carotene, 13-*cis*-β-carotene, ζ-carotene, 9-*cis*-β-carotene) in an orange–carrot juice. These results are consistent with those reported by Odriozola-Serrano *et al.* (2008c), who observed increasing amounts of lycopene in tomato juice for PEF-treatments of increasing intensity. Nevertheless, in a later publication, the same group reported different trends for each individual carotenoid as a consequence of PEF processing, thus leading to significant increases in the amounts of lycopene, β-carotene, and phytofluene, while the rest of the compounds remained without changes or slightly decreased (Odriozola-Serrano *et al.*, 2009b). These changes in the carotenoid profiles appear to be dependent on the intensity of the treatments. Oms-Oliu *et al.* (2009) showed that the amount of lycopene in a PEF-treated watermelon juice may be decreased or increased with respect to the untreated juice, depending on the process conditions set for the treatment. Thus, bipolar pulses applied at higher frequencies resulted into higher lycopene contents in the treated juice, which was probably due to the higher temperatures reached in the product. Results obtained by Zulueta *et al.* (2010) show that PEF treatments at 15 kV/cm caused a slight increase in the concentrations of some carotenoids in an orange juice–milk beverage, while slight decreases occurred when treating at electric field strengths of 40 kV/cm.

Information on the effect of PEF treatments on the phenolic composition of foods is scarce, especially with regard to qualitative analysis. Sánchez-Moreno *et al.* (2005a) did not report changes in the content of either total flavanones or individual flavanone glycosides and their aglycons hesperetin and naringenin in orange juice. Changes in the contents of the main phenolic acids (p-hydroxybenzoic acid, p-cumaric acid, and ellagic acid), flavonols (kaempferol, quercetin and myricetin), and anthocyanins (cyaniding-3-glucoside, pelargonidin-3-glucoside and pelargonidin-3-rutinoside) have been reported to be minimally affected by processing and subsequent storage (Odriozola-Serrano *et al.*, 2008d). Consistently, PEF-pasteurization did not result in significant changes in the phenolic content of a fruit juice–milk beverage. Zhang *et al.* (2007) reported an increase in the degradation of cyanidin-3-glucoside from red raspberry diluted in an aqueous-methanol solution as electric field strength and treatment time increased. In real food systems, though, the extent of these changes is rather limited and much less substantial than when a heat treatment is applied. Hence, total anthocyanin retention in HIPEF-treated strawberry juice ranged from 96.1 to 100.5% when exposed to electric field strengths from 20 to 35 kV/cm for up to 2,000 μs (Odriozola-Serrano *et al.*, 2008e).

The effects of PEF on health-related compounds of fluid foods in comparison to those obtained by heat treatment are summarized in Table 4.3.

There is no evidence that PEF treatments have any effects on other food compounds with health-related potential such as bioactive peptides or fatty acids.

Table 4.3 Some Examples of the Effects of Pulsed Electric Field and Heat Treatments on Bioactive Compounds in Fluid Foods

Compound	Food	PEF Processing		Heat Processing		Reference
		Treatment Conditions	% Retention	% Retention	Treatment Conditions	
Vitamins						
Vitamin A	Orange juice	E = 40 kV/cm, t = 130 μs, T < 69°C	90.1	85.4	T = 90°C, t = 20 s	Cortés *et al.* (2006b)
Vitamin C	Orange juice	E = 35 kV/cm, t = 750 μs, T < 50°C E = 40 kV/cm, t = 97 μs, T = 45°C	93.0 ~100	90.5 81	T = 90°C, t = 60 s T = 90°C, t = 90 s	Sánchez-Moreno *et al.* (2005a) Min *et al.* (2003a)
	Orange–carrot juice	E = 25 kV/cm, t = 280 μs, T = 65°C	90	83	T = 98°C, t = 21 s	Torregrosa *et al.* (2006)
	Carrot juice	E = 35 kV/cm, t = 1500 μs, T < 40°C	95.1	86.6	T = 90°C, t = 60 s	Quitão-Teixeira *et al.* (2009)
	Tomato juice	E = 35 kV/cm, t = 1500 μs, T < 40°C	86.5	79.2	T = 90°C, t = 60 s	Odriozola-Serrano *et al.* (2008a)
	Fruit juice–soymilk blend	E = 35 kV/cm, t = 1400 μs, T < 32°C	87-90	87-90	T = 90°C, t = 60 s	Morales-de la Peña *et al.* (2010)
Vitamins B1, B2, D and E	Skim milk	E = 27.1kV/cm, t = 400 μs, T < 55°C	~100	~100	T = 90°C, t = 10 min	Bendicho *et al.* (2002c)
Carotenoids						
Total content	Orange juice	E = 40 kV/cm, t = 130 μs, T < 69°C	92.2	87.4	T = 90°C, t = 20 s	Cortés *et al.* (2006b)
β-carotene	Orange–carrot juice	E = 30kV/cm, t = 340 μs, T < 64°C	154	~100	T = 98°C, t = 21 s	Torregrosa *et al.* (2005)

(Continued)

Table 4.3 (Continued)

Compound	Food	PEF Processing		Heat Processing		Reference
		Treatment Conditions	% Retention	% Retention	Treatment Conditions	
β-carotene	Carrot juice	E = 35 kV/cm, t = 1500 μs, T < 40°C	123	110	T = 90°C, *t* = 60 s	Quitão-Teixeira *et al.* (2009)
Lycopene	Tomato juice	E = 35 kV/cm, t = 1500 μs, T < 40°C	110	107	T = 90°C, *t* = 60 s	Odriozola-Serrano *et al.* (2009b)
Phenolic Compounds						
Anthocyanins	Strawberry juice	E = 35 kV/cm, t = 1700 μs, T < 40°C	96,5	91.6	T = 90°C, *t* = 60 s	Odriozola-Serrano *et al.* (2008d)
Ellagic acid	Strawberry juice	E = 35 kV/cm, t = 1700 μs, T < 40°C	97.8	91.1	T = 90°C, *t* = 60 s	Odriozola-Serrano *et al.* (2008d)
Chlorogenic acid	Tomato juice	E = 35 kV/cm, t = 1500 μs, T < 40°C	100	100	T = 90°C, *t* = 60 s	Odriozola-Serrano *et al.* (2009b)
Quercetin	Tomato juice	E = 35 kV/cm, t = 1500 μs, T < 40°C	98.0	96.5	T = 90°C, *t* = 60 s	Odriozola-Serrano *et al.* (2009b)

E = field strength, t *= treatment time, τ = pulse width, T = process temperature.*

Results obtained by Garde-Cerdán *et al.* (2007) in grape must support this statement, showing that even intense PEF treatments do not cause changes into the amino acid and fatty acid profiles.

There is very scarce information in scientific literature about the bioavailability of nutrients of fluid foods processed by PEF. Sánchez-Moreno *et al.* (2004, 2005b) studied the impact of consuming fruit and vegetable products treated by PEF on vitamin C bioavailability and biomarkers of oxidative stress and inflammation in a healthy human population. Specifically, the studies have been carried out with orange juice and a vegetable soup "gazpacho". In this sense, drinking two servings (500 ml) of PEF-treated orange juice or "gazpacho" daily maintained the vitamin C bioavailability and antioxidant properties of fresh products with a longer shelf-life, increasing plasma vitamin C and decreasing oxidative stress and biomarkers of inflammation in healthy humans.

4.10 SHELF-LIFE OF PULSED ELECTRIC FIELD-PROCESSED FLUID FOODS

Microbiological stability throughout storage of PEF-treated products may dramatically vary depending on processing conditions, the product itself, and the storage temperature. Min *et al.* (2003a) observed that, after processing orange juice with a commercial-scale PEF system (40 kV/cm, 97 μs), the counts of both total aerobic microorganisms and molds and yeasts were always below 10 cfu/ml throughout 112 days at 4°C. Elez-Martínez *et al.* (2006b) reported that a treatment of 35 kV/cm and 1000 μs combined with refrigerated storage (4°C) ensured the microbial stability of orange juice for at least 56 days. Shelf-life of PEF-treated apple juice has been also studied. A shelf-life of at least 3 weeks at either 4°C or 25°C was achieved when the juice was treated with 250 μs at 36 kV/cm (Qin *et al.* 1995b). Indeed, when processing apple juice at 35 kV/cm for 94 μs, shelf-life was at least 67 days under refrigerated storage (Evrendilek *et al.*, 2000). However, Min *et al.* (2003c) evaluated the refrigerated shelf-life of a tomato juice processed by PEF (40 kV/cm for 57 μs) in comparison to that of a heat-treated juice (92°C for 90 s). Both treatments provided a microbiological shelf-life of at least 112 days. Regarding microbial stability of milk, skim milk processed by PEF (40 kV/cm, 60 μs; 36 kV/cm, 84 μs) had a shelf-life of 14 days at 4°C with a level of acidity that was lower than that of heated milk (Fernández-Molina *et al.*, 2005c). In contrast, Odriozola-Serrano *et al.* (2006) reported 5 days of shelf-life after processing whole milk with PEF (35.5 kV/cm, 1000 μs) and no changes in acidity throughout storage for either HIPEF-treated or heated samples. Morales-de la Peña *et al.* (2010) maintained the marketability of a PEF-treated fruit juice–soymilk beverage (35 kV/cm, 1400 μs) stored at 4°C for 56 days (Fig. 4.3). Sampedro *et al.* (2009b) observed that an orange juice–milk beverage treated by PEF and stored at 8–10°C had a microbial shelf–life of 2.5 weeks. Refrigerated storage of liquid

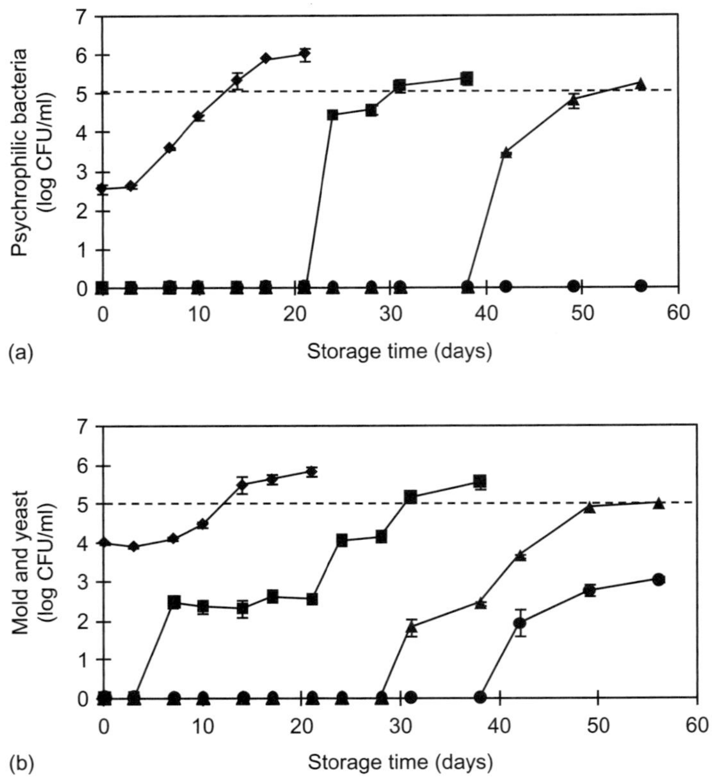

FIGURE 4.3

Microbial stability of a fruit juice-soymilk beverage throughout storage at 4°C (psychrophilic bacteria: a; mold and yeast: b). Fruit juice-soymilk beverages: (♦) untreated, (■) 800-μs-PEF treated, (▲) 1400-μs-PEF treated, (●) thermally treated. *[Reproduced with permission from Morales-de la Peña* et al. *(2010).]*

whole egg with 0.15% of citric acid and treated by PEF (89 μs of treatment time at 30 kV/cm) was prolonged for 20 days (Góngora-Nieto *et al.*, 2003).

Different studies reveal that no activation of quality-related enzymes take place during storage of fluid foods processed by PEF (Min *et al.*, 2003c; Elez-Martínez *et al.*, 2006b; Aguiló-Aguayo *et al.*, 2008c, 2009d, 2010b) (Fig. 4.4). In most cases, quality-related attributes, color, viscosity, acidity, flavor, and soluble solids, are better maintained during storage in PEF-processed fluid foods than in thermal-treated products (Min *et al.*, 2003c; Elez-Martínez *et al.*, 2006b; Aguiló-Aguayo *et al.*, 2008c, 2009d, 2010b; Morales-de la Peña *et al.*, 2010) (Fig. 4.5). Moreover, a higher retention of nutritional-related compounds (vitamin C, phenolics, carotenoids) is generally observed during storage of PEF-treated fluid foods compared to thermally processed foods (Odriozola-Serrano *et al.*, 2008acd, Sánchez-Moreno *et al.*, 2009) (Fig. 4.6).

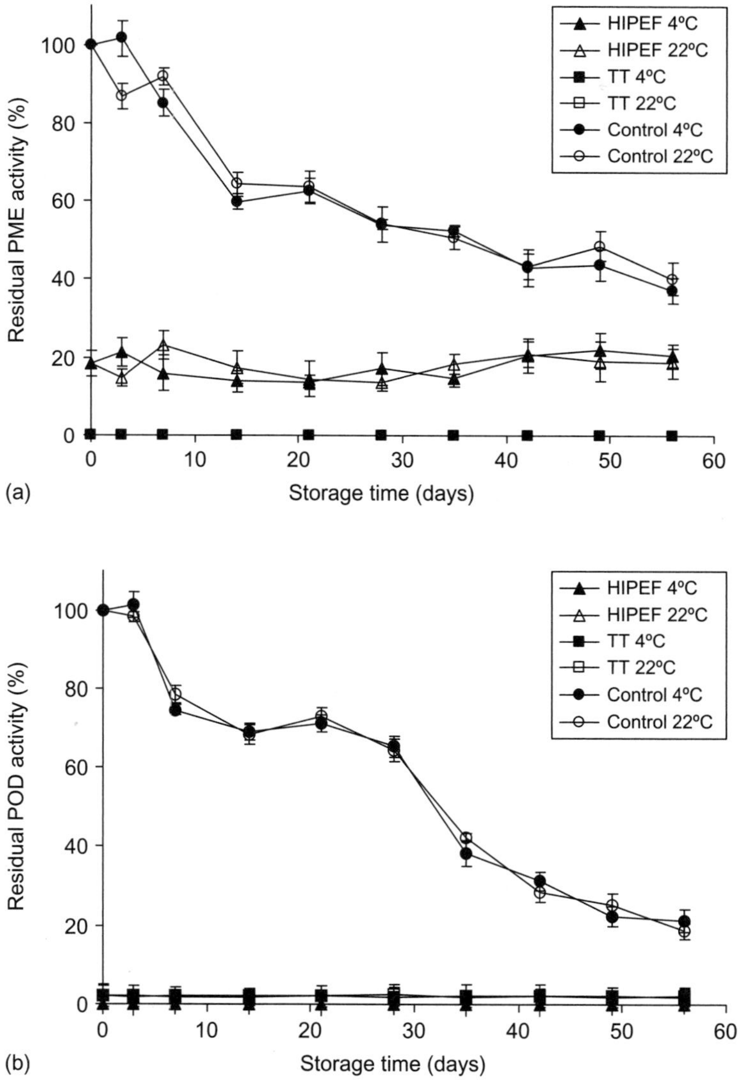

FIGURE 4.4

Effects of PEF treatment and heat pasteurization (TT) on the residual PME activity (a) and POD activity (b) of orange juice throughout storage at 4 and 22°C. PME activity of heat-pasteurized orange juice was under the limit of detection.

[Adapted with permission from Elez-Martínez et al. *(2006b).]*

However, some studies have been carried out studying the shelf-life of fluid foods processed by combining PEF treatments with some other processing technologies. Evrendilek *et al.* (2004b) reported that a PEF treatment achieved a longer shelf-life and better physical and sensory characteristics for a yogurt-based

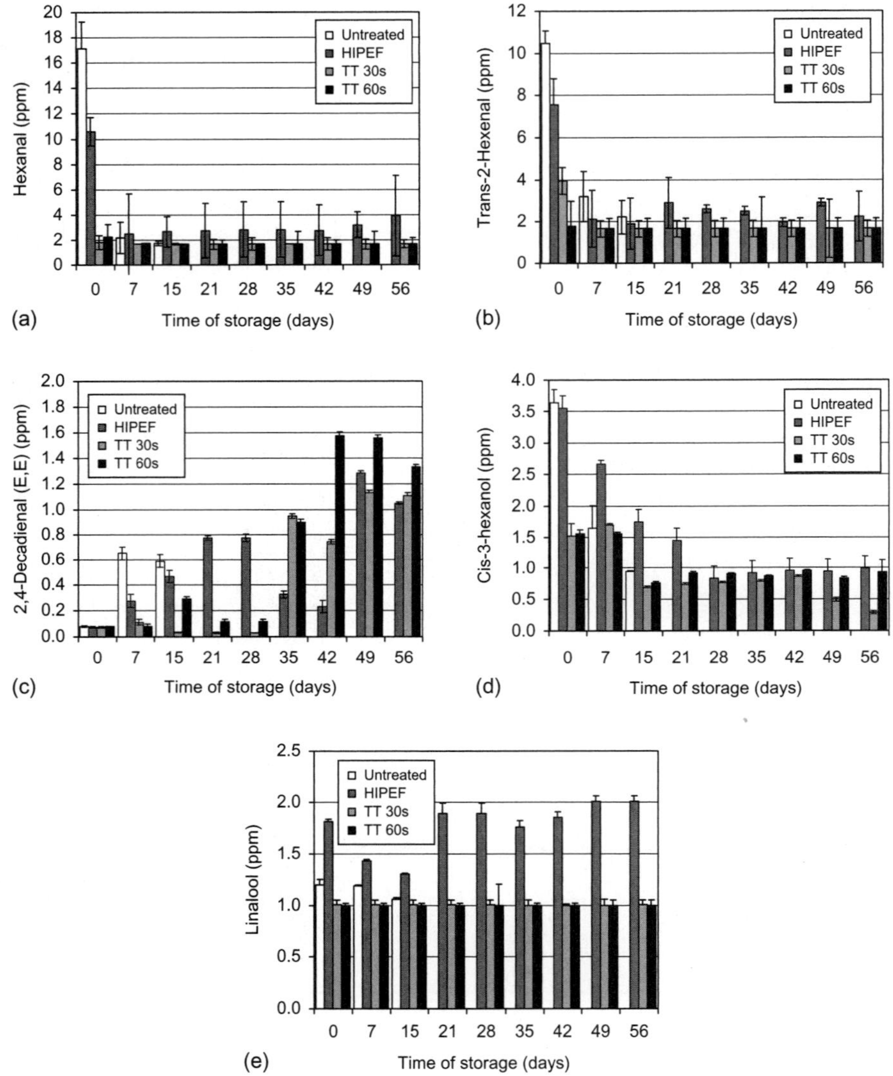

FIGURE 4.5

Quantitative comparison of (a–c) aldehydes, (d) cis-3-hexenol, and (e) linalool in fresh and PEF- and thermally processed tomato juices. PEF: 35 kV/cm, 1500 μs, 4-μs bipolar pulses, 100 Hz. TT: 90°C.

[Reproduced with permission from Aguiló-Aguayo et al. *(2010c).]*

drink compared to a control sample. The combination of PEF with natural antimicrobials was used by Mosqueda-Melgar *et al.* (2008d) to extend the shelf-life of fruit juices. In the latter study, it was observed that unprocessed and PEF-processed melon and watermelon juices were mainly limited by mesophilic

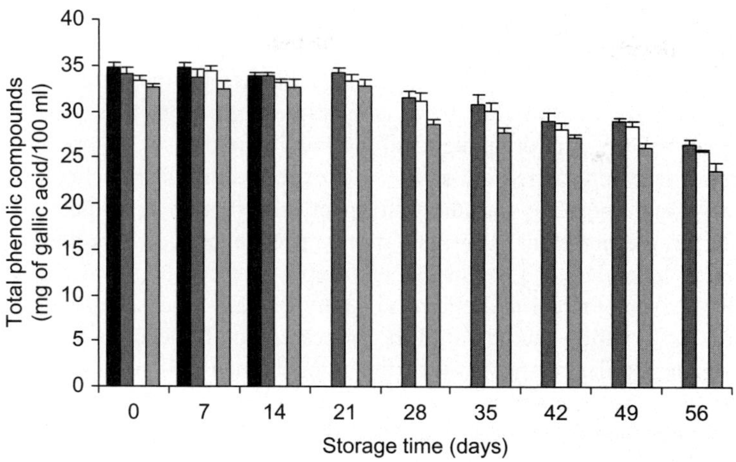

FIGURE 4.6

Effect of PEF and heat processing on total phenolic compounds of strawberry juices throughout storage at 4°C. Strawberry juices: (■) untreated, (▩) HIPEF treated at 35 kV/cm for 1700 μs in bipolar 4-μs pulses and 100 Hz, (□) heat treated at 90°C for 30 s and (▩) heat treated at 90°C for 60 s. Data shown are mean±standard deviation.

[Reproduced with permission from Odriozola-Serrano et al. *(2008d).]*

microorganisms at 5°C; whereas the natural microbiota in both juices was inhibited during more than 91 days of storage at 5°C by combining PEF with citric acid or cinnamon bark oil. Walkling-Ribeiro *et al.* (2009a) reported that the combination of thermosonication and PEF could represent a promising alternative to conventional pasteurization of orange juice in order to extend its shelf-life while maintaining its quality characteristics at 25°C. Fernández-Molina *et al.* (2005c) observed that a PEF treatment combined with a mild heat treatment (65°C for 21 s) prolonged the shelf-life of skim milk stored at 4°C to more than 30 days. In addition, Walkling-Ribeiro *et al.* (2009b) concluded that, by combining mild heat and PEF, a longer stabilization of microbiological shelf-life at 4°C was achieved compared to thermal pasteurization. A combination of PEF and thermal treatment (HTST) substantially improved microbial shelf-life stability of applesauce products without significant loss of physical and sensory attributes during storage at room temperature (27°C) (Jin *et al.*, 2009).

4.11 CURRENT STATUS AND FUTURE TRENDS

PEF processing has good prospects for being used in the food industry, since promising results have been obtained regarding the inactivation of undesirable microorganisms and enzymes in different foodstuffs, while maintaining the nutritional and sensory properties. PEF may be used to modify existing processes or to develop new, energy-efficient, waste-free technologies for the fluid food industry. This treatment may be an alternative to heat-pasteurization, thus providing safe, shelf-stable, healthy ,and organoleptically attractive juices.

With the research carried out so far, it is expected that PEF technology will be industrially implemented as a unique form of preservation or as part of a hurdle approach for preservation. However, more research and development activities are required to understand, optimize, and apply this complex process to its full potential. Microbiological and chemical safety of PEF-processed foods and demonstration of fresh-like quality of PEF-processed foods need to be validated in order to gain popularity and broaden the category of PEF-processed commercial products. Moreover, further investigation is required to better control the critical points and mechanisms involved in PEF processing of fluid foods in order to enable the scaling-up of the PEF technology for the food industry.

Furthermore, PEF systems with high production rates and identical or improved effectiveness of lab-scale equipment must be developed. In addition, a reliable assessment of the economic costs of the industrial implementation of PEF needs to be conducted to facilitate the transfer of the technology to the fluid foods processing industry.

References

Aguilar-Rosas, S.F., Ballinas-Casarrubias, M.L., Nevarez-Moorillon, G.V., Martín-Belloso, O., Ortega-Rivas, E., 2007. Thermal and pulsed electric fields pasteurization of apple juice: effects on physicochemical properties and flavour compounds. J. Food Eng. 83, 41–46.

Aguiló-Aguayo, I., Sobrino-López, A., Soliva-Fortuny, R., Martín-Belloso, O., 2008a. Influence of high-intensity pulsed electric field processing on lipoxygenase and β-glucosidase activities in strawberry juice. Innov. Food Sci. Emerg. 9, 455–462.

Aguiló-Aguayo, I., Odriozola-Serrano, I., Quintao-Texeira, L.J., Martín-Belloso, O., 2008b. Inactivation of tomato juice peroxidase by high-intensity pulsed electric fields as affected by process conditions. Food Chem. 107, 949–955.

Aguiló-Aguayo, I., Soliva-Fortuny, R., Martín-Belloso, O., 2008c. Comparative study con color, viscosity and related enzymes of tomato juice treated by high-intensity pulsed electric fields or heat. Eur. Food Res. Technol. 227, 599–606.

Aguiló-Aguayo, I., Soliva-Fortuny, R., Martín-Belloso, O., 2009a. Changes on viscosity and pectolytic enzymes of tomato and strawberry juices processed by high-intensity pulsed electric fields. Int. J. Food Sci. Technol. 44, 2268–2277.

Aguiló-Aguayo, I., Soliva-Fortuny, R., Martín-Belloso, O., 2009b. Effects of high-intensity pulsed electric fields on lypoxigenase and hydroperoxide lyase activities in tomato juice. J. Food Sci. 74, C595–C601.

Aguiló-Aguayo, I., Soliva-Fortuny, R., Martín-Belloso, O., 2009c. Avoiding non-enzymatic browning by high-intensity pulsed electric fields in strawberry, tomato and watermelon juices. J. Food Eng. 92, 37–43.

Aguiló-Aguayo, I., Oms-Oliu, G., Soliva-Fortuny, R., Martín-Belloso, O., 2009d. Flavor retention and related enzyme activities during storage of strawberry juices processed by high-intensity pulsed electric fields or heat. Food Chem. 116, 59–65.

Aguiló-Aguayo, I., Soliva-Fortuny, R., Martín-Belloso, O., 2010a. Impact of high-intensity pulsed electric field variables affecting peroxidase and lypoxigenase activities of watermelon juice. LWT-Food Sci. Technol. 42, 813–818.

Aguiló-Aguayo, I., Soliva-Fortuny, R., Martín-Belloso, O., 2010b. Color and viscosity of watermelon juice treated by high-intensity pulsed electric fields or heat. Innov. Food Sci. Emerg. 11, 299–305.

Aguiló-Aguayo, I., Soliva-Fortuny, R., Martín-Belloso, O., 2010c. Volatile compounds and changes in flavor-related enzymes during the cold storage of high-intensity pulsed electric field- or thermally-processed tomato juices. J. Sci. Food Agric. 90, 1597–1604.

Akin, E., Evrendilek, G.A., 2009. Effect of pulsed electric fields on physical, chemical, and microbiological properties of formulated carrot juice. Food Sci. Technol. Int. 15, 275–282.

Alkhafaji, S.R., Farid, M., 2007. An investigation on pulsed electric fields technology using new treatment chamber design. Innov. Food Sci. Emerg. 8, 205–212.

Álvarez, I., Raso, J., Palop, A., Sala, F., 2000. Influence of different factors on the inactivation of Salmonella senftenberg by pulsed electric fields. Int. J. Food Microbiol. 55, 143–146.

Aronsson, K., Rönner, U., 2001. Influence of pH, water activity and temperature on the inactivation of Escherichia coli and Saccharomyces cerevisiae by pulsed electric fields. Innov. Food Sci. Emerg. 2, 105–112.

Aronsson, K., Lindgren, M., Johansson, B.R., Rönner, U., 2001. Inactivation of microorganisms using pulsed electric fields: the influence of process parameters on Escherichia coli, Listeria innocua, Leuconostoc mesenteroides and Saccharomyces cerevisiae. Innov. Food Sci. Emerg. 2, 41–54.

Arroyo, C., Cebrián, G., Pagán, R., Condón, S., 2010. Resistance of Enterobater sakazakii to pulsed electric fields. Innov. Food Sci. Emerg. 11, 314–321.

Barbosa-Cánovas, G.V., Pothakamury, U.R., Palou, E., Swanson, B.G., 1998. Biological effects and applications of pulsed electric fields for the preservation of foods. Nonthermal Preservation of Foods. Marcel Dekker Inc., New York, USA, pp. 73–112.

Barsotti, L., Cheftel, J.C., 1999. Food processing by pulsed electric fields. II. Biological aspects. Food Rev. Int. 15, 181–213.

Bazhal, M., Ngadi, M.O., Raghavan, V.G.S., 2006. Kinetics of Escherichia coli in liquid whole egg using combined PEF and thermal treatments. LWT-Food Sci. Technol. 39, 420–426.

Bendicho, S., Barbosa-Cánovas, G.V., Martín, O., 2002a. Milk processing by high intensity pulsed electric fields. Trends Food Sci. Tech. 13, 195–204.

Bendicho, S., Estela, C., Giner, J., Barbosa-Cánovas, G.V., Martín, O., 2002b. Effects of high intensity pulsed electric field and thermal treatments on a lipase from Pseudomonas fluorescens. J. Dairy Sci. 85, 19–27.

Bendicho, S., Espachs, A., Arántegui, J., Martín, O., 2002c. Effect of high intensity pulsed electric fields and heat treatments on vitamins of milk. J. Dairy Res. 69, 113–123.

Bendicho, S., Barbosa-Cánovas, G.V., Martín, O., 2003. Reduction of protease activity in milk by continuous flow high-intensity pulsed electric field treatments. J. Dairy Sci. 86, 697–703.

Bendicho, S., Marsellés-Fontanet, R., Barbosa-Cánovas, G.V., Martín-Belloso, O., 2005. High intensity pulsed electric fields and heat treatments applied to a protease from Bacillus subtilis. A comparison study of multiple systems. J. Food Eng. 69, 317–323.

Castro, A.J., Swanson, B.G., Barbosa-Cánovas, G.V., Zhang, Q.H., 2001. Pulsed electric field modification of milk alkaline phosphatase activity. In: Barbosa-Cánovas, G.V., Zhang, Q.H. (Eds.), Pulsed Electric Fields in Food Processing. Fundamental Aspects and Applications. Technomic Publishing Company Inc., Lancaster, USA, pp. 65–82.

Cebrián, G., Sagarzazu, N., Pagán, R., Condón, S., Mañas, P., 2008. Resistance of Escherichia coli grown at different temperaturas to various environmental stresses. J. Appl. Microbiol. 105, 271–278.

Charles-Rodríguez, A.V., Nevárez-Moorillón, G.V., Zhang, Q.H., Ortega-Rivas, E., 2007. Comparison of thermal processing and pulsed electric fields treatment in pasteurization of apple juice. Trans. I. Chem. E. 85 (C2), 93–97.

Choi, J., Wang, D., Namihira, T., Katsuki, S., Akiyama, H., Lin, X., et al., 2008. Inactivation of spores using pulsed electric field in a pressurized flow system. J. Appl. Phys. 104, 1–6.

Cortés, C., Esteve, M.J., Frígola, A., Torregrosa, F., 2005. Quality characteristics of horchata (a Spanish vegetable beverage) treated with pulsed electric fields during shelf-life. Food Chem. 91, 319–325.

Cortés, C., Torregrosa, F., Esteve, M.J., Frígola, A., 2006a. Carotenoid profile modification during refrigerated storage in untreated and pasteurized orange juice and orange juice treated with high-intensity pulsed electric fields. J. Agric. Food Chem. 54, 6247–6254.

Cortés, C., Esteve, M.J., Rodrigo, D., Torregrosa, F., Frígola, A., 2006b. Changes of colour and carotenoids contents during high intensity pulsed electric field treatment in orange juices. Food Chem. Toxicol. 44, 1932–1939.

Cserhalmi, Z., Vidács, I., Beczner, J., Czukor, B., 2002. Inactivation of Saccharomyces cerevisiae and Bacillus cereus by pulsed electric fields technology. Innov. Food Sci. Emerg. 3, 41–45.

Cserhalmi, Z.S., Sass-Kiss, A., Tóth-Markus, M., Lechner, N., 2006. Study of pulsed electric field treated citrus juices. Innov. Food Sci. Emerg. 7, 49–54.

De Haan, S.W.H., Willcock, P.R., 2002. Comparison of the energy performance of pulse generator circuits for PEF. Innov. Food Sci. Emerg. 3, 349–356.

Donsi, G., Ferrari, G., Pataro, G., 2007. Inactivation kinetics of Saccharomyces cerevisiae by pulsed electric fields in a batch treatment chamber: the effect of electric field unevenness and initial cell concentration. J. Food Eng. 78, 784–792.

Dunn, J.E., 2001. Pulsed electric field processing: an overview. In: Barbosa-Cánovas, G.V., Zhang, Q.H. (Eds.), Pulsed Electric Fields in Food Processing: Fundamental Aspects and Applications. Technomic, Lancaster, PA, USA, pp. 1–30.

Dunn, J.E., Pearlman, J.S., 1987. Methods and apparatus for extending the shelf-life of fluid food products. U.S. Patent 4 695 472, Sep. 22.

Dutreux, N., Notermans, S., Witjez, T., Góngora-Nieto, M.M., Barbosa-Cánovas, G.V., Swanson, B.G., 2000. Pulsed electric fields inactivation of attached and free-living Escherichia coli and Listeria innocua under several conditions. Int. J. Food Microbiol. 54, 91–98.

Elez-Martinez, P., Escolá-Hernández, J., Soliva-Fortuny, R.C., Martín-Belloso, O., 2004. Inactivation of Saccharomyces cerevisiae suspended in orange juice using high intensity pulsed electric fields. J. Food Protect. 67, 2596–2602.

Elez-Martínez, P., Escolá-Hernández, J., Soliva-Fortuny, R.C., Martín-Belloso, O., 2005. Inactivation of Lactobacillus brevis in orange juice by high intensity pulsed electric fields. Food Microbiol. 22, 311–319.

Elez-Martínez, P., Aguiló-Aguayo, I, Martín-Belloso, O., 2006a. Inactivation of orange juice peroxidase by high-intensity pulsed electric fields as influenced by processing parameters. J. Sci. Food Agric. 86, 71–81.

Elez-Martínez, P., Soliva-Fortuny, R., Martín-Belloso, O., 2006b. Comparative study on shelf-life of orange juice processed by high intensity pulsed electric fields or heat treatment. Eur. Food Res. Technol. 222, 321–329.

Elez-Martínez, P., Suárez-Recio, M., Martín-Belloso, O., 2007. Modeling the reduction of pectin methyl esterase activity in orange juice by high intensity pulsed electric fields. J. Food Eng. 78, 184–193.

Elez-Martínez, P., Martín-Belloso, O., 2007. Effects of high intensity pulsed electric field processing conditions on vitamin C and antioxidant capacity of orange juice and gazpacho, a cold vegetable soup. Food Chem. 102, 201–209.

Elez-Martínez, P., Soliva-fortuny, R., Martín-Belloso, O., 2009. Impact of high-intensity pulsed electric fields on bioactive compounds in Mediterranean plant-based foods. Nat. Prod. Commun. 4, 741–746.

Espachs-Barroso, A., Barbosa-Cánovas, G.V., Martín-Belloso, O., 2003. Microbial and enzymatic changes in fruit juice induced by high-intensity pulsed electric fields. Food Rev. Int. 19, 253–273.

Evrendilek, G.A., Jin, Z.T., Ruhlman, K.T., Qiu, X., Zhang, Q.H., Ritcher, E.R., 2000. Microbial safety and shelf-life of apple juice and cider processed by bench and pilot scale PEF systems. Innov. Food Sci. Emerg. 1, 77–86.

Evrendilek, G.A., Dantzer, W.R., Streaker, C.B., Ratanatriwong, P., Zhang, Q.H., 2001. Shelf-life evaluations of liquid foods treated by pilot plant pulsed electric field system. J. Food Process. Pres. 25, 283–297.

Evrendilek, G.A., Li, S., Dantzer, W.R., Zhang, Q.H., 2004a. Pulsed electric field processing of beer: microbial, sensory, and quality analyses. J. Food. Sci. 69, M228–M232.

Evrendilek, G.A., Yeom, H.W., Jin, Z.T., Zhang, Q.H., 2004b. Safety and quality evaluation of a yogurt-based drink processed by a pilot plant PEF system. J. Food Process Eng. 27, 197–212.

Evrendilek, G.A., Zhang, Q.H., 2005. Effects of pulse polarity and pulse delaying time on pulsed electric fields-induced pasteurization of E. coli 0157:H7. J. Food Eng. 68, 271–276.

Fernández-Molina, J.J., Altunakar, B., Bermúdez-Aguirre, D., Swanson, B.G., Barbosa-Cánovas, G.V., 2005a. Inactivation of Pseudomonas fluorescens in skim milk by combinations of pulsed electric fields and organic acids. J. Food Protect. 68, 1232–1235.

Fernández-Molina, J.J., Barbosa-Cánovas, G.V., Swanson, B.G., 2005b. Skim milk processing by combining pulsed electric fields and thermal treatments. J. Food Process. Preserv. 29, 291–296.

Fernández-Molina, J.J., Fernández-Gutiérrez, S.A., Altunakar, B., Bermúdez-Aguirre, D., Swanson, B.G., Barbosa-Cánovas, G.V., 2005c. The combined effect of pulsed electric fields and conventional heating on the microbial quality and shelf-life of skim milk. J. Food Process. Preserv. 29, 390–406.

Fernández-Molina, J.J., Bermúdez-Aguirre, D., Altunakar, B., Swanson, G.G., Barbosa-Cánovas, G.V., 2006. Inactivation of Listeria innocua and Pseudomonas fluorescens by pulsed electric fields in skim milk: energy requirements. J. Food Process Eng. 29, 561–573.

Fox, M.B., Esveld, D.C., Mastwijk, H., Boom, R.M., 2008. Inactivation of L. plantarum in a PEF microreactor: the effect of pulse width and temperature on the inactivation. Innov. Food Sci. Emerg. 9, 101–108.

Gachovska, T.K., Kumar, S., Thippareddi, H., Subbiah, J., Williams, F., 2008. Ultraviolet and pulsed electric field treatment have additive effect on inactivation of E. coli in apple juice. J. Food Sci. 73, 412–417.

García, D., Gómez, N., Condón, S., Raso, J., Pagán, R., 2003. Pulsed electric fields cause sublethal injury in Escherichia coli. Appl. Microbiol. 36, 140–144.

García, D., Gómez, N., Mañas, P., Condón, S., Raso, J., Pagán, R., 2005. Occurrence of sublethal injury after pulsed electric field depending on the microorganism, the treatment medium pH and the intensity of the treatment investigated. J. Appl. Microbiol. 99, 94–104.

García, D., Mañas, P., Gómez, N., Raso, J., Pagán, R., 2006. Biosynthetic requirements for the repair of sublethal membrane damage in Escherichia coli after pulsed electric fields. J. Appl. Microbiol. 100, 428–435.

García, D., Gómez, N., Mañas, P., Raso, J., Pagán, R., 2007. Pulsed electric fields cause bacterial envelopes permeabilization depending on the treatment intensity, the treatment medium pH and the microorganism investigated. Int. J. Food Microbiol. 113, 219–227.

Garde-Cerdán, T., Arias-Gil, M., Marsellés-Fontanet, R., Ancín-Azpilicueta, C., Martín-Belloso, O., 2007. Effects of thermal and non-thermal processing treatments on fatty acids and free amino acids of grape juice. Food Control 18, 473–479.

Garde-Cerdán, T., Marsellés-Fontanet, A.R., Arias-Gil, M., Ancín-Azpilicueta, C., Martín-Belloso, O., 2008. Effect of storage conditions on the volatile composition of wines obtained from must stabilized by PEF during ageing without SO2. Innov. Food Sci. Emerg. 9, 469–476.

Gásková, D., Sigler, K., Janderová, B., Plásek, J., 1996. Effect of high-voltage electric pulses on yeast cells: factors influencing the killing efficiency. Bioelectroch. Bioener. 39, 195–202.

Giner, J., Gimeno, V., Barbosa-Cánovas, G.V., Martín, O., 2001. Effects of pulsed electric field processing on apple and pear polyphenoloxidase. Food Sci. Technol. Int. 7, 339–345.

Giner, J., Gimeno, V., Palomes, M., Barbosa-Cánovas, G.V., Martín, O., 2003. Lessening polygalacturonase activity in a commercial enzyme preparation by exposure to pulsed electric fields. Eur. Food Res. Technol. 217, 43–48.

Giner-Seguí, J., Elez-Martínez, P., Martín-Belloso, O., 2009. Modeling within the Bayesian framework, the inactivation of pectinesterase in gazpacho by pulsed electric fields. J. Food Eng. 95, 446–452.

Góngora-Nieto, M.M., Sepúlveda, D.R., Pedrow, P., Barbosa-Cánovas, G.V., Swanson, B.G., 2002. Food processing by pulsed electric fields: treatment delivery, inactivation level and regulatory aspects. LWT-Food Sci. Technol. 35, 375–388.

Góngora-Nieto, M.M., Pedrow, P.D., Swanson, B.G., Barbosa-Cánovas, G.V., 2003. Energy analysis of liquid whole egg pasteurized by pulsed electric fields. J. Food Eng. 57, 209–216.

Grahl, T., Märkl, H., 1996. Killing of microorganisms by pulsed electric fields. Appl. Microbiol. Biotechnol. 45, 148–157.

Ho, S.Y., Mittal, G.S., Cross, J.D., Griffiths, M.W., 1995. Inactivation of Pseudomonas fluorescens by high voltage electric pulses. J. Food Sci. 60, 1337–1340.

Ho, S.Y., Mittal, G.S., Cross, J.D., 1997. Effects of high field electric pulses on the activity of selected enzymes. J. Food Eng. 31, 69–84.

Hodgins, A.M., Mittal, G.S., Griffiths, M.W., 2002. Pasteurization of fresh orange juice using low-energy pulsed electrical field. J. Food Sci. 67, 2294–2299.

Huang, E., Mittal, G.S., Griffiths, M.W., 2006. Inactivation of Salmonella enteritidis in liquid whole egg using combination treatments of pulsed electric field, high pressure and ultrasound. Biosyst. Eng. 94, 403–413.

Huang, K., Wang, J., 2009. Designs of pulsed electric fields treatment chambers for liquid foods pasteurization process: a review. J. Food Eng. 95, 227–239.

IFT, 2001. Effect of preservation technologies and microbiological inactivation in foods. In: IFT (Ed.), Evaluation and Definition of Potentially Hazardous Foods. Comprehensive Reviews in Food Science and Food Safety, vol. 2. pp. 42–45.

Jaeger, H., Schulz, A., Karapetkov, N., Knorr, D., 2009. Protective effect of milk constituents and sublethal injuries limiting process effectiveness during PEF inactivation of Lb. rhamnosus. Int. J. Food Microbiol. 134, 154–161.

Jin, Z.T., Zhang, H.Q., Li, S.Q., Kim, M., Dunne, C.P., Yang, T., et al., 2009. Quality of applesauces processed by pulsed electric fields and HTST pasteurization. Int. J. Food Sci. Technol. 44, 829–839.

Kotnik, T., Pucihar, G., Rebersek, M., Miklavic, D., Mir, M.L., 2003. Role of pulse shape in cell membrane electropermeabilization. BBA-Biomembranes 1614, 193–200.

Li, Y.Q., Chen, Q., Liu, X.H., Chen, Z.X., 2008. Inactivation of soybean lipoxygenase in soymilk by pulsed electric fields. Food Chem. 109, 408–414.

Li, C., Hao, J., Zhong, H., Dang, M., Xie, B., 2009. Aroma components at various stages of litchi juice processing. J. Sci. Food Agric. 89, 2405–2414.

Lindgren, M., Aronsson, K., Galt, S., Ohlsson, T., 2002. Simulation of the temperature increase in pulsed electric field (PEF) continuous flow treatment chambers. Innov. Food Sci. Emerg. 3, 233–245.

Mabrook, M.F., Petty, M.C., 2003. Effect of composition on the electrical conductance of milk. J. Food Eng. 60, 321–325.

Mañas, P., Barsotti, L, Cheftel, J.C., 2001. Microbial inactivation by pulsed electric fields in a batch tratment chamber: effects of some electrical parameters and food constituents. Innov. Food Sci. Emerg. 2, 239–249.

Marquez, V.O., Mittal, G.S., Griffiths, M.W., 1997. Destruction and inhibition of bacterial spores by high voltage pulsed electric field. J. Food Sci. 62, 399–401.

Marselles-Fontanet, A.R., Martín-Belloso, O., 2007. Optimization and validation of PEF processing conditions to inactivate oxidative enzymes of grape juice. J. Food Eng. 83, 452–462.

Martín, O., Qin, B.L., Chang, F.J., Barbosa-Canovas, G.V., Swanson, B.G., 1997. Inactivation of Escherichia coli in skim milk by high intensity pulsed electric fields. J. Food Process Eng. 20, 317–336.

Martín-Belloso, O., Vega-Mercado, H., Qin, B.L., Chang, F.J., Barbosa-Cánovas, G.V., Swanson, B.G., 1997. Inactivation of Escherichia coli suspended in liquid egg using pulsed electric fields. J. Food Process. Preserv. 21, 193–208.

Martínez-Viedma, P., Sobrino-Lopez, A., Ben Omar, N., Abriouel, H., Lucas-Lopez, R., Valdivia, E., et al., 2008. Enhanced bactericidal effect of enterocin AS-48 in combination with high-intensity pulsed-electric field treatment against Salmonella enterica in apple juice. Int. J. Food Microbiol. 128, 244–249.

McManee, C., Noci, F., Cronin, D.A., Lyng, J.G., Morgan, D.J., Scannell, A.G.M., 2010. PEF based hurdle strategy to control Pichia fermentans, Listeria innocua and Escherichia coli k12 in orange juice. Int. J. Food Microbiol. 138, 13–18.

Michalac, S., Alvarez, V., Ji, T., Zhang, Q.H., 2003. Inactivation of selected microorganisms and properties of pulsed electric field processed milk. J. Food Process Preserv. 27, 137–151.

Min, S., Laura, R., Zhang, Q.H., 2002. Water activity and the inactivation of Enterobacter cloacae inoculated in chocolate liquor and a model system by pulsed electric field treatment. J. Food Process. Preserv. 26, 323–337.

Min, S., Zhang, Q.H., 2003. Effects of commercial-scale pulsed electric field processing on flavor and color of tomato juice. J. Food Sci. 68, 1600–1606.

Min, S., Jin, Z.T., Min, S.K., Yeom, H., Zhang, Q.H., 2003a. Commercial-scale pulsed electric field processing of orange juice. J. Food Sci. 68, 1265–1271.

Min, S., Min, S.K., Zhang, Q.H., 2003b. Inactivation kinetics of tomato juice lipoxygenase by pulsed electric fields. J. Food Sci. 68, 1995–2001.

Min, S., Jin, Z.T., Zhang, Q.H., 2003c. Commercial scale pulsed electric field processing of tomato juice. J. Agric. Food Chem. 51, 3338–3344.

Min, S., Evrendilek, G.A., Zhang, Q.H., 2007. Pulsed electric fields: processing system, microbial and enzyme inhibition, and shelf-life extension of foods. IEEE T. Plasma Sci. 35, 59–73.

Molinari, P., Pilosof, A.M.R., Jagus, M.J., 2004. Effect of growth phase and inoculum size on the inactivation of Saccharomyces cerevisiae in fruit juices, by pulsed electric fields. Food Res. Int. 37, 793–798.

Monfort, S., Gayán, E., Saldaña, G., Puértolas, E., Condón, S., Raso, J., et al., 2010. Inactivation of Salmonella Typhimurium and Staphylococcus aureus by pulsed electric fields in liquid whole egg. Innov. Food Sci. Emerg. 11, 306–313.

Morales-de la Peña, M., Salvia-Trujillo, L., Rojas-Graü, M.A., Martín-Belloso, O., 2010. Impact of high intensity pulsed electric field on antioxidant properties and quality parameters of a fruit juice–soymilk beverage in chilled storage. LWT-Food Sci. Technol. 43, 872–881.

Mosqueda-Melgar, J., Raybaudi-Massilia, R.M., Martín-Belloso, O., 2007. Influence of treatment time and pulse frequency on Salmonella Enteriditis, Escherichia coli and Listeria monocytogenes populations inoculated in melon and watermelon juices treated by pulsed electric fields. Int. J. Food Microbiol. 117, 192–200.

Mosqueda-Melgar, J., Raybaudi-Massilia, R.M., Martín-Belloso, O., 2008a. Non-thermal pasteurization of fruit juices by combining high-intensity pulsed electric fields with natural antimicrobials. Innov. Food Sci. Emerg. 9, 328–340.

Mosqueda-Melgar, J., Raybaudi-Massilia, R.M., Martín-Belloso, O., 2008b. Inactivation of Salmonella enterica ser. enteritidis in tomato juice by combining of high-intensity pulsed electric fields with natural antimicrobials. J. Food Sci. 73, M47–M53.

Mosqueda-Melgar, J., Elez-Martínez, P., Raybaudi-Massilia, R.M., Martín-Belloso, O., 2008c. Effects of pulsed electric fields on pathogenic microorganisms of major concern in fluid foods. Crit. Rev. Food Sci. 48, 747–759.

Mosqueda-Melgar, J., Raybaudi-Massilia, R.M., Martín-Belloso, O., 2008d. Combination of high-intensity pulsed electric fields with natural antimicrobials to inactivate pathogenic microorganisms and extend the shelf-life of melon and watermelon juices. Food Microbiol. 25, 479–491.

Noci, F., Walkling-Ribeiro, M., Cronin, D.A., Morgan, D.J, Lyng, J.G., 2009. Effect of thermosonication, pulsed electric field and their combination on inactivation of Listeria innocua in milk. Int. Dairy J. 19, 30–35.

Odriozola-Serrano, I., Bendicho-Porta, S., Martín-Belloso, O., 2006. Comparative study on shelf-life of whole milk processed by high-intensity pulsed electric field or heat treatment. J. Dairy Sci. 89, 905–911.

Odriozola-Serrano, I., Soliva-Fortuny, R., Martín-Belloso, O., 2008a. Changes of health-related compounds throughout cold storage of tomato juice stabilized by thermal or high intensity pulsed electric field treatments. Innov. Food Sci. Emerg. 9, 272–279.

Odriozola-Serrano, I., Soliva-Fortuny, R., Gimeno-Añó, V., Martín-Belloso, O., 2008b. Modeling changes in health-related compounds of tomato juice treated by high-intensity pulsed electric fields. J. Food Eng. 89, 210–216.

Odriozola-Serrano, I., Soliva-Fortuny, R., Hernández-Jover, T., Martín-Belloso, O., 2008c. Carotenoid and phenolic profile of tomato juices processed by high intensity pulsed electric fields compared with conventional thermal treatments. Food Chem. 112, 258–266.

Odriozola-Serrano, I., Soliva-Fortuny, R., Martín-Belloso, O., 2008d. Phenolic acids, flavonoids, vitamin C and antioxidant capacity of strawberry juices processed by high intensity pulsed electric fields or heat treatments. Eur. Food Res. Technol. 228, 239–248.

Odriozola-Serrano, I., Soliva-Fortuny, R., Gimeno-Añó, V., Martín-Belloso, O., 2008e. Kinetic study of anthocyanins, vitamin C, and antioxidant capacity in strawberry juices treated by high-intensity pulsed electric fields. J. Agric. Food Chem. 56, 8387–8393.

Odriozola-Serrano, I., Soliva-Fortuny, R., Martín-Belloso, O., 2009a. Impact of high-intensity pulsed electric field variables on vitamin C, anthocyanins and antioxidant capacity of strawberry juice. LWT-Food Sci. Technol. 42, 93–100.

Odriozola-Serrano, I., Soliva-Fortuny, R., Hernández-Jover, T., Martín-Belloso, O., 2009b. Carotenoid and phenolic profile of tomato juices processed by high intensity pulsed electric fields compared with conventional thermal treatments. Food Chem. 112 (1), 258–266.

Oms-Oliu, G., Odriozola-Serano, I., Soliva-Fortuny, R., Martín-Belloso, O., 2009. Effects of high-intensity pulsed electric field processing conditions on lycopene, vitamin C and antioxidant capacity of watermelon juice. Food Chem. 115, 1312–1319.

Otunola, A., El-Hag, A., Jarayam, S.H., Anderson, W.A., 2008. Effectiveness of pulsed electric fields in controlling microbial growth in milk. Int. J. Food Eng. 4, 1–14.

Picart, L., Dumay, E., Cheftel, C., 2002. Inactivation of Listeria innocua in dairy fluids by pulsed electric fields: influence of electric parameters and food composition. Innov. Food Sci. Emerg. 3, 357–369.

Pothakamury, U.R., Vega, H., Zhang, Q., Barbosa-Cánovas, G.V., Swanson, B.G., 1996. Effect of growth stage and processing temperature on the inactivation of E. coli by pulsed electric fields. J. Food Protect. 59, 1167–1171.

Puértolas, E., López, N., Condón, S., Raso, J., Álvarez, I., 2009. Pulsed electric fields inactivation of wine spoilage yeast and bacteria. Int. J. Food Microbiol. 130, 49–55.

Qin, B., Zhang, Q., Barbosa-Cánovas, G.V., Swanson, B.G., Pedrow, P.D., 1994. Inactivation of microorganisms by pulsed electric fields of different voltage waveforms. IEEE T. Dielect. El. In. 1, 1047–1057.

Qin, B.L., Barbosa-Cánovas, G.V., Swanson, B.G., Pedrow, P.D., Olsen, R.G., 1998. Inactivating microorganisms using a pulsed electric field continuous treatment system. IEEE T. Ind. Appl. 34, 43–50.

Qin, B.L., Zhang, Q., Barbosa-Cánovas, G.V., Swanson, B.G., Pedrow, P.D., 1995a. Pulsed electric field treatment chamber design for liquid food pasteurization using the finite element method. Trans. ASAE. 38, 557–565.

Qin, B.L., Chang, F.J., Barbosa-Cánovas, G.V., Swanson, B.G., 1995b. Non-thermal inactivation of Saccharomyces cerevisiae in apple juice using pulsed electric fields. LWT-Food Sci. Technol. 28, 564–568.

Qiu, X., Sharma, S., Tuhela, L., Jia, M., Zhang, Q.H., 1998. An integrated PEF pilot plant for continuous nonthermal pasteurization of fresh orange juice. Trans. Am. Soc. Agric. Eng. 41, 1069–1074.

Quitão-Texeira, L.J., Odriozola-Serrano, I., Soliva-Fortuny, R., Mota-Ramos, A., Martín-Belloso, O., 2009. Comparative study on antioxidant properties of carrot juice stabilised by high-intensity pulsed electric fields or heat treatments. J. Food Sci. Agric. 89, 2636–2642.

Rastogi, N.K., 2003. Application of high-intensity pulsed electrical fields in food processing. Food Rev. Int. 19, 229–251.

Reina, L.D., Jin, Z.T., Zhang, Q.H., Yousef, A.E., 1998. Inactivation of Listeria monocytogenes in milk by pulsed electric field. J. Food Prot. 61, 1203–1206.

Riener, J., Noci, F., Cronin, D.A., Morgan, D.J., Glyng, J., 2008a. Effect of high intensity pulsed electric fields on enzymes and vitamins in bovine raw milk. Int. J. Dairy Technol. 62, 1–6.

Riener, J., Noci, F., Cronin, D.A., Morgan, D.J., Glyng, J.G., 2008b. Combined effect of temperature and pulsed electric fields on apple juice peroxidase and polyphenoloxidase inactivation. Food Chem. 109, 402–407.

Riener, J., Noci, F., Cronin, D.A., Morgan, D.J., Glyng, J.G., 2009. Combined effect of temperature and pulsed electric fields on pectin methyl esterase inactivation in red grapefruit juice (Citrus paradisi). Eur. Food Res. Technol. 228, 373–379.

Rivas, A., Sampedro, F., Rodrigo, D., Martínez, A., Rodrigo, M., 2006a. Nature of the inactivaction of Escherichia coli suspended in an orange juice and milk beverage. Eur. Food Res. Technol. 223, 541–545.

Rivas, A., Rodrigo, D., Martínez, A., Barbosa-Cánovas, G.V., Rodrigo, M., 2006b. Effect of PEF and heat pasteurization on the physical–chemical characteristics of blended orange and carrot juice. LWT-Food Sci. Tech. 39, 1163–1170.

Rodrigo, D., Barbosa-Cánovas, G.V., Martínez, A., Rodrigo, M., 2003a. Pectin methyl esterase and natural microflora of fresh mixed orange juice and carrot juice treated with high intensity pulsed electric fields. J. Food Protect. 66, 2336–2342.

Rodrigo, D., Ruíz, P., Barbosa-Cánovas, G.V., Martínez, A., Rodrigo, M., 2003b. Kinetic model for the inactivation of Lactobacillus plantarum by pulsed electric fields. Int. J. Food Microbiol. 81, 223–229.

Rowan, N.J., MacGregor, S.J., Anderson, J.G., Fouracre, R.A., Farish, O., 2000. Pulsed electric field inactivation of diarrhoeagenic Bacillus cereus through irreversible electroporation. Lett. Appl. Microbiol. 31, 110–114.

Sagarzazu, N., Cebrián, G., Pagán, R., Condón, S., Mañas, P., 2010. Resistance of Campylobacter jejuni to heat and to pulsed electric fields. Innov. Food Sci. Emerg. 11, 283–289.

Saldaña, G., Puértolas, E., López, N., García, D., Álvarez, I., Raso, J., 2009. Comparing the PEF resistance and occurrence of sublethal injury on different strains of Escherichia coli, Salmonella Typhimurium, Listeria monocytogenes and Staphylococcus aureus in media of pH 4 and 7. Innov. Food Sci. Emerg. 10, 160–165.

Saldaña, G., Puértolas, E., Condón, S., Álvarez, I., Raso, J., 2010. Modeling inactivation kinetics and occurence of sublethal injury of a pulsed electric field-resistant strain of Escherichia coli and Salmonella Typhimurium in media of different pH. Innov. Food Sci. Emerg. 11, 290–298.

Sampedro, F., Rodrigo, M., Martínez, A., Rodrigo, D., Barbosa-Cánovas, G.V., 2005. Quality and safety aspects of PEF application in milk and milk products. Crit. Rev. Food Sci. 45, 25–47.

Sampedro, F., Rodrigo, D., Martínez, A., Barbosa-Cánovas, G.V., Rodrigo, M., 2006a. Review: application of pulsed electric fields in egg and egg derivatives. Food Sci. Technol. Int. 12, 397–406.

Sampedro, F., Rodrigo, D., Martínez, A., Barbosa-Cánovas, G.V., Rodrigo, M., 2006b. Review: application of pulsed electric fields in egg and egg derivatives. Food Sci. Technol. Int. 12, 397–406.

Sampedro, F., Rivas, A., Rodrigo, D., Martínez, A., Rodrigo, M., 2006c. Effect of temperature and substrate on PEF inactivation of Lactobacillus plantarum in an orange juice–milk beverage. Eur. Food Res. Technol. 223, 30–34.

Sampedro, F., Geveke, D.J., Fan, X., Zhang, H.Q., 2009a. Effect of PEF, HHP and thermal treatment on PME inactivation and volatile compounds concentration of an orange juice–milk based beverage. Innov. Food Sci. Emerg. 10, 463–469.

Sampedro, F., Geveke, D.J., Fan, X., Rodrigo, D., Zhang, H.Q., 2009b. Shelf-life study of an orange juice–milk based beverage after PEF and thermal processing. J. Food Sci. 74, S107–S112.

Sánchez-Moreno, C., Cano, M.P., De Ancos, B., Plaza, L., Olmedilla, B., Granado, F., et al., 2004. Pulsed electric fields-processed orange juice consumption increases plasma vitamin C and decreases F2-isoprostanes in healthy humans. J. Nutr. Biochem. 15, 601–607.

Sánchez-Moreno, C., Plaza, L., Elez-Martínez, P., De Ancos, B., Martín-Belloso, O., Cano, M.P., 2005a. Impact of high pressure and pulsed electric fields on bioactive compounds and antioxidant activity of orange juice in comparison with traditional thermal processing. J. Agric. Food Chem. 53, 4403–4409.

Sánchez-Moreno, C., Cano, M.P., De Ancos, B., Plaza, L., Olmedilla, B., Granado, F., et al., 2005b. Intake of Mediterranean vegetable soup treated by pulsed electric fields affects plasma vitamin C and antioxidant biomarkers in humans. Int. J. Food Sci. Nutr. 56, 115–124.

Sánchez-Moreno, C., De Ancos, B., Plaza, L., Elez-Martínez, P., Cano, M.P., 2009. Nutritional approaches and health-related properties of plant foods processed by high pressure and pulsed electric fields. Crit. Rev. Food. Sci. 49, 552–576.

Schilling, S., Schmid, S., Jäger, H., Ludwig, M., Dietrich, H., Toepfl, S., et al., 2008. Comparative study of pulsed electric field and thermal processing of apple juice with particular consideration of juice quality and enzyme deactivation. J. Agric. Food Chem. 56, 4545–4554.

Selma, M.V., Fernández, P.S., Valero, M., Salmerón, M.C., 2003. Control of Enterobacter aerogenes by high-intensity pulsed electric fields in horchata, a Spanish low-acid vegetable beverage. Food Microbiol. 20, 105–110.

Sepulveda, D.R., Góngora-Nieto, M.M., Guerrero, J.A., Barbosa-Cánovas, G.V., 2005. Production of extended-shelf-life milk by processing pasteurized milk with pulsed electric fields. J. Food Eng. 67, 81–86.

Shin, H.H., Pyun, Y., 1999. Cell damage and recovery characteristics of Lactobacillus plantarum by high voltage pulsed electric fields treatment. Food Sci. Biotechnol. 8, 261–266.

Shin, J.K., Lee, S.J., Cho, H.Y., Pyun, Y.R., Lee, J.H., Chung, M.S., 2010. Germination and subsequent inactivation of Bacillus subtilis spores by pulsed electric field treatment. J. Food Process. Preserv. 43, 43–54.

Smith, K., Mittal, G.S., Griffiths, M.W., 2002. Pasteurization of milk using pulsed electrical field and antimicrobials. J. Food Sci. 6, 2304–2308.

Sobrino-López, A., Raybaudi-Massilia, R., Martín-Belloso, O., 2006. High-intensity pulsed-electric field variables affecting Staphylococcus aureus in milk. J. Dairy Sci. 89, 3739–3748.

Sobrino-López, A., Martín-Belloso, O., 2008. Enhancing the lethal effect of high-intensity pulsed electric field in milk by antimicrobial compounds as combined hurdles. J. Dairy Sci. 91, 1759–1768.

Sobrino-López, A., Martín-Belloso, O., 2009. Review: potential of high-intensity pulsed electric field technology for milk processing. Food Eng. Rev. 2, 17–27.

Soliva-Fortuny, R., Balasa, A., Knorr, D., Martín-Belloso, O., 2009. Effects of pulsed electric fields on bioactive compounds in foods: a review. Trends Food Sci. Technol. 20, 544–556.

Somolinos, M., Garcia, D., Condon, S., Mañas, P., Pagán, R., 2007. Relationship between sublethal injury and inactivation of yeast cells by the combination of sorbic acid and pulsed electric fields. Appl. Environ. Microbiol. 73, 3814–3821.

Somolinos, M., Garcia, D., Condon, S., Mañas, P., Pagán, R., 2008a. Biosynthetic requirements for the repair of sublethally injured Saccharomyces cerevisiae cells after pulsed electric fields. J. Appl. Microbiol. 105, 166–174.

Somolinos, M., Garcia, D., Mañas, P., Condon, S., Pagán, R., 2008b. Effects of environmental factors and cell physiological state on pulsed electric fields resistance and repair capacity of various strains of Escherichia coli. Int. J. Food Microbiol. 124, 260–267.

Toepfl, S., Heinz, V., Knorr, D., 2005. Overview of pulsed electric field processing for food. In: Sun, D.W. (Ed.), Emerging Technologies for Food Processing. Elsevier Ltd., London, UK, pp. 69–97.

Toepfl, S., Heinz, V., Knorr, D., 2007. High intensity pulsed electric fields applied for food preservation. Chem. Eng. Process. 46, 537–546.

Torregrosa, F., Cortés, C., Esteve, M.J., Frígola, A., 2005. Effects of high-intensity pulsed electric fields processing and conventional heat treatment on orange–carrot juice carotenoids. J Agric. Food Chem. 53, 9519–9525.

Torregrosa, F., Esteve, M.J., Frigola, A., Cortés, C., 2006. Ascorbic acid stability during refrigerated storage of orange–carrot juice treated by high pulsed electric field and comparison with pasteurized juice. J. Food Eng. 73, 339–345.

Valappil, Z.A., Fan, X., Zhang, Q.H., Rouseff, R.L., 2009. Impact of thermal and nonthermal technologies on unfermented apple cider aroma volatiles. J. Agric. Food Chem. 57, 924–929.

Van Loey, A., Verachtert, B., Hendrickx, M., 2002. Effects of high electric field pulses on enzymes. Trends Food Sci. Technol. 12, 94–102.

Vega-Mercado, H., Martín-Belloso, O., Chang, F., Barbosa-Cánovas, G.V., Swanson, B.G., 1996. Inactivation of Escherichia coli and Bacillus subtilis suspended in pea soup using pulsed electric fields. J. Food Process. Preserv. 20, 501–510.

Vega-Mercado, H., Powers, J.R., Barbosa-Cánovas, G.V., Luedecke, L., Swanson, B.G., 2001. Change in susceptibility of proteins to proteolysis and the inactivation of an extracellular protease from Pseudomonas fluorescens M3/6 when exposed to pulsed electric fields. In: Barbosa-Cánovas, G.V., Zhang, H. (Eds.), Pulsed Electric Fields in Food Processing. Fundamental Aspects and Applications. Technomic Publishing Company Inc., Lancaster, USA, pp. 105–120.

Walkling-Ribeiro, M., Noci, F., Cronin, D.A., Lyng, J.G., Morgan, D.J., 2008. Inactivation of Escherichia coli in a tropical fruit smoothie by a combination of heat and pulsed electric fields. J. Food Sci. 73, 395–399.

Walkling-Ribeiro, M., Noci, F., Cronin, D.A., Lyng, J.G., Morgan, D.J., 2009a. Shelf-life and sensory evaluation of orange juice after exposure to thermosonication and pulsed electric fields. Food Bioprod. Process. 87, 102–107.

Walkling-Ribeiro, M., Noci, F., Cronin, D.A., Lyng, J.G., Morgan, D.J., 2009b. Antimicrobial effect and shelf-life extension by combined thermal and pulsed electric field treatment of milk. J. Appl. Microbiol. 106, 241–248.

Wouters, P.C., Smelt, J.P.P.M., 1997. Inactivation of microorganisms with pulsed electric fields: potential for food preservation. Food Biotechnol. 11, 193–229.

Yang, R., Li, S.Q., Zhang, Q.H., 2004a. Effects of pulsed electric fields on the activity and structure of pepsin. J. Agric. Food Chem. 52, 7400–7406.

Yang, R.J., Li, S.Q., Zhang, Q.H., 2004b. Effects of pulsed electric fields on the activity of enzymes in aqueous solution. J. Food Sci. 69, FCT241–FCT248.

Yeom, H.W., Streaker, C.B., Zhang, Q.H., Min, D.B., 2000a. Effects of pulsed electric fields on the quality of orange juice and comparison with heat pasteurization. J. Agric. Food Chem. 48, 4597–4605.

Yeom, H.W., Streaker, C.B., Zhang, Q.H., Min, D.B., 2000b. Effects of pulsed electric fields on the activities of microorganisms and pectin methyl esterase in orange juice. J. Food Sci. 65, 1359–1363.

Yeom, H.W., Zhang, Q.H., Chism, G.W., 2002. Inactivation of pectin methyl esterase in orange juice by pulsed electric fields. J. Food Sci. 67, 2154–2159.

Yeom, H.W., Evrendelik, G.A., Jin, Z.T., Zhang, Q.H., 2004. Processing of yogurt-based products with pulsed electric fields: microbial, sensory and physical evaluations. J. Food Process. Pres. 28, 161–178.

Zárate-Rodríguez, E., Ortegas-Rivas, E., Barbosa-Cánovas, G.V., 2000. Quality changes in apple juice as related to nonthermal processing. J. Food Quality 23, 337–349.

Zhang, Q., Barbosa-Cánovas, G.V., Swanson, B.G., 1995a. Engineering aspects of pulsed electric fields pasteurization. J. Food Eng. 25, 261–281.

Zhang, Q., Chang, F.J., Barbosa-Cánovas, G.V., Swanson, B.G., 1994a. Inactivation of microorganisms in a semisolid model food using high voltage pulsed electric fields. LWT-Food Sci. Technol. 27, 538–543.

Zhang, Q., Monsalve-Gonzalez, A., Qin, B.-L., Barbosa-Cánovas, G.V., Swanson, B.G., 1994b. Inactivation of S. cerevisiae in apple juice by square-wave and exponential-decay pulsed electric fields. J. Food Proc. Eng. 17, 469–478.

Zhang, Q., Monsalve-Gonzalez, A., Barbosa-Cánovas, G.V., Swanson, B.G., 1994c. Inactivation of E. coli and S. cerevisiae by pulsed electric fields under controlled temperature conditions. Trans. ASAE. 37, 581–587.

Zhang, Q., Qin, B.-L., Barbosa-Cánovas, G.V., Swanson, B.G., 1995b. Inactivation of E. coli for food pasteurization by high-strength pulsed electric fields. J Food Process Preserv. 19, 103–118.

Zhang, Y., Liao, X., Ni, Y., Wu, J., Hu, X., Wang, Z., et al., 2007. Kinetic analysis of the degradation and its color change of cyanidin-3-glucoside exposed to pulsed electric field. Eur. Food Res. Technol. 224, 597–603.

Zhao, W., Yang, R., 2008. Comparative study of inactivation and conformational change of lysozyme induced by pulsed electric fields and heat. Eur. Food Res. Technol. 228, 47–54.

Zhao, W., Yang, R., Lu, R., Wang, M., Qian, P., Yang, W., 2009. Effect of PEF on microbial inactivation and physical–chemical properties of green tea extracts. LWT-Food Sci. Technol. 41, 425–431.

Zhong, K., Hu, X., Zhao, G., Chen, F., Liao., X., 2005. Inactivation and conformational change of horseradish peroxidase induced by pulsed electric field. Food Chem. 92, 473–479.

Zhong, K., Wu, J., Wang, Z., Chen, F., Liao, X., Hu, X., et al., 2007. Inactivation kinetics and secondary structural change of PEF-treated POD and PPO. Food Chem. 100, 115–123.

Zimmermann, U., 1986. Electric breakdown,electropermeabilization and electrofusion. Rev. Physiol. Biochem. Pharmacol. 105, 196–256.

Zulueta, A., Esteve, M.J., Frasquet, I., Frígola, A., 2007. Fatty acid profile changes during orange juice–milk beverage processing by high-pulsed electric field. Eur J. Lipid Sci. Technol. 109, 25–31.

Zulueta, A., Barba, F.J., Esteve, M.J., Frígola, A., 2010. Effects on the carotenoid pattern and vitamin A of a pulsed electric field-treated orange juice–milk beverage and behavior during storage. Eur. Food Res. Technol. In press (DOI: 10.1007/s00217-010-1304-9).

CHAPTER

5 High-Pressure Processing of Fluid Foods

Rockendra Gupta[1] and V.M. Balasubramaniam[1,2]

[1]Department of Food Science and Technology, The Ohio State University, Columbus, Ohio

[2]Department of Food Agricultural and Biological Engineering, The Ohio State University, Columbus, Ohio

5.1 INTRODUCTION

The food industry is interested in developing alternative process technologies to accomplish a microbiological reduction in various fluid foods without compromising fresh-like product characteristics. High-pressure processing (HPP) is a method of food processing where food is subjected to elevated pressures (up to 87,000 pounds per square inch or approximately 600 MPa), with or without the addition of heat, to achieve microbial inactivation or to alter the food attributes in order to achieve consumer-desired qualities. The technology is also referred as *High Hydrostatic Pressure Processing* (HHP) and *Ultra High Pressure Processing* (UHP) in the literature. The technology has long been used in the material and process engineering industry for sheet-metal forming and isostatic pressing of advanced materials such as turbine components and ceramics.

The history of the use of high pressure to inactivate microorganisms in food dates back to 1899, when Hite demonstrated the application of high pressure in preserving milk and later to preserve fruits and vegetables (Hite, 1899). Engineering aspects of high-pressure processing were extensively investigated by Bridgman in the early 20th century (1909–1959) (Bridgman, 1964). However, its relevance in the food industry was explored only around 1990 in Japan. Depending on the operating conditions of temperature and pressure and the scale of operation, the cost of processing is typically around \$0.10 per pound more than thermally processed food. With a gradual increase in the volume of foods processed using HPP, advancements in HPP equipment and optimization of process parameters, the cost could be expected to drop further.

Novel Thermal and Non-Thermal Technologies for Fluid Foods. DOI: 10.1016/B978-0-12-381470-8.00005-0

5.2 TECHNOLOGY GOVERNING PRINCIPLES

During pressure treatment, the application of pressure is governed by the following basic principles. The Le Chatelier's principle states that any phenomenon such as phase change, change in molecular configuration, chemical reaction, etc., that is accompanied by a decrease in volume is enhanced by pressure and *vice versa* (Farkas and Hoover, 2000). The iso-static principle states that pressure is transmitted in a uniform and quasi-instantaneous manner throughout the whole sample, thus making the process independent of volume and geometry of the product. It has been generally accepted that the iso-static principle is assumed to be true for high-pressure food processing applications. However, deviations are possible for heterogenous large solid samples. Once the desired pressure is reached, it can be maintained for an extended period of time without any further energy input. The microscopic ordering principle states that at constant temperature, an increase in pressure increases the degree of ordering of molecules of a given substance (Heremans, 1995). Thus, it is important to consider the possibility of synergistic or antagonistic reactions when processing foods by combined pressure–heat treatment.

Unlike high-pressure homogenization. where the food is exposed to high velocity, turbulence, and shear forces, during HPP the food is subjected to iso-static pressure treatment. However, foods such as marshmallows or strawberries that contain large air packets are deformed during pressure treatment due to differences between the compressibility of air and the rest of the food.

5.2.1 Process Equipment

Fluid foods such as juices can be processed in batch or semi-continuous mode (Balasubramaniam and Farkas, 2008). The equipment typically consists the following:

1. Pressure vessel.
2. Top and bottom end closures.
3. Yoke (structure for restraining end closures).
4. High pressure pump and intensifier for generating target pressures.
5. Process control and instrumentation.
6. A handling system for loading and removing the product.

In the batch mode, liquid product is pre-packaged, preconditioned and pressure treated. A typical batch process cycle consists of loading the vessel with the pre-packaged, preconditioned product, and filling the remainder of the vessel with water, which acts as the pressure-transmitting fluid. The vessel is closed and the desired process pressure is achieved through the introduction of pressure-transmitting fluid in the vessel. After holding the product for the desired time at the target pressure and temperature, the vessel is decompressed by releasing the pressure-transmitting fluid. Because the pressure is transmitted uniformly (in all directions

Table 5.1 Examples of Commercial Fluid Food Products that can be Preserved by High-Pressure Processing

Country	Year	Product
Japan	1993	Rice wine
France	1994	Citrus juice
Mexico	2000	Citrus juice and smoothies
Lebanon	2001	Fruit juices
Portugal	2001	Apple and citrus blended juice
Czech Republic	2004	Broccoli, apple, beetroot, carrot juices
Northern Ireland	2006	Smoothies
USA	2005	Apple, strawberry, ginger juices
New Zealand	2009	Colostrum

Source: Tornello, 2011.

simultaneously), food retains its shape. In addition, because of minimal thermal effects, the sensory characteristics of the food are retained without compromising its microbial safety.

Semi-continuous pressure equipment employs two or more pressure vessels with free-floating pistons arranged to compress the liquid foods. A low-pressure transfer pump is used to fill the pressure vessel with the liquid food. After filling, the pressure vessel inlet valve is closed, and the pressure-transmitting fluid (usually water) is introduced to compress the liquid food. After the appropriate holding time, releasing the pressure on the pressure-transmitting fluid decompresses the system. The treated liquid food can then be filled aseptically into sterile containers. Three batch vessels in a semi-continuous system can be connected such that while one vessel discharges the product, the second vessel is being compressed, and the third vessel is being loaded. In this way a continuous output is obtained (Balasubramaniam and Farkas, 2008).

Juice and beverage companies using HPP are almost exclusively smaller companies processing bottled premium juices and smoothies for regional markets (Table 5.1). At the time of writing, continuous high-pressure equipment for liquid foods processing is not readily available. Use of batch or semi-continuous equipment constrains the throughput needs of some of the commodity-type beverages and limits the widespread technology implementation for processing such liquid foods.

5.2.2 Packaging

The packaging requirement for the pressure treatment varies depending upon the type of equipment (batch or semi-continuous) used. Semi-continuous systems are used in the case of pumpable liquid products which are aseptically packaged after pressure treatment. On the other hand, flexible or semi-rigid packaging with at least one flexible interface is best suited for batch processing. A variety of existing flexible packaging structures may be used (Balasubramaniam *et al.*, 2004). Since high-moisture foods compress by 15–20% in the range of 600 MPa (87,000 psi) at ambient temperature, packaging materials for pressure treatment must be able to accommodate volume reductions and then return to their original volume without loss of seal integrity or barrier properties. Accordingly, metal cans are generally not suited. High-barrier packaging materials with oxygen- and light-impermeable properties may be desired for extended refrigerated product storage. This can also help preserve the fresh color and flavor attributes of many pressure-treated products (Hogan *et al.*, 2005).

5.2.3 Pressure-Transmitting Fluid

Pressure-transmitting fluids are used to transmit the pressure to the food sample. Water, food-grade propylene glycol–water solutions, silicone oil, sodium benzoate solutions, ethanol solutions, and castor oil are some of the commonly used pressure-transmitting fluids (Balasubramanian and Balasubramaniam, 2003). Choice of pressure-transmitting fluids is in part dependent on their ability to seal under pressure, corrosion prevention properties, fluid viscosity changes under pressure, and heat of compression. The composition of the pressure-transmitting fluid, its thermal characteristics and ratio of fluid to sample play an important role in governing the thermal behavior of foods under pressure. The importance of considering the compression heating behavior of pressure-transmitting fluid on microbial inactivation studies has been documented (Balasubramanian and Balasubramaniam, 2003; de Heij *et al.*, 2003; Matser *et al.*, 2004; Robertson *et al.*, 2008; Otero and Sanz, 2003).

5.2.4 Food Material Response to Combined Temperature–Pressure Treatment

A typical pressure–temperature response of the sample during combined pressure–temperature processing is shown in Figure 5.1. During pressure processing, the food material is pressurized from atmospheric pressure (P_1) to a target pressure (P_2). The time interval between P_1 and P_2 represents the pressure come-up time. Typical commercial equipment may have pressure come-up time of approximately 2 min to reach 600 MPa. Once the target pressure is reached, the samples are processed for desired holding time (P_2–P_3). For economical justification and to minimize thermal effects on product quality, pressure holding times of less than 10 min may be desired. After processing, the samples are depressurized back to atmospheric pressure (P_4). Most of the commercial-scale high pressure equipment has short (<30 s) decompression times.

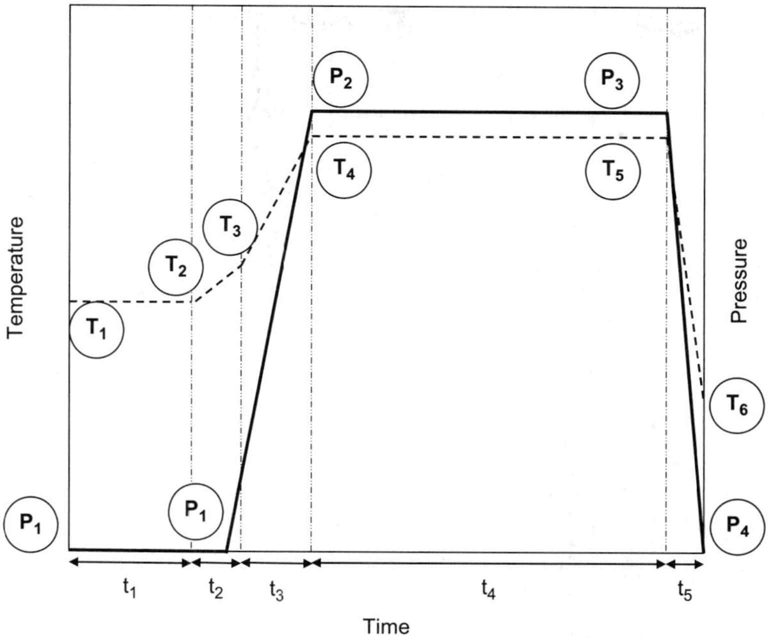

FIGURE 5.1

A typical pressure–temperature profile of a food sample during various stages (preheating, compression, pressure holding, and decompression) of combined pressure–temperature processing. t_1, t_3, t_4, and t_5 are the preheating, compression (come-up time), holding, and decompression times, respectively.

Depending on the processing objective (pasteurization or sterilization), the product will be preconditioned at the desired initial temperature. For example, during sterilization, the product may be preheated to a particular initial temperature (Figure 5.1, T_1–T_2). Subsequently the preheated (or chilled) product is loaded into the pressure vessel. Due to compression under pressure, the temperature of a food material increases (Ting *et al.*, 2002). The magnitude of this temperature change (Figure 5.1, T_3–T_4) depends on the compressibility of the substance, its thermal properties, initial temperature, and target pressure. The maximum product temperature at the target process pressure is independent of the rate of compression as long as heat transfer to the surroundings is negligible. During depressurization (T_5–T_6) the temperature may drop below the initial temperature if heat is lost from the sample to the environment during pressure holding.

Water is the major constituent present in most of the foods along with a complex mixture of proteins, carbohydrates, fats, minerals, vitamins, and salts. The properties of water under pressure are well documented after the pioneering work

Table 5.2 Properties of Water (25°C) at Different Pressures[1]

Properties/Pressure	0.1 MPa	100 MPa	200 MPa	500 MPa
Thermal conductivity (W/mK)	0.61	0.65	0.72	0.76
Specific heat (kJ/kgK)	4.18	3.99	3.83	3.79
Density (kg/m^3)	997	1038	1101	1150

[1]*Estimated values using NIST/ASME software.*

of Bridgman (Bridgman, 1912). Table 5.2 summarizes the data on thermal and physical properties of water obtained from the NIST database (Harvey *et al.*, 1996). Since water is the major constituent of fluid foods, thermophysical properties of liquid foods can be assumed to be similar to that of water.

During a high-pressure process, the food material undergoes physical compression. For example, a volume of 1 l of water may be reduced to 0.85 l while under 600 MPa pressure (~15% reduction). This physical compression also increases the temperature of the product and this is an unavoidable thermodynamic effect (Ting *et al.*, 2002). The reversible adiabatic temperature change of compression can be expressed using the following equation proposed by Zemansky (1957):

$$dT = \frac{(TV\alpha)}{C_p} dP \tag{5.1}$$

From the above equation it follows that the change in temperature as a result of physical compression depends on the thermal expansivity of the substance (α), temperature (T), specific volume (V), and specific heat capacity (Cp). Experimentally determined heat of compression data of selected foods is given in Table 5.3. In general, the heat of compression value of material depends on its phase and gases show highest compression, followed by liquids. Incompressible solids such as metals have the least heat of compression values (Ramaswamy *et al.*, 2005). Water and high-moisture-content foods exhibit the lowest heat of compression value, i.e. 3°C/100 MPa (at initial temperature of 25°C), whereas fats and oils show the highest heat of compression values of 6–9°C/100 MPa (Rasanayagam *et al.*, 2003; Patazca *et al.* 2007). Carbohydrates and proteins have similar heat of compression values to that of water. It is also worth noting that, while heat of compression values of water and high moisture foods are also dependent on the product's initial temperature, for fats and oils the heat of compression is fairly constant and does not depend on the product's initial temperature (Rasanayagam *et al.*, 2003; Patazca *et al.*, 2007) (Fig. 5.2). However, the heat of compression of incompressible materials such as the walls of the pressure vessel (mostly metals) is negligible (~0°C/100 MPa).

While the temperature of the polar compounds such as water readily follow the pressure curve, the temperature response of non-polar substances such as fats

Table 5.3 Heat of Compression Values of Selected Foods Determined at Initial Sample Temperature of 25°C

Food Sample	Temperature Increase (°C) per 100 MPa
Water	3.0
Orange juice, tomato salsa, 2% fat milk, and other water-like substances	2.6–3.0
Carbohydrates	3.6–2.6
Proteins	3.3–2.7
Linolenic acid	9.0–5.9
Soybean oil	9.1–6.2
Olive oil	8.7–6.3
Crude beef fat	4.4
Extracted beef fat	8.3–6.3
Beef ground	3.2
Gravy beef	3.0
Chicken fat	4.5
Chicken breast	3.1
Salmon	3.2
Egg albumin	3.0
Egg yolk	4.5–4.3
Egg whole	3.3
Mayonnaise	7.2–5.0
Whole milk	3.2
Tofu	3.1
Mashed potato	3.0
Yoghurt	3.1
Cream cheese	4.9–4.7
Hass avocado	4.1–3.7
Honey	3.2
Water/glycol (50/50)	4.8–3.7
Propylene glycol	5.8–5.1
Ethanol	10.6–6.8

Adapted from Rasanayagam et al., *2003; Kesavan* et al., *2002; Ramaswamy* et al., *2005.*

and oils exhibits a time lag of up to 30 s to reach the maximum temperature during the pressure holding time (Rasanayagam *et al.*, 2003). This could be attributed to the phase change of sample under pressure (Otero *et al.*, 2000).

Differences in heat of compression behavior of various materials also have interesting effects on thermal nonuniformity during high-pressure processing.

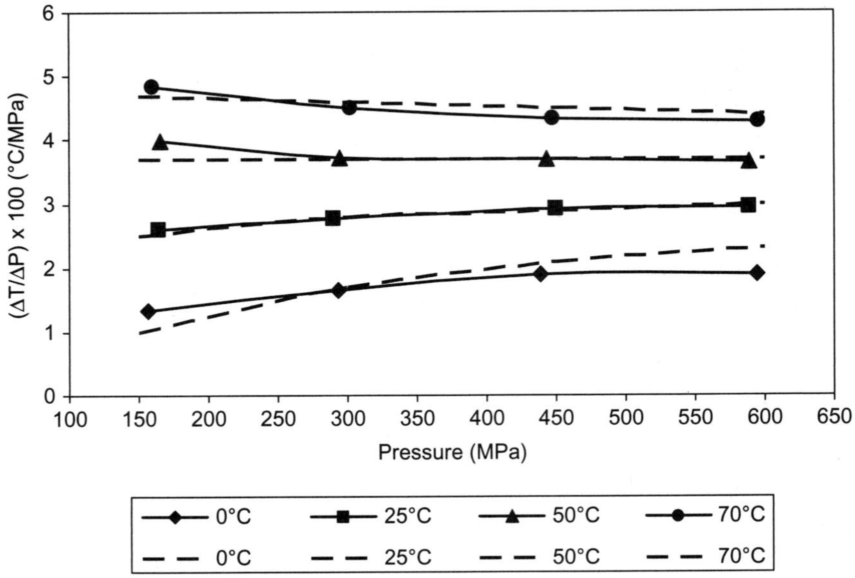

FIGURE 5.2

Heat of compression values of water (experimental: dashed lines) *vs* predicted (continuous lines estimated using NIST/ASME software) at different initial temperatures. *(Adapted with permission from Ramaswamy* et al., *2005.)*

These results are in transient temperature exchange between the food, packaging material, pressure-transmitting fluid, and the walls of the pressure chamber. Since the food material and pressure-transmitting fluid have higher heat of compression values than the pressure vessel, food material located near the vessel wall cools down and does not reach the same temperature as that at its center (de Heij *et al.*, 2003; Ting *et al.*, 2002). This non-uniformity of temperature distribution within the pressure vessel might influence the product microbial and enzyme inactivation. Further, for experiments involving elevated temperatures, if the pressure vessel is not properly insulated, the vessel tends to lose heat to the surroundings. Hence, it is necessary to control the vessel temperature at or near the desired final product temperature under process pressure.

Compression of a material decreases the average intermolecular distance between adjacent molecules. Compressibility, thus, is an intrinsic property of a material. A liquid contains molecules that occupy space in excess of that needed for closed packing. It is this free volume that is decreased during compression. Due to the decrease in compressible volume with increased pressure, the compressibility of a liquid significantly decreases at elevated pressures. Material compressibility can be estimated at constant temperature (isothermal compressibility) or constant entropy (adiabatic compressibility) (Rasanayagam *et al.*, 2003).

Heat is generated within the material due to the work of compression against intermolecular forces. Since food is a complex mixture of water, proteins, carbohydrates, and fats, each having different compressibility and specific-heat capacity, their heat of compression values are different. Since the water molecules are small and polar and more closely packed than the fat molecules by virtue of hydrogen bonds, they exhibit less compressibility values. Conversely, fats and oils are in general less closely packed than water. Hence, they exhibit compressibility values that are greater than that of water (Rasanayagam *et al.* 2003). Organic compounds such as propylene glycol and ethanol contain hydroxyl functional groups and can form hydrogen bonds between molecules. This might influence their thermal behavior under pressure.

5.2.5 pH Shift Under Pressure

High pressure is believed to cause ionic dissociation of water molecules with a corresponding decrease in pH (Cheftel, 1995; Hoover *et al.*, 1989). Studies conducted by Heremans (1995) indicate a lowering of pH in apple juice by 0.2 units per 100 MPa increase in pressure. pH plays an important role in phenomena such as gelation, enzyme activity, protein denaturation, and microbial inactivation kinetics and most microorganisms show increased susceptibility and inability to recover from sub-lethal injuries (Hoover *et al.* 1989). At the time of writing, reliable instruments for measurement of pH under pressure are not readily available. More research is needed to characterize transient pH shift under combined pressure–heat treatment as a function of various food compositions.

5.3 MICROBIOLOGICAL SAFETY

Fluid foods such as juices and beverages, with their high acidity and high a_w, can be pasteurized at ambient or chilled temperatures using pressures in the range of 400 to 600 MPa and process times generally under 10 min (Cheftel, 1995; Farkas and Hoover, 2000). High pressure pasteurization treatments inactivate pathogenic and spoilage bacteria, yeasts, molds and viruses. However, the treatment has limited efficacy against spores and enzymes. The extent of bacterial inactivation also depends on the type of microorganism, food composition, pH, and water activity. Cells in the exponential phase of growth have been found to be less resistant than those in the stationary phase. Gram-positive organisms are more resistant than gram negatives. Significant variations in pressure resistances can be seen among strains (Cheftel, 1995; Smelt, 1998). Water activity has a major influence on the rate of microbial inactivation.

Pressure treatment alone has little effect on the destruction of spores. It has been found in several studies that combined pressure–temperature treatment has a synergistic effect on spore inactivation (Cheftel, 1995; Hoover *et al.*, 1989; Rovere *et al.*, 1998, Ananta *et al.*, 2001). At ambient temperatures, bacterial spores can survive pressures above 1000 MPa (Farkas and Hoover, 2000). Elevated pressures (500–900 MPa; 72,500–130,500 psi) can be combined with several minutes of heat exposure (90–120°C) to sterilize low-acid foods (Ahn *et al.*, 2007; de Heij *et al.*, 2003; Matser *et al.*, 2004; Rajan *et al.*, 2006). During typical pressure-assisted thermal processing (PATP) (also referred to as "pressure-assisted thermal sterilization" or "PATS"), the food is subjected to a combination of elevated pressures and moderate heat for short duration. One of the unique advantages of PATP is its ability to provide a rapid and uniform increase in the temperature of treated food samples. Uniform compression heating and expansion cooling on decompression help to reduce the severity of thermal effects encountered with conventional processing techniques.

5.4 IMPACT OF COMBINED PRESSURE–HEAT TREATMENTS ON QUALITY ATTRIBUTES OF FOODS

Rastogi *et al.* (2007) reviewed combined pressure–heat effects on food enzymes. Pressure treatment has variable effects on enzyme inactivation. The effect of high pressure on certain enzymes may be reversible in the range 100–400 MPa. This may be attributed to conformational changes and subunit dissociation and association processes. Inactivation of selected enzymes in model buffers with treatment pressures ranging from 0.1 to 900 MPa ranked in order of ease of inactivation from easiest to the most difficult as follows: lipoxygenase, lactoperoxidase, pectinesterase, lipase, phosphatase, catalase, polyphenol oxidase, and peroxidase. Enzyme activity can be influenced by pressure-induced decompartmentalization. Pressure-induced damage of membranes facilitates enzyme substrate contact. The resulting reaction can either be accelerated or retarded by pressure. Combining pressure with a moderate temperature increase resulted in an increased enzyme inactivation rate. Thermal blanching prior to high-pressure processing can also be used as a means of controlling enzyme activity of high-pressure-processed foods. Combined pressure–heat effects on selected food enzymes are presented in Table 5.4.

High pressures at ambient or chilled temperatures have been employed for processing a number of liquid and semi-solid foods such as fruit juices, purees, smoothies, jellies, guacamole, etc. Several studies evaluated the beneficial effects of pressure treatment over conventional treatment on preserving quality attributes of foods (Table 5.5).

Small molecules such as vitamins, and flavor compounds remain unaffected by high pressure, while the structure of the large molecules such as proteins, enzymes, polysaccharides, and nucleic acid may be altered. Further, high pressure affects only non-covalent bonds (hydrogen, ionic, and hydrophobic bonds) and reduces the rate of browning reaction (Maillard reaction) (Rastogi *et al.*, 2007). High-pressure processing at low to moderate temperatures has very little effect on

Table 5.4 Combined Pressure–Temperature Effects on Selected Food Enzymes

Product	Temperature (°C)	Pressure (MPa)	Holding Time (min)	Results	References
PME, PG, PPO, POD, LOX in model systems, crude extracts and real foods	20–26	0.1–500	Variable	Most of the enzyme activity is retained except for LOX which can be completely inactivated using high pressure–low temperature combinations	Buggenhout *et al.*, 2006
Tomato PG and PME (enzyme extracts)	25	300–500 850	15 15	PG inactivated completely, PME still active. Only 50% PME inactivated	Rodrigo *et al.*, 2006
PPO in Victoria grape must	55–70 20–70	0.1 100–800	Variable	Synergistic effect of pressure and temperature on the inactivation of enzyme except in the high temperature–low pressure region where an antagonistic effect was observed	Rapeanu *et al.*, 2005
Purified tomato PG (pH 4.4)	25–65	100–500	Variable	High pressure shifts the optimum temperature of tomato PG activity to lower values as compared to studies at atmospheric pressure. Inhibition of PG activity increases with increasing pressure and the effect is higher at higher temperatures	Verlent *et al.*, 2004a
Purified tomato PME (pH 4.4 and 8.0)	20–65	0.1–600	Variable	Elevated pressures resulted in higher PME activities. At all pressures the PME activity was higher at pH 8.0 than at pH 4.4. Chemical de-esterification of pectin increases under pressure at pH 8.0 whereas no chemical de-esterification occurs at pH 4.4	Verlent *et al.*, 2004b

(*Continued*)

Table 5.4 (Continued)

Product	Temperature (°C)	Pressure (MPa)	Holding Time (min)	Results	References
PG and PME in tomato puree	20	300–700	2	70% inactivation of PG, 5- to 6-fold increase in PME activity >99% inactivation of PG, >95% inactivation of PME	Krebbers *et al.*, 2003
	90	700	0.5		
Carrot PME (carrot pieces, carrot juice, purified form)	10	800	36	90% reduction in PME activity. Purified PME more pressure- and temperature-stable at pH 6.0 than at 4.5 or 5.5. PME in carrot pieces more pressure- and temperature-stable than in purified form or in carrot juice	Balogh *et al.*, 2004
Lactoperoxidase in bovine milk and acid whey	69–73 20–65	0.1 700	Variable	First order inactivation kinetics at ambient pressure. Pressure and temperature exert antagonistic effects and pressure protects the enzyme at higher temperatures. Enzyme was more stable in milk than in whey	Ludikhuyze *et al.*, 2001
Orange PME (Orange juice)pH 3.2–3.7	20–25	100–400	0–720	Instantaneous pressure inactivation observed. Inactivation during holding phase followed first-order kinetics and increased with decreasing pH, increasing pressure, pressure hold time and total soluble solids	Basak and Ramaswamy, 1996
Lychee (PPO and POD)	20–60	200–600	10 or 20	PPO and POD activities increase at lower pressures (200–400 MPa). Rapid inactivation at 600 MPa (>50% POD; >90% PPO). Sugar syrup offered baroprotective effect	Phunchaisri and Apichartsrangkoon, 2005

Tomato puree (PPO, POD and PME) Hurdle technology	25	50–400	15	Hurdle technology employing citric acid (0–2%) and NaCl (0–0.8%) in combination with HPP gave optimal sensory and microbiological characteristics and lead to increased inactivation of PPO, POD and PME with increase in pressure and additive conc	Plaza *et al.*, 2003
Milk (Plasmin)	25–65	300–800	–	Synergistic effect of temperature and pressure in the inactivation of plasmin was observed in the range of 300–600 MPa at pH 6.7. However, at pressures higher than 600 MPa, antagonistic effect of pressure with temperature was observed	Borda *et al.*, 2004
Orange PE	45–85	0.1	Various preset time intervals	Influence of enzyme concentration, pH, Ca^{2+} concentration, and sucrose on the inactivation kinetics was studied. Ca^{2+} showed synergism with temperature treatments in inactivating the enzyme whereas decrease in Ca^{2+} increased enzyme inactivation during pressure treatments. Addition of acid showed enhanced inactivation of the enzyme under both temperature and pressure–heat treatments. The presence of sucrose increased temperature- and pressure-stability of orange PE	Van den Broeck *et al.*, 1999
	18–30	400 to 900			

LOX, lipoxygenase; POD, peroxidase; PPO, polyphenyloxidase; PE, pectinesterase.

Table 5.5 Combined Pressure–Temperature Effects on Sensory and Physico-Chemical Attributes of Selected Foods

Product	Temperature (°C)	Pressure (MPa)	Holding Time (min)	Results	References
Lycopene in tomato juice from two different cultivars	45.7 ± 0.56	700	10	All-*trans* lycopene was fairly stable to processing and *cis* isomer content of control (hot break juice) and processed juice did not differ significantly. During storage at 4, 25 and 37°C for 52 weeks, lycopene degradation varied as a function of cultivar type, processing method, storage temperature, and time	Gupta *et al.*, 2010
	101 ± 0.87	600	10		
	100	0.1	35		
Lycopene in hexane and Tomato puree	20 ± 1	100–600	12	Increase in pressure resulted in an increase in the 13-*cis* isomer of lycopene at the expense of trans form. Stability of lycopene in hexane was maximum for pressure treatments at 400 MPa and higher pressures resulted in significant losses. In tomato puree, maximum stability was observed at treatment pressure of 500 MPa followed by refrigerated storage at 4 ± 1°C	Qui *et al.*, 2006
White cabbage (Soluble and insoluble fiber)	20, 50, 80°C	400 and 500	10	Soluble fiber (SDF) decreases after high-pressure treatment and individual variations in pressure–temperature conditions might be required to achieve this effect in different cultivars. Effect of temperature on SDF and Insoluble Dietary Fibre (IDF) was reported to be lower at 500 MPa than at 0.1 or 400 MPa for one of the cultivars (Predikant)	Wennberg and Nyman, 2004

Guava Juice	25	600	15	Very similar flavor profile as compared to fresh juice. Heat processing at 95°C for 5 min results in significant decrease in flavor components. Residual enzyme activity causes degradation of volatile components during storage. However, storage at 4°C for 30 days after HPP results in similar flavor profile as that of fresh juice	Chin Yen *et al.*, 1999
Orange juice,		700	1	Cloud stability and shelf stability for 90 days under refrigerated conditions	Goodner *et al.*, 1999
	35–60	500–800	5	Higher retention of ascorbic acid and thus higher antioxidant potential as compared to juice processed in conventional way using heat	Fernandez *et al.*, 2001; Polydera *et al.*, 2003, 2004, 2005; Bull *et al.*, 2004
Lemon juice,		400–600	10–15	Very little effect on the physicochemical properties and flavor of these products without significantly affecting their ascorbic acid contents	Donsi *et al.*, 1998
Apple juice,					Novotna *et al.*, 1999
Guava puree	25				Gow and Hsin, 1996, 1998
Milk		400	3	Milk retains most of the vitamin B1 and B6 as compared to the original amount present in raw milk	Sierra *et al.*, 2000
Milk		200–600		Decrease in the rennet coagulation time as compared with untreated, unheated, and pasteurized milk	Needs *et al.*, 2000; Kolakowski *et al.*, 2000; Zobrist *et al.*, 2005; Huppertz *et al.*, 2005
				Decrease in the rennet coagulation time with increase in pressure. Increased gel strength of curd after HPP. Increase in the cheese yield by about 15% and decrease in the protein content of whey by 30%	

(*Continued*)

Table 5.5 (Continued)

Product	Temperature (°C)	Pressure (MPa)	Holding Time (min)	Results	References
				Increase in the curd firming rate of milk by HPP	
				Some effects could be due to the association of whey proteins with casein micelles under pressure	
				High-pressure treatments of milk up to 600 MPa results in denser network in rennet-catalyzed formation of gels than gels produced using non pressure-treated milk. They show higher storage modulus and casein micelles are found to be completely disrupted at pressures over 400 MPa. Whey proteins denature under high pressure and the overall rapid coagulation of pressure treated micelles reduced due to the interference of β-lactoglobulin (denatured whey protein). Syneresis from pressure-treated gels is lower at pressures below 400 MPa	
Whey		600		Denaturation of whey proteins leading to the decrease in amount of β- lactoglobulin in whey	Pandey and Ramaswamy, 1998; Brooker *et al.*, 1998
BSA and β-lactoglobulin	37	300–800	20	High-pressure treatment results in loss of sulphydryl groups in BSA and β- lactoglobulin and their 1:1 mixtures with dextran sulfate (42% loss in β- lactoglobulin and 55% in BSA). Protein aggregation partly assisted by	Galazka *et al.*, 1997

				the formation of disulfide bridges and changes in hydrophobicity are observed. Surface hydrophobicity increases by 40% for β- lactoglobulin and decreases by 41% for BSA. Dextran sulfate exhibits either a protective effect on β- lactoglobulin and prevents it from unfolding or enables it to regain some secondary structure	
Cheese from goats' milk	20	500	15	Higher concentration of organic acids than that prepared using pasteurized milk (72°C, 15 s). Factors affecting difference in organic acid profile could be due to differences in Salt/Moisture ratio, structural effects of high pressure and pasteurization methods and change in the microbial populations such as non-starter lactic acid bacteria, *Leuconostoc*, *Propionibacterium*, etc., responsible for the production of organic acids in cheese	Buffa *et al.*, 2003; Buffa *et al.*, 2004

BSA, bovine serum albumin.

the color pigments present in fruits and vegetables (Oey *et al.*, 2008). These include chlorophylls, anthocyanins, carotenoids, etc. However, prolonged storage after processing might result in undesirable chemical changes associated with partially active factors such as enzymes and microorganisms (Oey *et al.*, 2008). Texture modification during high-pressure processing mainly occurs due to cellular disruption and changes in cell wall polymers due to enzymatic and nonenzymatic reactions (Sila *et al.*, 2008). Changes in the hardness, softness and rheological characteristics of the food are dependent on the type of food product and the influence of pressure–temperature on its chemical composition.

Pressure–heat treatments have been reported to improve the retention of lycopene and rheology of tomato puree (Krebbers *et al.*, 2003). High-pressure sterilization of tomato puree at 700 MPa and 121°C showed color improvement and better retention of lycopene compared to 40% loss after thermal sterilization at 118°C for 20 min. The viscosity of high-pressure-sterilized tomato puree was found to be lower as compared to the thermally sterilized sample. High-pressure sterilization also resulted in 99% inactivation of polygalacturonase (PG) and pectinmethylesterase (PME) (Krebbers *et al.*, 2003). Butz *et al.* (2002) studied the effects of combined pressure and temperature on broccoli, carrots, tomatoes, and tomato pulp and found very little effect of high pressure on chlorophyll a and b in broccoli, and carotenoids and lycopene in tomatoes. Antioxidant capacity of the water-soluble fraction of tomatoes and carrots and antimutagenicity of high-pressure-treated tomato pulp was almost the same as that of the control. High-pressure-treated tomato pulp showed higher water-binding ability and increased glucose retardation index. Extractability of carotenoids from coarse carrot homogenate was found to decrease. Combined pressure–heat treatments between 500–700 MPa and 90–115°C were shown to affect the chemical pectin conversions (de Roeck *et al.*, 2009). Apple pectin solutions at pH 6.5 were subjected to increasing temperature treatments at atmospheric pressure (0.1 MPa) and combined pressure–temperature treatments. Whereas, a temperature-only effect resulted in higher rate constants for demethoxylation of pectin than beta elimination, an increase in temperature showed a higher rate of increase of beta elimination than demethoxylation. However, under combined pressure–heat conditions, demethoxylation was found to be stimulated and beta elimination was retarded.

5.5 COMBINED PRESSURE–HEAT TREATMENT EFFECTS ON NUTRIENT CONTENT OF FOODS

Consumer interest in health and wellness prompted the food industry to develop technologies for preserving functional food ingredients. The process conditions designed to inactivate bacteria may adversely impact the functional foods. Thus it is important to investigate combined pressure–heat effects on various nutrients. Several studies have reported the positive effects of pressure treatment at temperatures ≤60°C on maintaining the fresh-like nutrient content of foods (Sánchez-Moreno *et al.*, 2009). Vitamin C in different fruit and vegetable products is not significantly affected by high-pressure treatments. Similarly, thiamin, piridoxine, and other similar B vitamin compounds were not significantly affected by high-pressure treatments. However, when combined with heat, loss of vitamin C was reported. Temperature (>60°C) seems to be the cause of vitamin C degradation. Van den Broeck *et al.* (1998) studied the degradation kinetics of Vitamin C in an isobaric–isothermal treatment of orange and tomato juice at 850 MPa, 65–80°C up to 400 min holding time. Significant degradation of Vitamin C was observed and the degradation followed first-order kinetics. However, at 850 MPa and 50°C, no degradation of Vitamin C was observed during 60 min holding time. Pressure treatment in general was also reported to increase the antioxidant capacity of fruit and vegetables. Treatment efficacy varies with food composition and processing parameters (pressure, time and temperature). Also, the type of food matrix and duration of storage after pressure treatments influences the fate of Vitamin C (Sánchez-Moreno *et al.*, 2009).

While heat, light, and oxygen can decrease lycopene during food processing, high-pressure processing shows capability of preserving tomato lycopene content. A number of studies have reported increased extractability of lycopene from tomato products exposed to specific pressure–temperature treatments (Gupta *et al.*, 2010; Krebbers *et al.*, 2003; Sanchez-Moreno *et al.*, 2004; Sanchez-Moreno *et al.*, 2006). Patras *et al.*, (2009) reported that high-pressure treatment (400–600 MPa/15 min/20°C) better preserved the antioxidant activity (ARP), ascorbic acid, and carotenoids in tomato and carrot puree than in untreated or thermally processed (70°C/2 min) samples. Plaza *et al.* (2006) reported that pressure-treated (150–350 MPa, 60°C, 15 min) gazpacho (vegetable) soup stored under refrigerated conditions better preserved the carotenoid content than thermal treatment. Polydera *et al.* (2004) reported that high-pressure treatment resulted in higher overall antioxidant activity of orange juice during storage at 0–15°C compared to conventional pasteurization. In another study, the effects of pressure treatment of 350 MPa on orange juice carotenoids, β-carotene, ά-carotene, zeaxanthin, lutein, and β-cryptoxanthin, associated with pro-vitamin A and radical-scavenging capacity values, resulted in significant increases of 20–43% in the carotenoid content of fresh orange juice (de Ancos *et al.*, 2002). A study on pomegranate juice by Ferrari *et al.* (2010) showed that pressure treatment at room temperature improved its quality by increasing the intensity of red color and preserving the content of natural anthocyanins. In a study involving 5-methlytetrahydrofolic acid, Verlinde *et al.* (2009) reported that loss of water-soluble vitamin folate does occur during pressure–temperature treatments.

Most of the present studies are end-point in nature and more systematic kinetics studies are needed on degradation of nutrients in combined pressure–heat treatment. In addition, it is unclear if the comparisons presented in these studies are truly legitimate in terms of comparing equivalent thermal processes (Sastry and Balasubramaniam, 2009). More studies are needed to conduct a truly effective comparison between thermal and non-thermal processes. Also, further research is needed to accurately and reproducibly derive the correlation between safety of food processed using high-pressure treatments and corresponding changes in nutrients associated with such processing. Finally, changes to animal and plant tissue and nutrient bioavailability that occur in pressure processing need to be investigated (Balasubramaniam and Farkas, 2008).

5.6 REGULATORY ASPECTS

In the United States, the Food Drug and Cosmetic Act (FD&C Act) requires all foods to be processed, packaged, and held under sanitary conditions. FDA uses FD&C act to promulgate specific regulations. Like thermally pasteurized products, high-pressure pasteurized products are required to be processed under GMP conditions and relevant commodity-specific regulations (e.g., juice HACCP, Pasteurized Milk Ordinance (PMO), Sea Food HACCP, etc.) should be followed. It is important that processors conduct appropriate microbiological validation tests using relevant target pathogenic and surrogate organisms. The potential for temperature abuse during refrigerated storage and distribution has to be carefully evaluated and minimized. Processors must also work with equipment vendors to ensure that any part of the equipment, which may have incidental contact with the food, is made from approved materials.

5.7 CONCLUSIONS

Non-thermal processing methods such as high-pressure processing offer food processors exciting opportunities to develop a new generation of value-added foods that can preserve functional food ingredients. High-pressure treatments destroy harmful microorganisms at low or moderate temperatures without significantly changing the organoleptic and nutritional properties of food materials. The process has been successfully used in the commercial preservation of foods for more than 15 years and the number of commercial applications worldwide is growing. However, high capital expenditure and limited throughput are some current limitations of this technology for fluid-food processing. With wider industrial adaptation and technology innovation, it is expected that the cost of equipment will come down.

References

Ahn, J., Balasubramaniam, V.M., Yousef, A.E., 2007. Inactivation kinetics of selected aerobic and anaerobic spores by pressure-assisted thermal processing. Int. J. Food Microbiol. 113(3), 321–329.

Ananta, E., Heinz, V., Schlüter, O., Knorr, D., 2001. Kinetic studies on high-pressure inactivation of *Bacillus stearothermophilus* spores suspended in food matrices. Innovative Food Sci. Emerg. Technol. 2 (4), 261–272.

Balasubramaniam, V.M., Farkas, D., 2008. High-pressure food processing. Food Sci. Technol. Int. 14 (5), 413–418.

Balasubramaniam, V.M., Ting, E.Y., Stewart, C.M., Robbins, J.A., 2004. Recommended laboratory practices for conducting high-pressure microbial inactivation experiments. Innovative Food Sci. Emerg. Technol. 5 (3), 299–306.

Balasubramanian, S., Balasubramaniam, V.M., 2003. Compression heating influence of pressure transmitting fluids on bacteria inactivation during high pressure processing. Food Res. Int. 36, 661.

Balogh, T., Smout, C., Binh, L.N., Van Loey, A.M., Hendrickx, M.E., 2004. Thermal and high-pressure inactivation kinetics of carrot pectin–methylesterase: from model system to real foods. Innovative Food Sci. Emerg. Technol. 5 (4), 429–436.

Basak, S., Ramaswamy, H.S., 1996. Ultra high-pressure treatment of orange juice: a kinetic study on inactivation of pectin methylesterase. Food Res. Int. 29 (7), 601–607.

Borda, D., Indrawati, S.C., Van Loey, A.M., Hendrickx, M., 2004. High-pressure thermal inactivation kinetics of plasmin system. J. Dairy Sci. 87 (8), 2351–2358.

Bridgman, P.W., 1964. Collected Experimental Papers (Vol. I– VII). Harvard University Press, Cambridge, Massachusetts.

Bridgman, P.W., 1912. Water under pressure. Proc. Am. Acad. Arts Sci. 48, 441–558.

Brooker, B., Ferragut, V., Gill, A., and Needs, E., 1998. Properties of rennet gel formed from high-pressure treated milk. In: Autio, K. (Ed.), VTT symposium, Fresh Novel Foods by High Pressure. Helsinki, Finland, pp. 55–61.

Buffa, M., Guamis, B., Saldo, J., 2003. Changes in water binding during ripening of cheeses made from raw, pasteurized or high-pressure-treated goat milk. LAIT 83 (1), 89–96.

Buffa, M., Guamis, B., Saldo, J., Trujillo, A.J., 2004. Changes in organic acids during ripening of cheeses made from raw, pasteurized or high-pressure treated goat's milk. Lebensmittel-Wissenschaft-Und-Technologie 37 (2), 247–253.

Buggenhout, S., Messagie, I., der Plancken, I., Hendrickx, M., 2006. Influence of high-pressure low-temperature treatments on fruit and vegetable quality related enzymes. Eur. Food Res. Technol. 223 (4), 475–485.

Bull, M.K., Zerdin, K., Howe, E., Goicoechea, D., Paramanandhan, P., Stockman, R., et al., 2004. The effect of high pressure processing on the microbial, physical and chemical properties of Valencia and Navel orange juice. Innovative Food Sci. Emerg. Technol. 5 (2), 135–149.

Butz, P., Edenharder, R., Garcia, A.F., Fister, H., Merkel, C., Tauscher, B., 2002. Changes in functional properties of vegetables induced by high pressure treatment. Food Res. Int. 35, 295–300.

Cheftel, J.C., 1995. Review: high pressure, microbial inactivation and food preservation. Food Sci. Technol. Int. 1 (2-3), 75–90.

Chin Yen, G., Tang Lin, H., 1999. Changes in volatile flavor components of guava juice with high-pressure treatment and heat processing during storage. J. Agric. Food. Chem. 47, 2082–2087.

de Ancos, B., Sgroppo, S., Plaza, L., Cano, M.P., 2002. Possible nutritional and health-related value promotion in orange juice preserved by high-pressure treatment. J. Sci. Food Agric. 82 (8), 790–796.

de Heij, W.B.C., van Schepdael, L.J.M.M., Moezelaar, R., Hoogland, H., Master, A.M., van den Berg, R.W., 2003. High -pressure sterilization: maximizing the benefits of adiabatic heating. Food Technol. 57 (3), 37–41.

de Roeck, A., Duvetter, T., Fraeye, I., Van der Plancken, I., Sila, D.N., Van Loey, A., et al., 2009. Effect of high-pressure/high-temperature processing on chemical pectin conversions in relation to fruit and vegetable texture. Food Chem. 115 (1), 207–213.

Donsi, G., Ferrari, G., Matteo, M.d., Bruno, M.C., 1998. High-pressure stabilization of lemon juice. Ital. Food Beverage Technol. 14, 14–16.

Farkas, D.F., Hoover, D.G., 2000. High pressure processing. J. Food Sci. 65 (8 spec. suppl.), 47–64.

Fernandez, G.A., Butz, P., Bognar, A., Tauscher, B., 2001. Antioxidative capacity, nutrient content and sensory quality of orange juice and an orange lemon juice product after high-pressure treatment and storage in different packaging. Eur. Food Res. Technol. 213 (4/5), 290–296.

Ferrari, G., Maresca, P, Ciccarone, R., 2010. The application of high hydrostatic pressure for the stabilization of functional foods: pomegranate juice. J. Food Eng. 100 (2), 245–253.

Galazka, V.B., Ledward, D.A., Sumner, I.G., Dickinson, E., 1997. Influence of high pressure on bovine serum albumin and its complex with dextran sulfate. J. Agric. Food. Chem. 45 (9), 3465–3471.

Goodner, J.K., Braddock, R.J., Parish, M.F., Sims, C.A., 1999. Cloud stabilization of orange juice by high pressure processing. J. Food Sci. 64 (4), 699–700.

Gow, C.Y., Hsin, T.S., 1996. Comparison of high-pressure treatment and thermal pasteurization effects on the quality and shelf life of guava puree. Int. J. Food Sci. Technol. 31 (2), 205–213.

Gow, C.Y., Hsin, T.L., 1998. Effects of high pressure and heat treatment on pectic and related characteristics in guava juice. J. Food Sci. 63 (4), 684–687.

Gupta, Rockendra, Balasubramaniam, V.M., Schwartz, S.J., Francis, D., 2010. Storage stability of lycopene in tomato juice subjected to combined pressure-heat treatments. J. Agric. Food. Chem. 58 (14), 8305–8313.

Harvey, A.H., Peskin, A.P., and Klein, S.A., 1996. NIST Standard Reference Steam Database 10; U.S. Department of Commerce, National Institute of Standards and Technology, Boulder, CO. *version 2.2*.

Heremans, K., 1995. High pressure effects on biomolecules. In: Ledward, D.A., Johnston, D.E., Earnshaw, R.G., Hasting, A.P.M. (Eds.), High Pressure Processing of Foods. Nottingham University Press, Reading, UK, pp. 81–98.

Hite, B.H., 1899. The effect of pressure in preservation of milk. Morgantown Bull. W.V. Univ. Agric. Exp. Sta. 15–35.

Hogan, E., Kelly, A.L., Sun, D.W., 2005. High pressure processing of foods: an overview. In: Sun Da Wen (Ed.), Emerging Technologies for Food Processing. Elsevier Academic Press, CA, USA, pp. 3–31.

Hoover, D., Metrick, C., Papineau, A.M., Farkas, D.F., Knorr, D., 1989. Biological effects of high hydrostatic pressure on food microorganisms. Food Technol. 43 (3), 99–107.

Huppertz, T., Hinz, K., Zobrist, M.R., Uniacke, T., Kelly, A.L., Fox, P.F., 2005. Effects of high-pressure treatment on the rennet coagulation and cheese making properteis of heated milk. Innovative Food Sci. Emerg. Technol. 6 (3), 279–285.

Kesavan, V., Balasubramaniam, V.M., Adhikari, C., Narayanaswamy, G., and Parulekar, S.J., 2002. Compression heating characteristics of proteins during high pressure processing. In: IFT Annual Meeting (pp. Abstract No. 91E-6) Anaheim, CA.

Kolakowski, P., Reps, A., Fetlinski, A., 2000. Microbial qulaity and some physico chemical properties of high pressure processed cow milk. J. Food Nutr. Sci. 9/50 (4), 19–26.

Krebbers, B., Matser, A.M., Hoogerwerf, S.W., Moezelaar, R., Tomassen, M.M.M., van den Berg, R.W., 2003. Combined high-pressure and thermal treatments for processing of tomato puree: evaluation of microbial inactivation and quality parameters. Innovative Food Science & Emerging Technologies 4, 377–385.

Ludikhuyze, L.R., Claeys, W.L., Hendrickx, M.E., 2001. Effect of temperature and/or pressure on lactoperoxidase activity in bovine milk and acid whey. Journal of Dairy Research 68 (4), 625–637.

Matser, A.M., Krebbers, B., Van Den Berg, R.W., Bartels, P.V., 2004. Advantages of high pressure sterilisation on quality of food products. Trends in Food Science and Technology 15 (2), 79–85.

NC Hyperbaric, 2010. HPP juices and beverages in market. <http://www.nchyperbaric.com/descargas/Flyer_juices&beverages_2006.pdf/> (accessed 28.08.10).

Needs, E.C., Stenning, R.A., Gill, A.L., Ferragut, V., Rich, G.T., 2000. High-pressure treatment of milk: effects on caesin micelle structure and on enzyme coagulation. Journal of Dairy Research 67, 31–42.

Novotna, P., Valentova, H., Strohalm, J., Kyhos, K., Landfeld, A., Houska, M., 1999. Sensory evaluation of high pressure treated apple juice during its storage. Czech J. Food Sci. 17 (5), 196–198.

Oey, I., Lille, M., Van Loey, A., Hendrickx, M., 2008. Effect of high-pressure processing on colour, texture and flavour of fruit- and vegetable-based food products: a review. Trends in Food Science and Technology 19 (6), 320–328.

Otero, L., Sanz, P.D., 2003. High Pressure-Assisted and High Pressure-Induced Thawing: Two Different Processes. Jounal of Food Science 68 (8), 2523–2528.

Otero, L., Molina-García, A.D., Sanz, P.D., 2000. Thermal effect in foods during quasi-adiabatic pressure treatments. Innovative Food Sci. Emerg. Technol. 1 (2), 119–126.

Pandey, P. K., Ramaswamy, H. S., 1998. Effect of high pressure of milk on textural properties, moisture content and yield of Cheddar cheese. Anonymous Book of Abstracts of IFT Annual Meeting. Institute of Food Technologists, Atlanta, GA, pp. 173–174.

Patazca, E., Koutchma, T., Balasubramaniam, V.M., 2007. Quasi-adiabatic temperature increase during high pressure processing of selected foods. J. Food Eng. 80 (1), 199–205.

Patras, Ankit, Brunton, Nigel, Da Pieve, Sara, Butler, Francis, Downey, Gerard, 2009. Effect of thermal and high pressure processing on antioxidant activity and instrumental color of tomato and carrot purées. Innovative Food Sci. Emerg. Technol. 10 (1), 16–22.

Phunchaisri, C., Apichartsrangkoon, A., 2005. Effects of ultra-high pressure on biochemical and physical modification of lychee (Litchi chinensis Sonn.). Food Chem. 93 (1), 57–64.

Plaza, L., Munoz, M., Ancos, B.d., Cano, M.P., 2003. Effect of combined treatments of high pressure, citric acid and sodium chloride on quality parameters of tomato puree. Eur. Food Res. Technol. 216 (6), 514–519.

Plaza, L., Sanchez-Moreno, C., de Ancos, B., 2006. Carotenoid content and antioxidant capacity of Mediterranean vegetable soup (gazpacho) treated by high-pressure/temperature during refrigerated storage. Eur. Food Res. Technol. 223 (2), 210–215.

Polydera, A.C., Stoforos, N.G., Taoukis, P.S., 2003. Comparative shelf life study and vitamin C loss kinetics in pasteurised and high pressure processed reconstituted orange juice. J. Food Eng. 60, 21–29.

Polydera, A.C., Stoforos, N.G., Taoukis, P.S., 2004. The effect of storage on the antioxidant activity of reconstituted orange juice which had been pasteurized by high pressure or heat. International Journal of Food Science & Technology 39, 783–791.

Polydera, A.C., Stoforos, N.G., Taoukis, P.S., 2005. Effect of high hydrostatic pressure treatment on post processing antioxidant activity of fresh Navel orange juice. Food Chem. 91, 495–503.

Qui, W., Jiang, H., Wang, H., Gao, Y., 2006. Effect of high hydrostatic pressure on lycopene stability. Food Chem. 97 (3), 516–523.

Rajan, S., Ahn, J., Balasubramaniam, V.M., Yousef, A.E., 2006. Combined pressure–thermal inactivation kinetics of Bacillus amyloliquefaciens spores in egg patty mince. Journal of Food Protection 69 (4), 853–860.

Ramaswamy, R., Balasubramaniam, V. M., Sastry, S. K., 2005. Properties of food materials during high pressure processing. In: Encyclopedia of Agriculture, Food and Biological Engineering. Taylor and Francis, UK, pp. 1–6.

Rapeanu, G., Van Loey, A., Smout, C., Hendrickx, M., 2005. Thermal and high-pressure inactivation kinetics of polyphenoloxidase in Victoria grape must. J. Agric. Food. Chem. 53 (8), 2994.

Rasanayagam, V., Balasubramaniam, V.M., Ting, E., Sizer, C.E., Bush, C., Anderson, C., 2003. Compression heating of selected fatty food materials during high-pressure processing. J. Food Sci. 68 (1), 254–259.

Rastogi, N.K., Raghavarao, K.S.M.S., Balasubramaniam, V.M., Niranjan, K., Knorr., D., 2007. Opportunities and challenges in high pressure processing of foods. Critical Reviews in Food Science and Nutrition 47, 69–112.

Robertson, R.E., Carroll, T., Pearce, L.E., 2008. Bacillus spore inactivation differences after combined mild temperature and high pressure processing using two pressurizing fluids. Journal of Food Protection 71 (6), 1186–1192.

Rodrigo, D., Cortes, C., Clynen, E., Schoofs, L., Van Loey, A., Hendrickx, M., 2006. Thermal and high-pressure stability of purified polygalacturonase and pectinmethylesterase from four different processing varieties. Food Res. Int. 39, 440–448.

Rovere, P., Gola, S., Maggi, A., Scaramuzza, N., Miglioli, L., 1998. Studies on bacterial spores by combined high pressure–heat treatment: Possibility to sterilize low acid foods. In: Issacs, N.S. (Ed.), High pressure food science, bioscience and chemistry. Royal Society of Chemistry, Cambridge, England, pp. 354–363.

Sanchez-Moreno, C., Plaza, L., de Ancos, B., Cano, M.P., 2004. Effect of combined treatments of high-pressure and natural additives on carotenoid extractability and antioxidant activity of tomato puree (Lycopersicum esculentum Mill.). Eur. Food Res. Technol. 219 (2), 151.

Sanchez-Moreno, C., Plaza, L., de Ancos, B., Cano, M.P., 2006. Impact of high-pressure and traditional thermal processing of tomato puree on carotenoids, vitamin C and antioxidant activity. J. Sci. Food Agric. 86 (2), 171–179.

Sánchez-Moreno, C., de Ancos, B., Plaza, L., Elez-Martinez, P., Cano, M.P., 2009. Nutritional approaches and health-related properties of plant foods processed by high pressure and pulsed electric fields. Critical Reviews in Food Science and Nutrition 49 (6), 552.

Sastry, S.K., Balasubramaniam, V.M. (2009). Technology for delivery of heat-sensitive nutrients. In: Proceedings of the XX Convention of Food Scientists and Technologists, December 21–23, 2009, Bangalore, India.

Sierra, I., Vidal, V.C., Lopez, F.R., 2000. Effect of high pressure on the vitamin B1 and B6 content of milk. Milchwissenschaft-Milk Science International 55 (7), 365–367.

Sila, D.N., Duvetter, T., De Roeck, A., Velent, I., Smout, C., Moates, G.K., Hills, B.P., Waldron, K.K., Hendrickx, M., Van Loey, A., 2008. Texture changes of processed fruits and vegetables: potential use of high-pressure processing. Trends in Food Science & Technology 19 (6), 309–319.

Smelt, J.P.P.M., 1998. Recent advances in the microbiology of high pressure processing. Trends in Food science and Technology 9, 152–158.

Ting, E., Balasubramaniam, V.M., Raghubeer, E., 2002. Determining thermal effects in high-pressure processing. Food Technol. 56 (2), 31–35.

Tornello, C., 2011. Case Studies on High Pressure Processing of Foods. In: Zhang, Howard Q., Barbosa-Cánovas, Gustavo V., Balasubramaniam, V.M., Dunne, C. Patrick, Farkas, Daniel F., Yuan, James T.C. (Eds.), Nonthermal Processing Technologies for Food. Wiley, Hoboken, New Jersey, (Chapter 4).

Van den Broeck, I., Ludikhuyze, L., Weemaes, C., Van Loey, A., Hendrickx, M., 1998. Kinetics for isobaric – isothermal degradation of l-ascorbic acid. J. Agric. Food. Chem. 46 (5), 2001–2006.

Van den Broeck, I., Ludikhuyze, L.R., Van Loey, A.M., Weemaes, C.A., Hendrickx, M.E., 1999. Thermal and combined pressure–temperature inactivation of orange pectinesterase: Influence of pH and additives. J. Agric. Food. Chem. 47 (7), 2950–2958.

Verlent, I., Van Loey, A.M., Smout, C., Duvetter, T., Hendrickx, M.E., 2004a. Purified tomato polygalacturonase activity during hthermal and high pressure treatment. Biotechnology and Bioengineering 86 (1), 63–71.

Verlent, I., Van Loey, A., Smout, C., Duvetter, T., Nguyen, B.L., Hendrickx, M.E., 2004b. Changes in purified tomato pectinmethylesterase activity during thermal and high pressure treatment. Journal of Science of Food and Agriculture 84, 1839–1847.

Verlinde, P.H.C.J., Oey, I., Deborggraeve, W.M., Hendrickx, M.E., Van Loey, A.M., 2009. Mechanism and related kinetics of 5-methyltetrahydrofolic acid degradation during combined high hydrostatic pressure–thermal treatments. J. Agric. Food. Chem. 57 (15), 6803–6814.

Wennberg, M., Nyman, M., 2004. On the possibility of using high-pressure treatment to modify physico-chemical properties of dietary fiber in white cabbage (*Brassica oleracea* var. capitata). Innovative Food Sci. Emerg. Technol. 5 (2), 171–177.

Zemansky, M.W., 1957. Applications of thermodynamics to pure substances. In: Heat and Thermodynamics. McGraw–Hill Book Company, New York, pp. 243–279.

Zobrist, M.R., Huppertz, T., Uniacke, T., Fox, P.F., Kelly, A.L., 2005. High-pressure-induced changes in the rennet coagulation properties of bovine milk. Int. Dairy J. 15 (6–9), 655–662.

CHAPTER

Ultrasound Processing of Fluid Foods

6

B.K. Tiwari[1] and T.J. Mason[2]

[1]Manchester Food Research Centre, Manchester Metropolitan University, Manchester, UK

[2]Sonochemistry Centre, School of Science and the Environment, Coventry University, Coventry, UK

6.1 INTRODUCTION

The use of ultrasound within the food industry has been a subject of research and development for many years that can be broadly divided into two major applications: food analysis and food processing (McClements, 1995; Mason, 1998).

1. Analytical results are achieved using low-intensity ultrasound (with power levels, typically less than 1 Wcm^{-2}) generally within the frequency range of 5–10 MHz. This type of ultrasound is similar to Non-destructive Testing (NDT) that is used for flaw detection in material science where pulse–echo techniques are used. However, in food analysis the measurement techniques generally involve time of flight and using sound absorption which provide information on texture, composition, viscosity, or concentration. It is important to note that analytical ultrasound causes no physical or chemical alterations in the properties of the material through which the wave passes.
2. Processing is achieved using lower frequencies of ultrasound (20–100 kHz) at much higher intensities (typically in the range of 10–1000 Wcm^{-2}). These are the conditions traditionally used for cleaning and cutting but when applied to liquid foods this type of ultrasound can bring about significant changes in the food material through the effects of acoustic cavitation. Among the most promising of these is in the non-thermal processing for the preservation of food.

Novel Thermal and Non-Thermal Technologies for Fluid Foods. DOI: 10.1016/B978-0-12-381470-8.00006-2

Power ultrasound is used in a range of physical/mechanical operations in food processing producing improvements in:

- *Cooking and frying:* improved heat transfer gives shorter cooking times and more even cooking.
- *Cutting:* can be achieved rapidly with reduced pressure using ultrasonic vibrations on the edge of a cutting tool. The method can be used for frozen or soft foods.
- *Dispersal, mixing, and homogenization:* turbulent mixing is a feature of cavitation at liquid/liquid interfaces and has been used for many years in the production of sauces and mayonnaise.
- *Brining, pickling, and marination:* involving the penetration of liquids into food material that can be applied to both vegetable and meat products.
- *Drying and filtration:* air drying benefits arise from improved heat transfer and ultrasound has applications in filtration as it improves water flow through the filter cake.
- *Tenderizing meat:* this is normally achieved by pounding and ultrasound provides a mechanical vibration that can achieve similar results with less drastic mechanical damage.
- *Extraction:* the breakdown of cell walls in plants together with improved solvent penetration results in better and sometimes more targeted extract yields of flavorings and nutriceuticals from plants, herbs, and seeds.
- *Freezing and crystallization:* can be improved through cavitation initiating crystallization without seeding while also providing efficient heat transfer.
- *Degassing and defoaming:* is an important process for beverages before they are canned or bottled.

It has also been used in the initiation of oxidation/reduction reactions, enzyme inactivation and pasteurization (O'Donnell *et al.*, 2010; Soria and Villamiel, 2010; Roberts, 1993; Thakur and Nelson, 1997; Villamiel and de Jong, 2000).

The term applied to the process by which power ultrasound is introduced into a liquid is termed sonication and this has been shown to be effective in the inactivation of microbes at ambient temperatures for water disinfection (Phull *et al.*, 1997). When applied to food materials, it affords the possibility of antimicrobial treatment of heat-sensitive materials. Most ultrasound applications are limited to liquid foods, mainly fruit juices, smoothies, and milk. This chapter outlines the effects of ultrasound on inactivation of both microorganisms and enzymes in liquid foods along with quality and nutritional changes which may arise.

6.2 GENERATION OF POWER ULTRASOUND

Power ultrasound employed in food processing uses the lower-frequency ranges of 20–100 kHz with a sound intensity of between 10 and 1,000 W/cm^2. The vibrational energy is provided by ultrasonic transducers that convert electrical energy to sound energy of which there are two types in common usage, piezoelectric and magnetostrictive, with piezoelectric being the most common (Mason, 1998). Magnetostrictive transducers are less efficient in terms of electric power consumption but have the advantage of being of a much more robust construction and, for this reason, they are being looked at again for heavy-duty applications.

The driving force for the processing effects of sonication is acoustic cavitation. Cavitation itself is generally regarded as a problem in that it causes erosion where liquids move rapidly across metal surfaces, e.g., in pipes and impellors; however, when the energy of cavitation is harnessed it provides a remarkable energy source. This is essentially what sonication provides: controlled cavitation energy. The cavitation bubbles are generated by the ultrasound wave as it passes through the liquid. Like any sound wave, it is transmitted as a series of compression and rarefaction cycles affecting the molecules of the liquid. When the negative pressure of the rarefaction cycle exceeds the attractive forces between the molecules of the liquid, a void is formed. This void or cavity in the structure takes in a small amount of vapor from the solution so that on compression it does not totally collapse, but instead continues to grow in size in successive cycles to form an acoustic cavitation bubble. There are many thousands of such bubbles in a liquid, some of which are relatively stable but others expand further to an unstable size and undergo violent collapse to generate temperatures of about 5,000 K and pressures of the order of 50 MPa (see Fig. 6.1) (Vollmer *et al.*, 1998; Butz and Tauscher, 2002; Mason and Peters, 2002; Piyasena *et al.*, 2003). The pressure changes that occur from these implosions are the main mechanism for microbial cell disruption.

The cavitation bubbles formed in this way are loosely divided into two types:

- Stable cavition are bubbles that oscillate, often non-linearly, around some equilibrium size during many cycles of acoustic pressure. They exist for long enough to gather together and form large bubble clouds.
- Transient cavition, nowadays often called inertial cavitation, are bubbles which exist for a very short period, sometimes less than one cycle, and collapse violently (Abramov, 1998; Laborde *et al.*, 1998; Show *et al.*, 2007).

There is another way to generate cavitation via the Venturi effect of forcing a liquid at high pressure through a small orifice. On emerging from the orifice the liquid comes into ambient pressure conditions and the decompression generates cavitation bubbles. This is the driving force for the so-called liquid whistle, and this technology has been adopted by the food industry for mixing and emulsification for many years (Mason, 2005).

6.3 THE MEASUREMENT OF ULTRASONIC ENERGY (DOSIMETRY)

One of the most important facets of the use of power ultrasound in food technology is the need to find the appropriate amount of acoustic energy to apply to the

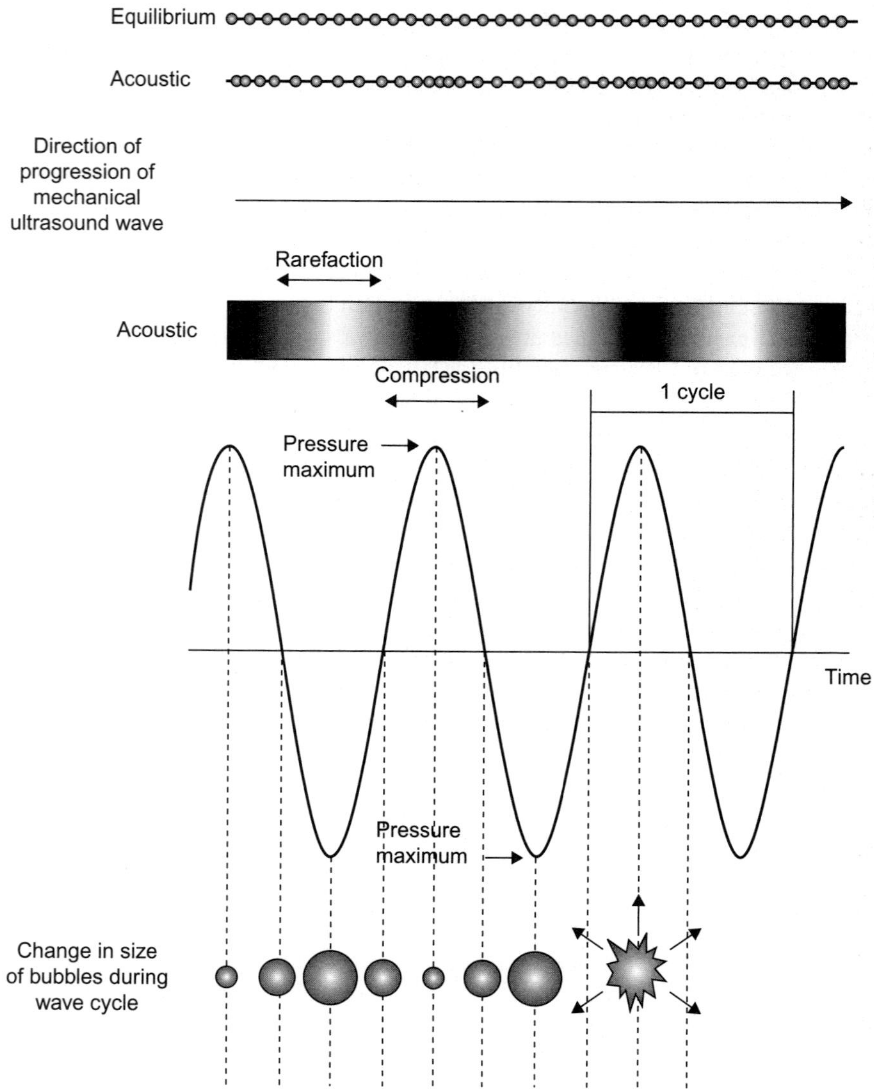

FIGURE 6.1

Ultrasonic cavitation.

[Reproduced with permission from Soria and Villamiel (2010).]

system. Almost all ultrasonic equipment has the capability to vary the power generated, i.e. the amplitude of vibration of the transducer. However, this parameter is not a measure of the absolute power that is actually entering the liquid being processed. Generally it is a measure of the energy used to drive the transducer. The normal measurement of acoustic power in a liquid is obtained using a hydrophone; however, at the intensities used in food processing, these instruments are rather fragile. Indeed it is almost impossible to obtain an accurate figure for the power because the sound energy is absorbed, reflected, and used in producing cavitation. If a reactor has no heat losses at all, then a good estimate is through calorimetry (see below); however, calorimetry is used as a rough guide as most reactors do have some losses. The level or energy transmitted to a food medium can be expressed as ultrasound intensity (W/cm^2) or acoustic energy density (Wcm^{-3} or W/ml). Each requires a measurement of power, and this can then be converted into intensity by dividing by the area of the emitting face of the transducer, or density by taking into account the total volume of the liquid. Generally the most useful of these is energy density (see below).

An alternative measurement may be required if it is the chemical effects of sonication that are of interest. In this case it is necessary to measure in some way the amount of radicals generated rather than the heating effect.

6.3.1 Calorimetry

Acoustic energy density is the amount of sound energy per unit volume. The advantage of this measure is that it can be used as a guide for scale up. It can be determined calorimetrically (Mason *et al.*, 1990) using Equations 6.1–6.3. The absolute ultrasonic power P is given as:

$$P = mC_p\left(\frac{dT}{dt}\right)_{t=0} \tag{6.1}$$

where, m is the mass, Cp is the specific heat capacity and (dT/dt) is the initial rate of change of temperature during sonication. This can be determined by fitting the data obtained for temperature rise (using a standard thermocouple) against time to a polynomial curve and extrapolating to time = 0.

The ultrasound intensity (UI) dissipated from an emitter of diameter D is given by (Mason *et al.*, 1990)

$$UI = \frac{4P}{\pi D^2}. \tag{6.2}$$

Acoustic Energy Density (AED) can be determined by dividing absolute ultrasound power by the volume (V) of the medium (cm^3 or ml):

$$AED = \frac{P}{V}. \tag{6.3}$$

6.3.2 Chemical Dosimetry

Cavitation collapse produces sufficient energy to break the chemical bonds of water to yield $H^{\cdot}$ and $OH^{\cdot}$ radicals. These radicals are very reactive and short-lived and can produce hydrogen peroxide (H_2O_2) as a product. The amount of peroxide is a measure of the radicals produced and so can be used as a chemical dosimeter for cavitation activity (Mead *et al.*, 1976; Alegria *et al.*,1989). The iodine dosimetry method is based upon the fact that H_2O_2 reacts quickly with I^- in solution to liberate molecular iodine which absorbs at 355 nm in the visible spectrum. Thus the yield of iodine can be estimated by the increase in absorbance at 355 nm on a UV/visible spectrophotometer which can be converted to H_2O_2 concentration by using stoichiometry and the Beer–Lambert Law (Ashokkumar *et al.*, 2008).

However, the determination of H_2O_2 generation during an ultrasound treatment in a food system is complex due to the presence of food components including ions and other colloidal components. To date, no reliable method of measuring cavitation activity in a food system has been developed (Raviyan *et al.*, 2005). Tsukamoto *et al.* (2004) reported that the measurement of the amplitude of vibration of the transducer is an indication of the ultrasonic cavitation and could thus be a reliable method for monitoring ultrasound power.

6.4 ULTRASOUND IN LIQUID FOOD STERILIZATION

Although ultrasound is generally regarded as an effective bactericidal agent, there have been reports of its use to enhance biocidal activity. The ability of ultrasound to stimulate biological activities has received very little attention; however, it is generally believed that the stimulatory impact detected is mainly due to increased cell membrane permeability and higher nutrient transfer rates. Thus, at low intensity, it has been reported to increase fermentation efficiency, in some food and beverage processes, e.g., *Bifidobacteria* and *Lactobacillus* in the fermentation of milk (Nguyen *et al.*, 2009). Şener *et al.* (2006) increased the rate of lactose hydrolysis in milk using ultrasound at an acoustic power level of 20 W, duty cycle of 10% and enzyme concentration of 1 ml/l, resulting in a minor loss (25%) of enzyme activity. Tentative efforts have also been made to employ ultrasound to stimulate the fermentation rates of yeast (Wood *et al.*, 1997). Possible applications in the improved performance of bioreactors have also been reviewed (Schläfer *et al.*, 2000).

For the sterilization of liquid foods, much higher acoustic energies are employed and the approaches can be classified as sonication alone, manosonication (pressure and ultrasound), themosonication (heat and ultrasound), or monothermosonication (heat, pressure, and ultrasound). Such approaches have been reported to be effective against several pathogenic and spoilage microorganisms for a range of liquid foods, including fruit juices, milk, smoothies, and liquid whole egg, with minimal changes in physicochemical characteristics. Power ultrasound has also been investigated for enzyme inactivation in fruit juices as well as dairy products.

6.4.1 Microbial Inactivation in Liquid Foods

The mechanisms through which ultrasound affects microbial inactivation are induced by acoustic cavitation which results in the weaking or disruption of bacterial cells through a number of processes (Joyce *et al.*, 2003). In simple terms, these may be grouped together into three classes:

- Bacteria cell wall damage, due to mechanical effects induced by pressure gradients generated during the collapse of cavitation bubbles within or near the bacteria.
- Shear forces induced by micro-streaming which occurs in the bacterial cell itself.
- Chemical attack due to formation of free radicals during cavitation which attack the cell wall structure leading to disintegration. In addition there will be the formation of a small amount of hydrogen peroxide via sonication, which itself is a bactericide.

It is difficult to establish the exact mechanism for inactivation. Ultrasound may inactivate microorganisms by physical (mechanical) effects and/or chemical effects through the formation of free radicals and hydrogen peroxide (Mason *et al.*, 1996; Raso *et al.*, 1998; Butz and Tauscher, 2002; Kadkhodaee and Povey, 2008; Piyasena *et al.*, 2004; Jiranek *et al.*, 2008). A more expanded list some of the mechanisms involved in inactivation of micro-organisms and enzymes is given in Table 6.1. Reported inactivation mechanisms are directly or indirectly dependent on processing variables such as ultrasound source and reactor geometry, frequency, and acoustic energy density. Media properties including treatment volume, temperature, viscosity, and gas concentration also affect the efficiency of inactivation (Sala *et al.*, 1995; Mason *et al.*, 1996; Kadkhodaee and Povey, 2008; Raso *et al.*, 1999).

One of the major bactericidal effects of ultrasound is attributed to intracellular cavitation; that is, micromechanical shocks that disrupt cellular structural and functional components up to the point of cell lysis. (Butz and Tauscher, 2002; Guerrero *et al.*, 2001; López-Malo *et al.*, 2005). Thinning of the cell membrane may occur due to localized heating and the production of free radicals by sonolysis of water (Butz and Tauscher, 2002). It is known that free radicals ($H^{\cdot}$, $O^{\cdot}$, $OH^{\cdot}$, $HO_2^{\cdot}$) and H_2O_2 are formed during the sonolysis of water molecules present in juice samples. These species contribute to microbial inactivation (Oyane *et al.*, 2009). A study conducted by Zenker *et al.* (2003) reported that ultrasound assisted treatments required a lower processing temperature compared to conventional thermal processing for inactivation of *Escherichia coli* and *Lactobacillus acidophilus* in various liquid food systems such as milk, fruit, and vegetable juices. Ugarte-Romero *et al.* (2006) reported a 5-log reduction in apple cider during sonication at an acoustic energy density of 0.46 W/ml and frequency of 20 kHz for sub-lethal temperatures (40°C). Similarly 5-log reductions for *Listeria monocytogens* at mild temperature were reported by Baumann *et al.* (2005) in apple cider.

Table 6.1 Ultrasonic Mechanisms for Inactivation

Theory	Mechanism	Application	Reference
Cavitation	Mechanical removal of attached or entrapped bacteria	Disinfection	Seymour *et al.* (2002)
	Localized hot spots (5,500°C) and high pressure (500 MPa)	Bactericidal Enzyme inactivation	Suslick (1988) Mason *et al.* (1996) Vercet *et al.* (2001)
	Increase in permeability of membranes or loss of selectivity	Bactericidal	Lehmann and Krusen (1954)
	Disruption (shear stress, localized heating) and chemical reactions within the cell of the microorganism	Bactericidal Inactivation of mold spores	Piyasena *et al.* (2003)
		Drying and dehydration	Fernandes *et al.* (2008)
Formation of free radicals	Sonolysis of water may produce OH^- and H^+ species and hydrogen peroxide	Antimicrobial	Mason *et al.* (1994); Suslick (1988)
		Denaturation of proteins and enzyme inactivation	Kadkhodaee and Powey (2008)
Intracellular micromechanical shocks	Disruption of cellular structural and functional components up to the point of cell lysis.	Inactivation of enzymes Cell lysis	Guerrero *et al.* (2001)
	Thinning of cell membranes	Antimicrobial	Sams and Feria (1991) Butz and Tauscher (2002); Fellows (2000)
Generation of mechanical energy	Cleaning action on surfaces	Bactericidal	Scherba *et al.* (1991); Sala *et al.* (1995); Raso *et al.* (1998)
Compressions and rarefactions or compression/expansion cycles	Acoustic microstreaming	Antimicrobial	Scherba *et al.* (1991) Floros and Liang (1994)
		Enzyme inactivation	Kashkooli *et al.* (1980)

Ugarte-Romero *et al.* (2006) studied the effects of thermal and ultrasound treatments on the morphology of *E. coli* K12 using environmental scanning electron microscopy (ESEM). They observed that the control cells exhibited a normal granular texture with external oligosaccharide strands (Fig. 6.2a) whereas thermal treatment (72°C) produced a non-granular surface and some minor pitting, but that the cell integrity was well maintained (Fig. 6.2b). In contrast, cells subject to sonication at 40°C for 3 min showed significant changes in cell morphology. Ugarte-Romero *et al.* (2006) also observed that sonication induces holes on the cell surface with approximate diameters of 200–300 nm (Fig. 6.2d). Such holes can be ascribed to the impact of high-velocity liquid jets generated during bubble collapse (Leighton, 1994) on the cell surface that perforate the cell wall and the cytoplasmic membrane. More prolonged ultrasound treatment may cause leakage of intracellular materials. Ugarte-Romero *et al.* (2006) also observed cell wall deformation and shrinkage of the cell surface (Fig. 6.2f) for sonication at 40°C whereas, sonication at a higher temperature (60°C) resulted in extensive damage and marked changes in cell morphology (Fig. 6.2h, f). The ESEM images revealed that sonication-induced cell damage at elevated temperatures is mainly mechanical. Similar observations were also reported by Raso *et al.* (1998) and others.

Cameron *et al.* (2008) studied the effect of sonication (20 kHz, 124 μm wave amplitude) on *E. coli, Saccharomyces cerevisiae* and *Lactobacillus acidophilus* in a saline salt solution and observed >99% inactivation for *E. coli, S. cerevisiae* and 72% for *L. acidophilus*. Transmission electron microscope (TEM) micrograph study of *E. coli*, *S. cerevisiae* and *L. acidophilus* indicated that sonication inflicts extensive external and internal damage leading to significant microbiocidal/microbiostatic effects. The authors also reported the formation of minute lipopolysaccharide vesicles formed by the ultrasound-induced emulsification of material liberated from the fragmented cell envelope in *E. coli*. As above the external and internal cell damage was shown by TEM in the micrographs (Fig. 6.3b, c). TEM also shows that sonication-induced damage is not uniform, as both intact and damaged cells were seen (Fig. 6.3d, b). Similarly an ultrasonic treatment of 5 min caused significant but variable damage to *L. acidophilus* cells (Cameron *et al.*, 2008). Figure (6.3h) shows intact *L. acidophilus* cells as well as empty cells, where the cell terminus had been "sheared off" due to cavitation (Fig. 6.3f, g). Uneven cell damage was clear in the case of *S. cerevisiae* cells sonicated for 2 min with a few cells devoid of content (Fig. 6.3j) compared to untreated controls (Fig. 6.3i).

Ciccolini *et al.* (1997) suggested that yeast cells could contain cavitation nuclei and sonication in combination with heat could cause internal cavitation as well as internal microstreaming. Such an effect would cause modification of the cellular structure. It has been reported that yeast cells undergo bactericidal effects during the early stages of sonication, and bactericidal and larger bacteriostatic effects appear with increasing time and intensity of sonication (Tsukamoto *et al.*, 2004).

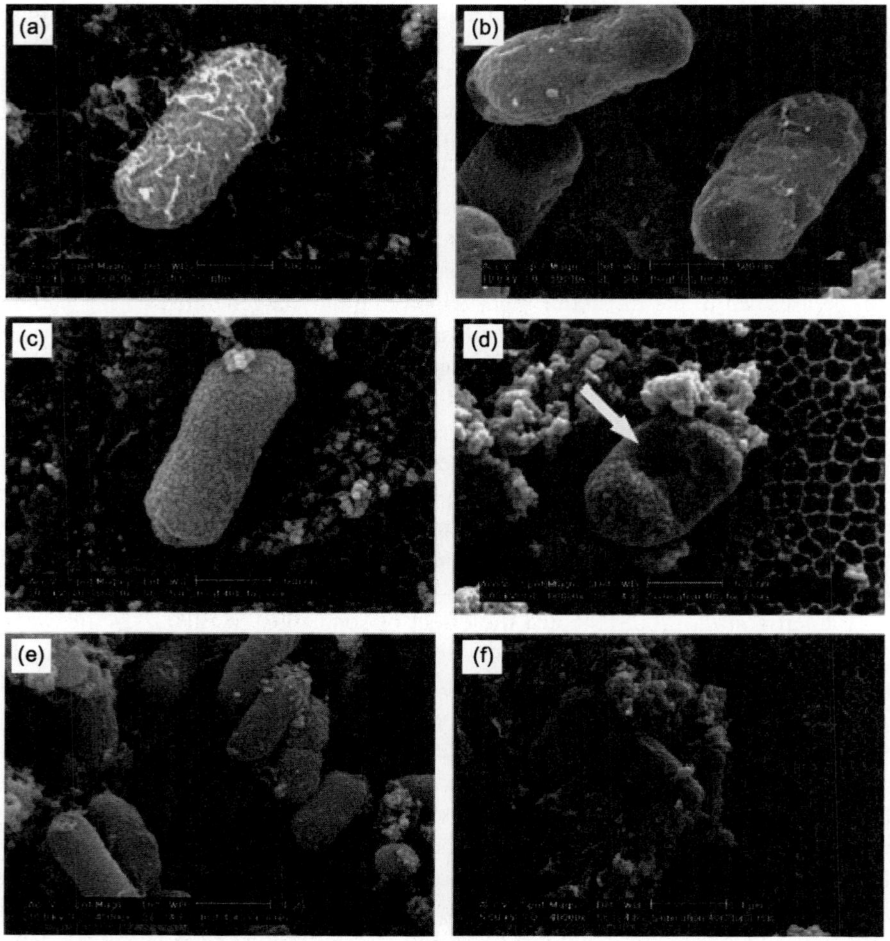

FIGURE 6.2

E. coli K12 cells observed with ESEM: (a) control; (b) heat treatment at 72°C for 30 s; (c) heat treatment at 40°C for 3 min (80,000 magnification); (d) sonication at 40°C for 3 min (80,000 magnification); (e) heat treatment at 40°C for 3 min (40,000 magnification); (f) sonication at 40°C for 3 min (40,000 magnification); (g) heat treatment at 60°C for 1 min (80,000 magnification); (h) sonication at 60°C for 1 min (80,000 magnification); (i) heat treatment at 60°C for 1 min (40,000 magnification); and (j) sonication at 60°C for 1 min (40,000 magnification) (Ugarte-Romero *et al.*, 2006).

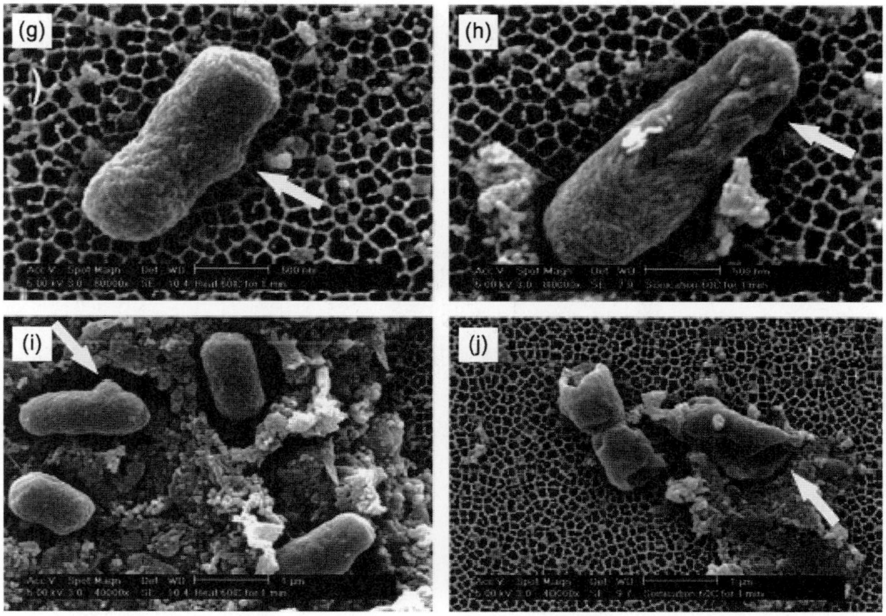

FIGURE 6.2

(*Continued*)

Microbial inactivation in fruit juices depends on both the type of microorganism and the processing parameters and the physicochemical properties of juice itself such as pH, soluble solids. For example, Baumann *et al.* (2005) reported increased inactivation rates for *L. monocytogenes* in apple cider (pH 3.4) compared to acidified saline solution when subjected to sonication at an acoustic energy density of 0.46 W/ml at 40–55°C. The effect of 500 kHz ultrasound was not found to be very effective on microorganisms in orange juice at ambient temperatures (Valero *et al.*, 2007). The presence of orange pulp increased the resistance of bacteria under these conditions and the authors suggested that a combination of sonication and other techniques, e.g., heating, might be more effective. Yeast inactivation in juice samples is possible at higher power levels (López-Malo *et al.*, 1999; Guerrero *et al.*, 2001; Adekunte *et al.*, 2010).

Sonication has been reported to be less effective on Gram-positive bacteria, such as *Staphylococcus aureus* and *Enterococci* (Ordóñez *et al.*, 1984). This can almost certainly be ascribed to their tougher cell-wall structure compared with Gram-negative bacteria. Ultrasound is also relatively less effective against spores due to their resistance to damage. However, Garcia *et al.* (1989) observed that the treatment of the spores with thermosonication (20 kHz, 150 W and 100°C) caused

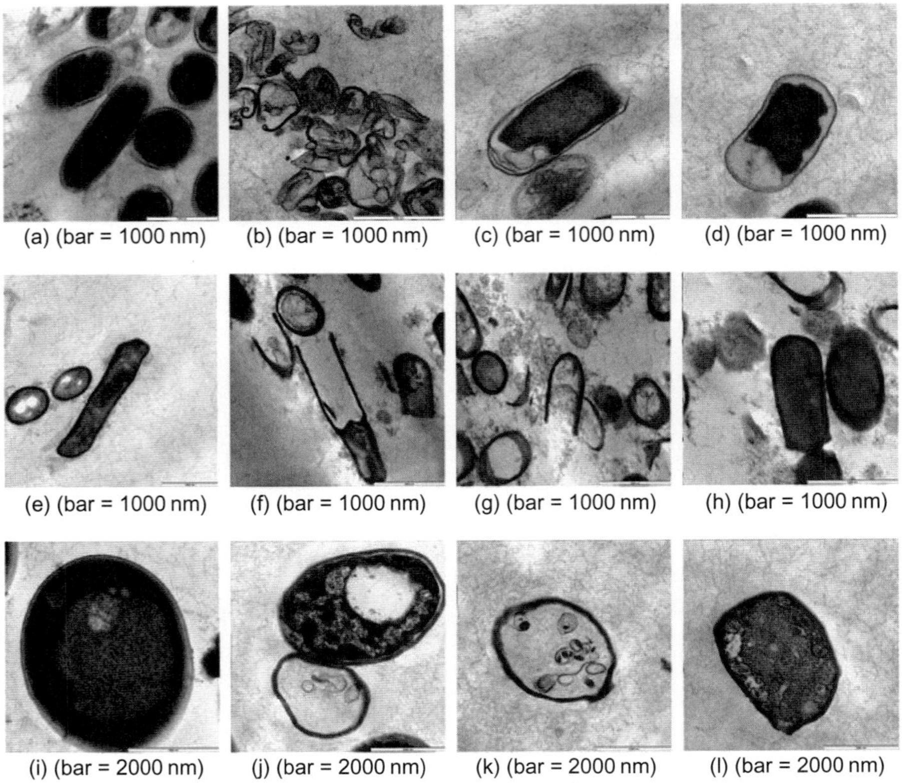

FIGURE 6.3

Transmission electron micrographs of *E. coli* (untreated (a) and ultrasonicated (b–d)), *L. acidophilus* (untreated (e) and ultrasonicated (f–h)) and *S. cerevisiae* (untreated (i) and ultrasonicated (j–l)) cells. (Cameron *et al.*, 2008)

a significant decrease in decimal reduction times for two strains of *B. subtilis* (niger-40 and ATCC 6051) for milk and glycerol: 63% (niger-40) and 74% (ATCC 6051) in glycerol and 79% (niger-40) and 40% (ATCC 6051) in milk.

Koda *et al.* (2009) studied the effect of sonication on microbial inactivation (*E. coli* and *S. mutans*) at 500 kHz, somewhat higher than the 20–40 kHz used by other workers, and concluded that the inactivation mechanism at the higher frequency was a combination of mechanical and chemical (via cavitation induced radicals) effects.

Yeast cells may also be resistant to the physical effects of sonication because they are relatively rigid and may not be disrupted by the action of microstreaming (Iida *et al.*, 2008). It is interesting to note that a Japanese group, Iida *et al.* (2008) proposed as a method for evaluating the physical (mechanical) effects of acoustic cavitation based on the rupture of yeast cells and the release of intercellular protein.

6.4.2 Enzyme Inactivation

It has been reported that, in general, the inactivation of enzymes using high-intensity ultrasound requires prolonged exposure periods. The inactivation is mainly the result of denaturation of protein either by chemical effects through free radicals produced during the sonolysis of water molecules ($H_2O \rightarrow OH^{\cdot} + H^{\cdot}$) or shear forces resulting from the formation or collapse of cavitating bubbles (Kadkhodaee and Povey, 2008).

Özbek and Ülgen (2000) reported that ultrasonic inactivation mechanisms are specific to the particular enzyme under investigation and depend on its amino acid composition and the conformational structure of the enzyme. For example manothermosonication is reported to inactivate peroxidase by splitting its prosthetic heme group, as for the mechanism of heat inactivation (López and Burgos, 1995a), whereas lipoxygenase appears to be inactivated by a free radical mediated mechanism (López and Burgos, 1995b) and possibly by denaturation of proteins (Mason, 1998). Some enzymes, such as catalase, yeast invertase, or pepsin are resistant to ultrasound (Sala *et al.*, 1995).

Nevertheless, critical factors governing microbial and/or enzymatic inactivation include the nature of the ultrasonic waves or amplitude, exposure time, microorganism or enzyme type, volume of food to be processed, composition of the food, and temperature. It is difficult to identify the specific enzyme inactivation mechanism during sonication which could be due to a singular or combination of several chemical and physical effects occurring simultaneously (Table 6.1).

6.4.3 Fruit Juices

Sonication has been identified as a potential technology to meet the US FDA's requirement of a 5-log reduction in pertinent microorganisms found in fruit juices (Salleh-Mack and Roberts, 2007). Various research groups have demonstrated the inactivation of pathogenic and spoilage microorganisms (*E. coli, Listeria*), spoilage enzymes (pectin methyl estrase, polyphenol oxidase) with high retentions of quality and nutritional parameters. Ultrasound is reported to have subtle adverse effects on the quality of fruit juices including: orange juice (Valero *et al.*, 2007), guava juice (Cheng *et al.*, 2007) and strawberry juice (Tiwari *et al.*, 2008a). Power ultrasound has been employed for inactivation of pathogenic and spoilage micro-organisms in apple cider (Baumann *et al.*, 2005), orange juice (Valero *et al.*, 2007, Patil *et al.*, 2009, Gómez-López *et al.*, 2010), guava juice (Cheng *et al.*, 2007) and tomato juice (Adekunte *et al.*, 2010). Ultrasound combined with mild heat treatment is reported to be effective against various food-borne pathogens including *E. coli* (Ugarte-Romero *et al.*, 2006) and *Listeria monocytogens* (Baumann *et al.*, 2005) in apple cider (Table 6.2). Thermosonication has shown

Table 6.2 Microbial Inactivation in Fruit Juices

Juice	Microorganism	Conditions	Inactivation	Reference
Model juice	*Escherichia coli*	24 kHz; maximum power 85 W/cm^2 and amplitude 100 μm	5 LR in 3–4 min	Salleh-Mack and Roberts (2007)
Apple cider	*Escherichia coli O157:H7*	40°C, 45°C, 50°C, 55°C, and 60°C 20 kHz; AED: 0.46 W/ml	AED 0.46 W/ml and 40°C 5.3 LR >5 LR E coli	Ugarte-Romero *et al.* (2006)
Apple cider	*Listeria monocytogenes*	(20 kHz, 457 mW • ml^{-1}) Sublethal (20, 30, and 40°C) Lethal (50, 55, and 60°C)	5 LR in combination with mild temperature	Baumann *et al.* (2005)
Orange juice	Total mesophilic aerobic counts total molds and yeast	120–600 W; 23 and 500 kHz; 15 min	APC (0.1–1.08 LR) Yeast and mold (~)	Valero *et al.* (2007)
Guava juice	Aerobic plate count	Sonication and carbonation 35 kHz (sonication bath); 30 min	APC (<1 LR)	Cheng *et al.* (2007)
Orange juice	*Listeria monocytogenes*	45°C, high-intensity ultrasound (600 W, 20 kHz, 95.2-μm wave amplitude), and the addition of different levels of vanillin (0, 1,000, 1,500, and 2,000 ppm), citral (0, 75, and 100 ppm)	TDT: 1.6 and 2.6 min	Ferrante *et al.* (2007)
Apple cider	*Escherichia coli O157:H7*	20 kHz; acoustic power 150 W and acoustic intensity 118 W/cm^2	6 LR with mild heat (57°C)	D'Amico *et al.* (2006)

to reduce inactivation treatment times for pathogenic microorganisms such as *L. monocytogenes* (Bermúdez-Aguirre *et al.*, 2009) and *S. Typhimurium* in skim milk and liquid whole egg (Wrigley and Llorca, 1992).

Manothermosonication is the simultaneous application of heat and high-energy ultrasonic waves under moderate pressure, and has been reported to enhance the bactericidal effect of ultrasound. Manothermosonication is reported to inactivate food-related enzymes and microorganisms at much higher rates than thermal treatments at comparable temperatures (Burgos, 1999). Raso *et al.* (1998) reported that during manothermosonication treatments at 30°C and 200 kPa, the increase in the amplitude of ultrasonic waves of 20 kHz from 21 to 150 μm exponentially decreased decimal reduction time values from 4 to 0.37 min for heat-resistant *Yersinia enterocolitica*. Manothermosonication is also reported to be effective against bacterial spores of *Bacillus subtilis* (Raso *et al.*, 1998). The pressure treatment is reported to increase the implosion power of cavitating bubbles and result in an enhanced killing (Lee *et al.*, 2009).

Ultrasound is reported to inactivate pectinmethylesterase (PME) in tomato juice and orange juice (Kuldiloke, 2002; López *et al.*, 1998; Vercet *et al.*, 1999, 2002a,b) in combination with heat and/or pressure. López *et al.* (1998) reported that the D-value of tomato PME was reduced from 45 min for thermal treatment to 0.85 min for manothermosonication at the same temperature (62.5°C). Raviyan *et al.* (2005) also reported a significant reduction in D value from 1571.4 min for thermal treatment to <80 min for thermosonication at the same temperature (50°C). The D value was further reduced from 240.6 min to 1.5 min with an increase in temperature from 50 to 61°C at a cavitation intensity of 0.007 $mgl^{-1}min^{-1}$ (Raviyan *et al.*, 2005). Wu *et al.* (2008) reported a reduction in D value for PME inactivation at 60 and 65°C compared with those for thermal inactivation. However, they did not observe this synergy at 70°C, where the D values for thermal and thermosonication treatment were similar. A number of studies have reported that sonication in combination with either heat or pressure has a synergistic effect on PME inactivation. Raviyan *et al.* (2005) reported increased inactivation of PME in sonicated tomato juice for a temperature range of 50–72°C compared with thermal treatment alone. Increased inactivation was dependent on cavitational intensity, which is reported to be temperature dependent. For example, simultaneous applications of heat (72°C) and ultrasound (frequency of 20 kHz and amplitude of 117 μm) under moderate pressure (200 kPa) increased the inactivation rate of orange juice PME by a factor of 25 in a buffer solution, and by more than a factor of 400 in orange juice (Vercet *et al.*, 1999). Higher inactivation rates in juice could be either due to the presence of co-solutes (substrates or other molecules that physically interact with enzymes) or loss of the protective effect of pectin in orange juice to which PME is bound (Vercet *et al.*, 1999). The effect of pectin on PME inactivation is also reported during orange juice ultrafiltration (Snir *et al.*, 1995). Raviyan *et al.* (2005) reported that the increase in enzyme inactivation during thermosonication is more pronounced at lower temperatures. One possible explanation for this could be that at higher temperatures, increased vapor pressure inside the bubbles introduces a cushioning effect and hence produces less effective bubble collapse (Mason, 1990). Tiwari *et al.* (2009a) concluded that sonication alone is not sufficient to inactivate PME. The maximum PME inactivation level reported for orange juice sonicated at the highest acoustic energy density of 1.05 W/ml for 10 min was 62%. The reduction of PME activity in sonicated lemon juice resulted in enhanced cloud stability during storage for 18 days at 4°C compared to thermally processed lemon juice (Knorr *et al.*, 2004). The improved cloud stability observed during storage could be due to the mechanical damage of the PME protein structure during sonication.

Enzymatic browning is one of the biggest problems faced during the processing of fruits and vegetables (Yemenicioglu and Cemeroglu, 2003). PPO is not an extremely heat stable enzyme, and short exposure to temperatures between 70 and 90°C is sufficient to inactivate it. Cheng *et al.* (2007) reported an increase in PPO in sonicated (35 kHz; for 30 min) guava juice compared to the control. They observed an increase in enzymatic activity possibly due to the processing conditions employed. A low ultrasound power level as in this case can enhance the disruption of biological cell walls to facilitate the release of their contents, indeed many ultrasonic horn systems were first marketed as cell disruptors (Mason *et al.*, 1996). Moreover, low power levels can induce stimulation of enzymes, whereas higher power levels inactivate enzymes due to denaturation. A synergistic effect of heat and pressure with ultrasound has been reported for the inactivation of PPO in model buffer systems (López *et al.*, 1994). They reported a linear decrease in log D values for an increase in ultrasound amplitude level over the range 35–145 μm. Heat- or pressure-assisted ultrasonic processing of juice can substantially reduce enzyme resistance and the heat treatment required for inactivation. As discussed earlier, the enzyme inactivation mechanism is complex and depends on several factors such as fruit juice composition, enzyme type, pH, and processing parameters. Manothermosonication treatment is reported to inactivate various enzymes relevant to fruit juices (lipoxygenase, peroxidase and polyphenol oxidase) in model buffer systems (López *et al.*, 1994). The inactivation mechanism of peroxidase under manothermosonication is reported to be the same as that for thermal processing namely the splitting of the prosthetic heme group (López and Burgos, 1995a). Thermal inactivation of peroxidase (POD) can be either by the dissociation of the prosthetic (heme) group from the haloenzyme (active enzyme system), through conformational changes in the protein or by modification or degradation of the prosthetic group (Lemos *et al.*, 2000). However, ultrasonic inactivation of POD results from conformational changes in protein and by splitting of prosthetic group from haloenzyme (López and Burgos, 1995a). Conversely lipoxygenase was reported to be ultrasonically inactivated by free radical formation (López and Burgos, 1995a) and possibly by the denaturation of enzymes and proteins.

Lipoxygenase (LOX) has been associated with quality deterioration because of its negative effects on pigments such as carotenes during storage, and its role in off-flavor and odor production (King and Klein, 1987; Aguiló-Aguayo *et al.*, 2008). However, in fruit juices a minimum LOX activity may be desirable for long storage periods (Min *et al.*, 2003). A 75–85% inactivation of LOX in soybeans has been achieved using ultrasound Thakur and Nelson (1997). The inactivation was strongly dependent on pH, treatment time and ultrasonic frequency. Similarly López and Burgos (1995a) reported that the resistance of LOX against heat and manothermosonication was also pH dependent during sonication over an amplitude range of 0–104 μm and a temperature range of 67.5–76.3°C. The dependency on pH is mainly due to the profound effects of pH on protein conformation since all enzymes having a maximum stability at an optimum pH.

6.4.4 Milk

Perhaps the four main applications of power ultrasound in dairy processing are in cleaning (McClements, 1995), microbial disinfection (Ordóñdez *et al.*, 1987), enzyme inactivation (Vercet *et al.*, 1997; O'Donnell *et al.*, 2010) and homogenization (Villamiel and de Jong, 2000). However a number of other applications have been reported for the use of ultrasound in the milk industry, e.g., extraction of chymosin (Kim and Zayas, 1991a; Kim and Zayas, 1991b) and β-galactosidase (Sakakibara *et al.*, 1994) and increasing the cheese yield (Müller, 1992).

The sonication of milk is reported to result in a diversity of physicochemical changes by reducing fermentation time during yoghurt preparation (Wu *et al.*, (2001) and improvement of yoghurt rheological properties (Vercet *et al.*, 2002b). Although many pathogenic and spoilage microorganisms are easily destroyed under standard heat treatments, many of them produce extracellular lipase and protease, which can withstand UHT treatment (Stead, 1986). These thermoresistant enzymes can reduce the quality and shelf-life of heat-treated milk and other dairy products. The simultaneous application of heat and ultrasound under pressure (manothermosonication) has been found to be more effective than heat treatment alone in the inactivation of heat-resistant protease and lipase secreted by *P. fluorescens* (Vercet *et al.*, 1997). The effect of ultrasound on enzymes involved in the coagulation of milk such as chymosin, pepsin, and several fungal enzymes, has been studied in model systems using batch processes. In general, ultrasonic treatment over a period of several minutes decreases the proteolytic activity of some enzymes. However, when a mixture of milk and chymosin was sonicated, hardly any enzyme inactivation was observed (Raharintsoa *et al.*, 1977, 1978). It has been reported that enzyme inactivation increases with an increase in solids content and decreases with an increase in enzyme concentration (Sala *et al.*, 1995; Villamiel and de Jong, 2000).

Despite early indications that sonication could provide several advantages as a preservation technique for milk, it has yet to be adopted as a standard procedure. This can almost certainly be ascribed to several "negative" factors resulting from work published using non-optimized sonochemical treatments that are now being addressed. Among these negative factors are concerns about the required energy consumption (Villamiel and de Jong, 2000) and the possible formation of oxidative compounds due to sonochemical reactions with milk fat. Riener *et al.* (2009) detected some weak and ill-defined "rubbery" aroma imparted during sonication of commercially pasteurized milk. They also reported the formation of n-alkenes such as hex-1-ene, hept-1-ene, oct-1-ene, and non-1-ene, which are also products of lipid oxidation. Similarly, Chouliara *et al.* (2010) investigated the combined effect of sonication and heat treatment and reported an increase in products of lipid oxidation with increased sonication and storage times.

However, other studies have shown that the ultrasonic processing of milk could be a promising milk-preservation technique, especially under mild conditions such as thermosonication. Ordoñez *et al.* (1984) studied the combined effect of ultrasound (20 kHz) and heat (5–62°C) for the inactivation of *Streptococus spp.* They observed that the combined treatment is more effective compared with treatment by either ultrasound or heat alone. In another study Ordoñez *et al.* (1987) investigated the effect of thermosonication on the survival of a strain of *Staphylococcus aureus* in ultra-heat-treated whole milk. They observed a 43% decrease in the decimal reduction times compared to the heat treatment alone. Villamiel and de Jong (2000) compared the treatment of raw milk with ultrasound alone under continuous flow with conventional heating at 48.6, 62 and 76°C. The initial total viable count of raw milk (5.3 log cfu ml^{-1}) was reduced by 0.2, 0.6 and 2.9 log cfu ml^{-1} after ultrasonication at 48.6, 62 and 76°C, respectively. Bermúdez-Aguirre *et al.* (2008) investigated the effect of fat content in milk on the inactivation of *Listeria innocua*; they observed that the rate of inactivation decreased with increasing fat content. In this case the inactivation rate decreased from 4.5 to 2.5 log reductions when the butter fat content of milk increased from 1% to 3.47% after 30 min sonication treatment at 63 ± 0.5°C and 120 μm amplitude level (Bermúdez-Aguirre *et al.*, 2008). It is well documented that the composition of food has an effect on microbial inactivation kinetics, e.g., Villamiel *et al.* (1999) observed higher inactivation of Salmonella in aqueous peptone (2.5 log) than in milk chocolate (0.1 log) after 10 min of treatment. This is because milk fat provides a protective effect to microbes and thereby reduces inactivation rates.

Villamiel and de Jong (2000) outlined the effect of ultrasound on native milk enzymes (Table 6.3). No effect on milk enzymes was observed when ultrasound was applied without thermal treatment. However inactivation effects were reported when sonication was carried out above 61°C. The resistance of enzymes to sonication was found to be both enzyme- and media-specific. Several studies have demonstrated that the effect of ultrasonic waves increases at higher total solids concentrations (Santamaria *et al.*, 1952; Sala *et al.*, 1995). In skim milk, the concentration of solids is lower than in whole milk resulting in a reduced ultrasonic effect. However, the concentration of enzymes in skim milk (alkaline phosphatase, AP and gamma-glutamyl transpeptidase, GGTP) is also lower than in whole milk, leading to a more pronounced effect, as these enzymes are linked to fat globules and can be liberated by the effects of ultrasound on the serum phase. Whereas, lactoperoxidase (LPO) is located in the whey, and the main cause of the higher decrease of enzyme activity in whole milk caused by sonication and heat (75.5°C; 102.3 s) compared with skim milk could be due to the higher concentration of solids in the former (Villamiel and Jong, 2000). LPO and AP enzymes have a significant function in dairy processing and a greater inactivation of with sonication at 40°C compared to 20°C has been reported (Ertugay *et al.*, 2003) (Table 6.3).

6.5 EFFECT ON QUALITY PARAMETERS

Given that sonication on its own or in combination with heat and/or pressure is an effective processing tool, it is of course essential that its use does not affect

Table 6.3 Enzymatic Inactivation of Selected Food Products

Food Product	Enzyme	Conditions	Residual Activity	Reference
Whole milk	Alkaline Phosphatase γ-Glutamyltranspeptidase Lactoperoxidase	Frequency: 20 kHz, Amplitude: 120 μm Residence time: 102.3 s	AP: 98.2% GGTP: 77.9% LPO: 85.6%	Villamiel and de Jong (2000)
Skim milk	Alkaline Phosphatase γ-Glutamyltranspeptidase Lactoperoxidase	Frequency: 20 kHz, Amplitude: 120 μm Residence time: 102.3 s	AP: 100.0% GGTP: 82.8% LPO: 100.0%	Villamiel and de Jong (2000)
Whole milk	Alkaline Phosphatase γ-Glutamyltranspeptidase Lactoperoxidase	Frequency: 20 kHz, Amplitude: 120 μm Residence time: 102.3 s Temperature: 61°C	AP: 0.0% GGTP: 0.0% LPO: 30.8%	Villamiel and de Jong (2000)
Skim milk	Alkaline Phosphatase γ-Glutamyltranspeptidase Lactoperoxidase	Frequency: 20 kHz, Amplitude: 120 μm Residence time: 102.3 s Temperature: 75°5 C	AP: 0.0% GGTP: 0.0% LPO: 47.2%	Villamiel and de Jong (2000)
Whole milk	Lactoperoxidase (LPO) and alkaline phosphatase (AP).	Amplitude level (0–80%) Temperature (20 and 40°C) and exposure times (30, 60, 90 and 120 s)	80% amplitude and 40°C (LPO: 6.875; AP: 3.813)	Ertugay *et al.* (2003)
Tomato juice	Pectinmethylesterase (PME) and polygalacturonase (PG)	24 kHz; Amplitude: 25, 50, and 75 μm 60°C, 65°C, and 70°C	D value 11.02–12.42 min (65°C; Amplitude 25–75 μm) D value 23.51 min at heat alone (65°C)	Wu *et al.* (2008)

food quality. It is certainly capable of achieving a desired 5-log for food-borne pathogens in fruit juices but there is some evidence that it could negatively modify some food properties including flavor, color, or nutritional value. Fruit juice color is a primary factor considered by the consumer in assessing juice quality and sensory acceptance and sonication is reported to have an effect on juice color. Differences in perceivable color can be analytically classified by total color difference (TCD) as very distinct (TCD $>$ 3), distinct (1.5 $<$ TCD $<$ 3) and small difference (TCD $<$ 1.5) (Tiwari *et al.*, 2008b). Small differences in perceivable color were reported for sonicated orange juice at lower intensity levels and treatment times. It must be borne in mind however that small differences observed by instrumental readings may not be registered by the consumer. Thus Gómez-López *et al.* (2010) observed significant instrumental color changes in orange juice color but these were not observed by the naked eye as indicated from sensory evaluation based on panel-tested color, aroma, and flavor even at extreme treatment conditions (89.25 μm for 10 min). Similar changes in color were reported for sonicated Guava juice (Cheng *et al.*, 2007). Valero *et al.* (2007) studied the effect of several ultrasonic treatments on limonin content and color of orange juices, as well as on brown pigments (BP) produced in non-enzymatic browning reactions. The authors reported that ultrasound treatment has almost no effect on the evolution of color scoring in the analyzed orange juices, and no effect on BP production in batch conditions. Tiwari *et al.*, 2008b found that sonication significantly influenced orange juice color at high amplitude levels and sonication times. Similar increases in lightness value were reported for apple cider (Ugarte-Romero *et al.*, 2006). Tiwari *et al.* (2008a) reported a slight increase (1–2%) in the anthocyanin content of sonicated strawberry juice at lower amplitude levels and treatment times which may be due to the extraction of bound anthocyanins from the suspended pulp. At higher amplitude levels and treatment times, a maximum of 5% anthocyanin degradation was observed. Weak ultrasonic irradiation was reported to promote an increase in the amount of phenolic compounds found in red wine (Masuzawa *et al.*, 2000). Some degradation of juice components may occur but can be minimal as is the degradation of the pelargonidin-3-glucoside content of strawberry juice (Tiwari *et al.*, 2008a) and the cyanidin-3-glucoside content of blackberry juice (Tiwari *et al.*, 2009b). The reported degradation of anthocyanins is mainly due to cavitation and the production of free radicals. Hydroxyl radicals produced by cavitation are involved in the degradation of anthocyanins by opening of rings and formation of chalcone (Sadilova *et al.*, 2007). The interaction of ascorbic acid with anthocyanin pigments results in mutual degradation and a decrease in color (Markakis *et al.*, 1957). This has also been reported for strawberry juice (Sondheimer and Kertesz, 1953).

Cloud stability is a critical juice quality parameter imparting characteristic flavor, color, and mouthfeel to fruit juices. Sonication of fruit juices is reported to improve the cloud value of orange juice (Tiwari *et al.*, 2008b). A significant increase in cloud value of 222% was reported for orange juice processed at an ultrasonic intensity of 8.61 W/cm^2 for 2 min. Cloud is attributed to a suspension of particles composed of a complex mixture of protein, pectin, lipids, hemicellulose, cellulose, and other minor components (Baker and Cameron, 1999). It has been reported that the degree of esterification of the pectin backbone necessary to cause cloud loss in orange juice is <36% (Krop and Pilnik, 1974). It is known that the reduction of PME activity in juices results in an enhanced cloud stability. However, PME residual activity of >95% was observed at ultrasonic intensity of 8.61 W/cm^2 for 2 min. The higher cloud retention observed at lower ultrasonic intensities could be due to a reduction in particle size, and breakdown of pectin molecules. This results in a reduced molecular weight and a weaker network formation (Seshadri *et al.*, 2003). Similarly, Lacroix *et al.* (2005) reported that the particle suspension stability of orange juice depends not only on PME activity but also on modifications of the pectin molecule. Structural damage to pectin may result from the microjets of liquid generated by the asymmetrical collapse of the cavitation bubble.

Viscosity is another important quality parameter of fruit juices. Tomato juice viscosity is highly dependent on pectic substances which form a colloidal suspension with other particles (Thakur *et al.*, 1996). The textural properties of tomato juice after manothermosonication were reported to result in higher product consistency and initial apparent viscosity when compared with unprocessed juice (Vercet *et al.*, 2002a). From a rheological perspective, tomato juices are pseudogels, whose flow properties depend on the interaction or entanglement of cell particles (mostly cell walls), soluble pectin concentration, and the chemical properties of the latter. Manothermosonication treatments of pure pectin solutions yield molecules with lower apparent viscosities due to a molecular weight reduction (López Buesa, unpublished observation). A similar study conducted by Tiwari *et al.* (2010) observed a significant reduction in the apparent viscosity of 2% w/v pectin dispersions. Seshadri *et al.* (2003) reported a similar influence of sonication on the gel strength properties of pectin.

Non-enzymatic browning of sonicated orange juice (Tiwari *et al.*, 2008b) has been reported. Such darkening of juice has been reported to be influenced by particulate fractions. Few sensory studies are reported in the literature. However, Mason *et al.* (2005) found minimal differences between manothermosonicated and unprocessed tomato juice.

Ultrasound treatment of fruit juices is reported to have a minimal effect on the ascorbic acid content during processing and results in improved stability during storage when compared to thermal treatment. This positive effect of ultrasound compared with heating is assumed to be due to the effective removal of occluded oxygen from the juice (Knorr *et al.*, 2004) and this is a critical parameter influencing the stability of ascorbic acid (Solomon and Svanberg, 1995). Tiwari *et al.* (2009c) reported a maximum degradation of only 5% in the ascorbic acid content of orange juice when sonicated at high acoustic energy densities (0.81 W/ml) and treatment times (10 min). During storage at 10°C, sonicated juice was found to have a higher retention of ascorbic acid compared with thermally processed and control samples. However, a higher reduction of ca. 15% was found in the ascorbic acid content of sonicated strawberry juice. Ascorbic acid degradation during sonication may occur either:

- in the bubble where it is subjected to the extreme conditions during collapse which is effectively thermolysis and this triggers a Maillard reaction; or
- in the bulk solution near to a collapsing bubble where it can be attacked by free radicals formed on bubble collapse (Portenlänger and Heusinger, 1992).

Thus sonication can be related to advanced oxidative processes since both pathways are associated with the production and use of hydroxyl radicals (Petrier *et al.*, 2007).

Bermúdez-Aguirre *et al.* (2009) studied the effect of thermosonication on the quality parameters of milk. They observed a slight increase in lightness in thermo-sonicated milk samples. Increase in the lightness value of sonicated milk is a result of the breakage of fat globules in milk because of the disruption of membranes generated by ultrasound, reducing the size of fat globules and regrouping them with the casein micelles in new complexes; also from the release of the triacylglycerols content in the globules, making the fat content easier for extraction and quantification (Bermúdez-Aguirre *et al.*, 2008, 2009). Bermúdez-Aguirre *et al.* (2009) observed a decrease in protein content of about 0.26% in thermosonicated milk samples with slight increases in fat content. Increased fat content is mainly due to the size reduction of fat globules and subsequent release into the medium, resulting in the triacylglycerols generating a higher fat content for the sonicated samples. Villamiel and de Jong (2000) observed a reduction of 73% in the size of the fat globule (with mean particles diameter of 0.64 μm) following the treatment of milk by a continuous-flow ultrasonic treatment at temperatures close to 62°C. The reduction in fat globule size through the homogenization effect is one possible advantage of using sonication (Wu *et al.*, 2001).

6.6 CURRENT STATUS AND CHALLENGES

Although the potential of power ultrasound has been investigated for many food applications, challenges remain prior to widespread adoption of the technology. One of the difficulties reported in the literature is the non-standardized reporting of methodology and control parameters. This is mainly due to the work being carried out in a number of different laboratories and mainly at laboratory scale. There can be no doubt that comparable reporting in terms of energy density, probe types, and sample volumes is required, and to some extent this has been achieved in other applications, e.g., sonochemistry. Ultrasound technology may be employed for many food applications, such as homogenization, crystallization, extraction, etc. However, the synergistic effects on enzymes or *vice versa* are generally not reported. Validation of the technique for enzyme or microbial inactivation needs to deal with the complex nature of food systems, in particular non-Newtonian fluids and particulate matter.

Despite the promising effects of sonication alone or in combination with heat or pressure on a laboratory scale, there are few reports of pilot or scale-up apparatus. This situation is changing as equipment manufacturers begin to realize the potential of the market. For application of power ultrasound on an industrial scale, it is essential to have energy-efficient processors and, if possible, design considerations and materials which mirror current trends in food processing.

6.7 CONCLUSIONS

Ultrasonic processing has been around for a long time, but applications that involve liquid foods are rather more recent, with the majority of investigations related to fruit juice and dairy production. From this chapter it can be concluded that ultrasound alone or in combination with mild heat or pressure treatments may be employed to meet the US FDA's mandatory 5-log reduction for microbes pertinent to fruit juices. The possible adverse effects involving key quality parameters such as color, anthocyanins, and ascorbic acid content are small and so not generally found to be problematic. Quality parameters such as cloud value may actually improve with mild sonication. In terms of dairy products, especially milk, the indications are again positive with ultrasound providing considerable benefit in lower-temperature pasteurization.

Further research is needed to investigate the combined effects of ultrasound with other non-thermal technologies in food processing and scale-up issues should now be a major target in liquid-food processing.

References

Abramov, O.V., 1998. High-intensity Ultrasonics: Theory and Industrial Applications. first ed. Gordon and Breach Science Publishers, Amsterdam.

Adekunte, A.O., Tiwari, B.K., Cullen, P.J., Scannell, A.G.M., O'Donnell, C.P., 2010. Effect of sonication on colour, ascorbic acid and yeast inactivation in tomato juice. Food Chem. 122 (3), 500–507.

Aguiló-Aguayo, I., Sobrino-López, Á., Soliva-Fortuny, R., Martín-Belloso, O., 2008. Influence of high-intensity pulsed electric field processing on lipoxygenase and β-glucosidase activities in strawberry juice. Innovative Food Sci. Emerg. Technol. 9 (4), 455–462.

Alegria, A.E., Lion, Y., Konda, T., Reisz, P., 1989. Sonolysis of aqueous surfactant solutions: probing the interfacial region of cavitation bubbles by spin trapping. J. Phys. Chem. 93, 4908–4913.

Ashokkumar, M., Sunartio, D., Kentish, S., Mawson, R., Simons, L., Vilkhu, K., et al., 2008. Modification of food ingredients by ultrasound to improve functionality: a preliminary study on a model system. Innovat. Food Sci. Emerg. Tech. 9 (2), 155–160.

Baker, R.A., Cameron, R.G., 1999. Clouds of citrus juices and juice drinks. Food Technol. 53, 64–69.

Baumann, A.R., Martin, S.E., Feng, H., 2005. Power ultrasound treatment of Listeria monocytogenes in apple cider. J. Food Prot. 68, 2333–2340.

Bermúdez-Aguirre, D., Corradini, M.G., Mawson, R., Barbosa-Cánovas, G.V., 2009. Modeling the inactivation of Listeria innocua in raw whole milk treated under thermosonication. Innovat. Food Sci. Emerg. Tech. 10 (2), 172–178.

Bermúdez-Aguirre, D., Mawson, R., Barbosa-Cánovas, G.V., 2008. Microstructure of Fat Globules in Whole Milk after Thermosonication Treatment. J. Food Sci. 73 (7), E325–E332.

Burgos, J., 1999. Manothermosonication. In: Robinson, R.K., Batt, C.A., Patel, P.D. (Eds.), Encyclopedia of Food Microbiology. Academic Press, New York, pp. 1462–1469.

Butz, P., Tauscher, B., 2002. Emerging technologies: chemical aspects. Food Res. Int. 35, 279–284.

Cameron, M., McMaster, L.D., Britz, T.J., 2008. Electron microscopic analysis of dairy microbes inactivated by ultrasound. Ultrason. Sonochem. 15 (6), 960–964.

Cheng, L.H., Soh, C.Y., Liew, S.C., Teh, F.F., 2007. Effects of sonication and carbonation on guava juice quality. Food Chem. 104, 1396–1401.

Choi, J.H., Kim, S.B., 1994. Effect of ultrasound on sulphuric acid-catalysed hydrolysis of starch. Korean J. Chem. Eng. 11 (3), 178–184.

Chouliara, E., Georgogianni, K.G., Kanellopoulou, N., Kontominas, M.G., 2010. Effect of ultrasonication on microbiological, chemical and sensory properties of raw, thermized and pasteurized milk. Int. Dairy J. 20 (5), 307–313.

Ciccolini, L., Taillandier, P., Wilhem, A.M., Delmas, H., Strehaiano, P., 1997. Low frequency thermo-ultrasonication of Saccharomyces cerevisiae suspensions: effect of temperature and of ultrasonic power. Chem. Eng. J. 65 (2), 145–149.

Coakley, W.T., Brown, R.C., James, C.J., 1973. The inactivation of enzymes by ultrasonic cavitation at 20 kHz. Arch. Biochem. Biophys. 159, 722–729.

Cocito, C., Gaetano, G., Delfini, C., 1995. Rapid extraction of aroma compounds in must and wine by means of ultrasound. Food Chem. 52, 311–320.

Cruz, R.M.S., Vieira, M.C., Silva, C.L.M., 2006. Effect of heat and thermosonication treatments on peroxidase inactivation kinetics in watercress (Nasturtium officinale). J. Food Eng. 72 (1), 8–15.

Cruz, R.M.S., Vieira, M.C., Silva, C.L.M., 2007. Modelling kinetics of watercress (Nasturtium officinale) color changes due to heat and thermosonication treatments. Innovative Food Sci. Emerg. Technol. 8, 244–252.

D'Amico, D.J., Silk, T.M., Wu, J., Guo, M., 2006. Inactivation of microorganisms in milk and apple cider treated with ultrasound. J. Food Prot. 69, 556–563.

De Guerrero, S., López-Malo, A., Alzamora, S.M., 2001. Effect of ultrasound on the survival of Saccharomyces cerevisiae: influence of temperature, pH and amplitude. Innovative Food Sci. Emerg. Technol. 2, 31–39.

Ertugay, M.F., Yuksel, Y., Sengul, M., 2003. The effect of ultrasound on lactoperoxidase and alkaline phosphatase enzymes from milk. Milchwissenschaft-Milk Sci. Int. 58 (11–12), 593–595.

Fellows, P., 2000. Food Processing Technology: Principles and Practice. second ed. CRC Press, New York.

Fernandes, F.A.N., Gallão, M.I., Rodrigues, S., 2008. Effect of osmotic dehydration and ultrasound pre-treatment on cell structure: melon dehydration. LWT — Food Sci. Technol. 41 (4), 604–610.

Ferrante, S., Guerrero, S., Alzamora, S.M., 2007. Combined use of ultrasound and natural antimicrobials to inactivate Listeria monocytogenes in orange juice. J. Food Prot. 70, 1850–1856.

Floros, J.D., Liang, H., 1994. Acoustically assisted diffusion through membranes and biomaterials. Food Technol. 48, 79–84.

Garcia, M.L., Burgos, J., Sanz, B., Ordoñez, J.A., 1989. Effect of heat and ultrasonic waves on the survival of two strains of Bacillus subtilis. J. Appl. Microbiol. 67 (6), 619–628.

Gómez-López, V.M., Orsolani, L., Martínez-Yepez, A., Tapia, M.S., 2010. Microbiological and sensory quality of sonicated calcium-added orange juice. LWT – Food Sci. Technol. 43 (5), 808–813.

Guerrero, S., López-Malo, A., Alzamora, S.M., 2001. Effect of ultrasound on the survival of Saccharomyces cerevisiae: influence of temperature, pH and amplitude. Innovative Food Sci. Emerg. Technol. 2, 31–39.

Henley, J.P., Sadana, A., 1985. Categorization of enzyme deactivation using series-type mechanism. Enzyme Microb. Technol. 7, 50–60.

Hunter, G., Lucas, M., Watson, I., Parton, R., 2008. A radial mode ultrasonic horn for the inactivation of Escherichia coli K12. Ultrason. Sonochem. 15, 101–109.

Iida, Y., Tuziuti, T., Yasui, K., Kozuka, T., Towata, A., 2008. Protein release from yeast cells as an evaluation method of physical effects in ultrasonic field. Ultrason. Sonochem. 15, 995–1000.

Jiranek, V., Grbin, P., Yap, A., Barnes, M., Bates, D., 2008. High power ultrasonics as a novel tool offering new opportunities for managing wine microbiology. Biotechnol. Lett. 30, 1–6.

Joyce, E., Mason, T.J., Phull, S.S., Lorimer, J.P., 2003. The development and evaluation of electrolysis in conjunction with power ultrasound for the disinfection of bacterial suspensions. Ultrason. Sonochem. 10 (4-5), 231–234.

Kadkhodaee, R., Povey, M.J.W., 2008. Ultrasonic inactivation of Bacillus α-amylase. I. effect of gas content and emitting face of probe. Ultrason. Sonochem. 15 (2), 133–142.

Kashkooli, H., Roony, J., Rooxby, R., 1980. Effects of ultrasound on catalase and malate dehydrogenase. J. Acoust. Soc. Am. 67, 1789–1801.

Kim, S.M., Zayas, J.F., 1991a. Comparative quality characteristics of chymosin extracts obtained by ultrasound treatment. J. Food Sci. 56 (2), 406–410.

Kim, S.M., Zayas, J.F., 1991b. Influence of ultrasound on the properties of chymosin and the ultrastructure of abomasum during chymosin extraction. J. Food Process. Pres. 15 (2), 89–100.

King, D.L., Klein, B.P., 1987. Effect of flavonoids and related compounds on soybean lipoxygenase-1 activity. J. Food Sci. 57 (1), 220–221.

Knorr, D., Zenker, M., Heinz, V., Lee, D., 2004. Applications and potential of ultrasonics in food processing. Trends Food Sci. Tech. 15 (5), 261–266.

Koda, S., Miyamoto, M., Toma, M., Matsuoka, T., Maebayashi, M., 2009. Inactivation of Escherichia coli and Streptococcus mutans by ultrasound at 500 kHz. Ultrason. Sonochem. 16 (5), 655–659.

Krop, J.J.P., Pilnik, W., 1974. Effect of pectic acid and bivalent cations on cloud loss of citrus juices. Lebensmittel Wissenschaft und Technologie 7, 62–63.

Kuldiloke, J., 2002. Effect of ultrasound, temperature and pressure treatments on enzyme activity and quality indicators of fruit and vegetable juices. Doctoral dissertation, Technical University of Berlin.

Laborde, J.L., Bouyer, C., Caltagirone, J.P., Gérard, A., 1998. Acoustic bubble cavitation at low frequencies. Ultrasonics 36, 589–594.

Lacroix, N., Fliss, I., Makhlouf, J., 2005. Inactivation of pectin methylesterase and stabilization of opalescence in orange juice by dynamic high pressure. Food Res. Int. 38, 569–576.

Lee, H., Zhou, B., Feng, H., Martin, S.E., 2009. Effect of pH on Inactivation of Escherichia coli K12 by Sonication, Manosonication, Thermosonication, and Manothermosonication. J. Food Sci. 74 (4), E191–E198.

Lehmann, J.F., Krusen, F.H., 1954. Effect of pulsed and continuous application of ultrasound on transport of ions through biologic membranes. Arch. Phys. Med. Rehabil. 35 (1), 20–23.

Leighton, T., 1994. The Acoustic Bubble. Academic Press, London.

Lemos, M.A., Oliveira, J.C., Saraiva, J.A., 2000. Influence of pH on the thermal inactivation kinetics of horseradish peroxidase in aqueous solution. LWT Food Sci. Technol. 33, 362–368.

López, P., Burgos, J., 1995a. Lipoxygenase inactivation by manothermosonication: effects of sonication physical parameters, pH, KCl. Sugars, glycerol and enzyme concentration. J. Agric. Food Chem. 43, 620–625.

López, P., Burgos, J., 1995b. Peroxidase stability and reactivation after heat treatment and manothermosonication. J. Food Sci. 60, 451–455.

López, P., Sala, F.J., de la Fuente, J.L., Condon, S., Raso, J., Burgos, J., 1994. Inactivation of peroxidase, lipoxygenase and polypohenol oxidase by manothermosonication. J. Agric. Food Chem. 42, 252–256.

López, P., Vercet, A., Sańchez, A.C., Burgos, J., 1998. Inactivation of tomato pectic enzymes by manothermosonication. Zeitschrift fur Lebensmittel-Untersuchung und -Forschung 207, 249–252.

Lopez-Malo, A., Guerrero, S., Alzamora, S.M., 1999. Saccharomyces cerevisiae, thermal inactivation kinetics combined with ultrasound. J. Food Prot. 62, 1215–1217.

López-Malo, A., Palou, E., Jiménez-Fernández, M., Alzamora, S.M., Guerrero, S., 2005. Multifactorial fungal inactivation combining thermosonication and antimicrobials. J. Food Eng. 67, 87–93.

Ly Nguyen, B., Van Loey, A.M., Smout, C., Eren Özcan, S., Fachin, D., Verlent, I., et al., 2003. Mild-heat and high-pressure inactivation of carrot pectin methylesterase: a kinetic study. J. Food Sci. 68, 1377–1383.

Markakis, P., Livingston, G.E., Fellers, C.R., 1957. Quantitative aspects of strawberry pigment degradation. Food Res. 22, 117–130.

Mason, T.J., 1990. Introduction. In: Mason, T.J. (Ed.), Chemistry with Ultrasound. Elsevier Science, New York, pp. 1–26.

Mason, T.J., 1990. Sonochemistry: The Uses of Ultrasound in Chemistry. Royal Society of Chemistry, Cambridge, England.

Mason, T.J., 1991. Practical Sonochemistry User's Guide to Application in Chemistry and Chemical Engineering. Ellis Horwood, New York.

Mason, T.J., 1998. Power ultrasound in food processing – the way forward. In: Povey, M.J.W., Mason, T.J. (Eds.), Ultrasound in Food Processing. Thomson Science, London, UK, pp. 105–126.

Mason, T.J., Lorimer, J.P., Baters, D.M., Zhao, Y., 1994. Dosimetry in sonochemistry: the use of aqueous terephthalate ion as a fluorescence monitor. Ultrason. Sonochem. 1, 91–95.

Mason, T.J., Paniwnyk, L., Lorimer, J.P., 1996. The uses of ultrasound in food technology. Ultrason. Sonochem. 3, 253–260.

Mason, T.J., Paniwnyk, L., Lorimer, J.P., 1996. The uses of ultrasound in food technology. Ultrason. Sonochem. 3, 253–260.

Mason, T.J., Peters, D., 2002. Practical Sonochemistry. Horwood Publishing Limited.

Mason, T.J., Riera, E., Vercet, A., López-Bueza, P., 2005. Application of ultrasound. In: Da-Wen Sun (Ed.), Emerging Technologies for Food Processing. Elsevier Ltd, Cambridge MA, pp. 323–351.

Masuzawa, N., Ohdaira, E., Ide, M., 2000. Effects of ultrasonic irradiation on phenolic compounds in wine. Jpn. J. Appl. Phys. 39, 2978–2979.

McClements, J., 1995. Advances in the application of ultrasound in food analysis and processing. Trends Food Sci. Tech. 6, 293–299.

Mead, E.L., Sutherland, R.G., Verrall, R.E., 1976. The effect of ultrasound on water in the presence of dissolved gases. Can. J. Chem. 54, 1114–1120.

Min, S., Min, S.K., Zhang, Q.H., 2003. Inactivation kinetics of tomato juice lipoxygenase by pulsed electric fields. J. Food Sci. 68, 1995–2001.

Müller, H., 1992. Die Käseausbeute als Kostenfaktor. Deutshe Milchwirtschaft 37, 1131–1134.

Nguyen, T.M.P., Lee, Y.K., Zhou, W., 2009. Stimulating fermentative activities of bifidobacteria in milk by highintensity ultrasound. Int. Dairy J. 19 (6–7), 410–416.

O'Donnell, C.P., Tiwari, B.K., Bourke, P., Cullen, P.J., 2010. Effect of ultrasonic processing on food enzymes of industrial importance. Trends Food Sci. Tech. 21 (7), 358–367.

Ordoñez, J.A., Aguilera, M.A., Garcia, M.L., Sanz, B., 1987. Effect of combined ultrasonic and heat treatment (thermoultrasonication) on the survival of a strain of Staphylococcus aureus. J. Dairy Res. 54 (01), 61–67.

Ordóñez, J.A., Sanz, B., Hernández, P.E., López-Lorenzo, P., 1984. A note on the effect of combined ultrasonic and heat treatment on the survival of thermoduric streptococci. J. Appl. Bacteriol. 56, 175–177.

Oyane, I., Takeda, T., Oda, Y., Sakata, T., Furuta, M., Okitsu, K., et al., 2009. Comparison between the effects of ultrasound and γ-rays on the inactivation of Saccharomyces cerevisiae: analyses of cell membrane permeability and DNA or RNA synthesis by flow cytometry. Ultrason. Sonochem. 16, 532–536.

Özbek, B., Ülgen, K., 2000. The stability of enzymes after sonication. Process Biochem. 35 (9), 1037–1043.

Pagán, R., Mañas, P., Alvarez, I., Condón, S., 1999. Resistance of Listeria monocytogenes to ultrasonic waves under pressure at sublethal (manosonication) and lethal (manothermosonication) temperatures. J. Food Microbiol. 16, 139–148.

Patil, S., Bourke, P., Kelly, B., Fras, J.M., Cullen, P.J., 2009. The effects of acid adaptation on Escherichia coli inactivation using power ultrasound. Innov. Food Sci. Emerg. Technol. 10 (4), 486–490.

Petrier, C., Combet, E., Mason, T., 2007. Oxygen-induced concurrent ultrasonic degradation of volatile and non-volatile aromatic compounds. Ultrason. Sonochem. 14, 117–121.

Phull, S.S., Newman, A.P., Lorimer, J.P., Pollet, B., Mason, T.J., 1997. The development and evaluation of ultrasound in the biocidal treatment of water. Ultrason. Sonochem. 4 (2), 157–164.

Piyasena, P., Mohareb, E., McKellar, R.C., 2003. Inactivation of microbes using ultrasound: a review. Int. J. Food Microbiol. 87, 207–216.

Portenlänger, G., Heusinger, H., 1992. Chemical reactions induced by ultrasound and γ-rays in aqueous solutions of L-ascorbic acid. Carbohydr. Res. 232, 291–301.

Raharintsoa, C., Gaulard, M.L., Alais, C., 1977. Etude de l'action des ultrasons cavitans sur quelques enzymes coagulantes. Lait 57, 631–645.

Raharintsoa, C., Gaulard, M.L., Alais, C., 1978. Effet des ultrasons cavitans sur la coagulation du lait par les enzymes. Lait 58, 559–574.

Raso, J., Manas, P., Pagan, R., Sala, F.J., 1999. Influence of different factors on the output power transferred into medium by ultrasound. Ultrason. Sonochem. 5, 157–162.

Raso, J., Pagan, R., Condon, S., Sala, F.J., 1998. Influence of temperature and pressure on the lethality of ultrasound. Appl. Environ. Microbiol. 64 (2), 465–471.

Raviyan, P., Zhang, Z., Feng, H., 2005. Ultrasonication for tomato pectinmethylesterase inactivation: effect of cavitation intensity and temperature on inactivation. J. Food Eng. 70, 189–196.

Riener, J., Noci, F., Cronin, D.A., Morgan, D.J., Lyng, J.G., 2009. Characterisation of volatile compounds generated in milk by high intensity ultrasound. Int. Dairy J. 19 (4), 269–272.

Roberts, R.T., 1993. High intensity ultrasonics in food processing. Chem. Ind. 15, 119–121.

Sadilova, E., Carle, R., Stintzing, F.C., 2007. Thermal degradation of anthocyanins and its impact on color and in vitro antioxidant capacity. Mol. Nutr. Food Res. 51, 1438–1439.

Sakakibara, M., Wang, D., Ikeda, K., Suzuki, K., 1994. Effect of ultrasonic irradiation on production of fermented milk with Lactobacillus delbrueckii. Ultrason. Sonochem. 1 (2), S107–S110.

Sala, F.J., Burgos, J., Condon, S., López, P., Raso, J., 1995. Effect of heat and ultrasounds on microorganisms and enzymes. In: Gould, G.W. (Ed.), New Methods of Food Preservation. Blackie, Glasgow, pp. 176–204.

Salleh-Mack, S.Z., Roberts, J.S., 2007. Ultrasound pasteurization: the effects of temperature soluble solids organic acids and pH on the inactivation of Escherichia coli ATCC 25922. Ultrason. Sonochem. 14, 323–329.

Sams, A.R., Feria, R., 1991. Microbial effects of ultrasonication of broiler drumstick skin. J. Food Sci. 56 (1), 247–248.

Santamaria, L., Castellani, A., Levi, F.A., 1952. Hyaluronidase inactivation by ultrasonic waves and its mechanism. Enzymologia 15 (5), 285–295.

Scherba, G., Weigel, R.M., O'Brien Jr, W.D., 1991. Quantitative assessment of the germicidal efficacy of ultrasonic energy. Appl. Environ. Microbiol. 57 (7), 2079–2084.

Schläfer, O., Sievers, M., Klotzbücher, H., Onyeche, T.I., 2000. Improvement of biological activity by low energy ultrasound assisted bioreactors. Ultrasonics 38 (1-8), 711–716.

Şener, N., Apar, D.K., Özbek, B., 2006. A modelling study on milk lactose hydrolysis and β-galactosidase stability under sonication. Process Biochem. 41 (7), 1493–1500.

Seshadri, R., Weiss, J., Hulbert, G.J., Mount, J., 2003. Ultrasonic processing influences rheological and optical properties of high-methoxyl pectin dispersions. Food Hydrocolloid 17, 191–197.

Seymour, I.J., Burfoot, D., Smith, R.L., Cox, L.A., Lockwook, A., 2002. Ultrasound decontamination of minimally processed fruits and vegetables. Int. J. Food Sci. Tech. 37, 547–557.

Show, K.Y., Mao, T.H., Lee, D.J., 2007. Optimisation of sludge disruption by sonication. Water Res. 41 (20), 4741–4747.

Snir, R., Koehler, P.E., Sims, K.A., Wicker, L., 1995. pH and cations influence permeability of marsh white grapefruit pectinesterase on polysulfone ultrafiltration membrane. J. Agr. Food Chem. 43, 1157–1162.

Solomon, O., Svanberg, U., 1995. Effect of oxygen and fluorescent light on the quality of orange juice during storage at 8°C. Food Chem. 53, 363–368.

Sondheimer, E., Kertesz, Z.I., 1953. Participation of ascorbic acid in the destruction of anthocyanin in strawberry juice and model systems. Food Res. 18, 475–479.

Soria, A.C., Villamiel, M., 2010. Effect of ultrasound on the technological properties and bioactivity of food: a review. Trends Food Sci. Tech. 21 (4), 323–331.

Stead, D., 1986. Microbial lipases—their characteristics, role in food spoilage and industrial uses. J. Dairy Res. 53 (3), 481–505.

Suslick, K.S., 1988. Ultrasounds: Its Chemical Physical and Biological Effects. VHC Publishers, New York.

Suslick, K.S., Hammerton, D.A., Cline, R.E., 1986. Thesonoche medical hot spot. J. Am. Chem. Soc. 108, 5641–5642.

Thakur, B.R., Nelson, P.E., 1997. Inactivation of lipoxygenase in whole soy flour suspension by ultrasonic cavitation. Nahrung 41 (5), 299–301.

Thakur, B.R., Singh, R.K., Nelson, P.E., 1996. Quality attributes of processed tomato products: a review. Food Rev. Int. 12, 375–401.

Tiwari, B.K., Muthukumarappan, K., O'Donnell, C.P., Cullen, P.J., 2008b. Effects of sonication on the kinetics of orange juice quality parameters. J. Agric. Food Chem. 56 (7), 2423–2428.

Tiwari, B.K., Muthukumarappan, K., O'Donnell, C.P., Cullen, P.J., 2008. Effects of sonication on the kinetics of orange juice quality parameters. J. Agr. Food Chem. 56, 2423–2428.

Tiwari, B.K., Muthukumarappan, K., O'Donnell, C.P., Cullen, P.J., 2008. Inactivation kinetics of pectin methylesterase and cloud retention in sonicated orange juice. Innovative Food Sci. Emerg. Technol. 10 (2), 166–171.

Tiwari, B.K., O'Donnell, C.P., Muthukumarappan, K., Cullen, P.J., 2008. Effect of ultrasound processing on the quality and nutritional properties of fruit juices. Stewart Postharvest Review 4 (5), 1–6.

Tiwari, B.K., O'Donnell, C.P., Cullen, P.J., 2009b. Effect of sonication on retention of anthocyanins in blackberry juice. J. Food Eng. 93 (2), 166–171.

Tiwari, B.K., O'Donnell, C.P., Muthukumarappan, K., Cullen, P.J., 2009c. Ascorbic and degradation kinetics of sonicated organge juice during storage and comparison with thermally pasteurised juice. LWT-Food Sci. Technol. 42 (3), 700–704.

Tiwari, B.K., O'Donnell, C.P., Patras, A., Cullen, P.J., 2008a. Anthocyanin and ascorbic acid degradation in sonicated strawberry juice. J. Agric. Food Chem. 56 (21), 10071–10077.

Tiwari, BK, Muthukumarappan, K, O'Donnell, CP, Cullen, PJ., 2010. Rheological properties of sonicated guar, xanthan and pectin dispersions. Int. J. Food Prop. 13 (2), 223–233.

Tsukamoto, I., Yim, B., Stavarache, C.E., Furuta, M., Hashiba, K., Maeda, Y., 2004. Inactivation of Saccharomyces cerevisiae by ultrasonic irradiation. Ultrason. Sonochem. 11, 61–65.

Ugarte-Romero, E., Feng, H., Martin, S.E., Cadwallader, K.R., Robinson, S.J., 2006. Inactivation of Escherichia coli with power ultrasound in apple cider. J. Food Sci. 71, 102–108.

Valero, M., Recrosio, N., Saura, D., Munoz, N., Martıc, N., Lizama, V., 2007. Effects of ultrasonic treatments in orange juice processing. J. Food Eng. 80, 509–516.

Van den Broeck, I., Ludikhuyze, L.R., van Loey, A.M., Hendrickx, M.E., 2000. Inactivation of orange pectinesterase by combined high-pressure and temperature treatments: a kinetic study. J. Agr. Food Chem. 48, 1960–1970.

Vercet, A, Sánchez, C, Burgos, J, Montanes, L, López-Buesa, P., 2002a. The effects of manothermosonication on tomato pectic enzymes and tomato paste rheological properties. J. Food Eng. 53, 273–278.

Vercet, A., Burgos, J., Crelier, S., López-Buesa, P., 2001. Inactivation of proteases and lipases by ultrasound. Innovative Food Sci. Emerg. Technol. 2, 139–150.

Vercet, A., López, P., Burgos, J., 1997. Inactivation of heat-resistant lipase and protease from Pseudomonas fluorescens by manothermoultrasonication. J. Dairy Sci. 80, 29–36.

Vercet, A., López, P., Burgos, J., 1999. Inactivation of heat-resistant pectinmethylesterase from orange by manothermosonication. J. Agr. Food Chem. 47, 432–437.

Vercet, A., Oria, R., Marquina, P., Crelier, S., López-Buesa, P., 2002b. Rheological properties of yoghurt made with milk submitted to manothermosonication. J. Agric. Food Chem. 50 (21), 6165–6171.

Villamiel, M., de Jong, P., 2000. Influence of high-intensity ultrasound and heat treatment in continuous flow on fat, proteins and native enzymes of milk. J. Agric. Food Chem. 48, 472–478.

Villamiel, M., van Hamerveld, E.H., de Jong, P., 1999. Review: effect of ultrasound processing on the quality of dairy products. Milchwissenschaft 54, 69–73.

Vollmer, A.C., Kwakye, S., et al., 1998. Bacterial stress responses to 1-megahertz pulsed ultrasound in the presence of microbubbles. Appl. Environ. Microbiol. 64 (10), 3927–3931.

Wood, B.E., Aldrich, H.C., Ingram, L.O., 1997. Ultrasound Stimulates Ethanol Production during the Simultaneous Saccharification and Fermentation of Mixed Waste Office Paper. Biotechnol. Prog. 13 (3), 232–237.

Wrigley, D.M., Llorca, N.G., 1992. Decrease of Salmonella typhimurium in skim milk and egg by heat and ultrasonic wave treatment. J. Food Prot. 55, 678–680.

Wu, H., Hulbert, G.J., Mount, J.R., 2001. Effects of ultrasound on milk homogenization and fermentation with yoghurt starter. Innovative Food Sci. Emerg. Technol. 211–218.

Wu, J., Gamage, T.V., Vilkhu, K.S., Simons, L.K., Mawson, R., 2008. Effect of thermosonication on quality improvement of tomato juice. Innovative Food Sci. Emerg. Technol. 9 (2), 186–195.

Yemenicioglu, A., Cemeroglu, B., 2003. Consistency of polyphenol oxidase (PPO) thermostability in ripening apricots (Prunus armeniaca L.): evidence for the presence of thermostable PPO forming and destabilizing mechanisms in apricots. J. Agric. Food Chem. 51 (8), 2371–2379.

Zenker, M., Hienz, V., Knorr, D., 2003. Application of ultrasound-assisted thermal processing for preservation and quality retention of liquid foods. J. Food Prot. 66, 1642–1649.

CHAPTER

Irradiation of Fluid Foods

7

Brendan A. Niemira[1] and Meixu Gao[2]

[1]*US Department of Agriculture, Agricultural Research Service, Wyndmoor, PA, USA*

[2]*Chinese Academy of Agricultural Sciences, Institute for Application of Atomic Energy, Beijing, China*

7.1 INTRODUCTION

Beverages and fluid foods present a unique challenge for non-thermal processing. Proper handling of the materials before, during, and after processing are the foundation of successful treatment. Among the various technologies proposed for the processing of liquids is irradiation, the use of ionizing radiation. Irradiation has been shown to be among the most flexible and effective antimicrobial processes applicable to foods (Smith and Pillai, 2004; Fan *et al.*, 2007). In recent years, irradiation has been considered with increasing interest by the producers and processors for improving the microbial safety of fresh and fresh-cut fruits and vegetables, beef carcasses, shellfish, processed meats, and other foods (Niemira and Zhang, 2009).

Thermal pasteurization remains the primary treatment for the majority of fruit juices sold in the US. Pasteurization results in the loss of essential oils and other juice components. Fresh juices are valued for premium flavor and aroma; however, these products have also been responsible for outbreaks of salmonellosis, enterohemorrhagic *E. coli* infection, and hemolytic uremic syndrome (FDA, 2009). *E. coli* O157:H7 was found to survive in fruit pulps during extended refrigerated storage, up to 30 days in the case of grape and up to 20 days in the case of passion fruit pulp (Marques *et al.*, 2001). Following a string of outbreaks, the US Food and Drug Administration (FDA) implemented a policy (Linton, 2001) requiring 5-log reductions in human pathogen load in fresh juices. Various non-thermal treatments for fresh juices have been evaluated (Sizer and Balasubramaniam, 1999), including irradiation, pulsed electric fields, minimal thermal processing, high-pressure processing, and ultraviolet (UV)

Novel Thermal and Non-Thermal Technologies for Fluid Foods. DOI: 10.1016/B978-0-12-381470-8.00007-4

radiation. Of these alternatives to pasteurization, UV has been the most widely adopted (Linton, 2001).

Juices, beverages, purees and other fluid foods are amenable to treatment with irradiation (Song *et al.*, 2005). However, as with any process, the specific protocols used for implementation determine the degree of applicability to any given commodity. Absorbed doses which are too low may not achieve the desired result: antimicrobial efficacy, extension of shelf-life, flavor or aroma modification, etc. In contrast, absorbed doses that are too high may result in undesirable sensory effects. This chapter will examine the microbiological and sensory impact of irradiation for a variety of liquid and pumpable commodities.

As a penetrating process, irradiation is typically viewed as being most applicable as a post-packaging treatment. For liquid foods, where "packaging" means bottling, pouch-fill, or similar operations, the geometry of the packaging becomes critical. Physical limitations arising from the different irradiation technologies can have a significant impact on the process. This aspect of irradiation processing of fluid foods will be examined in greater detail.

The logistics of irradiation as a processing technology operate within a framework of societal norms, economic realities, and regulatory structures. These factors vary from nation to nation. An overview of the varying situations in different markets is presented. Finally, this chapter concludes with a case study of irradiation used as a sensory enhancement process for fermented beverages. This draws on the experience of wine processors in China over the last 20 years.

7.2 IRRADIATION TECHNOLOGIES

Irradiation is a food-processing treatment endorsed by a variety of professional and governmental organizations. In reviewing the scientific literature on the process, the US Food and Drug Administration (CFR, 2000) the US General Accounting Office (GAO, 2000) the American Dietetic Association (ADA, 2000) and the UN World Health Organization (WHO, 1999) have endorsed the technology as safe and effective. This has recently been reaffirmed with studies showing that no detectable furan was produced in most fresh and fresh-cut fruits and vegetables following a dose of 5 kilogray (10 kGy = 1 Mrad), a very high dose in the context of food processing (Fan and Sokorai, 2008). In the most extreme cases in that study, irradiation-induced furan production was below the ppb limit of detection. These are in contrast to much higher levels ($\sim$100 ppb) of furan generated as a result of thermal processing. Therefore, at the doses practical for use with fruits and vegetables, irradiation is as safe as thermal processing. This supports irradiation's potential as one of several "hurdles" for those commodities (Smith and Pillai, 2004; Niemira and Deschenes, 2005).

Dose ranges may be characterized as low ($<$3 kGy), medium ($>$3 and $<$10 kGy), or high ($>$10 kGy). The irradiation treatment is based on the energy

FIGURE 7.1

The radura logo required for labeling on irradiated foods. The logo must be accompanied by the text "treated with radiation" or "treated by irradiation".

absorbed. For most foods, the controlling factors are the mass, bulk density, geometry, and thickness of the food product. For juices and liquid products, packaging size, shape, and orientation also become significant. This is discussed in more detail below.

In the United States, products which have been irradiated must be labeled with an indicative symbol called a "radura" (Fig. 7.1). Regulations for labeling vary among jurisdictions, as will be discussed later in this chapter. Three types of ionizing radiation used for commercial food processing are gamma rays, e-beam and X-rays (Table 7.1). All of the technologies are of sufficient energy to ionize molecules in the food target, leading to the generation of radicals, breakage of DNA, and other radiochemical effects. As with other food-processing technologies, temperature control during irradiation is an essential element of proper implementation.

Acceleration electron beams (e-beams) are produced by linear or cyclotron accelerators. In commercial food irradiators, energies of up to 10 MeV are used. The processing dose is delivered as a pulse of electrons, and the full dose is delivered quickly, taking typically <5 s. Short exposure times generally prevent any significant rise in temperature during processing; for some non-food applications, high doses are delivered by repeated exposure. e-Beam penetrability is lower than that of gamma rays or X-rays (Table 7.1). Single bags, relatively thin packages (<4 inches), and very low-bulk-density items have been irradiated successfully with e-beam systems. Packaging and product configuration must be carefully considered to avoid unacceptably uneven treatment.

X-rays are high-energy photons, generated when an e-beam is directed at a metal plate. Alloys of high-density metals, such as tungsten or tantalum, provide

Table 7.1 Major Irradiation Technologies: Advantages and Disadvantages

Factors	Electron Beam	X-Ray	Gamma
Source	**Electric power.** Electrons are generated using electronics and accelerated to high energy using magnetic fields, 5–10 MeV[a]. When accelerator is powered off, no radiation is emitted.	**Electric power.** Created when high energy electrons (up to 5 MeV) strike a metal plate (e.g., tungsten or tantalum alloys), typical conversion efficiency is 5–10%. When accelerator is powered off, no radiation is emitted.	**Radioisotopes.** Radioactive decay of ^{60}cobalt (2.5 MeV) or ^{137}cesium (0.51 MeV). Radioisotope source is always emitting radiation: shielding of source must be the default position.
Mechanism	**High-energy electrons** cleave water molecules, creating oxygen and hydroxyl radicals that damage DNA, membranes. Direct cleavage of DNA also occurs.	**High-energy photons** stimulate atoms within target to release high-energy electrons, which cleave water molecules into radicals. Direct cleavage of DNA also occurs.	**High-energy photons** stimulate atoms within target to release high-energy electrons, which cleave water molecules into radicals. Direct cleavage of DNA also occurs.
Infrastructure required	**Shielding:** >2 m concrete or <1 m steel/iron/lead. **Cooling:** extensive for high-voltage electronics and accelerator. **Ventilation:** for ozone removal while unit is operating.	**Shielding:** >2 m concrete or <1 m steel/iron/lead. **Cooling:** extensive for high-voltage electronics and accelerator. Additional cooling systems required for plate target. **Ventilation:** for ozone removal while unit is operating.	**Shielding:** depending on design, >5 m water or >2 m concrete or <1 m steel/iron/lead. **Cooling:** moderate for control equipment **Ventilation:** at all times for ozone removal when source is exposed to air.
Speed[b]	s	s	**Minutes** (depending on source strength)
Penetrability[c]	**6–8 cm**, suitable for relatively thin or low density products.	**30–40 cm**, suitable for all products.	**30–40 cm**, suitable for all products.

[a]*MeV = million electron volts.*
[b]*Speed of dose delivery. The desired dose will vary depending on the target organism and the commodity irradiated.*
[c]*Penetrability in food, average density approximately 1 g/cm^3. This figure will vary for individual commodities due to localized variations in density associated with bone, voids, fibrous matter, etc.*

the most efficient conversion rates. Electrons are absorbed by the metal atoms, stimulating the release of the X-rays. Depending on the specific alloy used in the metal target plate, the efficiency of this conversion is relatively low, typically 5–10%, with the balance lost as waste heat. Cooling subsystems in the metal plate are necessary for operation of X-ray irradiators. As photons, X-rays have greater penetration than e-beams. This makes them suitable for irradiating large or bulky food items, up to and including pallet- or crate-sized packages. Speed of processing is comparable to that of e-beam.

Gamma rays do not require significant inputs of electricity to create, as is the case with e-beam or X-rays systems. Gamma rays are high-energy photons produced by the radioactive isotopes cobalt-60 and, less commonly, cesium-137. Gamma rays have penetrability similar to X-rays, which makes these systems comparably suitable for irradiating large or bulky food items. The time required for processing is dependent on the dose desired and the strength of the source. Speed of processing is slower than X-ray of e-beam (Table 7.1).

Shielding for all three types of irradiators is comparable. The radioactive material used in gamma irradiators is continuously producing radiation, and must therefore be shielded for worker access all the time. Although e-beam and X-ray systems do not produce any ionizing radiation when the systems are powered down, while in operation they require control systems and shielding similar to that of gamma systems. This is typically concrete, steel, lead, or a combination of these.

7.3 GENERAL MODE OF ACTION

High-energy photons (gamma rays or X-rays) energize electrons within the atoms of the food. These electrons may leave the atom completely (ionization) or raise the energy of the electrons to a higher level (excitation). These processes yield free radicals, i.e. atoms with unpaired electrons on their outer shell. e-Beams interact with atoms to create these free radicals directly. Free radicals are very reactive, because their unpaired electrons pair up with outer shell electrons of the atoms in cells. In most foods, water makes up the bulk of mass. As a result, the majority of the absorbed energy from irradiation goes into the creation of hydrogen and hydroxyl radicals from water molecules (Diehl, 1995). Interaction of free radicals with organic molecules of the food is the main mode of action of irradiation. Under conditions of limited free water, such as in dried or frozen products, fewer radicals are produced per unit of energy applied. Also, as the radicals have reduced mobility in dry or frozen products, higher doses become necessary for microbial control in these commodities (Thayer and Boyd, 1995). This can have significance for juice concentrates with a reduced water activity and high osmotic potential, particularly if these concentrates are frozen as part of production and/or distribution.

7.4 FOOD SAFETY AND SHELF-LIFE OF IRRADIATED LIQUIDS

Historically, the main focus of irradiation research with respect to juices has been shelf-life extension, which means inactivation and reduction primarily of yeasts and molds. These are more resistant to irradiation than bacterial pathogens (Niemira and Fan, 2009). In a study of irradiated mango pulp, phytopathogenic fungi had D_{10} values (the dose in kGy necessary to reduce population by 90% or 1 $\log_{10}$) ranging from 0.39 to 1.11 kGy (El-Samahay *et al.*, 2000). Working with *Salmonella enteritidis* in reconstituted orange juice, Niemira (2001) reported a D_{10} value of 0.35 kGy. Buchanan *et al.* (1998) inactivated greater than 5 log cfu of *E. coli* O157:H7 in inoculated commercial apple juices with a dose of 1.0 kGy; D_{10} values were 0.12, 0.16, and 0.21 kGy for the three isolates tested. The resistance of four *Salmonella* isolates irradiated in orange juice also varied, with D_{10} values ranging from 0.35 kGy to 0.71 kGy (Niemira *et al.*, 2001). The authors of that study concluded that a 3.5-kGy dose would result in a 5-log reduction of *S. anatum*, the most resistant isolate tested. A comparable dose, 3.0 kGy, reduced *Salmonella* isolates as well as *L. monocytogenes* to undetectable levels in orange juice (Foley *et al.*, 2002). Following irradiation, the pathogen was able to be recovered through selective enrichment.

Carrot juice treated with 3 kGy showed a 7-log reduction of reduced aerobic and coliform bacteria, while kale juice similarly treated showed a 5-log reduction (Song *et al.*, 2005). In non-irradiated controls, microbial populations increased during storage, leading to spoilage and shortened shelf-life. Lee *et al.* (2009) showed near elimination of yeast and mold in tamarind juice after treatment with 3 kGy; aerobic bacteria were similarly reduced. Shelf-life of the product was extended without loss of antioxidant activity.

The antimicrobial efficacy of irradiation is influenced by several factors, including native variation in pathogen resistance among different isolates. Acid adaptation increased the radiation resistance of three strains of *E. coli* O157:H7 in apple juice by 54–67% (Buchanan, 1998). Turbidity of the juice also increased irradiation resistance of one strain in that study. Antioxidant capacity of the suspended solids was offered as a putative method. In contrast, D_{10} did not vary for *S. enteritidis* suspended in commercial citrus juices of varying composition (orange *vs* orange/tangerine, extra pulp *vs* pulp free, regular *vs* calcium enriched, etc.) and varying antioxidant power (Niemira, 2001). In culture solutions, chemical antioxidants are able to neutralize the radicals produced during irradiation, thus reducing the efficacy of the process. However, in many juices, these antioxidants are bound in pulps, cell debris, etc. Available metabolically after digestion, they are not as available for reaction in the free solution of the beverage. Therefore, the specific role of natural and artificial antioxidants in irradiation processing remains an area of research.

7.5 SENSORY AND NUTRITIVE PROPERTIES

In the production of juices, irradiation of fruits often increases juice yield during processing, due to the softening of internal tissues. When juices are irradiated, a portion of juice ascorbic acid is converted to the dehydro form; as in fruits, this conversion is expected to have little nutritional impact (Romani *et al.*, 1963; Thayer, 1994). Juices were the subject of a number of early studies of irradiation as a food-processing tool. A variety of juices (orange, guava, tomato, redcurrant, blackcurrant, apricot, peach, pear, and apple) evaluated by taste panels showed no reduction in flavor quality following treatment with 5.0 kGy; however, the quality of similarly irradiated grape juice was reduced (Fetter *et al.*, 1969). Grape, apple, and orange juices treated with sterilizing (2–80 kGy) doses of irradiation exhibited loss of anthocyanin (grape juice) and beta-carotene (orange juice) after 10 kGy. Browning was evident in apple juice after 5 kGy, and ascorbic acid was reduced in apple and orange juices at that level of treatment. Amendment with propyl gallate at 0.01% reduced the negative effects of high-dose (10 kGy) irradiation for orange juice, but did not similarly protect apple juice (Chachin and Ogata, 1969).

High-dose irradiation (10 kGy) of orange juice and mango pulp led to flavor degradation and browning; this sensory impact was neutralized by the addition of 0.1% sorbic acid pre-irradiation, resulting in acceptable products with extended shelf-life (Thakur and Singh, 1993). Irradiation of apple juice increased the antioxidant power and reduced brownness, rendering the juice lighter in color (Fan and Thayer, 2002). These effects were dose dependent up to 8.9 kGy. During subsequent refrigerated storage, the irradiated juice darkened more quickly than non-irradiated controls, but was still significantly lighter after 16 days of storage. In that study, processing temperature was determined to be of more significance than exclusion of oxygen or the level of suspended matter.

No changes in the appearance or aroma were evident in reconstituted orange juice irradiated to 2.5 kGy (Niemira *et al.*, 2001). Fresh juice from irradiated grapefruit was acceptable after 0.3 kGy, but declined in quality at 0.6 kGy (Miller and McDonald, 1996). Orange juice concentrate treated with 5.0 kGy and stored at room temperature (25°C) resulted in loss of fresh orange flavor and an increase in unacceptable off-flavors and aromas (Spoto *et al.*, 1997). The authors concluded that lower dose (2.5 kGy) combined with a fully maintained cold chain storage at 0 or 5°C would provide adequate processing. Aroma-related acyclic monoterpenes in orange juice were found to be sensitive to irradiation (Fan and Gates, 2001). Monoterpene concentrations were reduced immediately after irradiation in a dose-dependent manner. After refrigerated (7°C) storage for 21 days, there was no significant difference from the non-irradiated controls.

The color, taste and antioxidant capacity of tamarind juice was stable or slightly improved in irradiation treatments of up to 3 kGy (Lee *et al.*, 2009). Overall acceptability of treated tamarind juice was also stable up to 3 kGy, but declined with a 5-kGy dose. The authors of that study describe a slight brightening of the juice color, which they suggest could represent a value-added treatment. Incorporation of tamarind juice in industrial food formulations may be enhanced by improved color (Lee *et al.*, 2009). Total ascorbic acid content (ascorbic + dehydroascorbic) was stable in kale and carrot juice treated with 3 kGy (Song *et al.*, 2005). A taste panel determined the sensory acceptability of juices so treated were not different from controls immediately after treatment; during storage, the irradiated juices retained acceptable quality, while the non-irradiated controls deteriorated.

With respect to juice composition, the processing factors which may lead to undesirable sensory impacts (flavors, aromas, color, turbidity, etc.) resulting from irradiation are not fully established. These factors are known to be commodity- and blend-specific. As with whole fresh produce, temperature control during irradiation is paramount. At high doses, irradiation can lead to the formation of ozone in the headspace. Due consideration must be given to the packaging used in juices or other beverages to be irradiated. Plastics that are not approved for irradiation may be a source of migrated mono- or oligomers, or may lead to off-odors or flavors. As modern packaging systems use polylaminates, multiple types of inked layers, straw penetration ports (e.g., in the case of drink pouches), or other complex containers, each component must be evaluated for the durability and suitability for irradiation processing.

7.6 OPERATIONAL PARAMETER CASE STUDY: CONTAINER ORIENTATION

Engineering designs which apply irradiation to materials prior to containerization (bottling, pouch-fill, etc.) require aseptic packaging to preserve the antimicrobial effects of the irradiation. Treatment at this flow-through stage would involve irradiation of a devoted section of piping, a plate-style treatment area, or other suitable arrangements. Thermal process treatment equipment would guide the design of such irradiation treatments. However, as a penetrating process, one of the strongest advantages of irradiation as a food-processing technology is its application post-packaging.

For meats, poultry, seafood and produce applications, implementation is relatively straightforward. Packaged liquid foods, however, present distinct technical hurdles which must be addressed. As has been noted, the penetration of e-beam irradiation is significantly lower than that of gamma or X-ray irradiation. Thus, the operational parameter issues discussed in this section are more pronounced for e-beam systems.

Post-packaging, liquids are of uniform density, with no voids or heterogenous matter as would be found with other food products. The most significant issue of irradiation of liquids arises from edge effects. Such effects are also notable in meats, poultry, seafood, and produce irradiation, but the container materials of liquid foods are an additional complication. When considered along the centerline axis, the absorptive profile of an idealized container is uniform from top to bottom (Fig. 7.2). Irradiation from two or four sides will deliver a uniform dose across the product container. However, along the edge of the container, the irradiation must penetrate a smaller cross-section, which means less liquid is treated. The irradiation is less attenuated, and the liquid oriented edge-on to the beam path receives a higher dose than material oriented face-on.

For containers with a complex profile, these effects increase in significance. Figure 7.2 presents an idealized cylindrical container (left) and a schematic of a commercial-type beverage container (right). Note that the height of liquid is

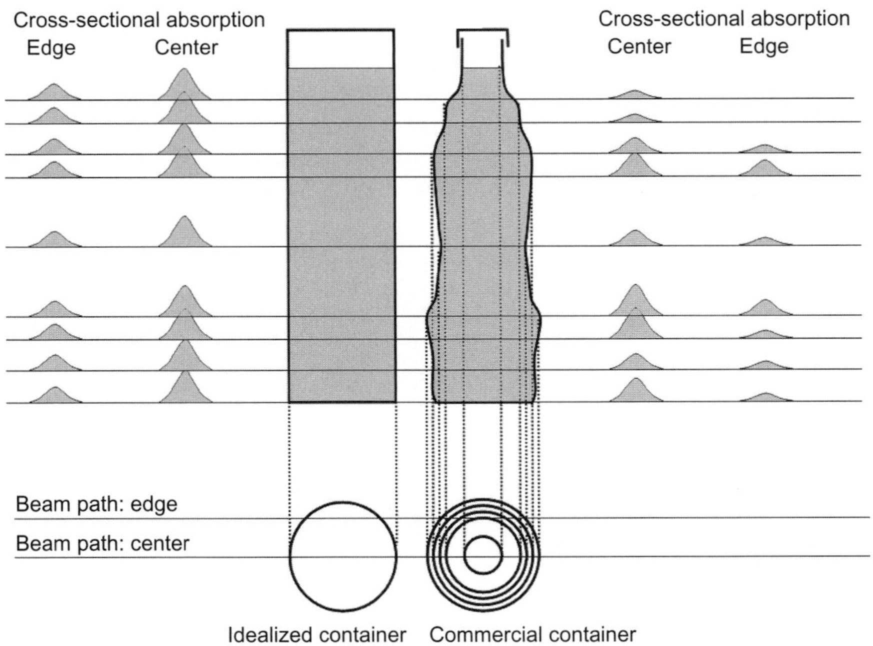

FIGURE 7.2

Schematic representation of irradiation dose absorption in bottle-type packaging. Left: idealized containers. Right: commercial-style containers.

identical in both containers. For the commercial container, the beam path along the center-line axis will cross a variable diameter from the top of the container to the bottom. Near the neck and the uppermost shoulder, only a relatively small amount of liquid is treated; at the waist is a more substantial amount, while the maximum diameter is seen at the upper and lower grip ridges. However, in a beam path along the edge of the container, the neck and uppermost shoulders are not treated at all. For this beam path, the waist and grip ridges will each receive a different absorbed dose.

These effects are observable in a scenario where individual containers are treated individually, as with a pass-through system on a conveyor. For containers that are arranged in a tray-pack, case-pack, or another bulk packaging arrangement, the edge effects and dose-absorption issues compound (Fig. 7.3). A perpendicular beam path which passes through multiple containers will see variable attenuation from each container in the path. Multi-pass processing will enhance the uniformity of the absorbed dose; the underlying complexity of the density profile in bulk packaging makes this a critical area. For beam paths which are not perpendicular to the product package edge, but are instead oriented at an oblique angle,

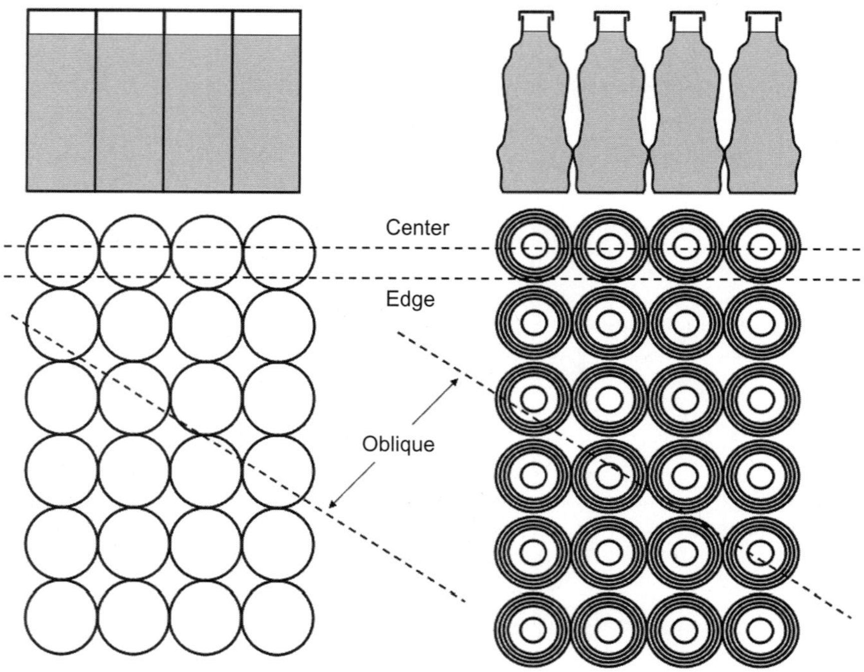

FIGURE 7.3

Schematic representation of irradiation dose absorption: in-line *vs* bulk-packaging. Left: idealized containers. Right: commercial-style containers.

establishing the absorbed dose across the rapidly changing density profiles is a challenge.

It should be noted that in this schematic, the commercial beverage container retains rotational symmetry along its own center axis. This means that the edge effects observed will be consistent irrespective of orientation of the container on a conveyor belt or in a case pack. However, an important recent trend is toward distinctively shaped containers. This increases the extent to which logos, grip ridges and other container features can complicate dose-mapping. Of course, pouch-style packages do not have rotational symmetry at all. Dose-absorption profiling for significantly asymmetric containers is an essential step in designing irradiation protocols.

Do these issues of container design and orientation present a fundamental barrier to post-packaging irradiation of liquid foods? It is important to remember that edge effects for liquid foods must be regarded differently from edge effects in solid foods such as meats, poultry, seafood, and produce. Unlike solid foods, as the liquid product is shipped, the material within the container mixes and equilibrates. The container as a whole will have a uniform taste and appearance. From this standpoint, irradiation can be a suitable non-thermal process so long as center-line dose absorption is sufficiently high to achieve the desired safety, shelf-life, or sensory effects, and so long as the edge effects are sufficiently limited so as to preserve the overall quality and marketability of the product. For foods which are only semi-liquid (yoghurts, condiments, thickened sauces, etc.) spontaneous homogenization after irradiation will be more limited. For these types of foods, proactive measures such as shaking, restacking, etc., may be warranted to gently mix the material inside the container.

7.7 REGULATIONS AND CONSUMER ATTITUDE OF FOOD IRRADIATION: CHINA

Like all food-processing technologies, irradiation must be applied within the guidelines of applicable regulatory controls. For domestic consumption, internal regulations apply, while food intended for export must meet the guidelines of the importing nation or economic zone. These governing regulations can vary widely. The applicable regulations for food irradiation in the United States are presented in Table 7.1. Guidelines for the packaging materials permitted for irradiation in the United States are presented in Tables 7.2 and 7.3. These guidelines are current as of 2010.

This section considers in detail the status of food irradiation in China, an important exporter of agricultural commodities and prepared foods, including juices, beverages, and other fluid foods. Chinese regulations encompass six general classes of foods approved nationally for irradiation. These are: (1) dried nuts and preserved fruits; (2) dried spices; (3) fresh fruits and vegetables; (4) frozen

Table 7.2 United States Code of Federal Regulations 21CFR179.26: Applications and Dose Limits for Irradiated Foods

Commodity and Purpose	Dose Limits
Control of *Trichinella* in pork	0.3–1.0 kGy
Suppression of growth and maturation in fresh foods	Maximum dose 1.0 kGy
Disinfestation of insect pests	Maximum dose 1.0 kGy
Antimicrobial treatment of dry enzymes	Maximum dose 10.0 kGy
Antimicrobial treatment of dry herbs and spices	Maximum dose 30.0 kGy
Control of pathogens in fresh and frozen raw poultry	Maximum dose 3.0 kGy
Sterilization of foods intended for use by NASA	Minimum dose 44.0 kGy
Control of pathogens and extension of shelf-life of refrigerated and frozen meats	Maximum dose 4.5 kGy (refrigerated) Maximum dose 7.0 kGy (frozen)
Control of *Salmonella* in fresh shell eggs	Maximum dose 3.0 kGy
Control of pathogens in seeds used to produce sprouts	Maximum dose 8.0 kGy
Control of *Vibrio* species and other food-borne pathogens in fresh or frozen molluscan shellfish	Maximum dose 5.5 kGy
Control of food-borne pathogens and extension of shelf-life in fresh Iceberg lettuce and fresh spinach	Maximum dose 4.0 kGy

packaged meat of livestock and poultry; (5) beans, grains and their products; and (6) cooked meat food of livestock and poultry. Also, three specific food items which have been approved for irradiation are: (1) pollen; (2) hog carcasses; and (3) sweet potato wine.

The commercial application of irradiation for sprout suppression, disinfestation, and shelf-life extension began in the 1990s, with spices, garlic, and grains as the major commodities treated. As of 2009, the total amount of irradiated foods in China is about 200,000 tons. Fermented foods have become the most important irradiated product since 2007. Irradiation can effectively prolong the shelf-life to 6 months; this extension serves to broaden the domestic market that Chinese manufacturers can economically reach.

Labeling regulations in China require that food products which have been irradiated be identified as such. However, these regulations are frequently not followed. In July 2009, it was established that manufacturers of dry noodles and other food products which were irradiated or contained irradiated components (such as dried vegetables) did not disclose this information on the label (Anonymous, 2009). Unlike in the United States and Europe, organized consumer groups opposed to food irradiation play a minimal role in China. Nevertheless,

Table 7.3 United States Code of Federal Regulations 21CFR179.45: Packaging Materials Approved for Irradiated Foods

Material	Maximum Dose
Nitrocellulose-coated or vinylidene chloride copolymer-coated cellophane	10 kGy
Glassine paper	10 kGy
Wax-coated paperboard	10 kGy
Films of polyolefin or polyethylene terephthalate. These may contain: 1. Sodium citrate, sodium lauryl sulfate, polyvinyl chloride* 2. Coatings comprising a vinylidene chloride copolymer containing a minimum of 85% vinylidene chloride with one or more of the following comonomers: acrylic acid, acrylonitrile, itaconic acid, methyl acrylate, and methyl methacrylate.	10 kGy
Kraft paper (only as a container for flour)	0.5 kGy
Polystyrene film	10 kGy
Rubber hydrochloride film	10 kGy
Vinylidene chloride–vinyl chloride copolymer film	10 kGy
Nylon 11	10 kGy
Ethylene–vinyl acetate copolymers	30 kGy
Vegetable parchments	60 kGy
Polyethylene film*	60 kGy
Polyethylene terephthalate film*	60 kGy
Nylon 6 films*	60 kGy
Vinyl chloride–vinyl acetate copolymer film*	60 kGy
Acrylonitrile copolymers*	60 kGy

This material may be amended with additional materials, listed in Table 7.4.

public reaction to the labeling issue led to investigations by the Ministries of Health and Agriculture.

The inquiries and dialog resulting from this had an educational outcome. In discussing irradiation, the presentation of background information on the technical nature of food irradiation, its strengths and weaknesses, served as a vehicle for education and outreach. In China, as in the United States, the European Union, and other parts of the world, the term "irradiation" is one that can provoke uncertainty in the consumer. In the extensive media coverage of food irradiation that followed the labeling violations, discussions and expert commentary by scientists in government and academia served to separate fact from fiction among Chinese consumers. In support of the consumers' right to know, increased regulatory compliance has resulted in food irradiation labeling by the two biggest instant-noodle companies in China. As of 2010, manufacturers of additional irradiated products, such as pollen and fermented foods, are more prominently labeled in the marketplace.

Table 7.4 United States Code of Federal Regulations 21CFR179.45: Adjuvants and Amendments Approved for Incorporation into Certain Packaging Materials Approved for Irradiated Foods

Adjuvant/Amendment	Limit (by Weight of Polymer)
Amides of erucic, linoleic, oleic, palmitic, and stearic acid.	1%
BHA (butylated hydroxyanisole)	1%
BHT (butylated hydroxytoluene)	1%
Calcium and sodium propionates	1%
Petroleum wax	1%
Mineral oil	1%
Stearates of aluminum, calcium, magnesium, potassium, and sodium	1%
Triethylene glycol	1%
Polypropylene, non-crystalline	2%

7.8 CASE STUDY: IRRADIATED WINE, CHINA

The use of irradiation for disinfestation (e.g., spices), sprout inhabitation (e.g., potato, garlic) are well known applications. However, irradiation has successfully been used to accelerate aging and flavor maturation in wine. As with conventional aging, wherein the taste and quality of wine improves with storage time, the irradiation-facilitated accelerated aging results in enhanced quality in a shorter amount of time.

Investigations of wine aging by irradiation started in the 1980s in China. Sweet potato wine is an inexpensive, widely consumed beverage across China. Aging for months or years is commonplace to improve the palatability. Market value of the wine is directly related to the time spent aging. In Sichuan province, sweet potato wine was irradiated and evaluated for marketability characteristics. The sensory analysis showed that the taste quality of irradiated wine was almost as good as the wine stored for 2–3 years. The content of total acid, ester and aldehyde could be increased in high-grade wine after a dose of 4.0 kGy. These studies also found that d-2,3-butanediol was formed in irradiated wine. Subsequent studies, including bioassays with *Tetrahymena pyriformis,* established that the concentrations of d-2,3-butanediol formed (10–1000 ppm, depending on dose) were well below levels of health concern (Lookchem, 2010). Peroxide was detected in irradiated wine, but the concentration of peroxide decreased with storage time. In 1994, a national standard was established in China especially for the irradiation of sweet potato wine. This specifies a peroxide content lower than 5 mg/l.

One commercial distillery in Sichuan initiated a program to irradiate wine, and successfully commercialized the technique. As of 1995, irradiated sweet potato wine enjoyed a very high reputation, commanding a premium price. The wine was notable as one of the national gifts of the central government in China. Although the product was successful in the marketplace, equipment failures led to a discontinuation of the company.

As of 2010, despite the demonstration of the irradiation as an artificial aging acceleration treatment for wine, the process is not currently used on a commercial scale. However, the successes noted for irradiated wine have encouraged further research in the field. Recent market shifts in China have led to a rising price for wines stored for 20 or 30 years. An ongoing research project is to study the irradiation effect on wine quality when the irradiation process is combined with aging. The goal of the research is to allow for aging for several months either before irradiation, or after irradiation (or both) in order to produce a product of comparable quality to that aged for several years. Pilot-scale tests at the Institute for Application of Atomic Energy in Beijing have shown a significant benefit from irradiation plus 2–3 months aging *vs* aging alone. These results suggest that commercialization of irradiated wine holds the potential for making a successful, high-value product.

7.9 CONCLUSIONS

Irradiation is a non-thermal process that can be applied to juices, beverages, and other fluid foods. Successful use of this technology is contingent on developing irradiation protocols that achieve the desired antimicrobial and food-safety goals while preserving (or improving) the sensory and nutritional value of the product. Beyond the microbiology and chemistry, issues of packaging, marketing, and regulations must be addressed as part of commercial adoption. As with any unfamiliar technology, these operational details will be established for each individual product as a normal part of process validation and optimization. As a flexible, broadly efficacious treatment, irradiation holds great potential for use with a variety of liquid commodities.

Acknowledgments

The authors would like to thank Drs X. Fan and J. Gurtler for their careful reviews of this manuscript. Mention of trade names of commercial products in this publication is solely for the purpose of providing specific information and does not imply recommendation or endorsement by the US Department of Agriculture.

References

ADA, 2000. Position of the american dietetic association: food irradiation. J. Am. Diet. Assoc. 100, 246–253.

Anonymous, 2009. Instant marking the beginning of a unified master "Radiation Sterilization". <http://www.chinafinancialdaily.com/financial/news/2009/08/12/9804/instant-marking-the-beginning-of-a-unified-master-radiation-sterilization.html/> (accessed 19.04.10).

Buchanan, R.L., Edelson, S.G., Snipes, K., Boyd, G., 1998. Inactivation of *Escherichia coli* O157:H7 in apple juice by irradiation. Appl. Envir. Microbiol 64 (11), 4533–4535.

Chachin, K., Ogata, K., 1969. Changes of chemical constituents and quality in some juice irradiated with the sterilizing dose level of gamma rays. Food Irradiation 4 (1), 85–90.

CFR, 2000. Irradiation in the production, processing and handling of food. Code of Federal Regulations, 21CFR179.

Diehl, J.F., 1995. Safety of Irradiated Foods. second ed. Marcel Dekker, Inc., New York, NY, 10016.

El-Samahay, S.K., Youssef, B.M., Askar, A.A., Swailam, H.M.M., 2000. Microbiological and chemical properties of irradiated mango. J. Food Safety 20, 139–156.

Fan, X., Gates, R.A., 2001. Degradation of monoterpenes in orange juice by gamma radiation. J. Agric. Food Chem. 49 (5), 2422–2426.

Fan, X., Thayer, D.W., 2002. [Gamma]-radiation influences browning, antioxidant activity and malonaldehyde level of apple juice. J. Agric. Food Chem. 50 (4), 710–715.

Fan, X., Sokorai, K.J.B., 2008. Effect of ionizing radiation on furan formation in fresh fruits and vegetables. J. Food Sci 73 (2), C79–C83.

Fan, X., Niemira, B.A., Prakash, A., 2007. Ionizing Irradiation of Fresh and Fresh-cut Fruits and Vegetables. Food Technol. 62 (3), 36–43.

FDA, 2009. Chapter IV. Outbreaks Associated with Fresh and Fresh-Cut Produce. Incidence, Growth, and Survival of Pathogens in Fresh and Fresh-Cut Produce. <http://www.fda.gov/Food/ScienceResearch/ResearchAreas/SafePracticesforFoodProcesses/ucm091265.htm/> (accessed 13.04.10).

Fetter, F., Stehlik, G., Kovacs, J., Weiss, S., 1969. Das flavourverhalten einger gamma-bestrahlter fruchtsaefte. Mitteilungen: -rube, -wein, -obstbau und fruechteverwertung 19, 140–151.

Foley, D.M., Pickett, K., Varon, J., Lee, J., Min, D.B., Caporaso, F., et al., 2002. Pasteurization of Fresh Orange Juice Using Gamma Irradiation: Microbiological, Flavor, and Sensory Analyses. J. Food Sci. 67 (4), 1495–1501.

GAO, 2000. Food Irradiation: Available Research Indicates that Benefits Outweigh Risks. U.S. General Accounting Office, GAO/RCED-00-217. <http://www.gao.gov/archive/2000/rc00217.pdf/> (accessed April 09.04.10).

Lee, J.W., Kim, J.K., Srinivasan, P., Choi, J.I., Kim, J.H., Han, S.B., et al., 2009. Effect of gamma irradiation on microbial analysis, antioxidant activity, sugar content and color of ready-to-use tamarind juice during storage. LWT—Food Sci. Technol. 42 (1), 101–105.

Linton, R., 2001. Controlling food safety using the HACCP approach and prerequisite programs <http://www.ces.purdue.edu/extmedia/FS/FS-13w.pdf/> (accessed 16.04.10).

Lookchem, 2010. 2,3-Butanediol <http://www.lookchem.com/2-3-Butanediol/> (accessed April 19.04.10).

Marques, P.A.H.F., Worcman-Barninka, D., Lannes, S.C.S., Landgraf, M., 2001. Acid tolerance and survival of Escherichia coli O157:H7 in fruit pulps stored under refrigeration. J. Food Prot. 64 (11), 1674–1678.

Miller, W.R., McDonald, R.E., 1996. Postharvest quality of GA-treated Florida grapefruit after gamma irradiation with TBZ and storage. Postharvest Biol. Technol. 7, 253–260.

Niemira, B.A., 2001. Citrus juice composition does not influence radiation sensitivity of Salmonella enteritidis. J. Food Prot. 64 (6), 869–872.

Niemira, B.A., Sommers, C.H., Boyd, G., 2001. Irradiation inactivation of four Salmonella species in orange juices with varying turbidity. J. Food Prot. 64 (5), 614–617.

Niemira, B.A., Deschenes, L., 2005. Ionizing radiation processing of fruits and fruit products. In: Barrett, D.M., Somogyi, L., Ramaswamy, H. (Eds.), Processing Fruits: Science and Technology, second ed. CRC Press, Boca Raton, FL, pp. 221–260.

Niemira, B.A., Fan, X., 2009. Ionizing radiation enhances microbial safety of fresh and fresh-cut fruits and vegetables while maintain product quality. In: Fan, X., Niemira, B.A., Doona, C.J., Feeherry, F., Gravani, R.B. (Eds.), Microbial Safety of Fresh Produce. Blackwell, Ames, IA, pp. 191–204.

Niemira, B.A, Zhang, H.Q., 2009. Advanced technologies for detection and elimination of pathogens. Ch. 17. In: Sapers, G.M., Solomon, E.S., Matthews, K. (Eds.), The Produce Contamination Problem—Causes and Solutions. Academic Press, New York, NY, pp. 425–443.

Romani, R.J., Van Kooy, J., Lim, L., Bowers, B., 1963. Radiation physiology of fruit — ascorbic acid, sulfhydryl and soluble nitrogen content of irradiated citrus. Radiat. Bot. 3, 58.

Sizer, C.E., Balasubramaniam, V.M., 1999. New intervention processes for minimally processed juices. Food Technol. 53 (10), 64–67.

Smith, J.S., Pillai, S., 2004. Irradiation and food safety and scientific status summary. Food Technol. 58 (11), 48–55.

Song, H.P., Byun, M.W., Jo, C., Lee, C.H., Kim, K.S., Kim, D.H., 2005. Effects of gamma irradiation on the microbiological, nutritional, and sensory properties of fresh vegetable juice. Food Control 18 (1), 5–10.

Spoto, M.H.F., Domarco, R.E., Walder, J.M.M., Scarminio, I.S., Bruns, R.E., 1997. Sensory evaluation of orange juice concentrate as affected by irradiation and storage. J. Food Process. Preserv. 21 (3), 179–191.

Thakur, B.R., Arya Singh, S., 1993. Effect of sorbic acid on irradiation-induced sensory and chemical changes in sweetened orange juice and mango pulp. Int. J. Food Sci. Technol. 28 (4), 371–376.

Thayer, D.W., 1994. Wholesomeness of irradiated foods. Food Technol. 48 (5), 132–136.

Thayer, D.W., Boyd, G., 1995. Radiation sensitivity of Listeria monocytogenes on beef as affected by temperature. J. Food Sci. 60, 237–240.

World Health Organization (WHO), 1999. High-Dose Irradiation: Wholesomeness of Food Irradiated with Doses above 10 KGy, Joint FAO/IAEA/WHO Study Group on High-Dose Irradiation, Geneva, 15–20 September 1997, WHO Technical Report Series 890.

CHAPTER

8 Ultraviolet and Pulsed Light Processing of Fluid Foods

Vicente M. Gómez-López[1], Tatiana Koutchma[2] and Karl Linden[3]

[1]*Instituto de Ciencia y Tecnología de Alimentos, Facultad de Ciencias, Universidad Central de Venezuela*

[2]*Agriculture and Agri-Food Canada, Guelph, Ontario, Canada*

[3]*Civil, Environmental, and Architectural Engineering, University of Colorado at Boulder, Boulder, Colorado, USA*

8.1 INTRODUCTION

Ultraviolet (UV) light is the part of the electromagnetic spectrum with wavelengths between 100 and 400 nm. The application of UV light for treatment of liquids on a municipal scale began in Marseille, France, in 1906 for disinfection of drinking water (Masschelein and Rice, 2002). About 100 years later, potential use of UV light to achieve the goals of food safety and food quality in liquid foods has expanded to apple cider (Koutchma *et al.*, 2004), orange juice (Tran and Farid, 2004), liquid egg derivates (Unluturk *et al.*, 2008), milk (Krishnamurthy *et al.*, 2007), and honey (Hillegas and Demirci, 2003), with increasing applications.

UV light has broad antimicrobial action, providing effective inactivation of viruses (Eischeid *et al.*, 2009), vegetative bacteria, bacterial spores, yeasts, conidia (Gómez-López *et al.*, 2005) and parasites (Hijnen *et al.*, 2006). The discovery of UV inactivation of the chlorine-resistant parasites *Cryptosporidium parvum* and *Giardia* sp. has catalyzed the use of UV light in the water industry (Hijnen *et al.*, 2006). UV can be utilized similarly in the disinfection of surfaces, foods, and other liquids. The application of UV light can also improve toxicological safety of foods through its ability to reduce levels of toxins such as patulin mycotoxin in fresh apple cider (Dong *et al.*, 2010), and possibly by decreasing allergenicity of products such as liquid peanut butter (Chung *et al.*, 2008).

Novel Thermal and Non-Thermal Technologies for Fluid Foods. DOI: 10.1016/B978-0-12-381470-8.00008-6

UV-light treatment of foods is a non-thermal method for microbial inactivation that is free of chemicals and waste effluents, which makes it ecologically friendly. It does not produce by-products. Even though the term "irradiation" is frequently used for this treatment, UV light is a non-ionizing radiation and it must not be associated with gamma irradiation. Some authors prefer using the term illumination to avoid consumer confusion. It is safe to use, although precautions must be taken to avoid human exposure to UV light and to evacuate ozone generated by vacuum and far-UV wavelengths (Gómez-López *et al.*, 2007).

UV-light treatment of liquid foods can be performed by use of continuous or pulsed UV sources. Continuous UV treatments are performed by mercury lamps that continuously emit UV photons and are called continuous-wave UV (CW UV) lamps in both monochromatic and polychromatic modes. Mercury-based UV lamps have been in use for over 100 years. Pulsed-light (PL) treatment is a more modern technology, in which short-time pulses of a high-power polychromatic flash, rich in UV-C emitted by a xenon lamp, are rapidly and repeatedly released on the sample surface. Its main advantage over CW UV light is that processing time can be dramatically reduced and PL sources do not need to be warmed-up, while its main disadvantage is the possibility of sample heating (Gómez-López *et al.*, 2007).

Due to low penetration capacity of UV light into low UV transmittance (UVT) or almost opaque substances such as liquid foods, it was initially intended to be used only for disinfection of food surfaces or packaging. Its use in fluid foods was originally limited to highly transparent liquids; however, development of new reactors with optimized hydraulics and thin-film design now allows delivery and exposure of target microorganisms to UV light in low-UVT fluids. An absolute requisite for UV-based microbial inactivation is the effective exposure of the UV photons to the target microorganisms. In solid foods, microorganisms located below the food surface are shielded from photon contact because the food surface will absorb photons intended for the microorganisms. In fluid foods, microorganisms located far from the UV source may be protected from photon contact by the highly absorbing fluid layer present between them and the UV source. The level of this protection is a function of the fluid UVT and the thickness of the layer. However, in contrast with solid foods, UV reactors for fluid treatment can relocate microorganisms, by hydraulic and reactor design, to a place where the fluid layer is not thick enough to absorb all the UV photons and consequently enable inactivation. In this regard, UV light is more suitable for fluids than for solid food treatment.

This chapter will provide an overview of the fundamental principles of continuous and pulsed UV light applications in low-UVT liquid foods, available designs and operation of UV systems, process and control parameters, mechanisms and efficacy of UV light microbial inactivation, examples of applications, effects on fluid food nutritional and quality parameters, and challenges and industrial scale-up for both CW UV and PL.

8.2 PRINCIPLES OF OPERATION

UV light is the portion of the electromagnetic spectrum that covers a range from 100 to 400 nm. The UV light is traditionally further subdivided into the following categories: UV-A range from 315 to 400 nm and are responsible for changes in human skin or tanning; UV-B range from 280 to 315 nm and can cause skin burning and possibly lead to skin cancer; UV-C range from 200 to 280 nm and is considered the germicidal range since it effectively inactivates bacterial, viral, and protozoan microorganisms; and the vacuum UV range from 100 to 200 nm and can be absorbed by almost all substances, and thus can be transmitted only in a vacuum. The UV spectrum from sunlight on the surface of the earth contains only wavelengths greater than 300 nm, due to absorption of lower wavelengths by the ozone layer. Short UV-C is absorbed in air within a few hundred meters. When vacuum UV and UV-C photons are absorbed by oxygen atoms, the energy exchange causes the formation of ozone.

8.2.1 Mechanisms of Ultraviolet Light Generation

Atoms and ions emit photons of light when electrons make a transition from a higher energy state, $E2$, to a lower energy state, $E1$. Each photon carries an energy, E (J), described by Equation 8.1, first stated by Planck.

$$E = E2 - E1 = hc/\lambda \tag{8.1}$$

where h is Planck's constant (6.23×10^{-34} Js), c is the speed of light (2.998×10^8 m/s), and λ is the wavelength of radiation (m). Energy levels of a given atom or ion are unique arising from the number of electrons, protons, and neutrons within that atom or ion, and their interaction with external force fields. As such, each element emits a unique spectrum of light. For some elements, the difference between energy levels results in the emission of photons in the germicidal UV range. A transition from a lower- to a higher-energy state requires an energy input. This energy may be derived from the collision of the atom with a photon of light of wavelength, λ, or by collision with other atoms, ions, or electrons. Energy transferred to the atom may result in an increase in the atom's kinetic energy, the transfer of electron to a higher energy level, or the removal of an electron from the atom. Removal of an electron from the atom is termed ionization and results in a positively charged cation and a negatively charged free electron. Recombination of a free electron and a cation may result in the emission of light. Since the free electron and cation may have a range of kinetic energies, the wavelength of emitting light will vary over a continuum or range. This wavelength (Equation 8.2) will be bound at one end by the ionization energy, E_o, of the atom and peak at a wavelength dependent on the temperature of the electrons and cations.

$$\lambda_{max} = hc/E_o \tag{8.2}$$

The photon may be viewed as the smallest discrete unit of radiation energy. For expressing the energy of a single photon, the unit of the electronvolt (eV) is used since the unit of the Joule (J) is rather large. One eV is defined as the energy gained by an electron in passing through a potential difference of 1 V which equals 1.6×10^{-19} J. For photochemical purposes, the photon energies are often expressed in kilocalories per Einstein. Note that 1 Einstein is defined as being equal to a mole of photons ($= 6.02 \times 10^{23}$ photons). Absorption of 1 Einstein can excite 1 mole of the absorbing substance (Equation 8.3).

$$E_E = h\frac{c}{\lambda}A \tag{8.3}$$

where E_E is energy per Einstein, A is Avogadro's number.

Radiation of the UV light and the adjacent visible spectral range as well as other less energetic types are summarily called non-ionizing radiation, as opposed to ionizing radiation. The latter is represented in the electromagnetic spectrum by X-rays and gamma-rays. Other kinds of ionizing radiation consist of ionizing particles (beta-rays, alpha-rays, protons). Ionizing in contrast to non-ionizing radiation is capable of ionizing many atoms and molecules. Absorption of non-ionizing radiation such as UV, however, leads solely to electronic excitation of atoms and molecules.

8.2.2 Gas Discharge

Gas discharges are responsible for the light emitted from UV lamps. A gas discharge is a mixture of non-excited atoms, excited atoms, cations, and free electrons formed when a sufficiently high voltage is applied across a volume of gas. The voltage required to start the gas discharge is typically higher than the ionization potential of the gas unless means are used to introduce electrons. Light is emitted from the gas discharge at wavelengths dependent upon the elemental composition of the gas discharge, and the excitation, ionization, and kinetic energy of those elements. When a voltage is first applied, free electrons and ions present in the gas are accelerated by the electric field formed between two electrodes. With sufficient voltage, the electrons are accelerated to high kinetic energies. Collisions of the free electrons with atoms result in a transfer of energy to the atoms and if the energy is sufficient, the atoms are ionized. This ionization results in a rapid increase in the number of electrons and cations, with a corresponding increase in lamp current, and a drop in the voltage across the lamp. Cations colliding with the electrodes cause electrons to be emitted. If sufficient electrons are emitted, a self-sustaining discharge occurs, termed a glow discharge. With an increase in current, the larger fraction of each electrode will emit electrons until the whole electrode is in use. To increase the current beyond that point, the voltage is increased to provide more kinetic energy to the cations. High-energy cations colliding with the electrode increase the electrode's temperature. At high-enough temperatures, the electrode begins to emit electrons, and a further increase in current reduces the voltage requirement. At this point the process is termed an arc discharge. Because voltage across the gas discharge is inversely related to current, the gas discharge has negative impedance and is intrinsically unstable. Therefore, ballast is placed in series with the gas discharge to provide positive impedance to the power supply.

The excitation energy provided by UV-photons is much higher than the energy of thermal motions of the molecules at physiological temperatures. The latter is of the order of Boltzman's constant times the absolute temperature which at 27°C amounts to 0.026 eV/molecule (0.60 kcal/mole) in contrast to the 3.3–6.5 eV/molecule (75 to 150 kcal/mole) available from UV absorption.

8.2.3 Pulsed Power Energization

PL is produced using a pulsed-power energization technique that multiplies the power many-fold, of the order of megawatts. Power is magnified by storing electricity in a capacitor over relatively long times (less than a second) and releasing it in a short time (millionths or thousandths of a second) through a high-speed switch. Electricity is discharged in a Xenon lamp that emits a high peak power flash consisting of wavelengths from 200 to 1100 nm (including UV), which can have a high repetition rate ($>$1 Hz).

8.2.4 Design Parameters

UV light emitted from the atoms and ions within the gas discharge of a UV lamp will propagate away from those atoms and ions. As UV light propagates, it interacts with the materials it encounters through absorption, reflection, refraction, and scattering. In a UV reactor, UV light emitted will interact with components of the UV reactor such as the lamp, the lamp sleeve, the reactor walls, as well as the liquid substance being treated. Each of these phenomena influences the intensity and wavelength of the UV light reaching the bacteria or chemical compound in the liquid.

Absorption of light is the transformation of energy of light photons to other forms of energy as it travels through a substance. *Reflection* is the change in the direction of propagation experienced by light deflected by an interface. *Scattering* is the phenomenon that includes any process that deflects electromagnetic radiation from a straight path through an absorber when photons interact with a particle. UV light scattered from particles is still capable of killing microbes. Much of the scattered light is in the forward direction and is a significant portion of the transmitted UV light. The scattering phenomenon plays an important role in disinfecting food liquids containing particles.

The UV absorbance of the liquid foods strongly influences UV-dose delivery. The Beer–Lambert law (Equation 8.4) is the linear relationship between absorbance (A), concentration of an absorber of electromagnetic radiation (*c*, mol/L) and extinction coefficient (ε, (L/mol)/cm) or molar absorptivity of the absorbing species, which is a measure of the amount of light absorbed per unit concentration absorbance or optical density, and path length of light (d, cm).

$$A = \varepsilon \times c \times d \tag{8.4}$$

If multiple species that absorb light are present in a liquid sample, the total absorbance at a given wavelength is the sum due to all absorbers (Equation 8.5):

$$A = (\varepsilon_1 \times c_1 \times d) + (\varepsilon_2 \times c_2 \times d) + ... \tag{8.5}$$

where the subscripts refer to the molar absorptivity and concentration of the different absorbing species that are present in the liquid. Molar absorptivity of apple juice, orange juice, and multifruit juice was determined at 253.7 nm and reported by Oteiza *et al.* (2005). The following absorptivities were reported: 0.0715, 0.3528, 0.7233 (L/mol)/cm for apple, orange, and multifruit juice, respectively.

The absorption coefficient (α) base e (α_e), called Naperian absorption coefficient, or base 10 (α_{10}), called the logarithmic coefficient, is also used in the calculations and is defined as the absorbance divided by the path length (m^{-1}) or (cm^{-1}). The absorption coefficient is a function of wavelength.

$$\alpha_e = 2.303A/d \tag{8.6}$$

Penetration depth (d_p) is the depth (cm) where the initial flux I_o drops by a specified percentage of its value at the quartz sleeve, for example, 95% or 99%. The penetration depth is defined by Equation (8.7)

$$d_p = 1/\alpha_e \tag{8.7}$$

Experimental measurements are usually made in terms of *transmittance* of a substance (*T*), which is defined as the ratio of the transmitted to the incident light irradiance (Equation 8.8). In practice, because of reflections at the quartz/air interfaces, I_o is the spectrophotometer reading with pure solvent in the cell and I_1 is the reading with the solution of interest in the cell (Koutchma, 2009).

$$T = I_1/I_o \tag{8.8}$$

As opposed to absorption coefficient that is a characteristic of the material only, the transmittance depends on thickness. The relationship between A and T is expressed by Equation 8.9.

$$A = -\log(T) = -\log(I_1/I_o) \tag{8.9}$$

From the Beer–Lambert Law, the per cent transmittance in a 10-mm path length ($\%T_{10}$) is related to absorption coefficient (α_{10}) by Equation 8.10.

$$\%T_{10} = 100 * 10^{-\alpha_{10}d} \tag{8.10}$$

where $d = 1$ cm and (α_{10}) is in cm^{-1}.

A convenient way of presenting information about UV transmittance of materials is to give the values of their absorption coefficient at various wavelengths, over a given depth (e.g., 1 cm). Knowing this, the transmittance for any particular depth and the depth of the liquid which will absorb 90% of the energy at 253.7 nm can be calculated from Equation 8.10.

The measurement techniques for low-UVT fluids include direct spectrophotometric measurement of absorbance at 253.7 nm using microcuvettes with a pathlength ranging from 0.05 to 1.0 mm and an integrating sphere. The absorption coefficient can be determined as a slope of the linear plot of the absorbance *vs* light path length (Equation 8.6). Oteiza *et al.* (2005) used dilutions of fruit juices in sterile water first and then the measurements were made using standard 1-cm light-path cuvettes. The molar absorptivity coefficient was obtained as a regression curve slope by plotting absorbance *vs* sample concentration. For the samples with suspended solids, the direct measurement dilution technique is controversial since suspended particles can scatter light out of the detector resulting in a high value of measured absorbance even though the light was not absorbed by the sample. Integrating sphere spectroscopy should be used to measure the absorbance of low-UVT fluids to compensate for this error.

It should be pointed out that UV absorbance and transmittance at 254 nm are important design parameters for UV reactors containing low-pressure mercury (LPM) lamps or low-pressure, high-output mercury lamps. However, for broadband UV lamps, such as medium pressure mercury (MPM) or PL lamps, it is important to measure the full scan of absorbance or transmittance in the germicidal region from 200 to 400 nm (Fig. 8.1). In addition, the molar absorption coefficient and its spectra should been collected for a variety of compounds. The examples of absorbance scans of four selected food fluids (milk, liquid egg, liquid protein, and apple juice) measured using integrated sphere and 1-mm path cuvettes are shown in Figure 8.2.

Other important terms to characterize UV light treatments in food fluid processing are fluence rate, and fluence. Fluence rate is "the total radiant power incident from all directions on to an infinitesimally small sphere of cross-sectional area dA, divided by dA", and it should be expressed in W/m^2, however mW/cm^2 is common in the literature (Bolton and Linden, 2003). The fluence rate (E) is affected when it passes through a substance according to the following equation.

$$E_1/E_0 = 10^{-\sum \varepsilon c d} = 10^{-\alpha_{10}d} = 10^{-A} = e^{-\alpha_e d} \tag{8.11}$$

where E_1 and E_0 and are the UV light irradiances (mW/cm^2) incident on the substance and transmitted through a length d (cm), respectively. As UV absorbance

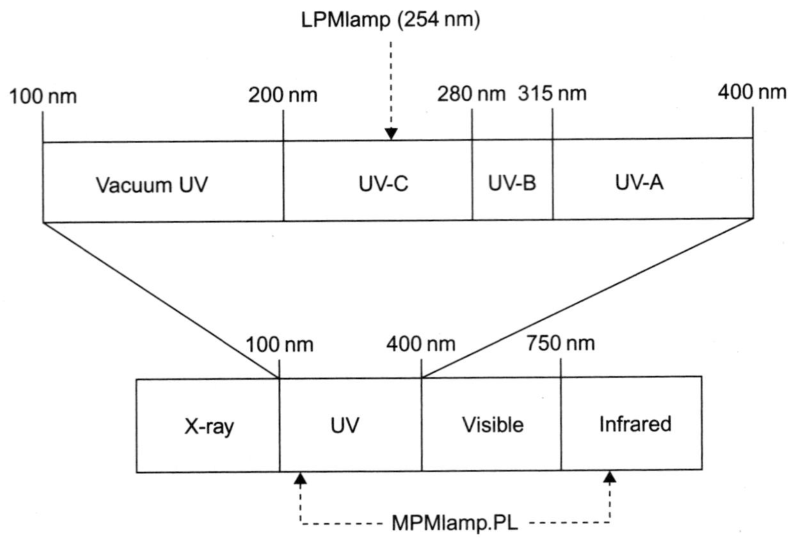

FIGURE 8.1

Position of the emission spectra or emission line in the electromagnetic spectrum of different UV lamps.

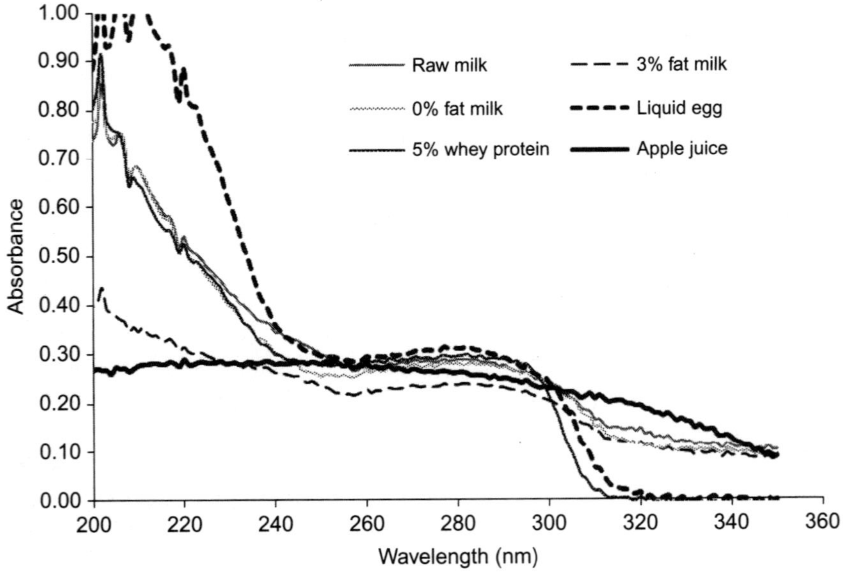

FIGURE 8.2

Absorbance scans of milk, liquid egg, liquid protein, and apple juice measured in 1-mm cuvettes using an integrated sphere attached to the spectrophotometer.

increases, the intensity throughout the product in the reactor decreases and results in a reduction in UV-dose delivery.

Fluence is defined as the fluence rate multiplied by the exposure time. Fluence is measured in J/m^2, although J/cm^2 or mJ/cm^2 can be also found in the literature. The term UV dose should be avoided as a synonym of fluence because dose refers in other contexts to absorbed energy, but only a small fraction of all incident UV light is absorbed by microorganisms (Bolton and Linden, 2003). In the case of PL, fluence is determined as energy per pulse multiplied by the number of pulses. If the broad side area of the solution, S, is exposed to the fluence H_o, which is the product of fluence rate and the time, $H_o = E_o t$, Equation 8.12 can be used to calculate absorbed fluence:

$$H_{abs} = (E_o - E)tS = E_o t(1 - e^{-A\ln 10})S = H_o(1 - e^{-A\ln 10})S \qquad (8.12)$$

The absorbed fluence indicates that radiant energy is available for driving the solution reaction. However, when UV light is absorbed by the solution, it is no longer available for inactivating the microorganisms. The remaining interactions including reflection, refraction, and scattering, change the direction of UV light but the light is still available for inactivation. The radiant energy delivered to the molecule or microorganism is called the effective or delivered germicidal UV dose. Microbial inactivation depends primarily on the effective dose.

8.3 PROCESS CONTROL PARAMETERS

8.3.1 Fluid Food Parameters

Three categories of fluid food products that are available commercially and can be treated by using UV light include liquids, emulsions, and liquids with particles or suspensions. The definitions and basic characteristics of each category of fluids are summarized in Table 8.1.

Fluid foods are characterized by a diverse range of chemical, physical, and optical properties. Optical properties are the major factors impacting UV light transmission and, consequently, microbial inactivation. The fluid can be transparent if $10\% < T < 100\%$, opaque if $T \sim 0\%$ or semitransparent if $0 < T < 10\%$. In a majority of cases, fluid foods will absorb UV radiation. For example, juices can be considered semi-transparent if they are clarified or opaque liquids if the juice contains suspended solids. Fluid chemical composition and concentration of dissolved and suspended solids determines whether the product is transparent, opaque, or semi-transparent. Chemical composition, pH, dissolved solids (°Brix), and water activity are also considered as hurdles that can significantly modify efficacy of UV inactivation.

The large variety of fluid foods represents different pH groups, °Brix levels, and varying viscosities. For instance, for fresh juices, apple and orange juices

Table 8.1 Categories and Definitions of Groups of Fluid Foods and Beverages Suitable for Treatment by Continuous and Pulsed Ultraviolet Light Technologies

Group of Fluid Foods	Definition and Characteristics	Key Properties			Examples
		pH	UV Transparency	Viscosity	
Liquids	Homogenous "pumpable" fluids with no particles or solids	High Acid $pH < 3.5$ Acid $3.5 < pH < 4.6$ Low acid $pH > 4.6$	Semi-transparent Semi-transparent Semi-transparent Nearly transparent	Newtonian Newtonian Newtonian Non-Newtonian	Fruit juices : Apple Grape Watermelon Iced tea Liquid sweeteners: sucrose, fructose, glucose
Liquid emulsions	Mixture of two unblendable liquids where one liquid is dispersed in other (continuous) phase	Low acid $pH > 4.6$	Opaque	Newtonian	Milk Liquid eggs Liquid white Whey protein
Liquid-particles	Heterogeneous liquids with suspended solids	High acid $pH < 3.5$	Opaque with particles	Non-Newtonian	Fruit and vegetable juices: Orange
	Density differences between particles and fluid are small, such that minimal settling occurs	Acid $3.5 < pH < 4.6$		Non-Newtonian	Pineapple
				Newtonian	Guava
	Particles should not be identified by individual size, but rather by dimensions and properties	Low acid $pH > 4.6$		Non-newtonian	Tomato Carrot

belong to the high-acid foods group (pH < 3.5). Pineapple juice (pH 3.96) is within the group of acid or medium acid foods (3.5 < pH < 4.6). Inactivation of vegetative pathogenic microorganisms of primary public health significance such as *Escherichia coli* O157:H7, *Salmonella* or *Listeria monocytogenes* and spoilage microorganisms in high-acid and medium-acid foods is a requirement for UV pasteurization. However, carrot juice (pH 5.75), watermelon (pH 5.19) and guava nectar (pH 6.32) are in the group of low-acid foods (pH > 4.5). Spore inactivation (*Clostridium botulinum*) is important for the safety of low-acid juices. Milk (pH 6.7) and liquid egg products (pH > 8) also belong to the group of low-acid foods. A 5-log $_{10}$ reduction of *Mycobacterium tuberculosis and Coxiella burnetii* and a negative alkaline phosphatase test are requirements for milk pasteurization. *Salmonella enteritidis* is a primary pathogen of concern for pasteurization of liquid egg products.

Physical properties such as viscosity and density influence the effectiveness of fluid momentum transfer and flow pattern in the reactor. In terms of viscosity, milk, apple, grape, guava, and watermelon juice represent a group of less viscous, Newtonian-fluid products, whereas carrot, orange, and pineapple juices, and liquid sweeteners are characterized by higher viscosity and Non-Newtonian behavior.

The examples of key product parameters for selected fluid foods and beverages are given in Table 8.2. As was noted above, the absorbance of fluid foods and ingredients, juices and beverages varies significantly from semi-transparent to opaque substances. For instance, liquid sucrose represents a nearly transparent case characterized by the absorption coefficient α_{10} of 0.45 cm $^{-1}$. Fructose solution has a higher UV absorbance (3–5 cm $^{-1}$) than sucrose solutions. Clarified

Table 8.2 pH, Viscosity, Absorption Coefficients and UVT at 253.7 nm of Water, Selected Juices, and Fluid Foods

Substance	pH	Viscosity, cP	Absorption Coefficient (cm $^{-1}$)	UVT, 1 cm (%)	UVT, 0.1 cm (%)	Penetration for 90% absorption (cm)
Water	7	1	0.01	97.72	99.77	100
Liquid sucrose	6–6.5	60–100	0.45	35	90	2.2
Liquid fructose	3.5–4.0	300–700	3.0–5.0	0.01	35	0.22
Clear apple juice	3.4–3.6	1.5	7.1 –39	~0	3.16	0.067
Apple cider	3.5–3.7	1.9	~40	~0	0.01	0.025
Orange juice	3.2	2–3	>60	~0	0.00	0.010
Egg white	>8	1.5–5	104	~0	~0	0.001
Raw milk	6.7	1.8–2.1	290	~0	0.00	0.003

apple juices are the examples of semi-transparent liquids. The α_{10} of four commercial brands of apple juice varied from 39.1 cm^{-1} to 7.1 cm^{-1} and a correlation with vitamin C contents was observed. In general, the larger values of vitamin C content resulted in the larger α_{10} of juice. The apple juice that had the highest α_{10} of 39.1 cm^{-1} was enriched with vitamin C at concentration of 0.3 mg/ml. Whereas the least absorptive apple juice with α_{10} of 7.1 cm^{-1} was not enriched with vitamin C and characterized by the lowest magnitude of color characteristics such as L (lightness), a (yellowness), and b (greenness). Orange juice with $\alpha_{10} > 60$ cm^{-1} is almost opaque. The presence of suspended solids leads to the apparent increase of α_{10} and needs to be considered in the UV-dose calculations. In addition, the observed differences show the importance of full characterization of liquid products to design a proper UV-light-based preservation process. These data also are necessary to determine optimal design of the UV reactor using mathematical modeling tools.

8.3.2 Ultraviolet Light Delivery

The performance of UV-light-processing systems is dependent upon dose delivery and assessed by dose-delivery monitoring, both of which are affected by system design and the liquid foods treated. System design affects transport phenomena relevant to treating the liquid according to its physical characteristics of viscosity, density, and irradiation transfer. Traditionally commercial UV systems are flow-through systems which have a distribution of residence time or exposure time (RTD) and light irradiance distribution (LID) due to attenuation following the Beer–Lambert Law. Consequently, there is a distribution of UV fluence levels (product of LID and RTD) that any population of microorganisms or chemical compound is exposed to. The dose received by any individual microorganism may be calculated as the integral of the irradiance history throughout its residence time within the irradiated zone (Equation 8.13)

$$Dose_i = \int_0^{T_i} I(t)dt \tag{8.13}$$

where $Dose_i$ is the radiation dose received by the ith particle (mJ/cm^2), $I\ (t)$ is the time-dependent radiation irradiance (mW/cm^2) and τ_i is the period of exposure for the ith particle.

Therefore, estimation of the UV fluence received by an individual microorganism requires knowledge of its trajectory through the irradiated zone using fluid dynamics. Understanding of UVT in the carrier fluid with low UVT in relation to inactivation of pathogenic microorganisms is needed in order to establish inactivation efficiency.

The magnitude of the level of irradiance I_r from the lamp surface depends on transmission (or absorbance) of UV light through the liquid. The Beer–Lambert law in the form of Equation 8.14 serves as the main mathematical basis for computation of the UV fluence rate distribution in a liquid medium.

$$I_r = I_o \frac{r_o}{r} exp[-\alpha_\lambda(r - r_o)] \tag{8.14}$$

where I_r is UV fluence rate at a distance r from the center (W/cm^2), I_o is UV fluence rate at the surface of the UV source (W/cm^2), r is radial distance between the center of the source to the point at which irradiance is measured, r_o is the radius of the UV lamp (cm) and α_λ is the absorption coefficient of the medium (cm^{-1}).

8.3.3 Fluids Flow Dynamics

UV fluence rate in low-UVT liquids is strongly related to the distance outward from the UV source as opposed to very-high-UVT liquids where the intensity of light striking microorganisms is less dependent on their position from the lamp. This signifies that accumulated or absorbed UV dose in a low-UVT liquid foods will be primarily dependent on the radial position of the liquid from the lamp and the time during which the liquid element resides at this position in the system. The flow pattern inside the UV system strongly influences the summed dose since the position and the residence time of the microorganisms in certain regions of the irradiation field can vary significantly. Another reason for establishing flow characteristics is to obtain an indication of the mixing behavior of the fluid and how it can effect inactivation.

Flow dynamics should be evaluated for all processed liquid products. Traditionally, velocity (v) is calculated as $v = Q/A$, where Q is a volumetric flow rate and A is the cross-sectional area of the tube. Reynolds numbers (Re) are then calculated as ($Re = vd\rho/\mu$), where d is characteristic dimension, ρ is density of fluid, and μ is dynamic viscosity. The Reynolds number is calculated for the range of flow rates and for each product if viscosity differs. For instance, when testing a UV coiled system for treatment of tropical fresh juices, Koutchma *et al.* (2007) reported the *Re* numbers that were calculated for all types of juices with variation of viscosities in a wide range. It was found that the magnitude of *Re* for pineapple and orange juice was less than 2,000 at the flow range from 3.5 to 7 gpm indicating that the hydraulic regime was laminar. However, for juices characterized as Newtonian liquids (guava, apple and watermelon), the flow dynamics were characterized as a turbulent flow with $Re > 2000$.

Flow simulations using a computational fluid dynamics (CFD) software program are a powerful tool recommended for the evaluation and optimization of flow regimes.

8.3.4 Utraviolet Light Sources

The UV light transfer phenomenon in liquid foods is defined by the emission characteristics of the UV source, considering long-term lamp aging and absorbance/scattering of the liquid. Consequently, the performance of the UV system and the success of UV technology depends on the correct matching of the UV source parameters to the optical behavior of a specific liquid and on the demands of the UV application. The available UV sources include LPM, MPM, excimer, and pulsed lamps. The LPM and excimer lamps are monochromatic sources whereas emission of MPM and PL is polychromatic.

Mercury-vapor lamps are highly developed and provide good efficiency, long life, and compact size for various applications. LPM lamps are easy to install and operate. They are readily available with their emission spectrum well established (single UV-C output at 253.7 nm) and quantified, at a comparatively low cost. MPM lamps have higher emission intensities in the UV-C range, however the source is polychromatic. The lamp source operates at high temperatures and at higher electrical potential. Aging of lamp material and enclosures of MPM lamps is faster than LPM. These two mercury lamps are the dominant sources for UV disinfection treatment. The efficacy of PL lamps is potentially greater due to high intensity and broader spectrum that may penetrate low UVT and opaque liquids but it has been scarcely evaluated for liquid food applications. PL lamp technologies are promising due to characteristics such as instant start, high intensity, and robust packaging with no mercury in the lamp, but more research is needed to establish them for food applications. Figure 8.3 shows the normalized spectra of

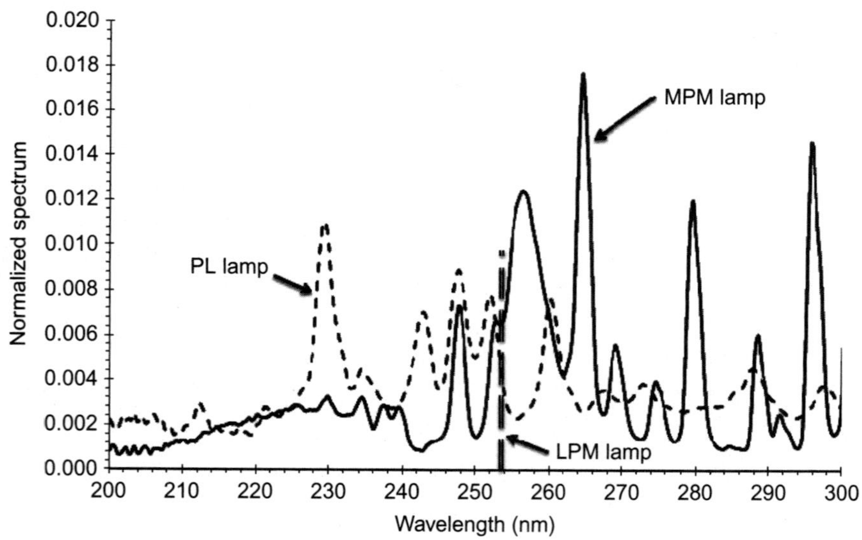

FIGURE 8.3

Emission spectra of common types of UV lamps including LPM vapor, MPM vapor, and surface discharge pulsed UV lamps (PL). MPM and PL spectra are presented as normalized to the total 200–300 nm, and LPM is monochromatic output at 253.7 nm, shown for comparison.

these three UV sources. Individual spectra are not comparable on a UV intensity basis but are comparable on a spectral basis regarding which wavelengths dominate the respective wavelength outputs. Table 8.3 provides a summary of some of the basic characteristics of common UV sources in commercial use and under development.

From this summary it is evident that no single lamp technology will represent the best source for all applications. Situation-specific requirements may dictate a clear advantage for a given process technology.

8.4 MICROBIAL INACTIVATION MECHANISM

The microbial inactivation mechanism by CW UV has been extensively studied through the years, while that of PL is still being reported in the literature. In both cases, DNA damage has been identified as the primary lethal event, but inactivation by PL has additional features that may explain its higher efficiency on a per fluence basis. For this reason, the mechanism of DNA damage by UV will be first discussed for both methods, and then special characteristics of the mechanism by PL will be considered. Virus inactivation is also important in fluid foods decontamination, and will also be treated in this section.

8.4.1 Photochemical Effect

The main lethal event caused by UV-C light on microorganisms is the formation of cyclobutane thymine dimmers in DNA, mainly thymine dimmers. When DNA is exposed to UV light, the electrons within specific bases become energized, which leads to formation of covalent links between adjacent bases The structural damage caused by the formation of these dimmers inhibits the formation of new DNA, resulting in the inactivation of the affected microorganism (Harm, 1980). As for PL, this reaction has been called the "photochemical effect". In order for a photochemical reaction to proceed, photons must be absorbed by the molecules and they must have enough energy to promote a reaction. LPM lamps have an almost monochromatic output at 253.7 nm, a wavelength which is close to the maximum absorbance of DNA, 260 nm; which explains the sensitivity of this macromolecule to exposure to the emission of that kind of lamp. The polychromatic emission spectra of medium-pressure mercury lamps and Xenon lamps include also other wavelengths that are absorbed by DNA. It can be observed in Figure 8.4 that the effectiveness of different wavelengths in causing the inactivation of microorganisms is microbe-specific, considering here *Bacillus subtilis* spores and *MS2* coliphage, compared to the absorbance spectrum of DNA. Similar data for *E. coli* is presented in Figure 8.5.

Some microorganisms have the capability to repair damage caused by UV light via two mechanisms. One mechanism needs light to work and it is called photoreactivation, the other does not need it and it is called dark repair.

Table 8.3 Summary of Ultraviolet Sources and their Basic Characteristics

UV radiation source	Electrical efficiency (%)	UV efficiency (%)	UV intensity (W/cm^2)	Lamp surface temperature	Lifetime (month)	Output spectrum
LPM	50	38	0.01	40	18–24	Monochromatic 253.7 nm
MPM	15–30	12	600	400–1000	0.5	Polychromatic (200–300 nm)
Pulsed	15–20	17	30,000	—	1	Polychromatic
Xenon and Surface Discharge*					NA	Polychromatic
Excimer	10–35	10–40	—	Ambient	>6	Monochromatic tunable
Sparker*		7	100,000		NA	Polychromatic

**Patented by and being commercialized by Phoenix Science &Technology, Inc. (MA, USA).*

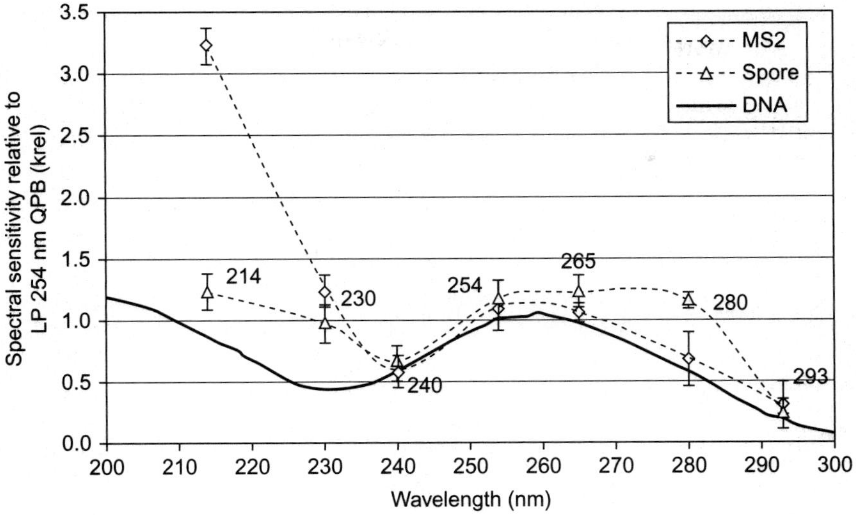

FIGURE 8.4

Spectral sensitivity of *Bacillus subtilis* spores and MS2 phage, relative to LPM lamp output of 253.7 nm. DNA absorbance is shown for reference.

(Reprinted with permission from Mamane-Gravetz et al., *2005)*

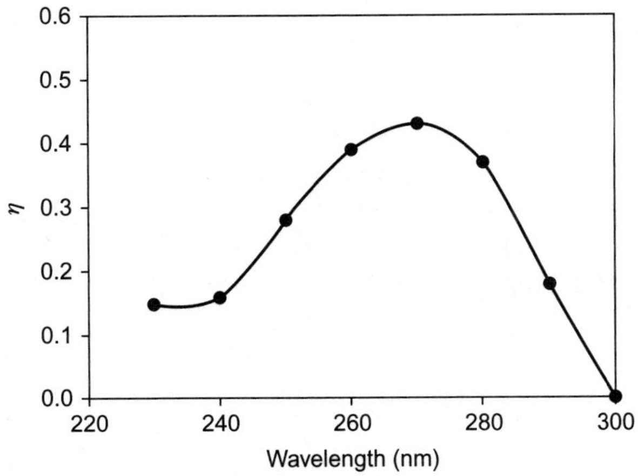

FIGURE 8.5

Germicidal efficiency η[$\log_{10}(N/N_0)$ per mJ/cm^2) for *E. coli* as a function of wavelength.

(Reprinted from Wang et al., *2005)*

Photoreactivation is a mechanism that cleaves the pyrimidine dimmer formed from UV exposure, restoring the individual pyrimidine components, by the enzyme photolyase, which harnesses blue or near-UV light energy to catalyze this reaction (Cleaver, 2003; Kao *et al.*, 2005). It has important implications for microbial inactivation efficacy, since some microorganisms in water and foods inactivated by UV light can become reactivated if exposed to light during storage. Dark repair does not need light; it is a process of nucleotide excision that involves several enzymes (Zimmer *et al.*, 2003), but is less efficient than photoreactivation and does not always lead to complete repair.

In many scientific disciplines, initial beliefs about novel technologies are confirmed, improved or discarded by new experiments that provide better insights. This is also true with respect to PL. Considerable work has been done to establish differences between inactivation mechanisms corresponding to different light sources. Refining dosimetry methods has allowed more trustworthy results, given the difficulties in comparing fluence supplied by continuous monochromatic and polychromatic lamps with pulsed-based polychromatic lamps.

PL has been shown to be more effective than continuous UV light in inactivating microorganisms in some cases. It can be twice as effective in inactivating bacteria and phages; and induces the formation of more thymine dimmers than LPM and MPM UV lamps per UV fluence (Bohrerova *et al.*, 2008).

The effect of PL is mainly, but not only, due to the UV-C portion of the spectrum, and differs between bacteria and viruses. This has been demonstrated by experiments on the effect of LPM and MPM lamps and PL on *E. coli* and phages T4 and T7 using light filters to discriminate effects of different wavelengths, as shown in Figure 8.6 (Bohrerova *et al.*, 2008). For *E. coli*, the inactivation caused by the UV-C portion of the PL spectrum is higher than that from MPM lamps, and wavelengths >295 nm are related to this higher germicidal efficiency. For the viruses (represented by phage), the inactivation due to wavelengths >295 nm fully explains the higher efficiency of PL, including effects from the visible light portion. The high intensity of wavelengths >400 nm might be responsible for some additional physical damage such as phage capsid rupture (Bohrerova *et al.*, 2008).

The higher efficiency of PL *vs* CW-UV seems to be unrelated to the high peak power of PL in the case of spores. Peak power is the pulse energy divided by the pulse width. An experiment on the inactivation of *B. subtilis* spores by two light sources, one a laser emitting at 248 nm and the other a low-power continuous lamp (254 nm) differing by a factor of 10^8 in peak power, showed no clear differences in spore survival (Rice and Ewell, 2001).

Early information stated that the damage inflicted by PL on microorganisms is so large that photoreactivation is impossible. Scientific evidence has, however, shown that photoreactivation occurs in flashed cells, which has been proved for *L. monocytogenes* seeded in agar (Gómez-López *et al.*, 2005) and for *S. enteriditis* on shell eggs (Hierro *et al.*, 2009). Nevertheless, the capacity of cells for photoreactivation after PL is weaker in comparison with that of CW UV light, which might be

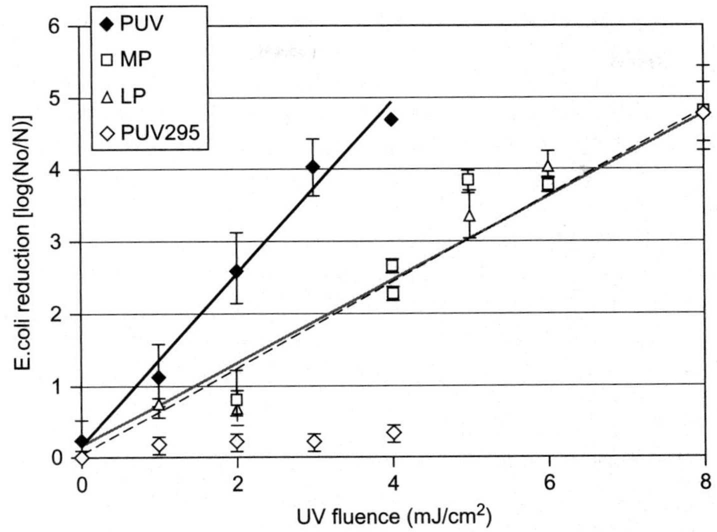

FIGURE 8.6

Escherichia coli inactivation using pulsed UV light (PUV), MPM lamp (MP), LPM lamp (LP) and 295-nm filtered PL (PUV295).

(Reprinted from Bohrerova et al.*, 2008)*

related to an effect of shorter wavelengths damaging the photoreactivation enzymes. Photoreactivation rates of *E. coli* cells were lower in flashed cells, although the maximal survival ratios were similar (Otaki *et al.*, 2003).

In order for a disinfection treatment to have long-term efficacy in the industry, microorganisms should not become resistant to it; otherwise resistant house microflora can develop. It has been shown that *L. monocytogenes* did not develop resistance to PL after successive cycles of treatment and survival culturing (Gómez-López *et al.*, 2005), however, information on other microorganisms of concern is not currently available.

8.4.2 Other Microbial Inactivation Mechanisms Specific to Pulsed Light

8.4.2.1 Photothermal Effect

The term "photothermal effect" has been used with two meanings, both having in common microbial inactivation by heat. One is when a large dosage of flash discharge absorbed by microorganisms heats them instantaneously, providing a sort of thermal sterilization, as proposed by Hiramoto (1984) to explain the inactivation of *Aspergillus niger* by PL. The other is heating a superficial layer of food product through absorption of light to a temperature effective to inactivate microorganisms. Heat may be localized at a very superficial layer and be quickly dissipated without raising the interior temperature of the food (Dunn *et al.*, 1989). In the first case, PL may still be considered a non-thermal method because heating is localized in the microorganism, while in the second case PL will be a kind of thermal treatment but without significantly heating the product overall.

The contribution of the photothermal effect may be considered minimal in liquid foods, because microorganisms can easily absorb PL energy in clear food materials, so that the treatment time will be short, and consequently the temperature increase should be minimal (Krishnamurthy *et al.*, 2009).

8.4.2.2 Photophysical Effect

A third inactivation mechanism has recently been identified by Krishnamurthy *et al.* (2009), which refers to the structural damage in microbial cells attributed to the constant disturbance caused by the high-energy pulses.

Structural damage caused by PL was identified in *Saccharomyces cerevisiae* cells, which was higher than that caused by a monochromatic UV lamp. This damage consisted of a loss of cell membrane permeability detected by a higher concentration of eluted protein and microscopic observation of raised and expanded vacuoles, cell membrane distortion, and change of cell shape, as seen in Figure 8.7 (Takeshita *et al.*, 2003). It has been observed in *Staphylococcus aureus* cells that damage due to PL consisted of cell leakage, lack of cell wall, cytoplasmic membrane shrinkage and collapse of internal structures (Krishnamurthy *et al.*, 2009).

8.4.3 Kinetics of Microbial Inactivation

Non linear models are currently applied to describe the kinetics of microbial inactivation by CW UV and PL, although some researchers have found that it could be modeled by a simple first-order approach. Inactivation curves should be reported in log counts or log reduction, as a function of UV fluence. Reporting results as a function of treatment time gives information on the length of the treatment but it does not allow comparisons between different reactors and difficulties scaling-up.

Inactivation curves in UV treatments can be composed of three phases: a shoulder, a log-linear portion typical of many inactivation curves, and a tail. The shoulder is an initial lag in the slope. The cell injury starts but no death occurs so that counts remain constant until threshold fluence is reached where the damage is enough such that inactivation begins. After the linear portion of the curve ends and the maximum amount of inactivation has occurred, a relatively small number of microorganisms remain alive. Tailing can result from suspended material that shields microorganisms from UV, cell aggregates where external cells are inactivated and shield

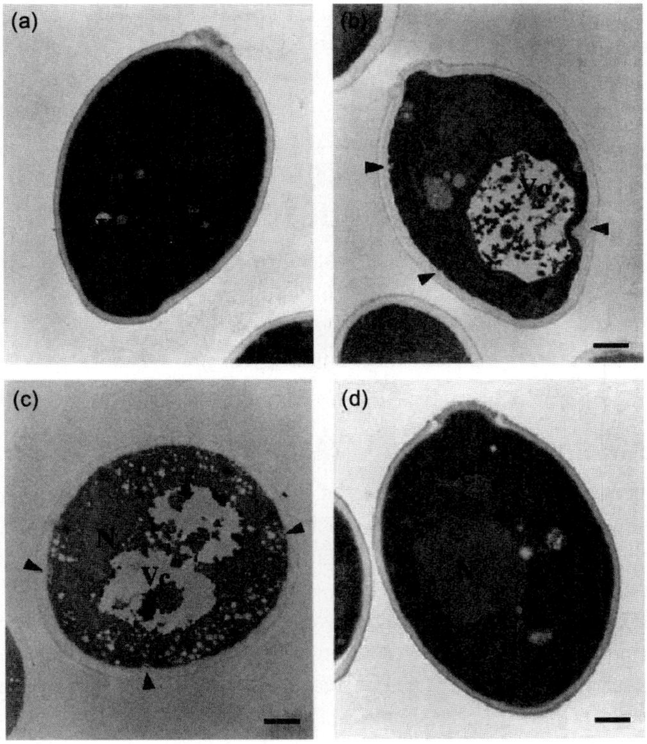

FIGURE 8.7

Transmission electron micrograph of *Saccharomyces cerevisiae*. (a) Untreated, (b) treated at 1.4 J/cm^2 and 2.1 J/cm^2, (c, d) treated with CW UV light (3 s at 60 mW/cm^2/s). Bar corresponds to 0.5 μm. N, nucleus; Vc, vacuole. Arrowheads indicate damaged/broken membranes. Magnification ×40,000.

(Reprinted from Takeshita et al.*, 2003)*

internal cells, non-homogeneity in illumination (Unluturk *et al.*, 2008), heterogeneity in UV resistance of microorganisms, varying abilities of cells to repair DNA damage, or multi-hit phenomena (Yaun *et al.*, 2003). Another point of view states that the probability of different targets being reached by photons is reduced with decreasing population density (McDonald *et al.*, 2000). Sometimes curves that do not show tailing can result from experiments with counting procedures of low sensitivity, where survivors from high fluences are simply not detectable.

On PL specifically, Sauer and Moraru (2009) found a very good fit of data to the Weibull model for the inactivation of *E. coli* O157:H7 and a non-pathogenic

E. coli strain inoculated in apple juice and apple cider, where tailing but not a shoulder could be observed.

8.5 ANTIMICROBIAL EFFECTS OF ULTRAVIOLET AND PULSED LIGHT IN SPECIFIC PRODUCTS

The results of several reports on inactivation of pathogenic microorganisms using CW UV light sources are summarized in Table 8.4. It must be stated that these data were obtained using different UV systems and approaches to determine UV fluence, and so reported results are not directly comparable.

8.5.1 Ultraviolet Inactivation of Microorganisms in Fresh Juice

The results of reported studies in terms of inactivation of non-pathogenic microorganisms in fresh juices are summarized in Table 8.5. Variations in UV fluence levels can be accounted for due to limitations in dosimetry and fluid-absorbance measurements.

Guerrero-Beltrán and Barbosa-Cánovas (2005) studied UV inactivation of *S. cerevisiae* in apple juice that was recirculated in two annular reactors with single LPM lamps in laminar flow. The reduction of 1.34-log after 30 min of UV treatment was reported for *S. cerevisiae.* The same authors also processed mango nectar and examined inactivation of *S. cerevisiae* and polyphenoloxidase (PPO) using the UV set-up described above (Guerrero-Beltrán and Barbosa-Cánovas, 2006). The maximum 3-log reduction was achieved after treatment for 30 min. In addition, shelf-life extension up to 20 days for mango nectar was reported by these researchers. UV inactivation of spoilage organisms in orange juice was also

Table 8.4 Inactivation Levels (log (N_o/N)) of Pathogenic Microorganisms in Fresh Juices Treated in Flow-Through Systems Continuous-Wave Ultraviolet Light Source at 253.7 nm

Fluid	Fluence (mJ/cm^2)	Test Microorganism	Inactivation	Reference
Apple cider	61	*Escherichia coli* O157:H7	5.4	Wright *et al.* (2000)
	14	*Cryptosporidium parvum*	>5	Hanes *et al.* (2002)
	14	*E. coli* O157:H7	>5	Basaran *et al.* (2004)
Goats' milk	16	*Listeria monocytogenes*	>5	Matak *et al.* (2005)

Table 8.5 Ultraviolet Inactivation of Non-Pathogenic Microorganisms in Fresh Juices

Juice	Type of UV reactor			Fluence (mJ/cm^2)	Test Microorganism	Log Inactivation (log (No/N))	Reference
	Flow Regime	Number /Type of UV Lamp/ Power	Gap Size (mm)				
Orange juice	Thin film laminar vertical	1/LPML/ 30 W	0.21–0.48	74	APC	0.53	Tran and Farid (2004)
					Yeasts	0.36	
Apple juice	Laminar	2/LPML/25 W	NA	45,000	*E. coli*	1.34	Guerrero-Beltrán and Barbosa-Cánovas (2005)
					APC*	4.29	
					Y&M**	5.10	
Mango nectar	Laminar	2/LPML/2 5 W	NA	45,000	APC	2.94	Guerrero-Beltrán and Barbosa-Cánovas (2006)
					Yeasts	2.71	
Model of tropical juices	Turbulent, Dean Flow	24/LPML/65 W	ID 10–12	21.5	Yeasts	Up to 6	Koutchma *et al.* (2007)
Orange juice					Moulds	1.5	
Guava					Moulds	1.2	
Carrot					APC	3.2	
Pineapple					Y&M	1.0	
Apple juice	Turbulent Re > 7500	1–10/LPML/ 100 W	NA	234	APC	> 3.50	Keyser *et al.* (2008)

(Continued)

Table 8.5 (Continued)

Juice	Type of UV reactor			Fluence (mJ/cm^2)	Test Microorganism	Log Inactivation (log (No/N))	Reference
	Flow Regime	Number /Type of UV Lamp/ Power	Gap Size (mm)				
					Y&M	> 2.99	
Guava-and-pineapple juice				1404	APC	3.31	
				468	Y&M	2.23	
Mango nectar				702	APC	0.40	
					Y&M	0.44	
Strawberry nectar				1404	APC	1.32	
					Y&M	2.45	
Orange juice				1634	APC	030	
					Y&M	0.30	
Cranberry juice	Laminar	2/LPML/25 W	NA	45,000	Yeast	0.53	Guerrero-Beltrán *et al.* (2009)
Grape juice						2.51	
Grapefruit juice						2.42	
Apple juice commercial naturally clouded	Laminar Dean vortices	1/LPML/9W	6.28 mm^2 and volume 24 ml	> 66	*Saccharomyces cerevisiae*	4	Franz *et al.* (2009)
Orange juice				1634	*E. coli*	0.30	
					Lactobacillus brevis	0.30	

**APC, aerobic plate count*
***Y&M, yeasts and moulds.*

reported by Tran and Farid (2004) using a vertical, single, LPM UV lamp thin-film reactor. The applied UV-fluence was in the range of 12 to 147.6 mJ/cm^2 and resulted in a 3-log reduction of aerobic bacteria, yeasts, and moulds. Koutchma *et al.* (2007) validated the performance of a coiled UV module (Salcor Inc., Fallbrook, CA) with 24 LPM lamps for pasteurization of fresh tropical juices with pulp such as orange, pineapple, guava nectar, and carrot. Approximately 1.2–1.6-log reductions of yeasts and up to 1.6-log reduction of moulds were achieved. Keyser *et al.* (2008) investigated the use of a turbulent flow system (PureUV) for treatment of apple, guava, pineapple, mango nectar, strawberry nectar, and orange juices. The spoilage microbial load of the juices was reduced successfully. Clear juice, i.e. the apple juice, needed a lower dose of UV radiation to achieve effective reduction whereas thicker juices, i.e. orange juice, with high levels of suspended solids, required higher UV doses. Furthermore, the UV treatment did not alter the flavor or color profiles of the juices. Franz *et al.* (2009) reported that a laboratory-scale UV reactor, UVivatec Lab from the Bayer Technology Services GmbH (BTS), where helically wound tubing was wrapped around a quartz glass tube containing an LPM lamp with an irradiation intensity of 60 W/m^2 was tested for its potential to inactivate bacteria in naturally cloudy apple juice. *Lactobacillus brevis* LMG 11438 was used as a typical representative of the lactic acid bacteria spoilage microflora of apple juice, while *S. cerevisiae* DSM 70478 was used as typical spoilage yeast. It was reported that the equipment was capable of reducing numbers of inoculated *E. coli* and *L. brevis* from an initial concentration of approximately 10^6 CFU/ml or 10^4 CFU/ml to below detectable limits in commercial naturally cloudy apple juice at a flow rate of 2 l/h, and to below 10^2 also at higher flow rates of 4 and 8 l/h. A 4-log reduction was easily achieved with a flow rate of 2 and 4 l/h. For *S. cerevisiae*, however, the same inactivation could not be obtained. Oteiza *et al.* (2010) studied the effect of yeast concentration on UV inactivation in five strains of *E. coli* O157:H7 inoculated both individually and simultaneously in orange juice. The presence of yeast cells in orange juice decreases the performance of UV radiation on *E. coli* inactivation. Absorption coefficients in the juice increased with increasing yeast concentration, and longer exposures were necessary to inactivate bacterial strains. All the tested individual strains showed higher resistance to the treatment when UV radiation was applied at 4°C in comparison with 20°C. It is possible that the lower temperature caused the bacterium to transfer to a different growth phase. The UV fluence of 2,190 mJ/cm^2 at 4°C (11 min of UV treatment) and 2,090 mJ/cm^2 at 20°C (10.55 min of UV treatment) was required to reach 5-log reduction for the most UV-resistant strain and maximum background yeast concentration. It was concluded that microbial inactivation tests using a cocktail of strains are particularly important in determining the performance of UV inactivation treatment. The UV fluence was calculated from the incident energy, and was not an integrated fluence through the fluid, and so did not account for the absorbance of the medium.

8.5.2 Ultraviolet Inactivation Case Studies in Dairy Products: Milk, Cheese Brines and Whey Protein

Bintsis *et al.* (2000) studied the effects of long-wave UVA (~365 nm) light and photosensitizing furocoumarines on pathogens such *as E. coli* O157:H7, *Listeria innocua, S. aureus* and yeasts.The purpose of the study was to eliminate pathogens but not the desired yeasts and lactic acid bacteria. In laboratory trials, UVA (45 W/m^2, 60 s) with prozalen was active against pathogens but less effective against yeasts from Feta cheese brines. However, only *Yarrowia lipolytica* survived the treatment. It was concluded that the treatment was not sufficiently selective in its action for use with cheese brines.

8.5.3 Antimicrobial Effects of Pulsed Light

Since PL systems are relatively new, literature still lacks reports of tests in flow-through systems although such devices are already commercially available. Nevertheless, encouraging results have been found in systems where a liquid sample with or without stirring is placed in a Petri dish or similar container and treated in a bench-top PL system. Counts of *E. coli* inoculated in apple juice have been reduced by 7.29-log CFU/ml after only 3 s of a treatment with a fluence amounting to 88,000 mJ/cm^2 (Sauer and Moraru, 2009).

Elmnasser *et al.* (2007) published a review summarizing the effects of PL on microflora including spoilage organisms in foods and model systems, mainly agar. The results from these pulsed tests can not be compared with other methods, since the fluences are reported in energy or energy per unit area, and the spectrum of the energy is not known. However the results may be used as an indication of the relative sensitivity of various organism types. The results have demonstrated that fungi are more resistant to PL than bacteria. PL treatment is more effective on solid surfaces than in liquids. Although the peak power of pulses is very high because of their short duration, it appears that PL does not penetrate very deeply into food products.

8.5.3.1 Dairy Products

The decontamination of bulk tank milk with a pulsed UV light excimer laser (248 nm) was investigated by Smith *et al.* (2002). One milliliter of bulk tank milk samples, in quartz cuvettes, was exposed to pulsed UV light at 25,000 mJ/cm^2. The pulsed UV light exposure resulted in total reduction of mesophilic aerobes. Moreover, the inactivation of seven different potential food-borne bacterial pathogens (*E. coli* O157:H7, *L. monocytogenes*, *Salmonella choleraesuis*, *Yersinia enterocolitica*, *S. aureus*, *Aeromonas hydrophila*, and *Serratia marcescens*) by pulsed UV light was studied under the same conditions as those for the treatment of mesophilic aerobes. No growth of any plated or subcultured samples was observed, even after incubation for 21 days. No initial concentrations were reported except for the indication that background microflora were present in the milk samples. Also, Krishnamurthy *et al.* (2007) reported 7.26-log reduction for *S. aureus* counts in milk using a flow-through reactor.

8.5.3.2 Liquid Sugars and Sweeteners, Honey

Liquid sugars are used extensively in the food and beverage industries. Sugar syrups (sucrose, fructose, honey) with high osmotic pressure can be subjected to microbial growth that can result in health problems and spoilage. In addition to the challenges of low UV transmission, liquid sugars have high viscosity characteristics. UV systems have been effectively designed and validated for manufacturing companies that produce soda, candy, honey, yoghurt, and a variety of other foods.

Hillegas and Demirci (2003) reported use of PL to inactivate *Clostridium sporogenes* spores in clover honey. PL was applied to the inoculated honey using the SteriPulse XL® 3000 PL Sterilization System. The energy released from the UV strobe was 5,600 mJ/cm^2 per pulse when the input voltage was 3,800 V. In order to study the effect of distance from the UV strobe and treatment time, inoculated honey was placed on the tray at three different distances from the UV lamp: 8, 13, 20 cm; and treated up to 135 s. The results showed an increase from 0% reduction at 15 and 135 pulses (5 and 45 s) to 89.4% reduction at 540 pulses (3 min). When the depth of honey was changed from 8 mm to 2 mm, an increase in reduction becomes evident; at the 8-mm depth, 135, 270, and 405 pulses demonstrated 0, 22.1, and 14.2% reduction, respectively. In comparison, at the 2-mm depth, 135, 270, and 405 pulses demonstrated 39.5, 28.9, and 73.9% reductions, respectively. Even though varying these three parameters enhanced the percentage inactivation of *C. sporogenes*, it failed to inactivate the spores completely. It was concluded that the PL has a limited penetration in the honey.

These results are not surprising, and do not provide quantitative understanding of the treatment of this organism in the medium of liquid honey. Honey undoubtedly absorbs UV light strongly, so that there will be an exponential decrease in intensity with depth, resulting in poor disinfection at the bottom of the sample. Similarly, increasing the distance from the lamp will reduce the amount of UV energy reaching the honey.

8.6 EFFECTS ON FLUID FOOD NUTRITIONAL AND QUALITY PARAMETERS

Formation of toxic by-products during processing is a concern in every food preservation technology. Furan is a possible human carcinogen induced by thermal processing and gamma irradiation of juices and sugar solutions. UV-C illumination induces furan formation in apple cider, and the major source of furan is apparently fructose. However, furan formation seems to be too low to cause alarm. At UV-C fluences that could inactivate 5-log of *E. coli* (2,500 mJ/cm^2), <1 ppb of furan is induced, which is lower than levels in some canned juices (Fan and Geveke, 2007). Foods rich in reducing sugars and proteins are susceptible to Maillard reaction, which reduces the nutritional value. No evidence of Maillard reaction in milk has been observed after 10 light pulses of 2,200 mJ/cm^2 each (Elmnasser *et al.*, 2008).

It is common knowledge that preservation treatments degrade nutrients, a phenomenon that has to be assessed and minimized. UV-C illumination decreases ascorbic acid content of juices at a similar level to that caused by thermal treatments. Ascorbic acid present in orange juice is degraded by UV-C following first-order kinetics at a rate of 0.175% per mJ/cm^2, with a reduction of 17.0% caused by 148 mJ/cm^2 (Tran and Farid 2004). Also for orange juice, a treatment of 299 mJ/cm^2 destroyed about 50% riboflavin and β-carotene, 17% vitamin C, 11% vitamin A, and did not affect folic acid or vitamin E (Anonymous, 1999). In apple juice, the reduction is reported to be from 5.4 to 4.0 mg/100 ml of juice (Walkling-Ribeiro *et al.*, 2008). Vitamin A present in milk is also degraded by a light pulse of 2,200 mJ/cm^2 (Elmnasser *et al.*, 2008).

Regarding PL specifically, care must be taken to avoid the sample overheating as a side effect, because degradation reactions associated with thermal treatment will occur. For example, in a study on the inactivation by a flow-through PL reactor of *S. aureus* inoculated in raw milk, the temperature rose from 0 to 38°C (Krishnamurthy *et al.*, 2007).

Besides the effect on micronutrients, the potential effect of UV light on antioxidant capacity and related compounds is also important to assess due to the beneficial effects of phytochemicals. The relative antioxidant capacity of fresh apple juice decreased by 18.9% and the concentration of total phenolic compounds changed from 22.74 to 15.98 mg gallic acid equivalents/l after 30 min of UV illumination (Noci *et al.*, 2008).

However, UV light should not be considered as a harmful treatment to the nutritional and antioxidant quality of fruit juices if it is evaluated from a broader point of view. UV treatment of fruits before juice extraction has been shown to yield excellent results to enhance phytochemical content. For example, UV-C illumination of grapes induces stilbene synthesis, especially that of *trans*-resveratrol, which will yield a phytochemical-enriched grape juice (González-Barrio *et al.*, 2009).

Changes in macromolecules, such as conformational modifications of proteins, can also lead to beneficial or deleterious alterations in food component functionality. Light pulses of 2,200 mJ/cm^2 promote milk protein aggregation by formation of disulfide bonds, with no change in amino acid composition, and no effect on small peptide fragments and on oxidation of proteins by lipids. Moreover, no oxidation of milk fatty acids has been observed after 22,000 mJ/cm^2 of PL treatment (Elmnasser *et al.*, 2008). However, 15.8 mJ/cm^2 of UV light treatment increased lipoxidation of goat milk (Matak *et al.*, 2005).

Color is a very important quality parameter in fruit juices. It is related to enzymatic and non-enzymatic browning, and PPO activity. As many fruit juices owe their color to carotenoids, which are very susceptible to UV light, it is predictable that some color degradation may occur during treatment of juices with UV light. The evidence shows, however, that UV light treatments do not affect instrumental color parameters of orange juice after fluences of 148 (Tran and Farid 2004) and 1,634 mJ/cm^2 (Keyser *et al.*, 2008), fresh apple juice (Noci *et al.*, 2008; Walkling-Ribeiro *et al.*, 2008) or apple juice treated with 1,404 mJ/cm^2 (Keyser *et al.*, 2008). Similar results have been also found for guava–passion fruit juice, "tropical juice", and strawberry and mango nectars treated, respectively, at 1,404, 1,634, 2,106 mJ/cm^2 (Keyser *et al.*, 2008). On the contrary, it has been demonstrated that UV light can "improve" the color of apple, peach and lemon juices previously heated to generate melanoidines due to photolysis of these polymeric compounds (Ibarz *et al.*, 2005).

The effect of UV light on the inactivation of enzymes related to food quality is diverse. One important factor in orange juice appearance is the "cloud" formed by pectin. Pectin methylesterase (PME) is an enzyme that tends to deesterify pectin, and consequently its inactivation is pursued. An UV treatment of 74 mJ/cm^2 only inactivates 5% of PME activity, which gives UV a serious disadvantage with respect to thermal treatments (Tran and Farid, 2004). PPO is the main agent responsible for enzymatic browning. While Noci *et al.* (2008) reported no effect of UV on apple PPO activity, Manzocco *et al.* (2009) reported about 80% inactivation of PPO at approximately 1,250 mJ/cm^2 of UV fluence. In mango nectar, an UV treatment of 44,633 mJ/cm^2 reduced PPO activity to 19% (Guerrero-Beltrán and Barbosa-Cánovas, 2006).

pH and °Brix of orange and apple juices are not affected by UV treatments (Tran and Farid, 2004; Noci *et al.*, 2008; Walkling-Ribeiro *et al.*, 2008). Only Ibarz *et al.* (2005) reported a minimum increase in °Brix for apple, peach and lemon juices, likely caused by evaporation due to sample heating.

As for any preservation treatment, the effect of UV-C illumination on the sensory quality of juices should be assessed. No matter how efficiently a novel method is for microbial inactivation, it is not practical if the juice is not accepted by consumers. It has been found that a UV fluence of 14 mJ/cm^2 of apple ciders did not change taste and preference, although UV-treated samples lost quality faster than pasteurized samples due to fermentation (Tandon *et al.*, 2003). UV light treatment of 15.8 mJ/cm^2 changed the odor of goats' milk, which was described after treatment as manure-like, stinky, barnyard, and goaty (Matak *et al.*, 2007). UV light treatment of guava juice at 8,561 mJ/cm^2 and passion fruit juice at 3,640 mJ/cm^2 caused detectable changes in sensory characteristics of these juices (Guevara *et al.*, in press).

8.7 INDUSTRIAL SCALE-UP AND CHALLENGES

8.7.1 Designs of Ultraviolet Reactors for Liquid Foods

A number of continuous flow systems were developed and validated for a variety of drinks and beverages ranging from raw milk, whey protein, exotic tropical juices and nectars, to the more common apple cider and apple juice. These UV reactors are commercially available or under development in order to be used to study UV treatment of liquids with low UVT. The reactor designs include traditional annular, thin film, static and dynamic mixers, and coiled tube devices.

8.7.1.1 Annular Reactors

In annular UV reactors (with singular or multiple lamps) liquid is pumped through the gap formed by two concentric stationary cylinders (Fig. 8.8). For example, the "Ultradynamics" UV reactor (Ye *et al.*, 2007) uses a single large quartz sleeve as the inner reactor wall with a surrounding metal cylinder as the outer reactor wall. The reports are available that annular type laminar reactor were used for treatment of apple juice and cider (Worobo *et al.*, 1999), mango nectar and cheese whey (Guerrero-Beltrán and Barbosa-Cánovas, 2006; Singh and Ghaly, 2007). The length and gap size can vary depending on the type of treated liquids or flow rate.

Thin film reactors form a thin fluid layer (less than 1 mm), either between fixed boundaries or by flowing fluid on a fixed surface, or by creating a "sheet" of fluid in air to decrease the path length and thus avoid problems associated with lack of penetration. Thin-film reactors are characterized by laminar flow with a parabolic velocity profile. Extensive research of the application of UV-light for

FIGURE 8.8

Single lamp annular UV-reactor manufactured by American Ultraviolet, Lebanon, MI, USA. *(Photo courtesy of NCFST, Summit, IL.)*

fresh apple cider by Worobo (1999) yielded a design and production model of a thin-film with 0.8 mm gap "CiderSure" UV reactor (Fig. 8.9). UV treatment of orange juice was reported by Tran and Farid (2004) using a vertical single UV lamp thin-film reactor. The thickness of the film was approximately $0.21 \sim 0.48$ mm.

Another commercial thin-film reactor is the PureUV/SurePure reactor that was used for treatment of apple juice, guava-and-pineapple juice, mango nectar, strawberry nectar, and two different orange and tropical juices (Keyser *et al.*, 2008). This reactor is a single-lamp system with a thin fluid film formed between the lamp surface and a surrounding rippled or undulating outer wall. The reactor consisted of inlet, outlet chambers and a corrugated spiral tube between the chambers. The tangential inlet created both a high velocity and turbulence ($Re > 7500$) even at a minimum flow-rate of 3800 l/h in the inlet chamber and liquid product was brought into contact with the UV radiation while flowing in a gap between sleeve and spiral tube.

FIGURE 8.9

Eight-lamp thin-film UV-reactor CiderSure 3500 manufactured by OESCO Inc, Conway, MA, USA.

(Photo courtesy of GFRC, Agriculture and Agri-Food Canada.)

8.7.1.2 Static Mixers

Latros of Dundee, Scotland, is the best-known example of the use of static mixers in UV treatment of low-UVT fluids. In this design, 10 or more "half-twist" static mixers are encased in a section of Teflon sleeve, with radiation applied by surrounding UV lamps. The static mixer consists of a series of alternating right- and left-hand helical elements with 180° rotations, each juxtaposed at 90° to the element preceding it. The static mixers achieve flow division, flow reversal, radial mixing, and axial differentiation of the fluid stream, which constantly changes the thin film at the inner wall of the pipe, thereby exposing more bacterial cells to the UV light during treatment. Tests of this reactor have been published in the academic literature. This unit reactor is repeated in series to achieve higher fluences, and they have a pilot unit with three of these sections "folded" within a single treatment zone. Altic *et al.* (2007) reported UV light inactivation of *Mycobacterium avium subsp. paratuberculosis* in Middlebrook 7H9 broth and whole and semi-skim milk using a laboratory-scale UV device.

Other types of static mixers are coiled-tube UV reactors used to increase liquid delivery by more mixing due to the Dean effect (Dean, 1927, 1928). Salcor Inc. has promoted a UV reactor in which Teflon tubes are coiled in a helix, with 12 lamps inside and 12 lamps outside the helix (Fig. 8.10). Fluid is pumped through the coiled tubing. The curved flow path can result in a pair of counter-rotating vortices

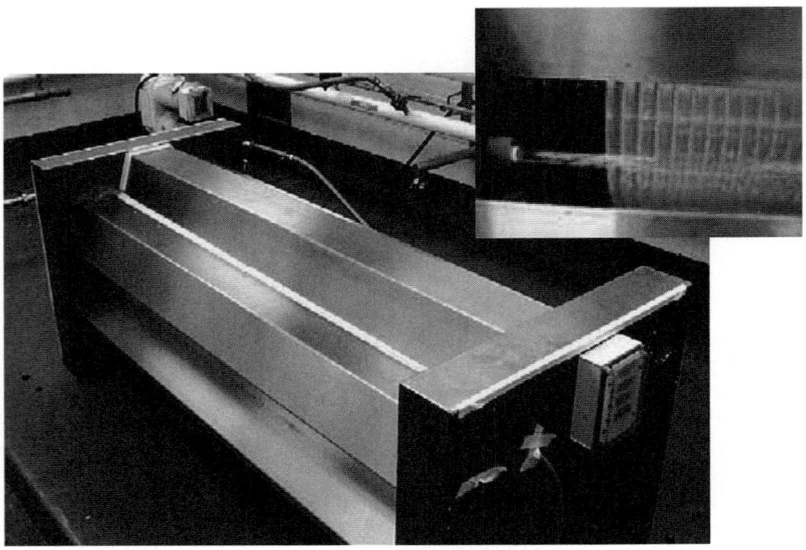

FIGURE 8.10

Twenty-four-lamp coiled-tube Salcor UV unit manufactured by Salcor, Inc, Fallbrook, CA, USA.

(Photo courtesy of NCFST, Summit, IL.)

with their axis along the length of the coil. Koutchma *et al.* (2007) validated the performance of a coiled UV module 420 model (Salcor Inc., Fallbrook, CA) for the pasteurization of fresh tropical juices. Geveke (2005, 2008) processed apple cider and liquid egg-white with a single-lamp UV reactor surrounded by a coil of UV-transparent Chemfluor tubing.

A UV coil reactor was designed and used for the online sterilization of cheese whey (Singh and Ghaly, 2007). Its microbial destruction efficiency was compared to the conventional annular UV reactor. Both reactors had the same geometry (840 ml volume and 17-mm gap size) and were tested at 11 flow rates. The flow was laminar in both the reactors. The phenomenon of Dean Flow was observed in the coil reactor and Dean vortices resulted in higher microbial destruction efficiency in the coil reactor in a shorter retention time. The results showed that, despite of its high turbidity, cheese whey could be sterilized using UV radiation if the proper reactor design and flow rate were used.

8.7.1.3 Dynamic Mixers

Ye (2007) investigated the inactivation efficiency of a Taylor–Couette flow reactor as an example of a dynamic mixer. Details of optimum performance for this reactor are discussed in Forney *et al.* (2008). This reactor (Fig. 8.11) has a stationary outer cylinder with a rotating inner cylinder. This results in a fluid

FIGURE 8.11

Pilot scale model of the Taylor–Couette UV reactor manufactured by Trojan UV Technologies Inc, London, ON, Canada.

(Photo courtesy of Agriculture and Agri-Food Canada, GFRC, ON, Canada.)

instability which causes counter-rotating vortices with the axis of the vortex along the circumference of the cylinders and brings fluid from the dark outer cylinder to the high-intensity inner-wall region.

The microbial inactivation results were reported in terms of logarithmic reduction ($\log_{10}$) of the number of target pathogenic organisms or total count while attempting to compare efficiency in terms of fluence (mJ/cm^2) or energy (mJ/l). However, more work is needed in the design of UV reactors capable of providing a sufficient and uniform UV fluence to all parts of the treated liquid. Quantitative tools and indicators need to be introduced to evaluate efficacy of UV treatments in food matrices.

8.7.2 Challenges and Future Perspectives

From an engineering perspective, the use of UV light to achieve proper inactivation must be designed effectively. Some of the challenges that exist in designing a system for fluid foods as noted above include the high-absorbance character of many of the food fluids, the particles present, the differences in viscosity, and the types of target pathogens or spoilage microorganisms. Although much experience can be borrowed from the water disinfection field, special care needs to be taken to adapt these systems to the needs of fluid foods.

Once the target level of inactivation of a specific pathogen is decided upon, the bench-scale experiments need to be validated at the flow-through level. The various systems described in this chapter are representative of the types of fluid food disinfection systems that have been designed. A validating authority or independent third party expert should oversee the testing. Various protocols for validation of UV disinfection systems are available and should be used as a basis for validating new designs. Such protocols include the NSF International Standard 55: Ultraviolet Microbiological Water Treatment Systems; the US EPA UV Disinfection Guidance Manual (US EPA, 2006); and the National Water Research Institute Ultraviolet Disinfection Guidelines for Water and Water Reuse (NWRI, 2003). All protocols share the fact that they require testing using a microbial surrogate appropriate for the target UV fluence range. A UV system will be validated over a range of UV absorbance levels, lamp power output, and flow rates such that it can be used for a variety of fluids. The validating authority may require addition of turbidity or require utilizing the specific fluid for the targeted application. In such cases, a surrogate microorganism can be spiked into the test fluid to quantify performance under actual operating conditions.

Although many systems used for disinfection of fluid foods are much smaller than UV disinfection for municipal water supplies, there is typically some scale-up required from the bench scale units. Currently, methods for scale-up of UV systems include computational modeling comprising fluence rate distribution modeling coupled with computational fluid dynamics (Bolton, 2000; Blatchley, 1997; Lyn and Blatchley, 2005). These approaches can be used to design a system with attributes that match the end-use relating to flow rate, target disinfection goals, and available footprint area. Eventually, the scaled-up unit will need to be validated before use.

8.8 CONCLUSIONS

UV-light processing using continuous and pulsed sources can be a viable non-thermal alternative for fluid foods, ingredients, drinks, and beverages based on sufficient evidence in the literature on the success of this method for eliminating or reducing the levels of most types of undesirable pathogenic and spoilage microorganisms. Compared to water, fluid foods have a range of optical and physical properties, diverse chemical compositions that influence UV light transmittance, dose delivery, momentum transfer and consequently microbial inactivation. Within a fluid food product several factors such as absorbance, pH, water activity, and food matrix can influence the delivery of the UV dose. The presence of particles can shield microorganisms from the UV irradiation. The flow pattern inside the UV reactor strongly influences the total UV dose, since the position and the residence time of the microorganisms in certain regions of the irradiance field can vary significantly. A number of UV-light continuous-flow apparatus that included annular laminar and turbulent flow reactors, thin-film devices, static and dynamic mixers were developed and validated for milk and a variety of beverages ranging from exotic tropical juices to the more common apple cider and apple juice for use in pasteurization. The correct choice of UV reactor design can reduce the interference of high UV absorbance and viscosity associated with some fluid products and therefore improves the inactivation efficiency. More work is needed on the design of UV systems capable of delivering sufficient UV doses to all parts of the treated liquid.

The effect of UV-light on quality of foods requires further study. UV light can adversely affect foods by generating free radicals in products by a wide variety of organic photochemical reactions. Possible undesirable effects include damage to vitamins and proteins, destruction of the antioxidants, oxidation of lipids, changes in color, and formation of off-flavours and aromas. In addition, the effects of UV light on the potential formation of chemical compounds in foods that may present a health threat should be evaluated to determine if there are any toxicological or chemical safety concerns associated with products that have undergone UV treatment.

Finally, the success of UV technology for low-UVT liquids depends on the correct alignment of the UV source parameters to the specific demands of the UV application. Low-pressure mercury lamps are currently the dominant sources for UV treatment of liquid foods and beverages since they were approved by US FDA and Health Canada. Special technology lamps like, for example, pulsed UV lamps, are promising due to the lack of mercury in the lamp, but more research is needed to establish their suitability for fluid food applications.

References

Altic, L.C., Rowe, M.T., Grant, I.R., 2007. UV light inactivation of *Mycobacterium avium* subsp. paratuberculosis in milk as assessed by FASTPlaque TB phage assay and culture. Appl. Environ. Microbiol. 73, 3728–3733.

Anonymous, 1999. A food additive petition for the use of ultraviolet light in the reduction of microorganisms on juice products. Submitted to FDA regarding CFR 21 179. Glendore Calif. California Day-Fresh Foods Inc, 1–117.

Basaran, N., Quintero-Ramos, A., Moake, M.M., Churey, J.J., Worobo, R.W., 2004. Influence of apple cultivars on inactivation of different strains of *Escherichia coli* O157:H7 in apple cider by UV irradiation. Appl. Environ. Microbiol. 70, 6061–6065.

Bintsis, T., Litopolou-Tzanetaki, E., Davies, R., Robinson, R.K., 2000. The antimicrobial effects of long-wave UV light and furocoumarins on some microorganisms that occur in cheese brines. Food Microbiol. 17, 687–695.

Blatchley III, E.R., 1997. Numerical modelling of UV intensity: application to collimated-beam reactors and continuous-flow systems. Water Res. 31, 2205–2218.

Bohrerova, Z., Shemer, H., Lantis, R., Impellitteri, C.A., Linden, K.G., 2008. Comparative disinfection efficiency of pulsed and continuous-wave UV irradiation technologies. Wat. Res. 42, 2975–2982.

Bolton, J.R., 2000. Calculation of ultraviolet fluence rate distributions in an annular reactor: significance of refraction and reflection. Water Res. 34, 3315–3325.

Bolton, J.R., Linden, K.G., 2003. Standardization of methods for fluence (UV dose) determination in bench-scale UV experiments. J. Environ. Eng. 129, 209–215.

Chung, S.-Y., Yang, W., Krishnamurthy, K., 2008. Effects of pulsed UV-light on peanut allergens in extracts and liquid peanut butter. J. Food Sci. 73, C400–C404.

Cleaver, J.E., 2003. Photoreactivation. DNA Repair 2, 629–638.

Dean, W.R., 1927. Note on the motion of fluid in a curved pipe. Philos. Mag. 4, 208–223, Sp. Iss. 7th Series.

Dean, W.R., 1928. The stream-line motion of fluid in a curved pipe. Philos. Mag. 5, 673–695, Sp. Iss. 7th Series.

Dong, Q., Manns, D., Feng, G., Yue, T., Churey, J.J., Worobo, R.W., 2010. Reduction of patulin in apple cider by UV radiation. J. Food Prot. 73, 69–74.

Dunn, J., Bushnell, A., Ott, T., and Clark, W., 1989. Methods for preservation of foodstuffs. US Patent 4,871,559.

Elmnasser, N., Guillou, S., Leroi, F., Orange, N., Bakhrouf, A., Federighib, M., 2007. Pulsed-light system as a novel food decontamination technology: a review. Can. J. Microbiol. 53, 813–821.

Elmnasser, N., Dalgalarrondo, M., Orange, N., Bakhrouf, A., Haertlé, T., Federighi, M., et al., 2008. Effect of pulsed-light treatment on milk proteins and lipids. J. Agric. Food Chem. 56, 1984–1991.

Eischeid, A.C., Meyer, J.N., Linden, K.G., 2009. UV disinfection of adenoviruses: molecular indications of DNA damage efficiency. Appl. Environ. Microbiol. 75, 23–28.

Fan, X., Geveke, D.J., 2007. Furan formation in sugar solution and apple cider upon ultraviolet treatment. J. Agric. Food Chem. 55, 7816–7821.

Forney, L.J., Ye, Z., Koutchma, T., 2008. UV Disinfection of *E. coli* between concentric cylinders: effects of the boundary layer and a wavy wall. Ozone Sci. Eng. 30, 405–412.

Franz, C., Specht, I., Cho, G.-S., Graef, V., Stahl, M.R., 2009. UV-C-inactivation of microorganisms in naturally cloudy apple juice using novel inactivation equipment based on Dean vortex technology. Food Control 20, 1103–1107.

Geveke, D., 2005. UV inactivation of bacteria in apple cider. J. Food Prot. 68 (8), 1739–1742.

Geveke, D., 2008. UV inactivation of *E. coli* in liquid egg white. Food Bioprocess Technol. 1 (2), 201–206.

Gómez-López, V.M., Devlieghere, F., Bonduelle, V., Debevere, J., 2005. Factors affecting the inactivation of microorganisms by intense light pulses. J. Appl. Microbiol. 99, 460–470.

Gómez-López, V.M., Ragaert, P., Debevere, J., Devlieghere, F., 2007. Pulsed light for food decontamination: a review. Trends Food Sci. Technol. 18, 464–473.

González-Barrio, R., Vidal-Guevara, M.L., Tomás-Barberán, F.A., Espín, J.C., 2009. Preparation of a resveratrol-enriched grape juice based on ultraviolet C-treated berries. Innovative Food Sci. Emerg. Technol. 10, 374–382.

Guerrero-Beltrán, J.A., Barbosa-Cánovas, G.V., 2005. Reduction of *Saccharomyces cerevisiae, Escherichia coli* and *Listeria innocua* in apple juice by ultraviolet light. J. Food Proc. Eng. 28, 437–452.

Guerrero-Beltrán, J.A., Barbosa-Cánovas, G.V., 2006. Inactivation of *Saccharomyces cerevisiae* and polyphenoloxidase in mango nectar treated with UV light. J. Food Prot. 69, 362–368.

Guerrero-Beltrán, J.A., Welti-Chanes, J., Barbosa-Cánovas, G.V., 2009. Ultraviolet-C Light processing of grape, cranberry and grapefruit juices to inactivate *Saccharomyces cerevisiae*. J. Food Proc. Eng 32, 916–932.

Guevara, M., Tapia, M.S., Gómez-López, V.M. (in press). Microbial inactivation and quality of guava and passion fruit nectars treated by UV-C light. Food Bioprocess Technol.

Hanes, D.E., Orlandi, D.H., Burr, M.D., Miliotis, M.G., Robi, J.W., Bier, G.J., et al., 2002. Inactivation of *Cryptosporidium parvum* oocysts in fresh apple cider using ultraviolet irradiation. Appl. Environ. Microbiol. 68, 4168–4172.

Harm, W., 1980. Biological Effects of Ultraviolet Radiation. Cambridge University Press, New York.

Hierro, E., Manzano, S., Ordóñez, J.A., de la Hoz, L., Fernández, M., 2009. Inactivation of *Salmonella enterica* serovar Enteriditis on shell eggs by pulsed light technology. Int. J. Food Microbiol. 135, 125–130.

Hijnen, W.A.M., Beerendonk, E.F., Medema, G.J., 2006. Inactivation credit of UV radiation for viruses, bacteria and protozoan (oo)cysts in water: a review. Wat. Res. 40, 3–22.

Hillegas, S.L., Demirci, A., 2003. Inactivation of *Clostridium sporogenes* in clover honey by pulsed UV-light treatment. CIGR-ejournal 5 (5).

Hiramoto, T., 1984. Method of sterilization. US Patent 4,464,336.

Ibarz, A., Pagán, J., Panadés, R., Garza, S., 2005. Photochemical destruction of color compounds in fruit juices. J. Food Eng. 69, 155–160.

Kao, Y.T., Saxena, C., Wang, L., Sancar, A., Zhong, D., 2005. Direct observation of thymine dimer repair in DNA by photolyase. Proc. Natl. Acad. Sci. USA 102, 16128–16132.

Keyser, M., Müller, I.A., Cilliers, F.P., Nel, W., Gouws, P.A., 2008. Ultraviolet radiation as a non-thermal treatment for the inactivation of microorganisms in fruit juice. Innovative Food Sci. Emerg. Technol. 9, 348–354.

Koutchma, T., 2009. Advances in ultraviolet light technology for non-thermal processing of liquid foods. Food Bioprocess Technol. 2, 138–155.

Koutchma, T., Keller, S., Chirtel, S., Parisi, B., 2004. Ultraviolet disinfection of juice products in laminar and turbulent flow reactors. Innovative Food Sci. Emerg. Technol. 5, 179–189.

Koutchma, T., Parisi, B., Patazca, E., 2007. Validation of UV coiled tube reactor for fresh fruit juices. J. Environ. Sci. Eng. 6, 319–328.

Krishnamurthy, K., Demirci, A., Irudayaraj, J.M., 2007. Inactivation of *Staphylococcus aureus* in milk using flow-through pulsed UV-light treatment system. J. Food Sci. 72, M233–239.

Krishnamurthy, K., Tewari, J.C., Irudayaraj, J., Demirci, A., 2009. Microscopic and spectroscopic evaluation of inactivation of *Stapylococcus aureus* by pulsed UV light and infrared heating. Food Bioprocess Technol. 3, 93–104.

Lyn, D.A., Blatchley III, E.R., 2005. Numerical computational fluid dynamics-based models of ultraviolet disinfection channels. J. Environ. Eng. 131, 838–850.

Mamane-Gravetz, H., Linden, K.G., Cabaj, A., Sommer, R., 2005. Spectral sensitivity of *Bacillus subtilis* spores and *MS2 coliphage* for validation testing of ultraviolet reactors for water disinfection. Environ. Sci. Technol. 39, 7845–7852.

Manzocco, L., Quarta, B., Dri, A., 2009. Polyphenoloxidase inactivation by light exposure in model systems and apple derivatives. Innovative Food Sci. Emerg. Technol. 10, 506–511.

Masschelein, W.J., Rice, R.G., 2002. Ultraviolet Light in Water and Wastewater Sanitation. Lewis Publishers, Boca Raton (p 2)

Matak, K.E., Churey, J.J., Worobo, R.W., Sumner, S.S., Hovingh, E., Hackney, C.R., et al., 2005. Efficacy of UV light for the reduction of *Listeria monocytogenes* in goat's milk. J. Food Prot. 68, 2212–2216.

Matak, K.E., Sumner, S.S., Duncan, S.E., Hovingh, E., Worobo, R.W., Hackney, C.R., et al., 2007. Effect of ultraviolet irradiation on chemical and sensory properties of goat milk. J. Dairy Sci. 90, 3178–3186.

McDonald, K.F., Curry, R.D., Clevenger, T.E., Brazos, B.J., Unklesbay, K., Eisenstark, A., et al., 2000. The development of photosensitized pulsed and continuous ultraviolet decontamination techniques for surfaces and solutions. IEEE Trans. Plasma Sci. 28, 89–96.

Noci, F., Riener, J., Walkling-Ribeiro, M., Cronin, D.A., Morgan, D.J., Lyng, J.G., 2008. Ultraviolet irradiation and pulsed electric fields (PEF) in a hurdle strategy for the preservation of fresh apple juice. J. Food Eng. 85, 141–146.

NSF Standard: < http://www.nsf.org/business/drinking_water_treatment/standards.asp >

NWRI, 2003. Ultraviolet Disinfection Guidelines for Drinking Water and Water Reuse. < http://www.nwri-usa.org/availablepubs.htm/ >.

Otaki, M., Okuda, A., Tajima, K., Iwasaki, T., Kinoshita, S., Ohgaki, S., 2003. Inactivation differences of microorganisms by low pressure UV and pulsed xenon lamps. Wat. Sci. Technol. 47, 185–190.

Oteiza, J., Peltzer, M, Gannuzzi, L, Zaritzky, N., 2005. Antimicrobial efficacy of UV radiation on *Escherichia coli* O157:H7 in fruit juices of different absorptivities. J. Food Prot. 68 (1), 49–58.

Oteiza, J.M., Giannuzzian, L., and Zaritzky, N., 2010. Ultraviolet treatment of orange juice to inactivate *E. coli* O157:H7 as affected by native microflora. Food Bioprocess Technol. 3, 603–614.

Rice, J.K., Ewell, M., 2001. Examination of peak power dependence in the UV inactivation of bacterial spores. Appl. Environ. Microbiol. 67, 5830–5832.

Sauer, A., Moraru, C.I., 2009. Inactivation of *Escherichia coli* ATCC 25922 and *Escherichia coli* O157:H7 in apple juice and apple cider, using pulsed light treatment. J. Food Prot. 72, 937–944.

Singh, J.P., Ghaly, A.E., 2007. Effect of flow characteristics on sterilization of cheese whey in UV reactors. Appl. Biochem. Biotechnol. 142, 1–16.

Smith, W.L., Lagunas-Solar, M.C., Cullor, J.S., 2002. Use of pulsed ultraviolet laser light for the cold pasteurisation of bovine milk. J. Food Prot. 65, 1480–1482.

Takeshita, K., Shibato, J., Sameshima, T., Fukunaga, S., Isobe, S., Arihara, K., et al., 2003. Damage of yeast cells induced by pulsed light irradiation. Int. J. Food Microbiol. 85, 151–158.

Tandon, K., Worobo, R., Churley, J., Padilla-Zakour, O., 2003. Storage quality of pasteurized and UV treated apple cider. J. Food Proc. Preserv. 27, 21–35.

Tran, M.T.T., Farid, M., 2004. Ultraviolet treatment of orange juice. Innovative Food Sci. Emerg. Technol. 5, 495–502.

Unluturk, S., Atilgan, M.R., Handan Baysal, A., Tari, C., 2008. Use of UV-C radiation as a non-thermal process for liquid egg products (LEP). J. Food Eng. 85, 561–568.

US EPA, 2006. Ultraviolet disinfection guidance manual for the final long term 2 enhanced surface water treatment rule. <http://www.epa.gov/ogwdw000/disinfection/lt2/pdfs/guide_lt2_uvguidance.pdf/>

Walkling-Ribeiro, M., Noci, F., Cronin, D.A., Riener, J., Lyng, J.G., Morgan, D.J., 2008. Reduction of *Staphylococcus aureus* and quality changes in apple juice processed by ultraviolet irradiation, pre-heating and pulsed electric fields. J. Food Eng. 89, 267–273.

Wang, T., MacGregor, S.J., Anderson, J.G., Woolsey, G.A., 2005. Pulsed ultra-violet inactivation spectrum of *Escherichia coli*. Wat. Res. 39, 2921–2925.

Worobo, R., 1999. Efficacy of the CiderSure 3500. Ultraviolet light unit in apple cider. CFSAN Apple cider food safety control workshop.

Wright, J.R., Sumner, S.S., Hackney, C.R., Pierson, M.D., Zoecklein, B.W., 2000. Efficacy of ultraviolet light for reducing *Escherichia coli* O157:H7 in unpasteurized apple cider. J. Food Prot. 63, 563–567.

Yaun, B.R., Sumner, S.S., Eifert, J.D., Marcy, J.E., 2003. Response of *Salmonella* and *Escherichia coli* O157:H7 to UV energy. J. Food Prot. 66, 1071–1073.

Ye, Z., 2007. UV disinfection between concentric cylinders. Ph.D. thesis, Georgia Institute of Technology.

Ye, Z., Koutchma, T., Parisi, B., Larkin, J., Forney, L.J., 2007. Ultraviolet inactivation kinetics of *E. coli and Y. pseudotuberculosis* in annular reactors. J. Food Sci. 72, E271–E278.

Zimmer, J.L., Slawson, R.M., Huck, P.M., 2003. Inactivation and potential repair of *Cryptosporidium parvum* following low- and medium-pressure ultraviolet radiation. Wat. Res. 37, 3517–3523.

CHAPTER

Ozone Processing of Fluid Foods

9

S. Patil and P. Bourke

School of Food Science and Environmental Health, Dublin Institute of Technology, Dublin, Ireland

9.1 INTRODUCTION

Ozone is a triatomic form of oxygen and is characterized by a high oxidation potential that conveys bacteriocidal and viricidal properties (Burleson, 1975; Horvath *et al.*, 1985; Kim *et al.*, 1999). It is a powerful broad-spectrum antimicrobial agent active against bacteria, fungi, viruses, protozoa, and bacterial and fungal spores (Khadre *et al.*, 2001). Ozone inactivates microorganisms through oxidization, and residual ozone spontaneously decomposes to non-toxic products (i.e. oxygen), making it an environmentally friendly antimicrobial agent for use in the food industry (Kim *et al.*, 1999). Ozone is widely used as a disinfectant in drinking water and wastewater treatment. The strong biocidal characteristics of ozone are due to a combination of its high oxidizing potential and its ability to diffuse through biological membranes (Hunt and Marinas, 1997). The effects of ozone on various microorganisms have received much attention because of its increasing use in water and sewage disinfection (Katzenelson and Biedermann, 1976; Boyce *et al.*, 1981; Glaze, 1987). Oxidation reactions of ozone in water follow two major pathways: direct oxidation by molecular ozone, and indirect oxidation by free-radical species formed from the autodecomposition of ozone, and from reactions between ozone and some inorganic and organic compounds (Hoigne and Bader, 1976, 1978; Forni *et al.*, 1982). Ozone is 1.5-times stronger than chlorine and is effective over a much wider spectrum of microorganisms than chlorine and other disinfectants (Xu, 1999). Ozonation was recognized as GRAS (generally recognized as a safe substance) by FDA for treatment of bottled water for drinking when used in accordance with good manufacturing practices (FDA, 1995). The affirmation as GRAS triggered broad usage of ozone gas in the food industry. The interest in ozone as an alternative to chlorine and other chemical disinfectants in cleaning and disinfection operations is based on its high

Novel Thermal and Non-Thermal Technologies for Fluid Foods. DOI: 10.1016/B978-0-12-381470-8.00009-8

biocidal efficacy, wide antimicrobial spectrum, absence of by-products that are detrimental to health and the ability to generate it on demand, *in situ*, without needing to store it for later use. Ozone also has the significant advantage of environmentally friendly technology that reduces the company's environmental costs and facilitates their compliance with statutory obligations (Pascual *et al.*, 2007). Ozone has also been used as a surface disinfecting agent (Greene *et al.*, 1993; Hampson, 2000; Moore *et al.*, 2000; Lagrange *et al.*, 2004). In 2001, ozone has been approved as an antimicrobial agent in foods in the US (US FDA, 2001). This approval of ozone as a direct additive to food triggered interest in ozone applications development, and industry guidelines for apple juice and cider were published by the USFDA in 2004, which also highlighted gaps in the scientific knowledge (US FDA, 2004). The oxidizing action of ozone for food application can be affected by some environmental factors, such as pH of medium, temperature, relative humidity and organic compounds in the medium (Kim *et al.*, 1999) and these factors are described in this chapter.

9.2 GENERATION OF OZONE

Ozone (O_3) results from the rearrangement of atoms when oxygen molecules are subjected to high-voltage electric discharge. The product is a bluish gas with a pungent odor and strong oxidizing properties (Muthukumarappan *et al.*, 2000). In nature, ozone generation occurs when oxygen molecules react in the presence of electrical discharges, e.g., lightning, and by the action of high energy electromagnetic radiation (Muthukumarappan *et al.*, 2009). Ozone is generated by reaction of free oxygen radicals with diatomic oxygen to form triatomic oxygen molecules. Generation of the free oxygen radical occurs by breakage of strong O–O bonds, requiring a significant energy input. Ultraviolet (UV) radiation and corona discharge methods can be used to initiate free radical oxygen formation and thereby generate ozone. In addition to photochemical (UV radiation) and electric discharge methods, ozone can be produced by chemical, thermal, chemonuclear and electrolytic methods (Kim *et al.*, 1999).

9.2.1 Electrical (Corona) Discharge Method

In this method, adequately dried air or O_2 is passed between two high-voltage electrodes separated by a dielectric material, which is usually glass. Air or concentrated O_2 passing through an ozonator must be free from particulate matter and dried to a dew point of at least −60°C to properly protect the corona discharge device. The ozone/gas mixture discharged from the ozonator normally contains 1–3% ozone when using dry air, and 3–6% ozone when using high-purity oxygen as the feed gas (Muthukumarappan *et al.*, 2000).

When a voltage is supplied to the electrodes, a corona discharge forms between the two electrodes, and the O_2 in the discharge gap is converted to ozone.

A corona discharge is a physical phenomenon characterized by a low-current electrical discharge across a gas-containing gap at a voltage gradient, which exceeds a certain critical value. First, oxygen molecules (O_2) are split into oxygen atoms (O), and then the individual oxygen atoms combine with remaining oxygen molecules to form ozone (O_3) (Muthukumarappan *et al.*, 2008).

9.2.2 Electrochemical (Cold Plasma) Method

In this method, an electrical current is applied between an anode and a cathode in an electrolytic solution containing water and a solution of highly electronegative anions. A mixture of oxygen and ozone is produced at the anode. The advantages associated with this method are use of low-voltage DC current, no feed gas preparation, reduced equipment size, possible generation of ozone at high concentration, and the generation in water (Muthukumarappan *et al.*, 2008).

9.2.3 Radiochemical Ozone Generation (RCOG)

High-energy irradiation of oxygen is bound to produce ozone. Although this technique is not yet used commercially in potable or waste water treatment, it is anticipated that in the future this will find much more widespread application (Muthukumarappan *et al.*, 2009). Even with the favorable thermodynamic yield of the process and the interesting use of waste fission isotopes, the cheminuclear ozone generation process has not yet become a significant application in water or waste water treatment due to its complicated process requirements (Pekarek, 2008).

9.2.4 Ultraviolet Method

In the UV method of O_3 generation, ozone is formed when O_2 is exposed to UV light of 140–190 nm wavelength. This splits the oxygen molecules into oxygen atoms, which then combine with other oxygen molecules to form O_3 (Muthukumarappan *et al.*, 2000).

Although a wide range of UV bulbs are available, an appropriate wavelength range varies from 180 to 254 nm. However, with currently available technology, these bulbs are still not an economical or efficient way to generate ozone, producing only up to 0.3–0.4% by weight (Sacco, 2009).

9.3 OZONE IN LIQUID FOOD PROCESSING

Ozone is a triatomic allotrope of oxygen, with greater solubility at lower temperatures and has the strong oxidation potential that conveys bacteriocidal and viricidal properties. Restaino *et al.* (1995) investigated the antimicrobial effects of ozonated water against food-related microorganisms and determined that ozone effectively killed Gram-positive bacteria such as *Listeria monocytogenes, Staphylococcus aureus, Bacillus cereus, Enterococcus faecalis*, and Gram-negative bacteria such as *Pseudomonas aeruginosa* and *Yersinia enterocolitica*. Ozone is an effective disinfectant for the reduction of protozoan parasites in water (Gyurek *et al.*, 1999; Rennecker *et al.*, 1999; Widmer *et al.*, 2002). Ozone was described as an effective alternative to chlorine for pathogen reduction in water (Thurston-Enriquez *et al.*, 2005). Ozone improves coagulation and is effective at controlling color, taste, and odor in water (AWWA, 1995). Ozone is also reported to inactivate yeast cells of *Candida albicans* and *Zygosaccharomyces bailli* and spores of *Aspergillus niger*.

Ozone has been reported to inactivate *E. coli* in orange juice and apple juice (Patil *et al.*, 2009a, 2010a). The effectiveness of ozone for reducing two strains of *L. monocytogenes* (ATCC 7644 and NCTC 11994) and *L. innocua* NCTC 11288 by 5-log cycles in orange juice was also reported by Patil *et al.* (2010b). Inactivation of *E. coli* O157:H7 and *Salmonella* in apple cider and orange juice treated with ozone in combination with antimicrobials (dimethyl dicarbonate (DMDC; 250 or 500 ppm) or hydrogen peroxide (300 or 600 ppm)) was evaluated by Williams *et al.* (2005). Their study demonstrated that combining ozone and antimicrobial treatments followed by refrigerated storage for 24 h can be used to achieve a 5-log cycle reduction of *E. coli* O157:H7 and *Salmonella* in apple cider and orange juice.

Ozonation of oil has gained attention recently. Ozonated vegetable oils have antibacterial and fungicidal effects, attributed with potential applications in food, pharmaceutical, and cosmetic industries. Antibacterial activity of ozonated sunflower oil (Sechi *et al.*, 2001) and antifungal activity of ozonated olive oil (Geweely, 2006) has been reported.

Ozone destruction of bacteria is accomplished by attack on the bacterial membrane glycoproteins and/or glycolipids. For most planktonic microorganisms (i.e. organisms suspended in water not attached to a surface) the order of disinfective effectiveness is (Montgomery, 1985):

$$O_3 > HOCl > ClO_2 > OCl^- > NH_2Cl > NHCl_2$$

9.4 FACTORS AFFECTING EFFICACY OF OZONE PROCESSING

There are different parameters that affect the disinfection ability of ozone in liquid processing treatment. Extrinsic and intrinsic parameters that affect the ozone efficacy include flow rate, ozone concentration, temperature, pH, and presence of solid contents (organic matter). These parameters are individually described below.

9.4.1 Extrinsic Parameters

9.4.1.1 Flow Rate

Depending on the gas flow rate applied for ozone production, different bubble sizes are produced. Bubble size has been shown to have an effect on ozone's solubilization rate and disinfection efficacy. Ahmad and Faroq (1985) reported that ozone mass transfer and disinfection efficacy increased as bubble size decreased (ozone bubble size was varied while all other factors were kept constant). The higher interfacial area available for mass transfer at the smaller bubble size may have been responsible for this effect. Decreasing the bubble diameter from 1 cm to 0.1 cm increases the contact area by 32 times (Ogden, 1970). Free suspended bacteria migrate toward the ozone bubbles due to their surface active properties and are preferentially inactivated by comparatively high ozone concentrations at the gas–liquid interface of the bubble (Hill and Spencer, 1974). Gong *et al.* (2007) showed a non-linear dependence of the ozone dissolution efficiency on the initial bubble size in a study of the mass transfer process of ozone dissolution in a bubble plume (bubble plumes are an interacting collection of bubbles formed by some event) inside a rectangular water tank. The dissolution efficiency varies rapidly when the initial bubble size reaches a certain value while the change in efficiency is much slower at other bubble sizes. A simple analysis of the mass transfer inside a bubble plume reveals that the mass transfer process is affected by a number of factors. They are the gross area of the contact surfaces between the gas and liquid phase, the contact time of bubbles with the liquid, and the mass transfer rate of individual bubbles. The gross contact area depends on the bubble diameter. At the same void fraction, the contact area increases as the size of the bubble decreases. The contact time depends mainly on the bubble diameter, in that it increases when the bubble size reduces, as smaller bubbles rise slower than larger ones. In general, reducing the diameter of ozone bubbles is useful for improving the efficiency of ozone's utilization. A longer distance or time of rising is important for larger bubbles to achieve a higher efficiency, since they need more time to transfer the dissoluble gas inside the bubbles (Gong *et al.*, 2007). Patil *et al.* (2009b) studied the effect of different flow rates on the inactivation of *E. coli* ATCC 25922 in a Tryptic Soy Broth (TSB) model system. The inactivation in a bubble column was found to be dependant on gas flow rate. Increased contact area results in enhanced disinfection efficacy allowing more microorganisms to come into direct contact with ozone bubbles.

9.4.1.2 Concentration

Ozone concentration present or available in the medium is another parameter determining ozone efficacy. Increased ozone concentration causes saturation and thus makes addition of further ozone to the reactor ineffective, resulting in longer times to achieve the same log-reduction values. This phenomenon was investigated by Patil *et al.* (2009b) when two flow rates (0.06 l/min and 0.12 l/min) were chosen for five different levels of ozone concentration (Table 9.1). For both

Table 9.1 Ozone Concentrations at Two Selected Gas Flow Rates

Flow rate (l/min)	Concentration (μg/ml)	Time required for 5-log reduction (min)
0.06	28–49	—
	60–70	—
	82–87	50.5
	103–114	42.2
	115–120	24.2
0.12	17–28	—
	33–40	107.04
	46–56	28.3
	67–70	24.5
	72–75	20.0

Adapted from Patil et al. *(2009b).*

flow rates, the highest concentration was the most effective to inactivate *E. coli* ATCC 25922 in TSB (Fig. 9.1). It is evident from Figure 9.1 that there is a stepwise decrease in the efficiency of *E. coli* inactivation at the lower concentration levels 1 and 2 (28–70 μg/ml and 17–28 μg/ml for flow rates of 0.06 and 0.12 l/min, respectively), where inactivation was insufficient to calculate D-values (Table 9.1). Similar findings of faster inactivation with shorter lag time and smaller D-values at the highest ozone concentrations were reported by Steenstrup and Floros (2004) during inactivation of *E. coli* O157:H7 in apple cider by ozone (Table 9.2). The concentrations applied were all greater than 1000 ppm.

In another study of ozone disinfection kinetics of *E. coli* in water, increased ozone efficiency was reported at higher ozone concentrations tested (Zuma *et al.*, 2009). At a flow rate of 2 l/min with an ozone concentration of 0.906 mg/l, *E. coli* populations of approximately 1×10^8 cells/ml were reduced by 4 log in 6 min, whereas with a higher ozone concentration of 4.724 mg/l the reduction was much higher, approximately 5 log in 4 min (Fig. 9.2).

9.4.1.3 Temperature

Ozone solubility in water is 13 times that of oxygen at 0–30°C and it is progressively more soluble in colder water (Rice, 1986). The solubility ratio for ozone increases as the temperature of water decreases (Bablon *et al.*, 1991). As temperature increases, ozone becomes less soluble and less stable, with an increase in the decomposition rate (Rice *et al.*, 1981). The mass transfer of ozone gas into the liquid phase is also influenced by temperature and pH. The inactivating capabilities of ozone are in line with the decreasing temperature (Farooq *et al.*, 1977a,

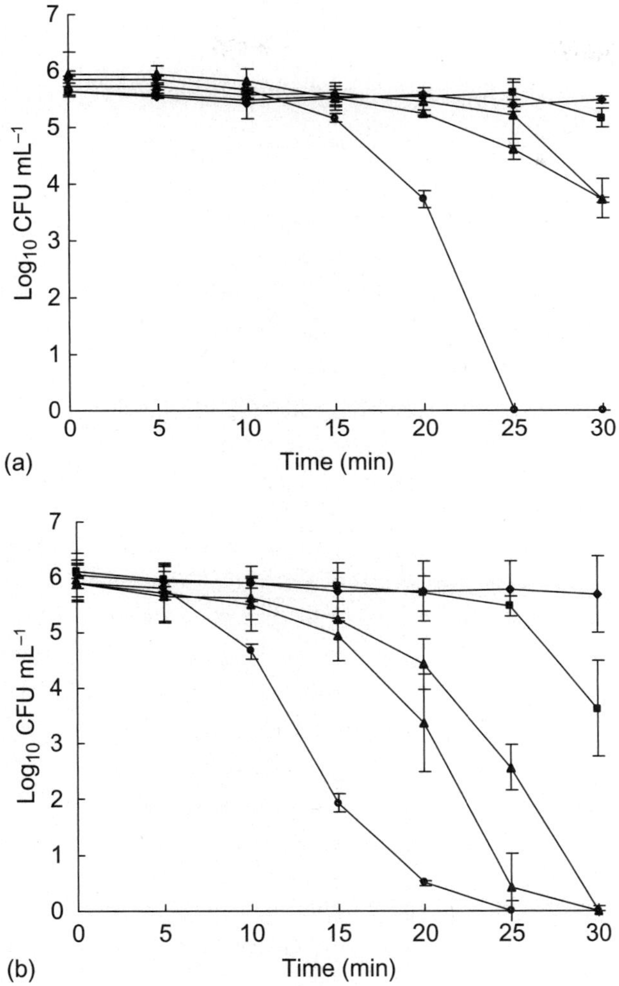

FIGURE 9.1

Effect of ozone concentration on the inactivation of *E. coli.* (a) At 0.06 l min^{-1} flow rate ozone concentrations: (♦) level 1, (■) level 2, (▲) level 3, (Δ) level 4, (o) level 5; and (b) at 0.12 l min^{-1} flow rate ozone concentrations: (♦) level 1, (■) level 2, (▲) level 3, (Δ) level 4, (o) level 5 respectively.

[Adapted from Patil et al. *(2009b).]*

Vaughn *et al.*, 1987, Hunt and Marinas, 1997). Patil *et al.* (2009b) examined the effects of four different temperatures on the efficacy of ozone inactivation on *E. coli* in TSB. Survival curves are shown in Figure 9.3. Ambient temperature (12–15°C) gave the best inactivation levels with a D-value of 20 min.

Table 9.2 Lag Times and D-Values for *E. coli* O157:H7 Treated with Selected Concentrations of Gaseous Ozone at 20°C in Apple Cider

Ozone Concentration	Lag Time (min)	D-Value (min)	Processing Time for 5-log reduction (Lag Time + 5D)
Low	2.7 ± 0.4	0.71 ± 0.07	6.25
Medium	2.6 ± 0.5	0.42 ± 0.09	4.70
High	1.8 ± 0.1	0.33 ± 0.01	3.45

Adapted from Steenstrup and Floros (2004).

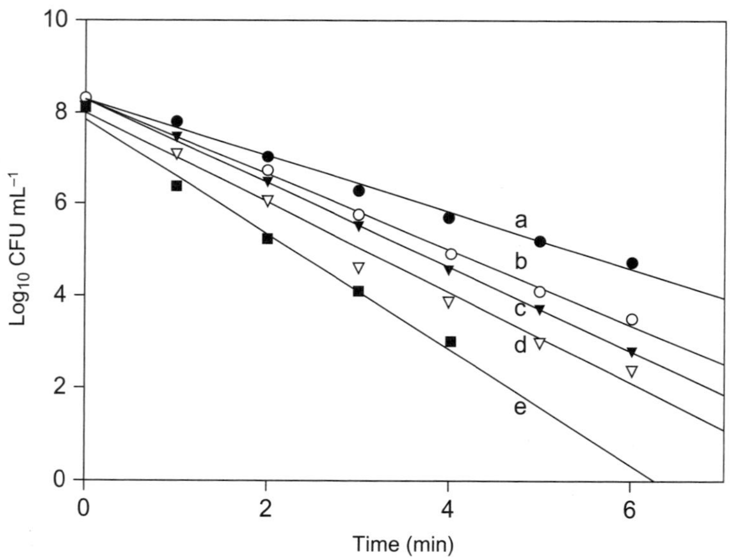

FIGURE 9.2

Effect of applied voltage on *E. coli* inactivation rate. Plot of log CFU/ml *vs* time. Flow rate of 2 l/min. Settings for curves are as follows: a, 40 V; b, 50 V; c, 60 V; d, 70 V; and e: 80 V. [Ozone]/(mg/l) for curves are as follows: a, 0.91; b, 1.78; c, 2.78; d, 3.44; and e, 4.72.

[Adapted from Zuma et al. (2009).]

Steenstrup and Floros (2004) also reported that the processing times for inactivation of *E. coli* in apple cider varied from 6.5 to 13.4 min, and generally increased with decreasing temperature. A higher rate of ozone inactivation with increasing temperature was obtained for *B. Subtilis* spores and *Cryptosporidium*

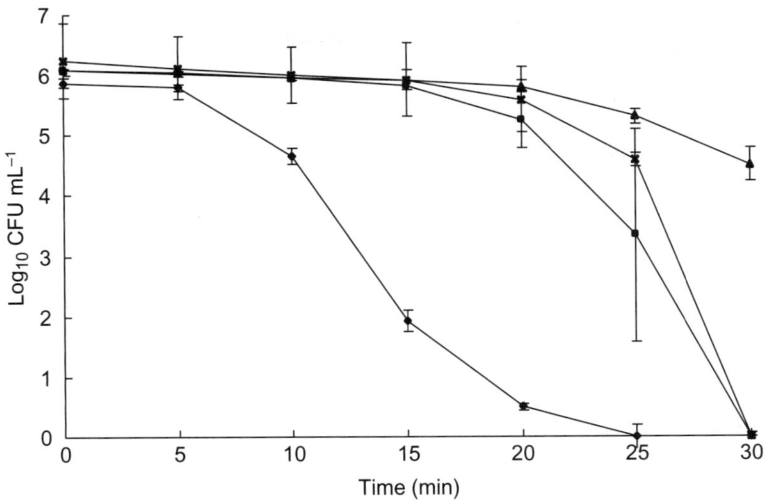

FIGURE 9.3

Effect of temperature on the efficacy of ozone inactivation of *E. coli*. (♦) ambient, (■) 20°C, (▲) 25°C, (×) 30°C.

[Adapted from Patil (2009b).]

parvum oocysts (Corona-Vasquez, 2002; Larson and Marinas, 2003) (Fig. 9.4). Dow *et al.* (2006) indicated that increasing temperature over 7–22°C had the strongest influence on the inactivation rate of *B. subtilis* spores in oxidant-demand-free phosphate buffer and also led to a corresponding decrease in the lag phase and CT2 log. Driedger *et al.* (2001) reported a reduced ozone inactivation rate of *C. parvum* with decreasing temperature. Finally, there is no clear consensus on the effect of temperature on the biocidal efficacy of ozone; for example, a drop in the temperature of the aqueous medium increases ozone solubility and stability, augmenting its availability in the medium and consequently efficacy rises. The simultaneous contribution of these two factors (solubility/stability and reactivity) to ozone efficacy can vary with experimental conditions, making it difficult to predict the influence of temperature on a particular application (Pascual *et al.*, 2007).

9.4.2 Intrinsic Parameters

9.4.2.1 pH

The effect of pH on ozone inactivation is mainly attributed to the fact that ozone decomposition rate changes substantially with changes in pH (Farooq *et al.*, 1977b; Roy *et al.*, 1980). Patil *et al.* (2010a) investigated the efficacy of gaseous

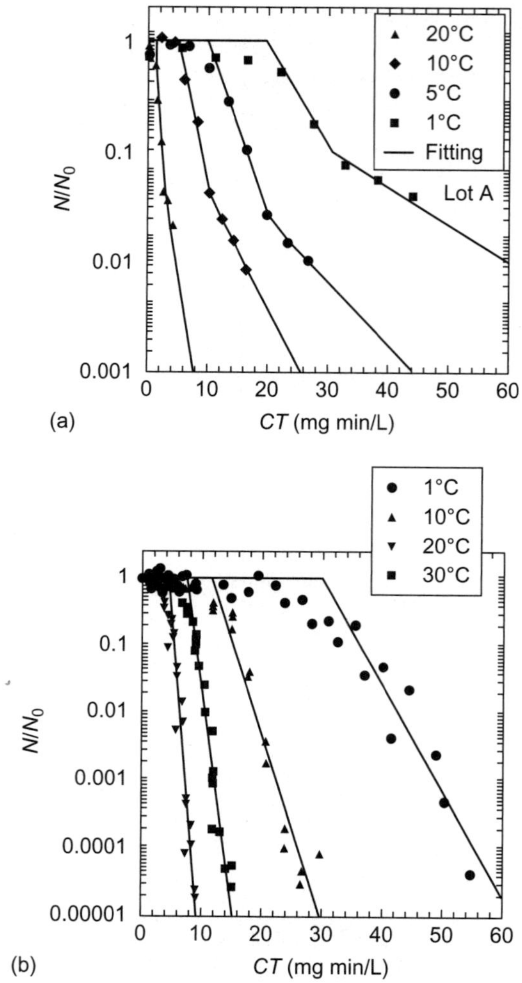

FIGURE 9.4

Effect of temperature on (a) the inactivation kinetics of *B. subtilis* spores (Stock 2) with ozone at pH 7.0; and (b) the kinetics of primary inactivation of *C. parvum* oocysts with ozone at pH 7.0.

[Adapted from Larson and Marinas (2003) and Corona-Vasquez (2002).]

ozone (0.048 mg/min per ml) on the inactivation of *E. coli* strains (ATCC 25922 and NCTC 12900) in apple juice of a range of pH levels (3.0, 3.5, 4.0, 4.5, and 5.0). In this work, it was observed that ozone inactivation of *E. coli* was much faster at the lower pH. The ozone treatment duration required for achieving a

5-log reduction was 4 min at the lowest pH and 18 min at the highest pH studied. Similar results by Zuma *et al.* (2009) showed ozone-mediated disinfection kinetics of *E. coli* in water. They investigated the effect of different control parameters and pH (4.93–9.16) on the disinfection rate of *E. coli*, and reported that the kinetics of inactivation of *E. coli* were much faster in acidic than basic medium. The inactivation of *E. coli* (10^8 CFU/ml) at a flow rate of 2 l/min with an ozone concentration of 0.906 mg/l at $25 \pm 2°C$ resulted in a higher rate constant of 2.209 min at a pH value of 4.93 than at pH 9.16 with a rate constant of 1.126 min, indicating a 56% decrease in disinfection rate (Zuma *et al.*, 2009). Lin and Wu (2006) studied the potential of ozone for inactivation of intra- and extracellular enterovirus 71 as affected by a number of parameters; flow rate (100, 80, 60 mg/h), a constant flow rate of 80 mg/h for a pH of 6.7, and variable pH (3, 5, 7, 9, and 11). The results demonstrated that EV71 inactivation by ozone was related to the kinetics of ozone solubility. However, the inactivation was dependent on the ozone input flow rate and was positively enhanced at acidic pH. Vaughn *et al.* (1987) and Harakeh and Butler (1984) also reported that the efficacy of ozone decreased at alkaline pH for rotavirus and Poliovirus type 1 in ozone-demand-free phosphate carbonate buffer and effluent, respectively. Farooq *et al.* (1977b) reported an increased survival rate of *Mycobacterium fortuitum* during ozone treatment when pH was increased from 5.7 to 10.1. Leiguarda *et al.* (1949) reported that bacteriocidal efficiency of ozone on *E. coli* and *C. perfringens* was slightly greater at pH 6.0 than at pH 8.0. Foegeding (1985) studied ozone inactivation of *Bacillus* and *Clostridium* spores at different pH values and found that acidic pH values enhanced the lethality of ozone. The authors attributed this increased survival to a smaller ozone residual as the pH of water increased. At high pH, the chain reactions of ozone decomposition result in the formation of numerous radical species with high oxidative capabilities. Normally, under acidic conditions ($pH < 4$), the direct pathway occurs; above pH 10 it changes to indirect pathway (Gottschalk *et al.*, 2010).

$$O_3 + OH^- \rightarrow HO_2^- + O_2$$

$$O_3 + HO_2^- \rightarrow {}^{\bullet}OH + O^{\bullet}{}_2 + O_2$$

It has not been well established whether molecular ozone or the radical species are responsible for inactivation of microorganisms. Some researchers have suggested that direct reaction with molecular ozone is the predominant mechanism of inactivation of microorganisms (Finch *et al.*, 1992; Labatiuk *et al.*, 1994; Hunt and Marinas, 1997), while others suggest indirect reactions with radicals to be responsible for inactivation (Bancroft *et al.*, 1984; Dahi, 1996). It is likely that the relative importance of direct and indirect reactions with ozone in determining microbial inactivation responses will vary between microorganisms (Blatchley and Nimrata, 2002).

9.4.2.2 Organic Matter

Ozone demand can be caused by certain organics, inorganics, or suspended solids. The inhibiting effect depends on the types of turbidity rather than the turbidity level. Dissolved organic matter reduces the disinfection activity by consuming ozone to produce compounds with little or no microbiocidal activity, thereby reducing the concentration of active species available to react with microorganisms. Inactivation of viruses either in sewage effluent or in secondary effluent was reduced in the presence of increased level of organics or in non-ozonated wastewater (Majumdar, 1973; Harakeh and Butler, 1984). Williams *et al.* (2005) studied the inactivation of *E. coli* in orange juice, and found that the efficacy of ozonation was reduced in the presence of ascorbic acid and organic matter. Patil *et al.* (2009a) investigated the inactivation of *E. coli* in orange juice and reported that efficacy of ozone inactivation depends on the juice type, specifically showing a faster inactivation rate achieved in juice with either low or nil pulp content. In wastewater, proteins, carbohydrates, lipids, and organic amines will elevate the concentration of dissolved organic carbon. Oxidizing disinfectants like ozone will lose bacteriocidal strength through reaction with organic matter. The reaction products will generally have weak or no bacteriocidal activity.

Ozone-driven inactivation of microorganisms in wastewaters required longer contact times and larger doses than inactivation in a demand-free system. Hunt and Marinas (1999) reported a slower rate for the ozone inactivation of *E. coli* in the presence of humic acid than in the absence of natural organic matter due to faster decomposition of dissolved ozone and, thus, the lower exposure of *E. coli* to the disinfectant ozone. Farooq *et al.* (1977b) reported the interference of organic matter present in the water with the inactivation of *C. parapsilosis* by exerting an ozone demand. Hence the applied ozone dosage had to be sufficient to overcome this demand before a free residual concentration was available for the inactivation of the microbial organisms. Schuchmann and Von Sonntag (1989) explained ozone effectiveness in reducing the load of organic matter (added D-glucose) in raw water purification. They found that a direct mode of reaction by ozone predominated at high-glucose-concentration; however, the $^{\bullet}OH$ pathway predominated at low-glucose-concentration, especially at higher pH (e.g., 9.0). Janex *et al.* (2000) reported the biggest influence of organic matter on the ozone demand of the effluents during wastewater disinfection by ozone. Dissolved organic or inorganic matter affect the disinfection rate because they compete with the microorganisms for disinfectant. Macauley *et al.* (2006) stated that the inactivation of bacteria was ineffective up to an ozone dose of 20 and 10 mg/l in wastewater samples obtained at two different swine-production facilities located in central Missouri. It implies that at a low ozone dose, the majority of ozone was preferentially consumed by natural organic materials (non-bacterial) present in lagoon samples. At the ozone dose of 100 mg/l, the efficiency of bacteria inactivation could reach 3.3- and 3.9-log for Lagoon A and Lagoon B samples, respectively (Fig. 9.5).

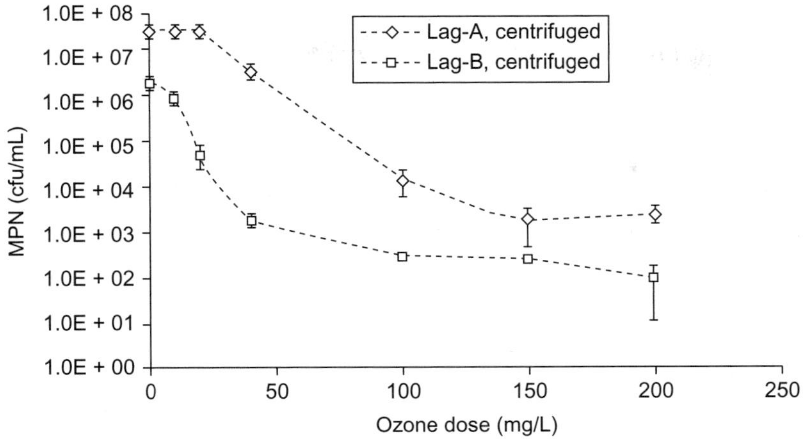

FIGURE 9.5

Effect of ozone dose on bacteria inactivation. Data are from three-tube most probable number (MPN) assays that were run twice (mean of MPN ± standard deviation). Experimental conditions: 10 mM KH_2PO_4, pH 7.7, 2.5 h disinfection time (on ice). *[Adapted from Macauley* et al. *(2006).]*

9.5 MICROBIAL INACTIVATION

The bacteriocidal effects of ozone have been studied in a variety of organisms, including Gram-positive and Gram-negative bacteria as well as spores and vegetative cells (Ishizaki *et al.*, 1987; Restaino *et al.*, 1995) (Table 9.3)

Ozone has been shown to reduce populations of *E. coli* O157:H7 in phosphate buffer (Byun *et al.*, 1998) while its preservation efficacy has also been evaluated in a variety of food products including milk, gelatin, albumin, casein, and meat products (Kim *et al.*, 1999). Selma *et al.* (2007) reported that ozone treatment at levels of 1.6 and 2.2 ppm for 1 min decreased *Shigella sonnei* population in water by 3.7- and 5.6-log CFU/ml, respectively.

Franz and Gagnaux (1971) found ozone effective in the inactivation of coliforms and spore-forming bacteria at 0.1–0.2 mg/l and 1.6–3.2 mg/l ozone during 8 min treatment, respectively.

Ozone (5–35 mg/l) decreased the microbial population of skim milk in a treatment time of 5–25 min and pressurized ozone was effective for decreasing psychrotrophic counts by 2.4-logs (Rojek *et al.*, 1995). Treatment of whey and apple juice also produced favorable microbial reduction. Greene *et al.* (1993) reported that ozone treatment at 0.5 ppm for 10 min eliminated more than 99% of the population of milk spoilage bacteria, such as *P. fluorescens* and *Alcaligenes faecalis* on stainless steel plates. Sobsey (1989) reviewed studies to inactivate

Table 9.3 Inactivation of Bacteria, Viruses, Yeasts, and Protozoa by Ozone

Organism	Per Cent Reduction or $\log_{10}$ Reduction	Time (min)	Concentration (mg/l)	pH	Temperature (°C)	Medium	References
Escherichia coli	99.99	1.67	0.23–0.26	7	24	O_3 demand free water	Farooq and Akhlaque (1983)
L. pneumophila	99.997	20	0.32	7	24	Sterile distilled water	Edelstein *et al.* (1982)
Mycobacterium fortuitum	90	1.67	0.23–0.26	7	24	O_3 demand free water	Farooq and Akhlaque (1983)
Salmonella typhimurium	99.995	1.67	0.23–0.26	7	24	O_3 demand free water	Farooq and Akhlaque (1983)
E. coli	17	19	Initial 0.85 Residual 0	7.5	16	Raw wastewater	Joret *et al.* (1982)
E. coli	97	19	Initial 1.4 Residual 0	7.5	16	Raw wastewater	Joret *et al.* (1982)
Bacillus cereus	>2.0	5	0.12		28	O_3 demand free water	Broadwater *et al.* (1973)
B. cereus spores	>2.0	5	2.29		28	O_3 demand free water	Broadwater *et al.* (1973)
Leuconostoc mesenteroides	1.3 to ~7	0.5	0.3–3.8 μg/ml	5.9	25		Kim and Yousef 2000
Listeria monocytogenes	0.7 to ~7	0.5	0.2–1.8 μg/ml	5.9	25		Kim and Yousef 2000
E. coli ATCC 25922	5.0	11.86	75–78 μg/ml	3.5–4.0	15–18	Unfiltered orange juice	Patil *et al.* (2009a)

E. coli ATCC 25922	5.0	4.76	75–78 μg/ml	3.5–4.0	15–18	Filtered orange juice without pulp	Patil *et al.* (2009a)
E. coli NCTC 12900	5.0	11.30	75–78 μg/ml	3.5–4.0	15–18	Unfiltered orange juice	Patil *et al.* (2009a)
E. coli NCTC 12900	5.0	4.90	75–78 μg/ml	3.5–4.0	15–18	Filtered orange juice without pulp	Patil *et al.* (2009a)
E. coli	5.0	0.5	75–78 μg/ml	3.0	15–18	Model orange juice	Patil *et al.* (2009a)
Shigella sonnei	3.7	1	1.6 ppm			Ozonated deionized water	Selma *et al.* (2007)
S. sonnei	5.6	1	2.2 ppm			Ozonated deionized water	Selma *et al.* (2007)
E. coli O157:H7 (five strain mixture)	6.0	45	0.9 g/h	3.8	50	Apple cider	Williams *et al.* (2004)
E. coli O157:H7 (five strain mixture)	6.0	75	0.9 g/h	3.8	50	Orange juice	Williams *et al.* (2004)
Salmonella (five–serovar mixture)	4.8	15	0.9 g/h	3.8	50	Apple cider	Williams *et al.* (2004)
Salmonella (five–serovar mixture)	5.0–6.0	15	0.9 g/h	3.8	50	Orange juice	Williams *et al.* (2004)
E. coli O157:H7 (five strain mixture)	4.8	240	0.9 g/h	3.8	4	Apple cider	Williams *et al.* (2004)
E. coli O157:H7 (five strain mixture)	5.4	240	0.9 g/h	3.8	4	Orange juice	Williams *et al.* (2004)

(*Continued*)

Table 9.3 (Continued)

Organism	Per Cent Reduction or log_{10} Reduction	Time (min)	Concentration (mg/l)	pH	Temperature (°C)	Medium	References
Salmonella (five-serovar mixture)	4.5	240	0.9 g/h	3.8	4	Apple cider	Williams *et al.* (2004)
Salmonella (five-serovar mixture)	4.2	240	0.9 g/h	3.8	4	Orange juice	Williams *et al.* (2004)
E. coli O157:H7 (five strain mixture)	<5.0	240	0.9 g/h	3.8	20	Apple cider	Williams *et al.* (2004)
E. coli O157:H7 (five strain mixture)	<5.0	240	0.9 g/h	3.8	20	Orange juice	Williams *et al.* (2004)
Salmonella (five-serovar mixture)	<5.0	240	0.9 g/h	3.8	20	Apple cider	Williams *et al.* (2004)
Salmonella (five-serovar mixture)	<5.0	240	0.9 g/h	3.8	20	Orange juice	Williams *et al.* (2004)
Viruses							
Poliovirus type 1 (Mahoney)	99.7	1.67	0.23–0.26	7	24	O_3 demand free water	Farooq and Akhlaque (1983)
CoxSackie virus B5	90	4.9	0.18	7.2	20	Activated sludge	Harakeh and Butler (1984)
Poliovirus type 1	99.99	2.5	0.40	7.2	20	Reactor effluent	Harakeh and Butler (1984)
Enteric virus	92	19	Initial 1.4 Residual 0	7.5	16	Raw wastewater	Joret *et al.* (1982)
Enteric virus	>98	19	Initial 4.10 Residual 0.02	7.8	18	Raw wastewater	Joret *et al.* (1982)

Human rotavirus	80	10	0.31	7.2	20	Activated sludge	Harakeh and Butler (1984)
Poliovirus type 1	99	9	0.2	4	20	Activated sludge	Harakeh and Butler (1984)
H5N1	4 log	10	0.5 (residual ozone level)		22±2	Treated water	Lénès *et al.* (2010)
H1N1	4 log	10	0.5 (residual ozone level)		22±2	Treated water	Lénès *et al.* (2010)
H5N1	4 log	10	1.0 (residual ozone level)		22±2	Treated water	Lénès *et al.* (2010)
H1N1	4 log	10	1.0 (residual ozone level)		22±2	Treated water	Lénès *et al.* (2010)
Yeasts							
Candida parapsilosis	99.8	1.67	0.23–0.26	7	24	O_3 demand free water	Farooq and Akhlaque (1983)
C. tropicalis	99	0.30–0.08	0.02–1.0	7.2	20	O_3 demand free water	Kawamura *et al.* (1986)
Protozoa							
Giardia lamblia	2.0 log	1.1	0.7 mg/l	7	5	Water	Wickramanayake *et al.* (1984)
G. muris	2.0	2.8	0.5	7	5	Water	Wickramanayake *et al.* (1984)
Naegleria gruberi	2.0 log	2.1	2.0	7	5	Water	Wickramanayake *et al.* (1984)
Cryptosporidium parvum	>1.0	5	1.0	7	25	O_3 demand free water	Korich *et al.* (1990)

Modified from Weavers and Wickramanayake (2001).

health-related microorganisms in water by several disinfectants and concluded that Gram-positive bacteria, including *S. aureus* and *Bacillus* spp., and the *Mycobacteria* were more resistant than were Gram-negatives. Lee and Deniniger (2000) observed the dominance of Gram-positive bacteria among the surviving microorganisms in ozonated drinking water. Gram-negative bacteria were substantially more sensitive to ozone in pure water than the Gram-positive bacteria tested including *L. monocytogenes* (Restaino *et al.*, 1995). However, Kim and Yousef (2000) and Kim *et al.* (1999) found the resistance of food-borne spoilage and pathogenic bacteria treated with ozone under identical conditions followed this descending order: *E. coli* O157:H7, *P. fluorescens*, *Leuconostoc mesenteroides*, and *L. monocytogenes*. *Alicyclobacillus acidocaldarius* vegetative cells and *Z. bailii* ascospores that commonly spoil fruit juices were inactivated rapidly with aqueous ozone.

Khadre and Yousef (2001) compared the effects of ozone and hydrogen peroxide against food-borne *Bacillus* spp. spores. It has been found that ozone was more effective than hydrogen peroxide. Steenstrup and Floros (2004) studied the inactivation of *E. coli* O157:H7 in apple cider and reported a 5-log reduction within 4–14 min of treatment, depending on temperature and ozone concentration. Dosti *et al.* (2005) investigated the efficacy of ozone against food spoilage bacteria in synthetic broth. Exposure of fresh 24-h bacterial cultures of *P. fluorescens, P. fragi, P. Putida, Enterobacter aerogenes, E. Cloacae*, and *B. Licheniformis* to ozone at 0.6 ppm for 10 min caused the highest bacterial population reduction, with a mean reduction of 7.3 log units. Patil *et al.* (2009a) reported the ozone inactivation of two strains of *E. coli* (ATCC 25922 and NCTC 12900) in orange juice at a concentration of 75–78 μg/ml. Reduction of *E. coli* cells by 5 log cycles was reported in 5–12 min, depending on the type of juice. Fast inactivation rates (reduction by 6.0 log cycles in 60 s) were reported in a model orange juice system. Ozone (75–78 μg/ml) inactivation of *Listeria* species by 5 log cycles was reported between 5.08 and 8.44 min of treatment in orange juice (Patil *et al.*, 2010b). Lezcano *et al.* (2001) reported ozone inactivation of Gram positive bacteria such as *Streptococcus faecalis, S. aureus* and yeast; *C. albicans* at concentrations from 0.3 to 2.5 mg of dissolved ozone per liter of water. Their study found the *S. aureus* isolate to be more resistant than *S. faecalis* or the *C. albicans* strain, and the longest time required to achieve total inactivation was 10 min. Exposure of *P. fluorescens*, *E. coli* O157:H7, *L. mesenteroides*, and *L. monocytogenes* to ozone at 2.5 ppm for 40 s caused a 5- to 6-log decrease (Kim and Yousef, 2000). Lin and Wu (2006) reported a significant reduction of intracellular virus EV 71 titer (≥99%, $p < 0.01$) after 45 or 60 min exposure of ozone (60 mg/h) but with low cell viability. Ozone has also been shown to be effective against a number of enveloped and non-enveloped viral species (Akey and Walton, 1985; Herbold *et al.*, 1989; Carpendale and Freeberg, 1991; Wells *et al.*, 1991; Arimoto *et al.*, 1996; Khadre and Yousef, 2002; Shin and Sobsey, 2003). Ozone is very effective for inactivating feline calicivirus (FCV) and adenovirus type 40 (AD40) in treated water and at low ozone concentrations

(Thurston-Enriquez *et al.*, 2005). Other researchers (Kim *et al.*, 1980; Hall and Sobsey, 1993) also reported the sensitivity of the bacteriophages f2, and MS2 to ozone. Rapid inactivation of bacteriophage f2 by 5- and 7-logs at 0.09 and 0.8 mg/l ozone concentrations during the first 5 s of exposure was reported. Hall and Sobsey (1993) reported complete inactivation of bacteriophage MS2 and hepatitis A virus with ozone residuals between 0.3 and 2.0 mg/l and inactivations ranged from >3.9-log to >6.0-log occurred within 5 s.

9.6 MICROBIAL INACTIVATION MECHANISM

Inactivation of bacteria by ozone is a complex process because ozone attacks numerous cellular constituents including proteins, unsaturated lipids, and respiratory enzymes in cell membranes, peptidoglycans in cell envelopes, enzymes and nucleic acids in the cytoplasm, and proteins and peptidoglycan in spore coats and virus capsids (Khadre *et al.*, 2001). Ozone is very unstable both in the gaseous phase and in solution, decomposing into hydroxyl (OH^-), hydroperoxy ($\bullet HO_2$) and superoxide radicals ($\bullet O_2^-$). The reactivity of ozone is attributed to the oxidizing power of these free radicals (Manousaridis *et al.*, 2005), which are responsible for microbial inactivation. Microorganisms are inactivated by disruption of the cell envelope or disintegration leading to cell lysis. Both molecular ozone and the free radicals produced by its breakdown play a part in this inactivation mechanism but there is no consensus on which is more decisive. The resultant disruption or lysis associated with ozone is a faster inactivation mechanism than that of other disinfectants, which require the disinfectant agent to permeate through the cell membrane in order to be effective. Generally, with regard to the spectrum of microbial action, each microorganism has an inherent sensitivity to ozone. Bacteria are more sensitive than yeasts and fungi. Spores are more resistant than vegetative cells. Due to the mechanism of ozone action, which destroys the microorganism through cell lysis, the development of resistance is not found (Pascual *et al.*, 2007).

The inactivation of bacteria by ozone can be considered as an oxidation reaction (Bringmann 1954). The bacterial membrane seems to be the first site of the attack with proteins and unsaturated lipids in the cell membrane being the primary targets (Christensen and Giese, 1954; Scott and Lesher, 1963; Pryor *et al.*, 1983). Ozone causes alteration in *E. coli* cell membrane permeability leading to leakage of cell contents and eventually causing lysis (Scott and Lesher, 1963). Dave (1999) found that treatment of *S. enteritidis* with aqueous ozone disrupted the cell membranes. Thanomsub *et al.* (2002), also supported the proposed mechanism of bacteria inactivation by ozone, namely cell membrane destruction and finally cell lysis reaction (Fig. 9.6).

Reactions of ozone with proteins and amino acids showed that the susceptibility to oxidation is in the order: cysteine = tryptophan = methionine > histidine

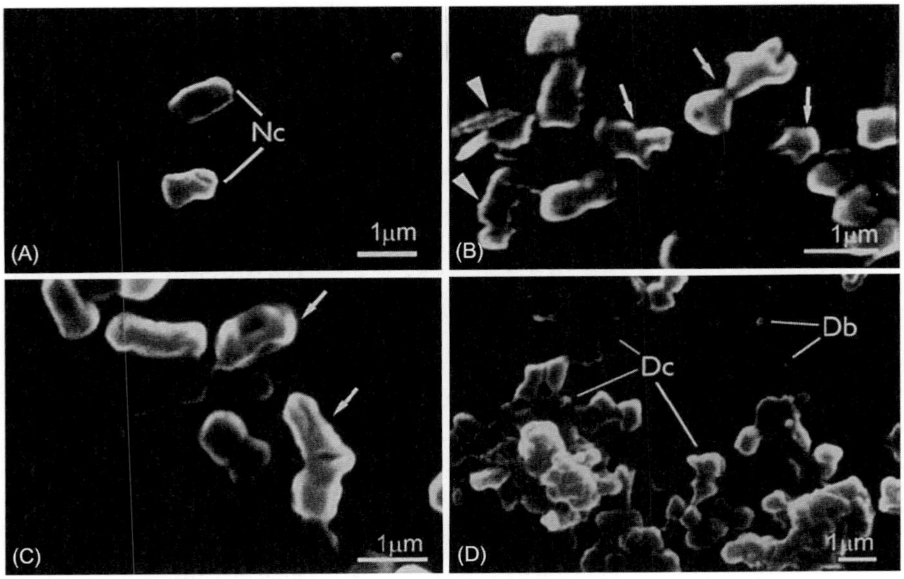

FIGURE 9.6

Scanning electron micrograph of *Escherichia coli* treated with ozone. After ozone exposure of 30–90 min, ultrastructural changes and damage to surface morphology were observed. A: Normal cell (Nc) of control group. B, C: Bacteria treated with ozone for 30 and 60 min, respectively, showed deformity (arrows) and destruction (arrowheads) of cells. D: Bacteria treated with ozone for 90 min showed destroyed cell (Dc) characteristics and some exploded debris (Db).

[Adapted from Thanomsub et al. *(2002).]*

(Mudd *et al.*, 1969, Pryor *et al.*, 1984). Proteins requiring any of these residues for catalytic activity are inactivated by ozone (Knight and Mudd, 1984; Berlett *et al.*, 1991). Inactivation of enzymes by ozone is probably due to oxidation of sulfhydryl groups in Cysteine residues (Chang, 1971).

Vrochinskii (1963) noted that ozone-treated bacteria lost their ability to degrade sugars and produce gases. Foegeding (1985) found that *B. cereus* spores with coat proteins removed, were rapidly inactivated by ozone compared to intact spores. Komanapalli *et al.* (1997) studied the effect of ozone on metabolic activities of *E. coli* K-12. The cell viability was unaffected with short-time ozone exposure. Glyceraldehyde-3-phosphate dehydrogenase showed the greatest susceptibility to ozone followed by oxidation of glutathione, a non-protein sulfhydryl, and total sulfhydryl compounds. Effects on malate dehydrogenase, lactate dehydrogenase, and glutathione disulfide reductase were negligible. Their results also indicated that the mutant (DNA repair deficient) and the wild type strains of *E. coli* were equally sensitive to ozone, suggesting that RecA DNA repair system may not play an important or vital role in ozone resistance. Fisher *et al.* (2000) reported that both catalase and superoxide dismutase protected listerial cells from ozone attack, with superoxide dismutase being more important than catalase in this protection. Takamoto *et al.* (1992) observed that ozone decreased enzyme activity in *E. coli* to a greater degree in the case of cytoplasmic β-galactosidase than in the case of the periplasmic alkaline phosphatase. Hinze *et al.* (1987) studied effects of ozone on ATP, cytosolic enzymes, and permeability of *Saccharomyces cerevisiae*. They reported the most drastic inactivation for glyceraldehyde-3-phosphate dehydrogenase and to lesser extents, pyruvate decarboxylase, phosphofructokinase-1 and nicotinamide adenine dinucleotide (NAD) alcohol dehydrogenase. The ATP, NAD and protein also accumulated in the medium indicating the permeabilization of yeast cells by ozone treatment. Ozone also affects both purines and pyrimidines in nucleic acid (Christensen and Giese, 1954; Scott and Lesher, 1963). Ozone produces single-strand breaks in DNA which, if unrepaired, cause extensive breakdown of DNA in E. coli, resulting in loss of cell viability (Hamelin *et al.*, 1977, 1978). Ishizaki *et al.* (1987) reported that the plasmid DNA present in *E. coli* cells is converted from closed circular DNA to open circular DNA by ozone. Komanapalli and Lau (1996) determined the effects of short- and long-interval ozone exposure on *E. coli* membrane permeability, total proteins, and plasmid DNA. They found that short-term exposures of *E. coli* K-12 to ozone gas had no effect on cell viability and tryptophan oxidation. Conversely, a significant leakage of protein, followed by malondialdehyde and nucleic acid components in the supernatant indicated an increase in the membrane permeability but did not affect viability, which progressively decreased with longer exposure. Ozone induces mutations in *E. coli* by penetrating into the cell and genetically altering cytoplasmic contents before destruction of the cell membrane (Hamelin and Chung, 1974). The mechanism of inactivation of bacteriophage f2 ribonucleic acid (RNA) was studied by Kim *et al.* (1980) who found that RNA was released from phage particles after the phage coat was broken into many protein subunit pieces. They suggested that ozone breaks the protein capsid, liberating RNA and disrupting virus adsorption to the host pili, and that the naked RNA may be secondarily inactivated but at a rate less than that for the RNA within the intact phage. Ozone inactivates f2 and T4 bacteriophage by attacking the protein capsid, liberating nucleic acid and inactivating the nucleic acid (Sproul *et al.*, 1982). Kim *et al.* (1984) confirmed the results of Sproul and Kim (1980) concerning bacteriophage T4; they found that ozone randomly destroyed the head, collar, contractile sheath, end plate, and tail fibers and liberated the DNA from the head. Shriniki *et al.* (1988) posed an inactivation hypothesis against Tobacco Mosaic Virus (TMV), that ozone attacks protein coat and RNA. The damaged RNA cross-links with amino acids of the coat protein subunits. TMV loses its infectivity because of its inability to uncoat. Yoshizaki *et al.* (1988) also found that aqueous ozone caused the coat proteins subunits of TMV to aggregate with each other and cross-link with the viral RNA. Ozone exposure caused damage to viral RNA of poliovirus 1 by altering two of the four polypeptide chains in the poliovirus protein coat leading to poliovirus inactivation (Roy *et al.*, 1981). Enveloped viruses such as hepatitis A virus (HAV) are expected to be much more resistant to ozone compared to non-enveloped viruses such as poliomyelitis (Khadre *et al.*, 2001). The inactivation of human immunodeficiency virus (HIV) by ozone indicated that it could exert several antiviral effects, including viral particle disruption, reverse transcriptase inactivation, and/or a perturbation of the ability of the virus to bind to its receptor on target cells (Wells *et al.*, 1991). Lin and Wu (2006) reported a significant effect of inactivation of intracellular EV71 in the case of 45 or 60 min ozone exposure which correlated with high cell death rate. Ozone inactivation of viruses occurs primarily in two ways: by lipid peroxidation and by protein peroxidation (Carbonneau *et al.*, 1991; Dianzani, 1993; Friedman and Stromberg, 1993). Murray *et al.* (2008) studied the inactivation of enveloped and non-enveloped viruses including herpes simplex virus type-1 (HHV-1, strainMcIntyre), vesicular stomatitis Indiana virus (VSIV), vaccinia virus (VACV, strain Elstree), adenovirus type-2 (HAdV-2), and the PR8 strain of influenza A virus (FLUAVA/PR/8/34/H1N1; FLUAV) using an ozone–oxygen delivery system. The results of the study showed that ozone exposure reduced viral infectivity by lipid peroxidation and subsequent lipid envelope and protein shell damage.

9.7 MYCOTOXINS

Ozone is reported to be effective for the detoxification and degradation of commonly occurring mycotoxins including aflatoxin, patulin, cyclopiazonic acid, secalonic acid D, ochratoxin A, and ZEN (McKenzie *et al.*, 1997; Lemke *et al.*, 1999). Concentrated ozone gas has been used to degrade and detoxify mycotoxins *in vitro* (McKenzie *et al.*, 1997). Ozone either completely degrades mycotoxins or causes chemical modifications, thereby reducing their biological activity (McKenzie *et al.*, 1998; Lemke *et al.*, 1999). However degradation or chemical modification is specific to the structure of mycotoxins under investigation. McKenzie *et al.* (1997) noted a greater resistance of aflatoxin B2 and G2 compared to B1 and G2. The presence of double bonds at C8–C9 position for B1 and G1 and the tendency of ozone to react at olefinic positions indicate the possibility for greater sensitivity towards ozone. Apart from the degradation of double-bonded aliphatic or polycyclic aromatic hydrocarbons, the presence of chlorinated ring structures (e.g., Ochratoxin) and nitrogen heterocycles (e.g., Fumonisin) in mycotoxins are also subject to ozone attack, resulting in free chlorine or amino acids (Lemke *et al.*, 1999). Therefore, the biological toxicity of certain mycotoxins may significantly decrease following ozone treatment due to the formation of new products.

McKenzie *et al.* (1997) reported the formation of Fumonisin B1 derivatives due to ozonation of Fumonisin B1 in aqueous solution. Similarly, Young *et al.* (2006) studied the degradation of trichothecene mycotoxins

(3-acetyldeoxynivalenol, 15 acetyldeoxynivalenol, diacetoxyscirpenol, 4-deoxynivalenol, fusarenon X, HT-2 toxin, 15-monoacetoxyscirpenol, neosolaniol, T-2 triol, and verrucarol) by aqueous ozone. They observed a complete degradation of these mycotoxins to simple products at ozone concentrations of 25 ppm and intermediate products at ozone concentrations of <25 ppm due to partial degradation or chemical modification. Levels of pH were also important for oxidation of trichothecenes in an ozone-saturated water system. At pH 4–6, DON (Deoxynivalenol), 3-acetyldeoxynivalenol (3-ADON), 15-acetyldeoxynivalenol (15-ADON), diacetoxyscirpenol (DAS), 15-monoacetoxyscirpenol (MAS), fusarenon-X (FUS), neosolaniol (NEO), and VER were degraded completely. Intermediate compounds (keto trichothecenes or de-epoxy trichothecenes) formed after ozone treatment of trichothecenes are reported to be less toxic than their parent compounds (Sundstøl Eriksen *et al.*, 2004; Beyer *et al.*, 2009; Könings *et al.*, 2009). Mammalian studies show that ozone effectively detoxifies mycotoxins. Gaseous ozone was effective for reduction of aflatoxin B_1 (Inan *et al.*, 2007; Zorlugenç *et al.*, 2008). Maeba *et al.* (1988) confirmed the destruction and detoxification of aflatoxins B_1 and G_1 with ozone.

9.8 EFFECT ON FLUID FOOD QUALITY

Microbial studies to date typically show mandatory 5-log reductions of spoilage and potentially pathogenic species most commonly associated with fruit and vegetable juices achieved using ozone. The effects of ozone treatment on quality and physiology of various foods are reported in Table 9.4. Applying ozone at doses leading to effective decontamination may impact the sensory qualities of food. Ozone is not universally beneficial and in some cases may promote oxidative spoilage in foods. Richardson (1994) reported that ozone helps to control odor, flavor, and color while disinfecting wastewater. Ozone is used for disinfection and for other various reasons at disinfecting water-treatment plants. Application of ozone results in beneficial effects such as color, taste, and odor removal. Oxidation of undesirable or unwanted organic and inorganic compounds (iron, manganese, nitrite, cyanide, hydrogen sulphide) by application of ozone is rapid (Rakness, 2005). Dock (1995) reported no detrimental change in quality attributes of apple cider when it was treated with ozone. Williams *et al.* (2004) reported orange juice was darker and apple cider was lighter in color after ozone treatment for up to 240 min. Choi and Nielsen (2005) noted greater sedimentation, lower sucrose content, and a decrease in soluble solids in apple juice samples when ozonated at 860 ppm for 28 min. Segovia Bravo *et al.* (2007) concluded that ozone treatment (7.27×10^{-4} g O_3/h) for 24 h caused rapid destruction of most of the polyphenols present in green table olive solutions (GTOS) and ozone bubbling for a further 72 h was necessary to reduce the remaining tyrosol content in the solution. The sugar content in GTOS remained unaltered after ozone treatment. Ozonated water and ozone exposure are reported to be effective for

Table 9.4 Effect of Ozone Processing on Quality Parameters

Fruit Juice	Ozone Application	Quality and Nutritional Attributes	Reference
Orange juice	Ozone gas (pumped into juice)	Ascorbic acid (↓), color (×)	(Angelino *et al.*, 2003)
Orange juice	Ozone gas (bubble column reactor)	Color (↓), non-enzymatic browning (NEB) (~), cloud value (~), pH(~), titratable acidity (TA) (~), ascorbic acid (↓)	Tiwari *et al.*, 2008a, 2008b
Apple cider	Ozone gas (pumped into juice)	Sediments (↑), color (×)	Choi and Nielsen, 2005
Strawberry juice	Ozone gas (bubble column reactor)	Color (↓), pH(~), TA (~), ascorbic acid (↓), anthocyanins (↓)	Tiwari *et al.*, 2009b
Blackberry juice	Ozone gas (bubble column reactor)	Color (↓), anthocyanins (↓)	Tiwari *et al.*, 2009a
Tomato juice	Ozone gas (bubble column reactor)	Color (↓), NEB (~), cloud value (~), pH (~), TA (~), ascorbic acid (↓)	Tiwari *et al.*, 2009d

(×): Significant difference; (↑): increases ; (↓): decreases; and (~): no change.
Adapted from Cullen et al. *(2009).*

inactivating microorganisms in milk biofilm and skim milk (Greene *et al.*, 1993; Rojek *et al.*, 1995; Dosti *et al.*, 2005). However, a negative effect of ozone on sensory quality of milk powder due to lipid oxidation was reported (Ipsen, 1989). Ozone treatment for fruit juices and liquid food products was developed by Sander (1985), which minimizes possible quality deterioration. Ozone has also been used in milk and its possibility to oxidize major milk components was tested. Ozone decreased the fat content in condensates (80–230 mg/l) by 96–98% and completely eliminated turbidity (Loorits *et al.*, 1975). Pastair, a Sweden-based company claimed that ozone gas was effective for killing pathogens but did not impact the active and healthy components in dairy products (www.foodproductiondaily.com). Much research has focused on the fruit juice industry and the effect of ozone on fruit juice quality and nutrition are outlined in detail in the following sections.

9.8.1 Physical Quality

9.8.1.1 Juice Color

Ozonation of fruit juices, including apple cider (Choi and Nielsen, 2005; Williams *et al.*, 2005), orange juice (Angelino *et al.*, 2003; Tiwari *et al.*, 2008a), blackberry juice (Tiwari *et al.*, 2009a), and strawberry juice (Tiwari *et al.*, 2009b), is reported to result in a significant loss of color pigment leading to color loss. Ozone has a high oxidation potential (2.07 V) resulting in the degradation of most organic compounds. The oxidizing ability of ozone is derived from the nascent oxygen atom. However, it has been reported that ozonation of organic dyes results in loss of color due to the oxidative cleavage of the chromophores (Nebel, 1975) due to attack on conjugated double bonds. Similarly, the chromophore of conjugated double bonds of carotenoids is responsible for orange juice color. Carotenoid pigments which contribute to yellow, orange, or red color in orange juice contain one or more aromatic rings (Melendez-Martinez *et al.*, 2007). The ozone and hydroxyl radicals (OH^-) generated in the aqueous solution may open these aromatic rings and lead to partial oxidation of products such as organic acids, aldehydes, and ketones.

9.8.2 Nutritional Quality

9.8.2.1 Bioactive Compounds

Ozone is used for washing of fruits and vegetables (Karaca and Velioglu, 2007). Ozone treatments were reported to have minor effects on anthocyanin contents in strawberries (Perez *et al.*, 1999) and blackberries (Barth *et al.*, 1995). Anthocyanin content in blackberries stored in air and at 0.1 ppm ozone remained stable, but fluctuated in the 0.3-ppm ozone-treated samples during storage. Nevertheless, ozonation of fruit juices rich in anthocyanins such as strawberry and blackberry juice causes a significant reduction in these color pigments. A significant reduction of 98.2% in the pelargonidin-3-glucoside content of strawberry juice was reported at an ozone concentration of 7.8% w/w processed for 10 min (Tiwari *et al.*, 2009b). Reductions of >90% in the cyanidin-3-glucoside content of blackberry juice was reported under similar treatment conditions (Tiwari *et al.*, 2009a). Tiwari *et al.* (2009c) reported that both color and anthocyanin content were significantly affected during ozone processing of grape juice.

The degradation of anthocyanins or ascorbic acid during ozone treatment could be due to direct reaction with ozone or indirect reactions of secondary oxidators such as •OH, HO_2^-, •O_2^-, and •O_3^-. Such secondary oxidators may lead to electrophilic and nucleophilic reactions occurring with aromatic compounds that are substituted with an electron donor (e.g., OH^-) having high electron density on the carbon compounds in ortho- and para-positions. A direct reaction is described by the Criegee mechanism (Criegee, 1975), where ozone molecules undergo 1–3 dipolar cyclo addition with double bonds present, leading to the formation of ozonides (1,2,4-trioxolanes) from alkenes and ozone with aldehyde or ketone oxides as decisive intermediates, all of which have finite lifetimes. This leads to the oxidative disintegration of ozonide and the formation of carbonyl compounds, while oxidative work-up leads to carboxylic acids or ketones. Ozone attacks OH radicals, preferentially the double bonds in organic compounds leading to the formation of unstable ozonide which subsequently disintegrates. The degradation mechanism for anthocyanin in strawberry juice was proposed by Tiwari *et al.* (2009b). The temperature during juice extraction influences the stability of anthocyanins by affecting the enzymatic activity of polyphenoloxidase (PPO), which accelerates the rate of degradation of anthocyanins. Inactivation of this enzyme prior to juice depectination could reduce losses of anthocyanins during juice extraction and subsequent storage. Anthocyanin stability is also influenced by the presence of various other compounds such as ascorbic acid and other polyphenols.

9.8.3 Enzyme Inactivation

Ozonation of fruit juices is reported to have no effect on non-enzymatic browning, cloud value, pH, °Brix and titratable acidity of fruit juices, including orange and tomato juice (Tiwari *et al.*, 2008a; Tiwari *et al.*, 2008b; Tiwari *et al.*, 2009d) and apple cider (Choi and Nielsen, 2005). However, Choi and Nielsen, (2005) reported a significant decrease in the °Brix of ozonated apple cider during a storage period of 21 days. They also reported no significant change in cloud value of apple cider after ozone treatment but they observed higher levels of sedimentation after 21 days. Similarly no significant changes in cloud value of orange juice was reported by Tiwari *et al.* (2008b). Fruit juice cloud is mainly caused by suspended particles in colloidal form maintained by pectin molecules. Cloud loss or clarification occurs mainly due to the enzymatic activity of pectinesterase (PE), which de-esterifies pectin. PE attacks a methyl ester group adjacent to a free carboxyl group and cleaves these methyl esters, producing methanol and polygalacturonic acid (Baker and Cameron, 1999). As consumers associate cloud loss with spoilage and quality degradation, no reduction in cloud value is desirable. Enzymes such as polyphenoloxidase, peroxidise, and β-glucosidase have been implicated in the degradation of anthocyanins. Enzymatic degradation of anthocyanins by β-glucosidase is mainly due to the loss of the glycosidic moiety leading to the formation of aglycon (anthocyanidin) consequently affecting juice color (Barbagallo *et al.*, 2007). Browning is a common problem encountered during juice processing which results from the action of a group of enzymes called polyphenol oxidases (PPO), often referred to as enzymatic browning. Ozone treatment is reported to be effective against PPO and peroxidases in fresh-cut lettuce which could be due to the high oxidation potential of ozone (Rico *et al.*, 2006). Fruit treated with ozone in combination with organic acids had less browning and reduction of polyphenol oxidase (Whangchai *et al.*, 2006). Nonenzymatic browning may result from the condensation of a carbonyl group with amino acids (Maillard reaction). However, sugars and ascorbic acid also undergo browning reactions in the absence of free amino acids (caramelization), and many of the products formed are similar to those resulting from the Maillard reaction (Yeom *et al.*, 2000) and also due to particulate fractions (Zarate-Rodríguez *et al.*, 2000). Ozone is a strong oxidizing agent and browning is an oxidative reaction process (McEvily and Lyengar, 1992) caused by the oxidation of phenols (Chubey and Nylund, 1969).

However, ozone has been reported to prevent browning reactions in fresh cut potatoes (Beltran *et al.*, 2005).

9.9 CURRENT STATUS AND FUTURE TRENDS

Ozone has been found to be an effective antimicrobial agent against many bacteria, spores, viruses, yeasts and protozoa. The effectiveness of ozone against microorganisms present in food systems depends on several factors, including the amount of ozone applied, the residual ozone in the medium, various environmental factors such as medium pH, temperature, humidity, additives (surfactants, sugars, etc.), and the amount of organic matter surrounding the cells (Pascual *et al.*, 2007). Application of ozone at higher doses that cause inactivation of the challenge organism may alter the nutritional and sensory qualities of food products. The role of organic matter in ozone consumption increases the need for the additional supply of ozone, which is also required for elimination of microorganisms embedded within a food product (Muthukumarappan *et al.*, 2009). Hence, it is necessary to use suitable application methods to ensure the direct contact of ozone with the target microbial population. A detailed study of the influence of food ingredients on both the inactivation and quality degradation kinetics will be required to account for the complexity of food systems. It is difficult to predict the reaction of ozone in the presence of organic material (Kim *et al.*, 1999). Reactions involving ozone and food components will be complex, involving simultaneous degradation and formation of new products. Ascorbic acid may be a key reactant for ozone reactivity in juices, as in plants and human respiratory tract (Thiele *et al.*, 1997). Fast ozone inactivation of *Cryptosporidium* oocysts in orange juice due to the high amount of ascorbic acid was reported (Kniel, 2002). Various radicals were produced due to reaction of ozone with ascorbic acid. It was hypothesized that oocyst cell wall proteins were oxidized by the reactive oxygen species generated from the decomposition of ozone (Kniel, 2002).

Ozone is applied either directly or indirectly for processing of foods. Ozonated water has been applied for decontamination and sanitization processes. It has been reported to extend the shelf-life of fresh-cut lettuce (Beltran *et al.*, 2005). Similarly, Selma *et al.* (2007) reported that ozone can be an alternative treatment to chlorine for disinfection of wash water and for reduction of microbial population on fresh produce as it decomposes to non-toxic products. Wang *et al.* (2004) reported that ozone (aqueous) achieved the highest overall quality of cilantro during storage and also maintained typical cilantro aroma. Thus, for many vegetables and fruits, decontamination aqueous ozone has been applied. Although, ozone solubility is greater in water; its applications in aqueous systems require techniques to overcome the gas/liquid interface. Therefore, most of the methods of mixing ozone with water involve passing the ozone-containing gas into liquid (Muthukumarappan *et al.*, 2000). Another parameter that should be considered is the large amount of water used for washing or decontamination. Also, the half-life of ozone is short; therefore immediate usage of aqueous ozone needs to be done so as to avoid loss of efficacy. Hence, direct ozone delivery for target food products to attain desirable microbial reduction and quality could be useful. To achieve the target inactivation, optimization of process parameters (temperature, ozone concentration, etc.) is also very important.

Ozone in an aqueous solution may act on various compounds, either by direct reaction with the molecular ozone or indirect reaction with the radical species that are formed during ozone decomposition. The possibility that interaction of ozone with the types of organic material present in the food product results in differing radical production trends that can lead to microbial inactivation should be studied. Therefore, studying the mechanism of the reaction of ozone with organic materials will contribute to establishing the impact of specific radical species on target microorganisms. Further research is required to ascertain the interaction of food constituents with ozone and role of resulting compounds in the inactivation process. Optimization of process parameters may also follow more specific process-related studies on mechanisms of action.

References

Ahmad, M., Farooq, S., 1985. Influence of bubble sizes on ozone solubility utilization and disinfection. Water Sci. Technol. 17, 1081–1090.

Akey, D.H., Walton, T.E., 1985. Liquid-phase study of ozone inactivation of Venezuelan equine encephalomyelitis virus. Appl. Environ. Microbiol. 50, 882–886.

Angelino, P.D., Golden, A., Mount, J.R., 2003. Effect of ozone treatment on quality of orange juice. IFT Annual Meeting Book of Abstract. Institute of Food Technologists, Chicago, IL.

Arimoto, M., Sato, J., Maruyama, K., Mimura, G., 1996. Effect of chemical and physical treatment on the inactivation of striped jack nervous necrosis virus. Aquaculture 143, 15–22.

AWWA, 1995. Water Treatment: Principles and Practices of Water Supply Operations. second ed. American Water Works Association, Denver.

Bablon, G., Bellamy, W.D., Bourbigot, M.M., Daniel, F.B., Dore, M., Erb, F., et al., 1991. Fundamental aspects. In: Langlais, G., Reckhow, D.A., Brink, D.R. (Eds.), Ozone in Water Treatment: Application and Engineering. Lewis Publishers, Inc., Chelsea, Mich., USA, pp. 11–132.

Baker, R.A., Cameron, R.G., 1999. Clouds of citrus juices and juice drinks. Food Technol. 53, 64–69.

Bancroft, K., Chrostowski, P., Wright, R.L., Suffet, I.H., 1984. Ozonation and oxidation competition values. Water Res. 18, 473–478.

Barbagallo, R.N., Palmeri, R., Fabiano, S., Rapisarda, P., Spagna, G., 2007. Characteristic of β-glucosidase from Sicilian blood oranges in relation to anthocyanin degradation. Enz. Microb. Technol. 41 (5), 570–575.

Barth, M.M., Zhou., C., Mercier, J., Payne, F.A., 1995. Ozone storage effects on anthocyanin content and fungal growth in blackberries. J. Food Sci. 60 (6), 1286–1288.

Beltran, D., Selma, M.V., Tudela, J.A., Gil, M.I., 2005. Effect of different sanitizers on microbial and sensory quality of fresh-cut potato strips stored under modified atmosphere or vacuum packaging. Postharvest Biol. Technol. 37, 37–46.

Berlett, B.S., Omar, O.H.M., Sahakian, J.A., Levine, R.L., Stadtman, E.R., 1991. Oxidation of proteins by ozone. FASEB J. 6691 (abstract).

Beyer, M., Ferse, I., Mulac, D., Würthwein, E.U., Humpf, H.U., 2009. Structural elucidation of T-2 toxin thermal degradation products and investigations toward their occurrence in retail food. J. Agri. Food Chem. 57 (5), 1867–1875.

Blatchley, E.R., Nimrata, K., 2002. Ozone disinfection in drinking water. In: Lingireddy, S. (Ed.), Control of Microorganisms in Drinking Water. American Society of Civil Engineers, Water Supply Engineering, pp. 241–270.

Boyce, D.S., Sproul, O.J., Buck, C.E., 1981. The effect of bentonite clay on ozone disinfection of bacteria and viruses in water. Water Res. 15, 759–767.

Bringmann, G., 1954. Determination of the Lethal Activity of Chlorine and Ozone on *E. coli*. Z. f., Hygiene 139, 130–139.

Broadwater, W.T., Hoehn, R.C., King, P.H., 1973. Sensitivity of 3 selected bacterial species to ozone. Appl. Microbiol. 26, 391–393.

Burleson, G.R., Murray, T.M., Pollard, M., 1975. Inactivation of viruses and bacteria by ozone with and without sonication. Appl. Microbiol. 29, 340–344.

Byun, M.W., Kwon, O.J., Yook, H.S., Kim, K.S., 1998. Gamma irradiation and ozone treatment for inactivation of *Escherichia coli* O157: H7 in culture media. J. Food Prot. 61 (6), 728–730.

Carbonneau, M.A., Peuchant, E., Sess, D., Canioni, P., Clerc, M., 1991. Free and bound malondialdehyde measured as thiobarbituric acid adduct by HPLC in serum and plasma. Clin. Chem. 37, 1423–1429.

Carpendale, M.T., Freeberg, J.K., 1991. Ozone inactivates HIV at noncytotoxic concentrations. Antiviral Res. 16, 281–292.

Chang, S.L., 1971. Modern concept of disinfection. J. Sanitation Eng. Div. 97, 689–707.

Choi, L.H., Nielsen, S.S., 2005. The effects of thermal and nonthermal processing methods on apple cider quality and consumer acceptability. J. Food Qual. 28 (1), 13–29.

Christensen, E., Giese, A.C., 1954. Changes in absorption spectra of nucleic acids and their derivatives following exposure to ozone and ultraviolet radiations. Arch. Biochem. Biophys. 50, 209–216.

Chubey, B.B., Nylund, R.E., 1969. Surface browning of carrots. Can. J. Plant. Sci. 49, 421–426.

Corona-Vasquez, B., Samuelson, A., Rennecker, J.L., Marinas, B.J., 2002. Inactivation of Cryptosporidium parvum oocysts with ozone and free chlorine. Water Res. 36 (16), 4053–4063.

Criegee, R., 1975. Mechanism of ozonolyisis. Angew. Chem.-Int. Edit. Engl. 14 (11), 745–752.

Cullen, P.J., Tiwari, B.K., O'Donnell, C.P., Muthukumarappan, K., 2009. Modelling approaches to ozone processing of liquid foods. Trends Food Sci. Technol. 20, 125–136.

Dahi, E., 1996. Physiochemical aspects of disinfection of water by means of ultrasound and ozone. Water Res. 10, 677–684.

Dave, S.A., 1999. (MSc thesis) Efficacy of Ozone Against *Salmonella Enteritidis* in Aqueous Suspensions and on Poultry Meat. Ohio State University, Columbus, Ohio, USA, pp. 26–68

Dianzani, M.U., 1993. Lipid peroxidation and cancer. Crit. Rev. Oncol. Hematol. 15, 125–147.

Dock, L.L., 1995. (MSc thesis) Development of Thermal and Non-Thermal Preservation Methods for Producing Microbially Safe Apple Cider. Purdue University, West Lafayette, IN, USA

Dosti, B., Guzel-Seydim, Z., Greene, A.K., 2005. Effectiveness of ozone, heat and chlorine for destroying common food spoilage bacteria in synthetic media and biofilms. Int. J. Dairy Technol. 58, 19–24.

Dow, S.M., Barbeau, B., von Gunten, U., Chandrakanth, M., Amy, G., Hernandez, M., 2006. The impact of selected water quality parameters on the inactivation of *Bacillus subtilis* spores by monochloramine and ozone. Water Res. 40 (2), 373–382.

Driedger, A.M., Rennecker, J.L., Marinas, B.J., 2001. Inactivation of *Cryptosporidium parvum* oocysts with ozone and monochloramine at low temperature. Water Res. 35 (1), 41–48.

Edelstein, P.H., Whittacker, R.E., Kreiling, R.L., Howell, C.L., 1982. Efficacy of ozone in eradication of *Legionella pneumophila* from hospital plumbing fixtures. Appl. Env. Microbiol. 44, 1330–1334.

Farooq, S., Akhlaque, S., 1983. Comparative response of mixed cultures of bacteria and virus to ozonation. Water Res. 17, 809–812.

Farooq, S., Chian, E.S.K., Engelbrecht, R.S., 1977b. Basic concepts in disinfection with ozone. Journal WPCF 49 (8), 1818–1831.

Farooq, S., Engelbrecht, R.S., Chian, E.S.K., 1977a. Influence of temperature and UV light on disinfection with ozone. Water Res. 11, 737–741.

FDA, 1995. Beverages: bottled water; final rule, food and drug admin., Fed. Regul. 60, 57075–57130.

Finch, G.R., Yuen, W.C., Uibel, B.J., 1992. Inactivation of *Escherichia coli* using ozone and ozone hydrogen peroxide. Environ. Technol. 13, 571–578.

Fisher, C.W, Lee, D., Dodge, B.A., Hamman, K.M., Robbins, J.B., Martin, S.E., 2000. Influence of catalase and superoxide dismutase on ozone inactivation of *Listeria monocytogenes*. Appl. Env. Microbiol. 66 (4), 1405–1409.

Foegeding, P.M., 1985. Ozone inactivation of Bacillus and Clostridium spore populations and the importance of the spore coat to resistance. Food Microbiol. 2 (2), 123–134.

Forni, L., Bahnemann, D., Hart, E.J., 1982. Mechanism of the hydroxide ion initiated decomposition of ozone in aqueous solution. J. Phys. Chem. 86, 255–259.

Franz, J., Gagnaux, A., 1971. Sterilization with ozone under extremely high sterilization requirements. Wasser Luft Betrieb 15, 393–396.

Friedman, L.I., Stromberg, R.R., 1993. Viral inactivation and reduction in cellular blood products. Rev. Fr. Transfus. Hemobiol. 36, 83–91.

Geweely, N.S.I., 2006. Antifungal activity of ozonized olive oil (Oleozone). Int. J. Agri. Biol. 8 (5), 670–675.

Glaze, W.H., 1987. Drinking water treatment with ozone. Environ. Sci. Technol. 21, 224–230.

Gong, X., Takagi, S., Huang, H., Matsumato, Y., 2007. A numerical study of mass transfer of ozone dissolution in bubble plumes with an Euler–Lagrange method. Chem. Eng. Sci. 62 (4), 1081–1093.

Gottschalk, C., Libra, J.A., Saupe, A., 2010. Reaction mechanism. Ozonation of Water and Wastewater a Practical Guide to Understanding Ozone and its Application. Wiley–VCH, Germany, p. 20

Greene, A.K., Few, B.K., Serafini, J.C., 1993. A comparison of ozonation and chlorination for the disinfection of stainless steel surfaces. J. Dairy Sci. 76, 3617–3620.

Gyurek, L.L., Hanbin, L., Belosevic, M., Finch, G.B., 1999. Ozone inactivation kinetics of Cryptosporidium in phosphate buffer. J. Environ. Eng. 125, 913–924.

Hall, R.M., Sobsey, M.D., 1993. Inactivation of hepatitis A virus and MS2 by ozone and ozone–hydrogen peroxide in buffered water. Wat. Sci. Technol. 27, 371–378.

Hamelin, C., Sarhan, F., Chung, Y.S., 1977. Ozone induced DNA degradation in different DNA polymerase I mutants of *Escherichia coli* K12. Biochem. Biophys. Res. Commun. 77, 220–224.

Hamelin, C., Sarhan, F., Chung, Y.S., 1978. Induction of deoxyribonucleic acid degradation in *Escherichia coli* by ozone. Experentia 34, 1578–1579.

Hamelin, C., Chung, Y.S., 1974. Optimal conditions for mutagenesis by ozone in *Escherichia coli* K12. Mut. Res. 24 (3), 271–279.

Hampson, B.C., 2000. Use of ozone for winery and environmental sanitation. Practical WineryandVineyard January–February, 27–30.

Harakeh, M., Butler, M., 1984. Inactivation of human rotavirus, SA11 and other enteric viruses in effluent by disinfectants. J. Hyg. Camb. 93, 157–163.

Herbold, K., Flehmig, B., Botzenhart, K., 1989. Comparison of ozone inactivation, in flowing water, of hepatitis A virus, poliovirus 1, and indicator organisms. Appl. Environ. Microbiol. 55, 2949–2953.

Hill, A.G., Spencer, H.T., 1974. Mass transfer in a gas sparged ozone contactor. In: Presented in the First International Symposium on Ozone for Water and Wastewater Treatment. Washington, D.C., pp. 367–380.

Hinze, H., Prakash, D., Holzer, H., 1987. Effect of ozone on ATP, cytosolic enzymes and permeability of *Saccharomyces cerevisiae*. Arch. Microbiol. 147, 105–108.

Hoigne, J., Bader, H., 1976. The role of hydroxyl radical reactions in ozonation processes in aqueous solutions. Water Res. 10, 377–386.

Hoigne, J., Bader, H., 1978. Ozone initiated oxidations of solutes in wastewater: a reaction kinetic approach. Prog. Wat. Technol. 10, 657–671.

Horvath, M., Bilitzky, L., Huttner, J., 1985. Ozone. Elsevier, New York, 350 p.

Hunt, N.K., Marinas, B.J., 1999. Inactivation of Escherichia coli with ozone: chemical and inactivation kinetics. Water Res. 33 (11), 2633–2641.

Hunt, N.K., Marinas, B.J., 1997. Kinetics of *Escherichia coli* inactivation with ozone. Water Res. 31, 1355–1362.

Inan, F., Pala, M., Doymaz, I., 2007. Use of ozone in detoxification of aflatoxin B1 in red pepper. J. Stored Products Res. 43, 425–429.

Ipsen, R., 1989. Factors affecting the storage of whole milk powder. Scand. Dairy Ind. 3, 24–26.

Ishizaki, K., Miura, K., Shinriki, N., 1987. Effect of ozone on plasmid DNA of *Escherichia coli* in situ. Water Res. 21, 823–827.

Janex, M.L., Savoye, P., Roustan, M., Do-Quang, Z., Laine, J.M., Lazarova, V., 2000. Wastewater disinfection by ozone: influence of water quality and kinetics modeling. Ozone Sci. Eng. 22 (2), 113–121.

Joret, J.C., Block, J.C., Hartemann, P., Richard, Y., 1982. Wastewater disinfection: elimination of fecal bacteria and enteric viruses by ozone. Ozone Sci. Eng. 4, 91–99.

Karaca, H., Velioglu, Y.S., 2007. Ozone applications in fruit and vegetable processing. Food Rev. Int. 23 (1), 91–106.

Katzenelson, E., Biedermann, N., 1976. Disinfection of viruses in sewage by ozone. Water Res. 10, 629–631.

Kawamura, K., Kanesko, M., Hirata, T., Taguchi, K., 1986. Microbial indicators for the efficiency of disinfection process. Water Sci. Technol. 18 (10), 175–184.

Khadre, M.A., Yousef, A.E., Kim, J.G., 2001. Microbiological aspects of ozone applications in food: a review. J. Food Sci. 66 (9), 1241–1252.

Khadre, M.A., Yousef, A.E., 2001. Sporicidal action of ozone and hydrogen peroxide: a comparative study. Int. J. Food Microbiol. 71 (2–3), 131–138.

Khadre, M.A., Yousef, A.E., 2002. Susceptibility of human rotavirus to ozone, high pressure, and pulsed electric field. J. Food Prot. 65, 1441–1446.

Kim, C.K., Gentile, D.M., Sproul, O.J., 1980. Mechanism of ozone inactivation of bacteriophage f2. Appl. Environ. Microbiol. 39, 210–218.

Kim, C.K., Kim, C.H, Mheen, T.I., 1984. Destructive action of ozone on the capsid and deoxyribinucleic acid of bacteriophage T4. Sanop Misaengmul Hakhoechi 12 (2), 105–109.

Kim, J.G., Yousef, A.E., Dave, S., 1999. Application of ozone for enhancing the microbiological safety and quality of foods: a review. J. Food Prot. 62 (9), 1071–1087.

Kim, J.G., Yousef, A.E., 2000. Inactivation kinetics of foodborne spoilage and pathogenic bacteria by ozone. J. Food Sci. 65, 521–528.

Kniel, K.E., 2002. Evaluation of Chemical Treatments and Ozone on the Viability of *Cryptosporidium Parvum* Oocysts in Fruit Juices. (PhD thesis)

Knight, K.L., Mudd, J.B., 1984. The reaction of ozone with glyceraldehyde–3-phosphate dehydrogenase. Arch. Biochem. Biophys. 229, 259–269.

Komanapalli, I.R., Lau, B.H.S., 1996. Ozone-induced damage of *Escherichia coli* K-12. Appl. Microbiol. Biotechnol. 46, 610–614.

Komanapalli, I.R., Mudd, J.B., Lau, B.H.S., 1997. Effect of ozone on metabolic activities of *Escherichia coli* K-12. Toxicol. Lett. 90, 61–66.

Königs, M., Mulac, D., Schwerdt, G., Gekle, M., Humpf, H.U., 2009. Metabolism and cytotoxic effects of T-2 toxin and its metabolites on human cells in primary culture. Toxicology 258 (2–3), 106–115.

Korich, D.G., Mead, J.R., Madore, M.S., Sinclair, N.A., Sterling, C.R., 1990. Effects of ozone, chlorine dioxide, chlorine, and monochloramine on *Cryptosporidium parvum* oocyst viability. Appl. Environ. Microbiol. 56, 1423–1428.

Labatiuk, C.W., Belosevic, M., Finch, G.R., 1994. Inactivation of Giardia muris using ozone and ozone hydrogen peroxide. Ozone Sci. Eng. 16, 67–78.

Lagrange, F., Reiprich, W., Hoffmann, M., 2004. CIP-cleaning and disinfection with ozone water. Journal article 84 (2), 112–114.

Larson, M.A., Mariñas, B.J., 2003. Inactivation of *Bacillus subtilis* spores with ozone and monochloramine. Water Res. 37, 833–844.

Lee, J.Y., Deniniger, R.A., 2000. Survival of bacteria after ozonation. Ozone Sci. Eng. 22, 65–75.

Leiguarda, R.H., Peso, O.A., Palazzolo, A.Z., 1949. Bactericidal action of ozone. Water Pollution (Abstract) 22, 268.

Lemke, S.L., Mayura, K., Ottinger, S.E., McKenzie, K.S., Wang, N., Fickey, C., et al., 1999. Assessment of the estrogenic effects of zearalenone after treatment with ozone utilizing the mouse uterine weight bioassay. J. Toxicol. Environ. Health—Part A 56 (4), 283–295.

Lénès, D., Deboosere, N., Ménard-Szczebara, F., Jossent, J., Alexandre, V., Machinal, C., et al., 2010. Assessment of the removal and inactivation of influenza viruses H5N1 and H1N1 by drinking water treatment. Water Res. 44 (8), 2473–2486.

Lezcano, I., Pérez Rey, R., Gutiérrez, M.S., Baluja, C., Sanchez, E., 2001. Ozone inactivation of microorganisms in water: gram positive bacteria and yeasts. Ozone Sci. Eng. 23 (2), 183–187.

Lin, Y.C., Wu, S.C., 2006. Effects of ozone exposure on inactivation of intra- and extracellular enetrovirus 71. Antiviral Res. 70, 147–153.

Loorits, K.A., Munter, R.R., Siirde, E.K., Lisenkova, L.L., 1975. Use of ozone for oxidation of major milk components in effluent. Molochn. Promst. 4, 27–30.

Macauley, J.J., Qiang, Z., Adams, C.D., Surampalli, R., Mormile, M.R., 2006. Disinfection of swine wastewater using chlorine, ultraviolet light and ozone. Water Res. 40 (10), 2017–2026.

Maeba, H., Takamoto, Y., Kamimura, M., Miura, T., 1988. Destruction and detoxification of aflatoxins with ozone. J. Food Sci. 53, 667–668.

Majumdar, S.B., Ceckler, W.H., Sproul, O.J., 1973. Inactivation of polioviruses in water by ozonation. J. Water Pollut. Control Fed. 45, 2433–2443.

Manousaridis, G., Nerantzaki, A., Paleologos, E.K., Tsiotsias, A., Savvaidis, I.N., Kontominas, M.G., 2005. Effect of ozone on microbial, chemical and sensory attributes of shucked mussels. Food Microbiol. 22 (1), 1–9.

McEvily, A.J., Lyengar, R., 1992. Inhibition of enzymatic browning in foods and beverages. Crit. Rev. Food Sci. Nutr. 32, 253–273.

McKenzie, K.S., Kubena, L.F., Denvir, A.J., Rogers, T.D., Hitchems, G.D., Bailey, R.H., et al., 1998. Aflatoxicosis in turkey poults is prevented by treatment of naturally contaminated corn with ozone generated by electrolysis. Poultry Sci. 77, 1094–1102.

McKenzie, K.S., Sarr, A.B., Mayura, K., Bailey, R.H., Miller, D.R., Rogers, T.D., et al., 1997. Oxidative degradation and detoxification of mycotoxins using a novel source of ozone. Food Chem. Toxicol. 35 (8), 807–820.

Melendez-Martinez, A.J., Vicario, I.M., Heredia, F.J., 2007. Review: analysis of carotenoids in orange juice. J. Food Compos. Anal. 20 (7), 638–649.

Montgomery, J.M., 1985. Water Treatment Principles and Design. John Wiley and Sons, New York, NY, p. 273

Moore, G., Griffith, C., Peters, A., 2000. Bactericidal properties of ozone and its potential application as a terminal disinfectant. J. Food Protect. 63 (8), 1100–1106.

Mudd, J.B., Leavitt, R., Ongun, A., McManus, T.T., 1969. Reactions of ozone with amino acids and proteins. Atmos. Environ. 3, 669–682.

Murray, B.K., Ohmine, S., Tomer, D.P., Jensen, K.J., Brent Johnson, F., Kirsi, J.J., et al., 2008. Virion disruption by ozone-mediated reactive oxygen species. J. Virol. Methods 153, 74–77.

Muthukumarappan, K., O'Donnell, C.P., Cullen, P.J., 2008. Ozone utilization. Encyclopedia of Agricultural, Food, and Biological Engineering. Taylor and Francis Group, New York, USA (DOI: 10.1081/E-EAFE-120044004)

Muthukumarappan, K., Halaweish, F., Naidu, A.S., 2000. Ozone. In: Naidu, A.S. (Ed.), Natural Food Anti-Microbial Systems. CRC Press, Boca Raton, FL, pp. 783–800.

Muthukumarappan, M., O'Donnell, C.P., Cullen, P.J., 2009. Ozone treatment of food materials. In: Jun, S., Irudayaraj, J.M. (Eds.), Food Processing Operations Modeling Design and Analysis. CRC Press Taylor and Francis Group, Boca Raton, Florida, pp. 263–280.

Nebel, C. 1975. Ozone decolorization of secondary dye laden effluents. In: Second Symposium on Ozone Technology Montreal.

Ogden, M. 1970. "Ozonation Today", Industrial Water Engineering, June 1970.

Pascual, A., Llorca, I., Canut, A., 2007. Use of ozone in food industries for reducing the environmental impact of cleaning and disinfection activities. Trends Food Sci. Technol. 18, S29–S35.

Patil, S., Bourke, P., Frias, J.M., Tiwari, B.K., Cullen, P.J., 2009a. Inactivation of *Escherichia coli* in orange juice using ozone. Innov. Food Sci. Emerg. Technol. 10, 551–557.

Patil, S., Cullen, P.J., Kelly, B., Frias, J.M., Bourke, P., 2009b. Extrinsic control parameters for ozone inactivation of *Escherichia coli* using ozone bubble column. J. Appl. Microbiol. 107 (3), 830–837.

Patil, S., Valdramidis, V.P., Cullen, P.J., Frias, J., Bourke, P., 2010a. Inactivation of *Escherichia coli* by ozone treatment of apple juice at different pH levels. Food Microbiol. 27 (6), 835–840.

Patil, S., Valdramidis, V.P., Cullen, P.J., Frias, J., Bourke, P., 2010b. Ozone inactivation of acid stressed *Listeria monocytogenes* and *Listeria innocua* in orange juice using a bubble column. Food Control. (doi: 10.1016/j.foodcont.2010.04.031)

Pekarek, S., 2008. DC corona ozone generation enhanced by TiO2 photocatalyst. European Physics. J. D 50, 171–175.

Perez, A.G., Sanz, C., Rios, J.J., Olias, R., Olias, J.M., 1999. Effects of ozone treatment on postharvest strawberry quality. J. Agric. Food Chem. 47 (4), 1652–1656.

Pryor, W.A., Dooley, M.M., Church, D.F., 1983. Mechanisms of the reactions of ozone with biological molecules: the source of the toxic effects of ozone. In: Mustafa, M.G., Mehlman, M.A. (Eds.), Advances in Modern Environmental Toxicology. Ann Arbor Science, Ann Arbor, MI.

Pryor, W.A., Giamalva, D.H., Church, D.F., 1984. Kinetics of ozonation. 2. Amino acids and model compounds in water and comparisons to rates in nonpolar solvents. J. Am. Chem. Soc. 106, 7094–7100.

Rakness, K.L., 2005. Applications in drinking water treatment. Ozone in Drinking Water Treatment: Process Design, Operation and Optimization. American Water Works Association, Denver, CO, pp. 17–74

Rennecker, J.L., Marinas, B.J., Owens, J.H., Rice, E.W., 1999. Inactivation of *Cryptosporidium parvum* oocysts with ozone. Water Res. 33, 2481–2488.

Restaino, L., Frampton, E.W., Hemphill, J.B., Palnikar, P., 1995. Efficacy of ozonated water against various food-related microorganisms. Appl. Environ. Microbiol. 61 (9), 3471–3475.

Rice, R.G., 1986. Application of ozone in water and waste water treatment. In: Rice, R.G., Browning, M.J. (Eds.), Analytical Aspects of Ozone Treatment of Water and Waste Water. Syracuse, The institute, New York, pp. 7–26.

Rice, R.G., Robson, C.M., Miller, G.W., Hill, A.G., 1981. Uses of ozone in drinking water treatment. J. Am. Water Works Ass. 73 (1), 44–57.

Richardson, S.D., 1994. Drinking water disinfection by-products. In: Meyers, R.A. (Ed.), The Encyclopedia of Environmental Analysis and Remediation. John Wiley and Sons, New York, pp. 1398–1421.

Rico, D., Martin-Diana, A.B., Frias, J.M., Gary, T.M.H., Barry-Ryan, C., 2006. Effect of ozone and calcium lactate treatments on browning and texture properties of fresh-cut lettuce. J. Sci. Food Agric. 86 (13), 2179–2188.

Rojek, U., Hill, A., Griffiths, M., 1995. Preservation of milk by hyperbaric ozone processing. J. Dairy Sci. 78 (Suppl. 1), 125.

Roy, D., Engelbrecht, R.S., Wong, P.K.Y., Chian, E.S.K., 1981. Inactivation of enteroviruses by ozone. Prog. Water Technol. 12, 819–836.

Sacco, A. 2009. Introduction to Ozone Generation Techniques—Corona, UV and Electrochemistry in <http://ezinearticles.com/?Introduction-to-Ozone-Generation-Techniques—Corona,-UV-and-Electrochemistryandid = 1686734> (Accessed 18 March 2010).

Sander, M. 1985. Mild ozone treatments of liquids such as fruit juices, milk, liquid dairy products, wine, oils, liquid medicaments, blood and/or similar products. German Federal Republic Patent Application no. DE3325568A1.

Schuchmann, M.N., Von Sonntag, C., 1989. Reactions of ozone with D-glucose in oxygenated aqueous solution-direct action and hydroxyl radical pathway. JWSRT-Aqua 38, 311–317.

Sechi, L.A., Lezcano, I., Nunez, N., Espim, M., Dupre, I., Pinna, A., et al., 2001. Antibacterial activity of ozonated sunflower oil (Oleozon). J. Appl. Microbiol. 90, 279–284.

Segovia Bravo, K.A., Arroyo Lopez, F.N., Garcia Garcia, P., Duran Quintana, M.C., Garrido Fernandez, A., 2007. Treatment of green table olive solutions with ozone. Effect on their polyphenol content and on *Lactobacillus pentosus* and *Saccharomyces cerevisiae* growth. Int. J. Food Microbiol. 114 (1), 60–68.

Scott, D.B.M., Lesher, E.C., 1963. Effect of ozone on survival and permeability of *Escherichia coli*. J. Bacteriol. 85, 567–576.

Selma, M.V., Beltran, D., Allende, A., Chacon-Vera, E., Gil, M., 2007. Elimination by ozone of Shigella sonnei in shredded lettuce and water. Food Microbiol. 24, 492–499.

Shin, G.A., Sobsey, M.D., 2003. Reduction of Norwalk virus, poliovirus 1, and bacteriophage MS2 by ozone disinfection of water. Appl. Environ. Microbiol. 69, 3975–3978.

Shriniki, N., Ishizaki, K., Yoshizaki, T., Miura, K., Ueda, T., 1988. Mechanism of inactivation of tobacco mosaic virus with ozone. Water Res. 22 (7), 933–938.

Sobsey, M.D., 1989. Inactivation of health-related microorganisms in water by disinfection processes. Wat. Sci. Technol. 21, 179–195.

Sproul, O.J., Kim, C.K. 1980. The mechanism of ozone inactivation of waterborne viruses. Report (1980) (W81-00315, OWRT-B-013-ME/2), Order No. PB81-127334). From Government Rep Announce-Index (U.S.)1981, 81(5), 861.

Sproul, O.J., Pfister, R.M., Kim, C.K., 1982. The mechanism of ozone inactivation of water borne viruses. Water Sci. Technol. 14, 303–314.

Steenstrup, L.D., Floros, J.D., 2004. Inactivation of *Escherichia coli* O157:H7 in apple cider by ozone at various temperatures and concentrations. J. Food Process. Preserv. 28 (2), 103–116.

Sundstøl Eriksen, G., Pettersson, H., Lundh, T., 2004. Comparative cytotoxicity of deoxynivalenol, nivalenol, their acetylated derivatives and de-epoxy metabolites. Food Chem. Toxicol. 42 (4), 619–624.

Takamoto, Y., Maeba, H., Kamimura, M., 1992. Changes in survival rates of enzyme activities in *Escherichia coli* with ozone. Appl. Microbiol. Biotechnol. 37, 393–395.

Thanomsub, B., Anupunpisit, V., Chanphetch, S., Watcharachaipong, T., Poonkhum, R., Srisukonth, C., 2002. Effects of ozone treatment on cell growth and ultrastructural changes in bacteria. J. Gen. Appl. Microbiol. 48, 193–199.

Thiele, J.J., Traber, M.G., Tsang, K., Cross, C.E., Packer, L., 1997. In vivo exposure to ozone depletes vitamin C and E and induces lipid peroxidation in epidermal layers of murine skin. Free Rad. Biol. Med. 23, 385–391.

Thurston-Enriquez, J.A., Haasb, C.N., Jacangeloc, J., Gerba, C.P., 2005. Inactivation of enteric adenovirus and feline calicivirus by ozone. Water Res. 39, 3650–3656.

Tiwari, B.K., Muthukumarappan, K., O'Donnell, C.P., Cullen, P.J., 2008b. Kinetics of freshly squeezed orange juice quality changes during ozone processing. J. Agric. Food Chem. 56 (15), 6416–6422.

Tiwari, B.K., Muthukumarappan, K., O'Donnell, C.P., Cullen, P.J., 2008a. Modelling color degradation of orange juice by ozone treatment using response surface methodology. J. Food Eng. 88 (4), 553–560.

Tiwari, B.K., O'Donnell, C.P., Brunton, N.P., Cullen, P.J., 2009d. Degradation kinetics of tomato juice quality parameters by ozonation. Int. J. Food Sci. Technol. 44 (6), 1199–1205.

Tiwari, B.K., O'Donnell, C.P., Patras, A., Brunton, N., Cullen, P.J., 2009b. Effect of ozone processing on anthocyanins and ascorbic acid degradation of strawberry juice. Food Chem. 113 (4), 1119–1126.

Tiwari, B.K., O'Donnell, C.P., Patras, A., Brunton, N., Cullen, P.J., 2009c. Anthocyanins and color degradation in ozonated grape juice. Food Chem. Toxicol. 47 (11), 2824–2829.

Tiwari, B.K., O'Donnell, C.P., Muthukumarappan, K., Cullen, P.J., 2009a. Anthocyanin and color degradation in ozone treated blackberry juice. Innovative Food Sci. Emerg. Technol. 10 (1), 70–75.

United States Food and Drug Administration (USFDA), 2001. Hazard analysis and critical point (HACCP); procedures for the safe and sanitary processing and importing of juice; final rule. Fed. Reg. 66, 6137–6202.

USFDA. 2004. FDA Guidance to Industry, 2004: Recommendations to Processors of Apple Juice or Cider on the Use of Ozone for Pathogen Reduction Purposes. Available online: <http://www.fda.gov/Food/GuidanceComplianceRegulatoryInformation/GuidanceDocuments/Juice/ucm072524.htm> (Accessed 29 April 2010).

Vaughn, J.M., Chem, Y.S., Lindburg, K., Morales, D., 1987. Inactivation of human and simian rotaviruses by ozone. Appl. Env. Microbiol. 53, 2218–2221.

Vrochinskii, K.K., 1963. Experimental datta on water decontamination with ozone. Hyg. Sanit. 28, 3.

Wang, H., Feng, H., Luo, Y., 2004. Microbial reduction and storage quality of fresh-cut cilantro washed with acidic electrolyzed water and aqueous ozone. Food Res. Int. 37 (10), 949–956.

Weavers, L.K., Wickramanayake, G.B., 2001. Disinfection and sterilization using ozone. In: Block, S.S (Ed.), Disinfection, Sterilization and Preservation. Lippincott Williams and Wilkins, Philadelphia, USA, pp. 205–214.

Wells, K.H., Latino, J., Gavalchin, J., Poiesz, B.J., 1991. Inactivation of human immunodeficiency virus type 1 by ozone in vitro. Blood 78, 1882–1890.

Whangchai, K., Saengnil, K., Uthaibutra, J., 2006. Effect of ozone in combination with some organic acids on the control of postharvest decay and pericarp browning of longan fruit. Crop protection 25, 821–825.

Wickramanayake, G.B., Rubin, A.J., Sproul, O.J., 1984. Inactivation of *Naegleria* and *Giardia* cysts in water by ozonation. J. Water Pollut. Control Fed. 56, 983–988.

Widmer, G., Clancy, T., Ward, H.D., Miller, D., Batzer, G.M., Pearson, C.B., et al., 2002. Structural and biochemical alterations in Giarida lamblia cysts exposed to ozone. J. Parasit. 88, 110–1106.

Williams, R., Sumner, S., Golden, D., 2004. Survival of *Escherichia coli* O157:H7 and *Salmonella* in apple cider and orange juice as affected by ozone and treatment temperature. J. Food prot. 67 (11), 2381–2386.

Williams, R., Sumner, S., Golden, D., 2005. Inactivation of *Escherichia coli* O157:H7 and Salmonella in apple cider and orange juice treated with combinations of ozone, dimethyl dicarbonate, and hydrogen peroxide. J. Food Sci. 70 (4), 197–201.

Xu, L., 1999. Use of ozone to improve the safety of fresh fruits and vegetables. Food Technol. 53 (10), 58–63.

Yeom, H.W., Streaker, C.B., Zhang, Q.H., Min, D.B., 2000. Effects of pulsed electric fields on the quality of orange juice and comparison with heat pasteurization. J. Agric. Food Chem. 48, 4597–4605.

Yoshizaki, T., Miura, K., Ueda, T., 1988. Mechanism of inactivation of tobacco mosaic virus with ozone. Water Res. 22 (7), 933–938.

Young, J.C., Zhu, H., Zhou, T., 2006. Degradation of trichothecene mycotoxins by aqueous ozone. Food Chem. Toxicol. 44 (3), 417–424.

Zarate-Rodríguez, E., Ortega-Rivas, E., Barbosa-Cánovas, G.V., 2000. Quality changes in apple cider as related to nonthermal processing. J. Food Qual. 23, 337–349.

Zorlugenç, B., Kıroglu, F., Öztekin, S., Bülend Evliya, I., 2008. The influence of gaseous ozone and ozonated water on microbial flora and degradation of aflatoxin B1 in dried figs. Food Chem. Toxicol. 46, 3593–3597.

Zuma, F., Lin, J., Jonnalagadda, S.B., 2009. Ozone-initiated disinfection kinetics of *Escherichia coli* in water. J. Environ. Sci. Health Part A 44, 48–56.

CHAPTER

10

Dense-Phase Carbon Dioxide Processing of Fluid Foods

Giovanna Ferrentino[1] and Murat O. Balaban[2]

[1]*Department of Materials Engineering and Industrial Technologies, University of Trento, Trento, Italy*

[2]*Department of Chemical and Materials Engineering, University of Auckland, Auckland, New Zealand*

10.1 PRINCIPLES

10.1.1 Introduction to Dense-Phase Carbon Dioxide Technology

The need for a food preservation method that is safe, inexpensive, and that preserves heat-sensitive compounds resulted in the use of pressurized carbon dioxide (CO_2) as a food preservation method. CO_2 is used because of its safety, low cost, and high purity. Dense-phase carbon dioxide (DPCD) treatment has attracted great interest in the non-thermal treatment of liquid foods or liquid model solutions. DPCD has been shown to inactivate microorganisms as well as conventional heat pasteurization without the loss of nutrients or quality changes that may occur due to thermal effects. In the DPCD process, food is contacted with pressurized sub- or supercritical CO_2 for a period of time in a batch, semi-batch or continuous manner. The CO_2 pressures can range from 7.0 to 40.0 MPa. These levels are much lower than those of ultra-high pressure processes, for example, making the management and control of pressure during the process easier. Process temperatures can range from 20 to 60°C, and therefore the temperature levels that may cause thermal damage to foods are avoided. The temperature increase induced by the pressure build-up is negligible. The treatment times can range from about 3 to 9 min for continuous, or from 120 to 140 min for semi-continuous or batch DPCD processes.

Carbonation to preserve food started with the study by Brown *et al.* (1939) with apple cider. Carbonation was shown to preserve the cider for up to 3 months at 21°C with no change in flavor. Carbonation was also investigated for its use in soft drinks as a preservative. Even at low gas pressures (3 l CO_2 per 1 beer)

Novel Thermal and Non-Thermal Technologies for Fluid Foods. DOI: 10.1016/B978-0-12-381470-8.00010-4

sterility was achieved on the 20th day depending on the °Brix of the beverage (Insalata, 1952). Carbonation was not effective for some bacteria: Molin (1983) examined the inhibitory effect of CO_2 on a variety of food related bacteria at a variety of pressures. Inhibition was only partially successful. Although 100% CO_2 slowed the growth of all organisms some were affected less than others. CO_2 had a 75% inhibitory affect on *Bacillus cereus, Brochothrix thermosphacta*, and *Aeromonas hydrophila*, and a 53–29% inhibitory effect on *Escherichia coli* and *Streptococcus faecalis*. Inhibitory rates for anaerobic bacteria were even lower. This discovery proved that carbonation of foods alone would not inactivate all food-related bacteria.

10.1.2 Carbon Dioxide Properties

Pressurized CO_2 dissolves in water to form carbonic acid (H_2CO_3), which dissociates into bicarbonate (HCO_3^-), carbonate (CO_3^{2-}) and hydrogen (H^+) ions, lowering the pH of the food according to the following equilibria:

$$CO_2(g) \leftrightarrow CO_2(aq) \tag{10.1}$$

$$CO_2(aq) + H_2O \leftrightarrow H_2CO_3 \tag{10.2}$$

$$H_2CO_3 \leftrightarrow H^+ + HCO_3^- \tag{10.3}$$

$$HCO_3^- \leftrightarrow H^+ + CO_3^{2-} \tag{10.4}$$

The dissociation constants for Equations 10.1–10.4 are:

$$[CO_2]_{aq} = H_K \cdot p_{CO_2} \quad H_K = 3.3 \cdot 10^{-2}\ \text{mol}/(1 \cdot \text{atm})\ at\ 25°\text{C}$$

$$K_b = \frac{[H_2CO_3]}{[CO_2]_{aq}} = 1.7 \cdot 10^{-3}\ \text{at } 25°\text{C and 1 atm} \tag{10.5}$$

$$K_c = \frac{[H^+] \cdot [HCO_3^-]}{[H_2CO_3]} = 2.5 \cdot 10^{-4}\ \text{mol}/lat25°\text{C and 1 atm} \tag{10.6}$$

$$K_d = \frac{[CO_3^{2-}] \cdot [H^+]}{[HCO_3^-]} = 5.61 \cdot 10^{-11}\ \text{mol/l at } 25°\text{C and 1 atm} \tag{10.7}$$

Subscripts (g) and (aq) refer to the states of gas and aqueous solution, respectively. At atmospheric pressure, the concentration of dissolved CO_2 *(aq)* can be related to the external partial pressure of CO_2 (p_{CO_2}) and Henry's law constant (H_K). Henry's law requires ideal gas behavior and a dilute solution. Meyssami *et al.* (1992) predicted the pH of water, orange juice, and model liquid foods

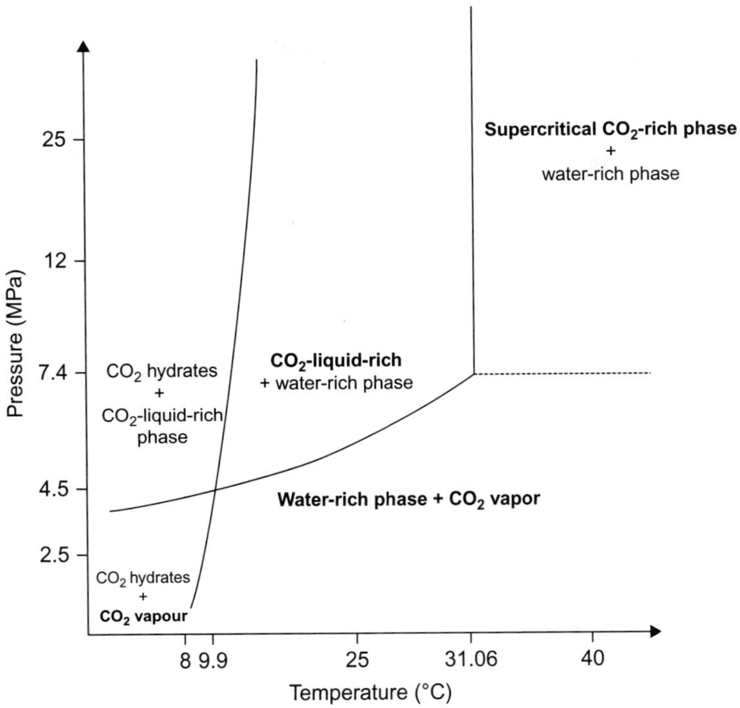

FIGURE 10.1

Phase diagram of CO_2–H_2O system (Ballestra *et al.*, 1996).

under DPCD, obtaining good correlations with measured pH. Other studies showed that dissolved acids and salts reduced CO_2 solubility and also demonstrated the lowering of pH by DPCD (Calix *et al.*, 2008).

In water, 98% of CO_2 is hydrated to form aqueous molecular CO_2 that generates clathrates or hydrates with water. Water in clathrates or hydrates is linked by hydrogen bonds in a crystalline structure in which CO_2 molecules are entrapped (Ikeda and Mae, 1998). The CO_2–H_2O system composition and the state of CO_2 change with pressure and temperature, as shown in Figure 10.1.

Different molecular species such as hydrates, the water-rich phase, the vapor CO_2-rich phase, and the liquid CO_2-rich phase can form. At low pressure, CO_2 is a vapor. As pressure increases, CO_2 goes into the liquid state. Two phases, one rich in water and one rich in liquid CO_2 co-exist. Their proportions change with pressure and temperature. At the CO_2 critical point (7.4 MPa and 31.06°C) the two phases merge into a single state. As pressure and temperature increase above the critical point, the CO_2 is in its supercritical state with a high diffusivity, low viscosity, and a strong solvent power. The

CO_2. density influences its solvent power. CO_2 in supercritical state can dissolve certain solutes and is used as an excellent extractor of natural products (Rizvi *et al.*, 1986).

10.1.3 Engineering Aspects of Dense-Phase Carbon Dioxide Technology

Several batch, semi-continuous, and continuous systems have been developed for DPCD applications. In a batch system, CO_2 and the solution to be treated are stationary in a container during treatment. A semi-continuous system allows a continuous flow of CO_2 through the chamber; however, the liquid remains stationary. A continuous system allows flow of both CO_2 and the liquid food through the system.

Early studies used batch systems. A typical batch system (Fig. 10.2) has a CO_2 gas cylinder, a pressure regulator, a vessel, a water bath or heater, and a CO_2 release valve (Hong and Pyun, 1999).

The sample is placed into the vessel and the temperature is set to the desired value. Then CO_2 is introduced into the vessel until the sample is saturated at the desired pressure and temperature. The sample is left in the vessel for a period of time and then the CO_2 outlet valve is opened to release the gas. Some systems contain an agitator to decrease the time to saturate of the sample with CO_2.

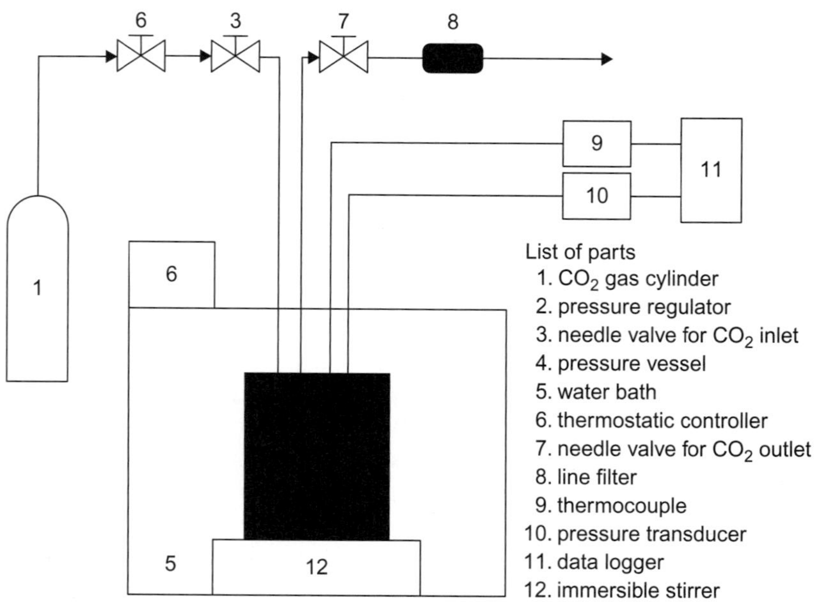

FIGURE 10.2

A typical batch dense phase CO_2 system (Hong and Pyun, 1999).

Spilimbergo and Mantoan (2005) developed a multi-batch DPCD apparatus to determine the rate of microbial inactivation in foods. The apparatus is shown in Figure 10.3. It consisted of six identical reactors with magnetic stirrers, connected in parallel and submerged in a water bath to assure uniform temperature in all reactors. Each reactor was connected to an on–off valve for independent depressurization. Each run provided six experimental data taken in identical process conditions but different treatment times.

Ishikawa *et al.* (1995) developed a semi-continuous system using a filter to micro-bubble CO_2 into the vessel (Fig. 10.4). They could achieve three times more inactivation of enzymes using this micropore filter than without it. Using the filter increased the concentration of dissolved CO_2 in the sample from 0.4 to 0.92 mol/l at 25 MPa and 35°C without mixing. A continuous micro-bubble system, very effective in inactivating microorganisms, was developed where liquid

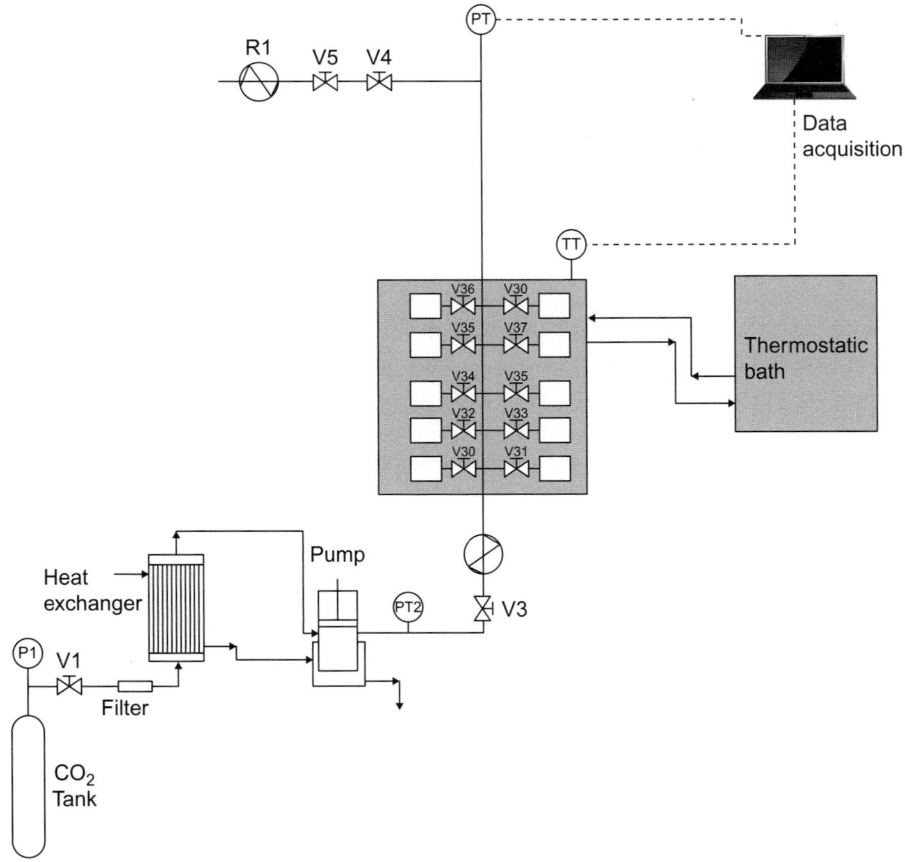

FIGURE 10.3

Conceptual diagram of the multi-batch system (Spilimbergo and Mantoan, 2005).

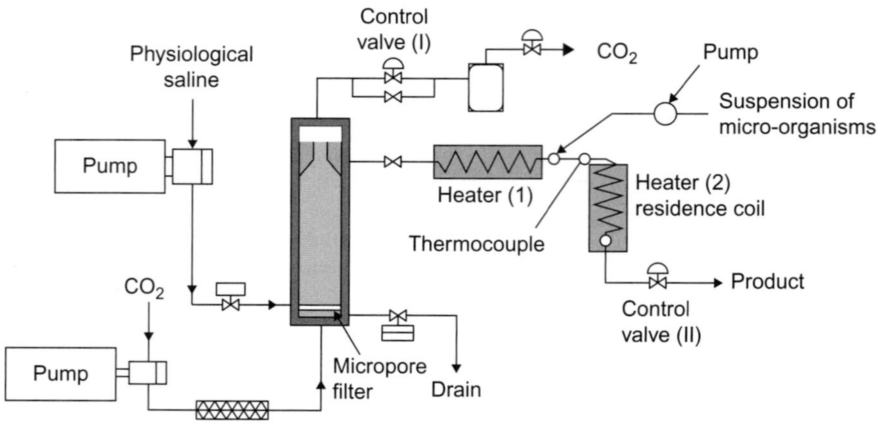

FIGURE 10.4

A continuous micro-bubble dense-phase CO_2 system (Shimoda *et al.*, 2001).

CO_2 and a saline solution were pumped through a vessel. CO_2 was dispersed into the saline solution through a stainless steel mesh filter with 10-μm pore size.

The CO_2 micro-bubbles moved upward while dissolving into the solution. Then, the saturated CO_2 solution passed through a heater to reach the desired temperature and a suspension of microorganisms was pumped into it. Another coil with a heater was used to adjust the residence time (Shimoda *et al.*, 2001).

A semi-continuous apparatus was built by Spilimbergo *et al.* (2003) as shown in Figure 10.5. Liquid CO_2 is continuously fed into a thermally insulated 30-ml vessel by a high-performance liquid chromatography (HPLC) pump with a maximum flow rate of 25 ml/min. A 5-mm porous metallic filter, placed at the bottom of the vessel, resulted in CO_2 micro-bubbles. The vessel had a resistance temperature probe located inside, a temperature on–off control, and an electric resistance. The outlet micrometric valve that regulated the CO_2 pressure was heated by resistance to prevent freezing during CO_2 expansion. Pressure and temperature were continuously recorded by a real-time acquisition data system (LabView™).

In 1999, Praxair (Chicago, IL, USA.) developed a continuous-flow DPCD system (Fig. 10.6). CO_2 and the products were pumped and mixed before passing through the high-pressure pump, which increased the pressure to the process levels. Product temperature was controlled in holding coils. Residence time was adjusted by setting the flow rate of the product through the coils. An expansion valve released CO_2 from the mixture at the end of the holding coils. Residual CO_2 in the food could be removed using a vacuum tank. This system has been effective in killing pathogens and spoilage bacteria in a short time (Damar and Balaban, 2006; Kincal *et al.*, 2005; Lecky, 2005; Lim *et al.*, 2006).

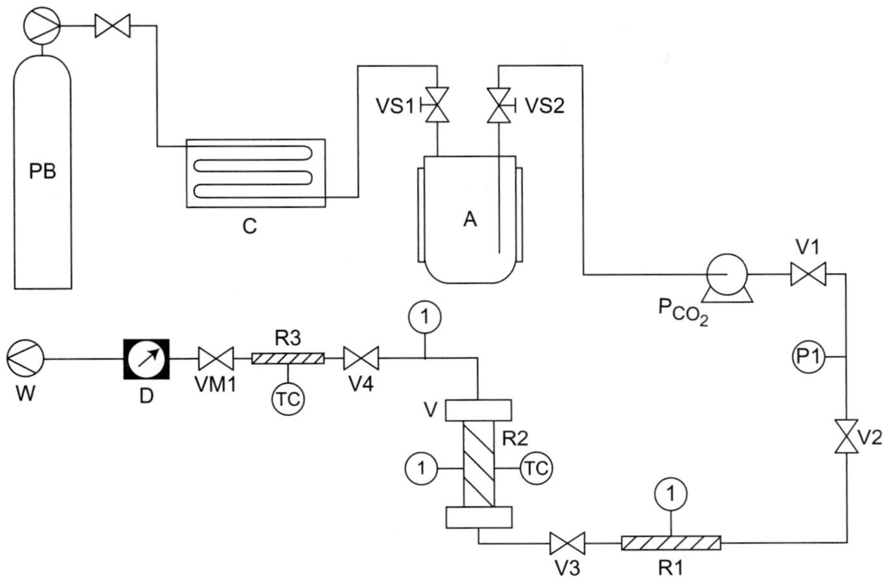

FIGURE 10.5

A semi-continuous dense phase CO_2 system (Spilimbergo *et al.*, 2002).

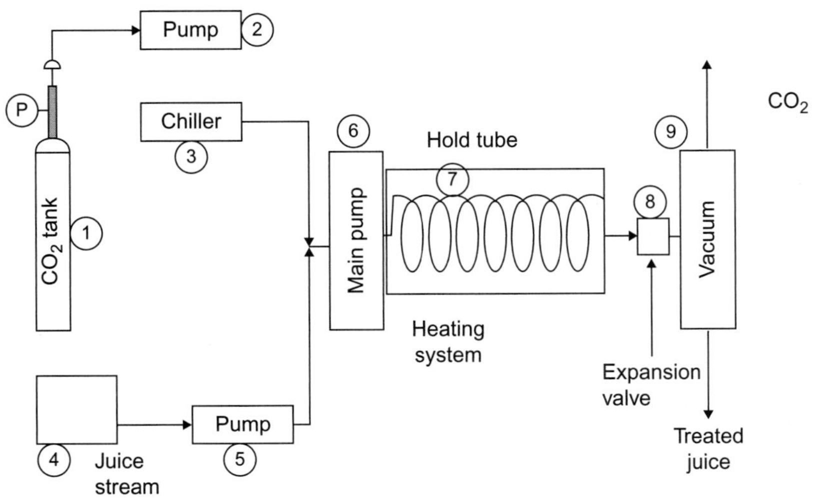

FIGURE 10.6

A continuous flow dense phase system (Damar and Balaban, 2006).

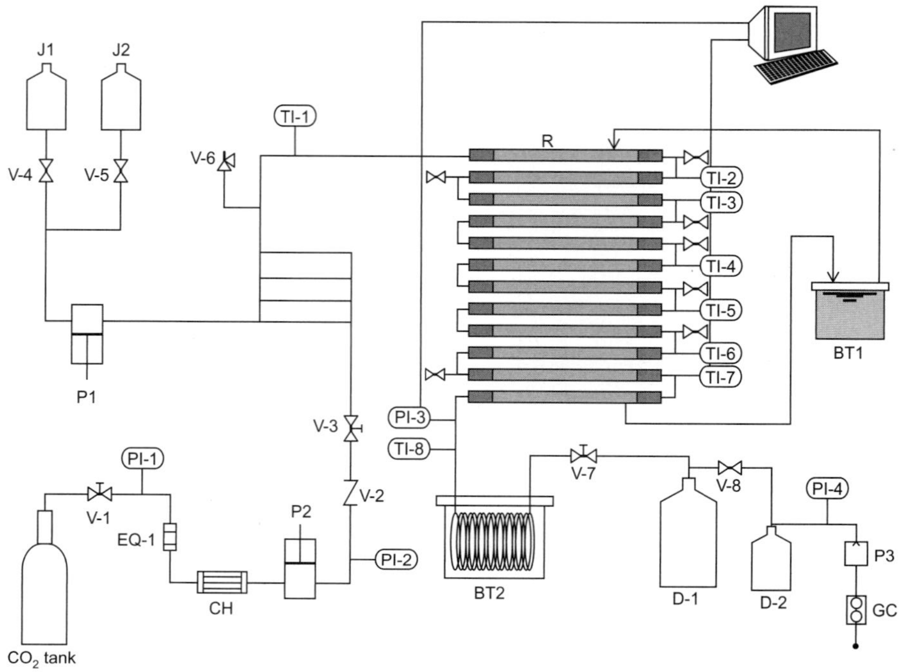

FIGURE 10.7

Diagram of the continuous pilot plant for high pressure CO_2 inactivation (Parton *et al.*, 2007).

A continuous high-pressure CO_2 system has been developed by Parton *et al.* (2007) to process 1 l/h of liquid at 40.0 MPa. The sample liquid was stored in two 5-l high-density polyethylene containers, both connected to the pump (Fig. 10.7).

CO_2 passed through an in-line 0.5-μm filter and a cooling system, then pumped to four mixing points. The pressurized CO_2 was mixed with the liquid and the mixture went to a temperature-controlled holding tube. Several valves along the holding tube allowed for sampling at different residence times. The treated liquid depressurized through a capillary tube inside a thermostatic bath. The liquid was degassed in two containers.

10.2 PROCESS CONTROL PARAMETERS

10.2.1 Pressure and Temperature

Pressure and temperature are the main control parameters for DPCD. They significantly affect microbial inactivation, and they influence the CO_2 physical state (i.e. liquid, gas, or supercritical) and its properties such as viscosity and diffusivity.

These parameters, coupled with the composition of the medium, influence CO_2 solubility in the liquid food.

A wide range of temperatures has been employed for DPCD treatment ranging from ambient to 90°C depending on the type of microorganisms to be inactivated. In general, inactivation rate increases with increasing temperature. Higher temperatures increase the diffusivity of CO_2 and the fluidity of cell membrane making CO_2 penetration easier (Lin *et al.*, 1993; Hong *et al.*, 1997; Hong and Pyun, 1999). The cell membrane is modified which makes CO_2 diffusion and accumulation into the phospholipid layer easier. Different authors reported the effect of increasing temperature on vegetative cells and spores. Oule *et al.* (2006) reported inactivation of *E. coli* cells by CO_2 at different temperatures. At a pressure of 25 MPa for 30 min, an increase in temperature from 25 to 40°C led to 5 log-cycles inactivation increase. A review of the process conditions during the treatment of vegetative microbial cells is presented in Table 10.1.

The process temperature should not be far above the CO_2 critical temperature because the density of the CO_2, and hence its solvent capacity decrease quite rapidly as temperature increases (Lucien and Foster, 1999). Thus, the increase in CO_2 penetration by temperature can in part be counteracted by its reducing effect on CO_2 solubility. Also, DPCD should not be performed at high temperatures to avoid the deterioration of the food quality. Many studies investigated the effect of DPCD temperature on spores showing that a range from 20 to 40°C failed in the inactivation process, although the pressure applied was up to 30 MPa (Kamihira *et al.*, 1987; Enomoto *et al.*, 1997; Spilimbergo and Bertucco, 2003; Watanabe *et al.*, 2003). Generally DPCD required temperatures higher than 60°C to induce spore inactivation. Higher temperatures were necessary to stimulate spore germination and then inactivation to the extent required for food sterilization. Spilimbergo *et al.* (2002) conducted experiments in a semi-continuous equipment with a 5-μm porous filter to generate microbubbles. They observed no significant reduction in the number of *B. subtilis* spores even though the treatment was performed at 54°C, 20 MPa for 60 min. A study which needs to be mentioned for low-temperature spore inactivation is that of Ishikawa *et al.* (1997). They reported 3.5–4 log-cycles reduction after a treatment of 30 MPa, 40°C and 30 min for different *Bacillus* spores with microbubbles of CO_2. The increased inactivation was attributed to an increase in CO_2 concentration in the spore's suspension, and hence to the enhanced amount of CO_2 sorbed by the spores due to the microbubble method. A summary of the literature, including different species of bacterial spores, treatment conditions and inactivation levels, is reported in Table 10.2.

In general, increasing the pressure accelerates microbial inactivation. At higher pressure a shorter exposure time is needed to achieve the same inactivation level. Pressure controls the solubility of CO_2 as well as its dissolution rate in the suspending medium. By increasing the pressure, CO_2 solubility is enhanced, facilitating the acidification of the external medium as well as its contact with the cells. Hong and Pyun (1999) showed that the inactivation of *Lactobacillus plantarum* cells was accelerated with increasing CO_2 pressure from 50 to 80 kg/cm^2 at

Table 10.1 Summary of Some Studies on Vegetative Forms Inactivated by Dense-Phase CO_2

Medium	Inoculated Microorganism	Treatment	T (°C)	System	Maximum Reduction	Reference
Yeast malt peptone broth	*S. cerevisiae*	7–20 MPa 15–7 min	35	Batch	7 log	Lin *et al.* (1992a)
Physiological saline	*Candida utilis* *S. cerevisiae* *Kluyveromyces fragilis*	10 MPa 5 min	33	Batch	6 log	Isenchmid *et al.* (1995)
de Man, Rogosa and Sharpe broth	*L. monocytogenes*	21 MPa 14 min	35	Batch	9 log	Lin *et al.* (1994)
Physiological saline	*L. brevis* *S. cerevisiae*	25 MPa 15 min 25 MPa 30 min	35	Semi-cont. System	6 log	Ishikawa *et al.* (1995)
Physiological saline	*E. coli*	5 MPa 20 min	35	Batch	6 log	Ballestra *et al.* (1996)
Sterilized water	*S. cerevisiae*	4 MPa 24 h	35	Batch	8 log	Enomoto *et al.* (1997)
Sterilized water	*S. cerevisiae*	15 MPa 1 h	40	Batch	4 log	Kumagai *et al.* (1996)
Brain heart infusion broth	*S. aureus*	8 MPa 60 min	25	Batch	7 log	Erkmen (1997)
Hydroph. filter disk	*E. coli* *S. cerevisiae* *E. faecalis*	5 MPa 290 min	23	Batch	5 log	Debs-Louka *et al.* (1999)

Different buffer	*L. plantarum*	7 MPa 120 min	30	Batch	8 log	Hong *et al.* (1999)
de Man, Rogosa and Sharpe broth		14 MPa 30 min			6 log	
Growth media and biodegradable polymers	*S. aureus*	20.5 MPa 4 h	40	Batch	9 log	Dillow *et al.* (1999)
	B. cereus		60		8 log	
	L. innocua	20.5 MPa 0.6 h	34		9 log	
	P. aeruginosa	20.5 MPa 4 h	40		8 log	
	Salmonella salford				9 log	
	P. vulgaris	20.5 MPa 0.6 h	34		8 log	
	L. dummifii	20.5 MPa 1.5 h	40		4 log	
	E. coli	20.5 MPa 0.5 h	34		8 log	
Physiological saline	*L. monocytogenes*	6 MPa 60 min	35	Batch	6 log	Erkmen (2000b)
Physiological saline	*B. thermosphacta*	6 MPa 10 min	35	Batch	6 log	Erkmen (2000a, b, c)
Brain heart infusion broth	*S. typhimurium*	6 MPa 15 min			7 log	
		6 MPa 110 min				
Physiological saline	*L. monocytogenes*	6 MPa 40 min			7 log	
Brain heart infusion broth		6 MPa 120 min				

(*Continued*)

Table 10.1 (Continued)

Medium	Inoculated Microorganism	Treatment	T (°C)	System	Maximum Reduction	Reference
Nutrient broth	*E. coli*	10 MPa 50 min	30	Batch	6 log	Erkmen (2001b)
Physiological saline	*S.cerevisiae*	8 MPa 0.7 min	38	Continuous plant	4 log	Shimoda *et al.* (2001)
Physiological saline	*B. subtilis* *P. aeruginosa*	7.5 MPa 2.5 min	38	Semi-continuous system	7 log	Spilimbergo *et al.* (2002)
Potato dextrose broth	*S. cerevisiae*	2.5–10 MPa 18 h–125 min	30	Batch	Total inactivation	Erkmen (2003)
Physiological saline	*B. subtilis*	11 MPa 0.05 gCO_2/gH_2O 30 min	36	Continuous plant	Total inactivation	Parton *et al.* (2007)
Sabouraud dextrose broth	*S. cerevisiae*	9 MPa 1.8 ml 15 min	38	Multi-batch system	>6.87	Parton *et al.* (2007)

Table 10.2 Summary of Some Studies on Spores Inactivated by Dense-Phase CO_2

Medium	Inoculated Microorganism	Treatment	T (°C)	System	Maximum Reduction	Reference
Sterile water	*B. subtilis*	2 MPa 2 h	35	Batch	0.3 log	Kamihira *et al.* (1987)
Sterile water	*B. megaterium*	5.8 MPa 30 h	60	Batch	8 log	Enomoto *et al.* (1997)
Nutrient broth	*B. polymixa* *B. cereus* *B. subtilis*	30 MPa 60 min	45 50 55	Batch	6 log	Ishikawa *et al.* (1997)
de Man, Rogosa and Sharpe broth	*Lactobacillus spores*	7 MPa 250 min	30	Batch	6 log	Hong *et al.* (1997)
Sterile ringer solution	*B. subtilis*	5 MPa 60 min	80	Batch	4 log	Ballestra and Cuq (1998)
	B. fulva	5 MPa 85 min	80		90%	
	A. niger conidia	5 MPa 11 min	50		90%	
Heart brain infusion	*B. subtilis*	7 MPa 24 h	75	Semi-continuous system	7 log	Spilimbergo *et al.* (2002)
Carb. physiological saline	*A. niger*	10 MPa 1 min	52	Batch	7 log	Shimoda *et al.* (2002)
Nutrient broth	*G. stearothermopl.*	30 MPa 100 min	95	Batch	4 log	Watanabe *et al.* (2003)
Sterile water	*B. coaugulans* *B. licheniformis*	6.5 MPa 120 min	35	Batch	10% 20%	Furukawa *et al.* (2004)
Diluted aqueous H_2O_2 solution	*B. pumilus*	27.5 MPa 4 h	60	Batch	6.28 log	Zhang *et al.* (2006)

30°C. It took 150 min to reduce the cells by 5 log-cycles at 50 kg/cm^2, and less than 50 min at 70 kg/cm^2. However, the inhibiting effect of CO_2 is limited by its saturation solubility in the suspending medium (Sims and Estigarribia, 2003). Spilimbergo *et al.* (2002) demonstrated that above 10 MPa, the solubility of CO_2 was a weak function of pressure. An increase in pressure from 10 to 30 MPa at 55–60°C did not appreciably influence the solubility of CO_2 in water. Ferrentino *et al.* (2010a) also demonstrated that the medium in which microbial cells are suspended is also important. *S. cerevisiae* cells in a 0.4-M sodium phosphate monobasic solution did not show an appreciable increase in the lethality, even by increasing the pressure from 7.5 to 13.0 MPa for treatment times ranging from 5 to 120 min. The effect is linked to the saturation solubility of CO_2 in the suspending medium. However, pressure is a parameter to control from an economic perspective, considering that increasing the pressure will increase both the operating costs and the capital costs (Spilimbergo *et al.*, 2005).

10.2.2 CO_2 Physical States

Depending on pressure and temperature, CO_2 physico-chemical properties change from subcritical (liquid or gas) to supercritical. In supercritical conditions, CO_2 exists as a single phase and exhibits a more liquid-like density with mass transport properties such as viscosity and diffusivity closer to that of a gas. The liquid-like density allows a higher solvating and extracting power. The gas mass transport properties enhance the diffusion rate compared to the liquid state. Also the very low surface tension of supercritical CO_2 facilitates its penetration into cells and extracting intracellular components, resulting in an increased disruption of biological systems (Lucien and Foster 1999). Lin *et al.* (1992a) reported experiments on the inactivation of yeast cells induced by CO_2 at subcritical and supercritical temperatures over the pressure range of 1000–3000 psi (6.8–20.4 MPa). Results showed that a maximum of 15 min, depending on pressure, was sufficient to inactivate yeast cells by 7 log-cycles under supercritical conditions while subcritical carbon dioxide was found to be less effective.

Oule *et al.* (2006) reported the effect of the physical state of CO_2 on the cells by plating and by observation using transmission and scanning electron microscopy. Vapor phase CO_2 generated a bacteriostatic effect while CO_2 in liquid or supercritical state provided a bactericidal effect. Scanning electron microscopy images showed that during vapor CO_2 treatment, the cells seemed to be sustaining a stress translated by a slight depression of the cellular envelope. Due to its hydrophobic and liposoluble nature, CO_2 diffuses into and accumulates in the cellular membrane. Presence of CO_2 increases the disordering of the hydrocarbon lipid chains and the fluidity of the membrane, leading to the rupture of the lipid–protein interactions according to the so called anesthetic effect (Isenschmid *et al.*, 1995). The accumulated CO_2 in the lipid phase of the membrane can induce a decrease in membrane viscosity because the phospholipids are rendered soluble. The alterations in the cell membrane were visible as surface modifications generated by the cell wall collapse and caused stress to the cells. It is believed that this effect is reversible and as soon as the contact between cells and CO_2 is broken, proteins to repair damage are synthesized by the cells to continue cellular division and growth.

When CO_2 is in liquid or supercritical state, its effect is not limited to the cellular membrane since it can penetrate the cell to cause irreversible damage and its effect becomes bactericidal. In the liquid state, the mechanism of cell inactivation of CO_2 involves two stages (Ballestra *et al.*, 1996). The first stage consists of stressing the cells, making them more sensitive to the treatment, blocking the biological functions of the cells, and altering the membrane which allows the absorption of nutrients. Then, CO_2 diffusion across the biological membrane causes a collapse and formation of depressions in the cell wall with a retraction of the cytoplasm and precipitation of the cytoplasmic contents. In inactivation kinetics terms, this stage during the process represents the lag phase corresponding to the time during which the number of cells remain constant before inactivation. After the first stage, the inactivation phase starts. The cells are flaccid and emptied of their cytoplasmic content with holes on the surface of their cellular envelope, completely sunken and wrinkled.

In supercritical conditions, the effect of CO_2 was drastic and the cells did not pass through a stressed step. CO_2 diffused rapidly inside the cells to exercise its bactericidal effects, which involve the supercritical extraction of intracellular substances and the rupture of the cytoplasmic membrane, causing loss of vital cellular constituents and the fragmentation of the cell envelope. The cells were completely destroyed and the wall fragments were found in the suspending medium.

10.2.3 CO_2 Solubility

10.2.3.1 Effects of Solubility on Microbial Inactivation

Solubility of a solute is the analytical composition of a saturated solution, expressed in terms of the proportion of the designated solute in a designated solvent (Karukstis and Van Hecke, 2000) and has been studied since the early 19th century (Battino and Clever, 1965; King *et al.*, 1992). The knowledge of CO_2 solubility is becoming increasingly important considering that the microbial inactivation during the DPCD process is linked to the concentration of CO_2 dissolved in the microbial suspension. Considering the following equation:

$$[CO_2]_{aq} = f(p, T, p_{CO_2}|_{gas}) \tag{10.8}$$

the concentration of dissolved CO_2 is a function of pressure, temperature, and CO_2 partial pressure.

Solubility of CO_2 in water depends on its pressure and temperature in equilibrium with the solution (Dodds *et al.*, 1956). Pressure has a direct effect on CO_2 solubility: increasing pressure increases CO_2 solubility. However, with increasing temperature, solubility of CO_2 decreases. The presence of other substances and the water content in the liquid food can have a positive or negative effect on solubility (Descoins *et al.*, 2006; Meyssami *et al.*, 1992).

The inactivation rate is strongly affected by the composition of the suspending medium and nature of foods. Factors such as acidity, fat or oil content have a direct influence on the CO_2 concentration dissolved in the liquid food and subsequently on the CO_2 amount in contact with the microorganisms. Lin *et al.* (1992b) noticed an increased resistance to the treatment for vegetative cells suspended in complex media with lipids. They proposed that the presence of fat in growth and suspending media probably led to decreased CO_2 penetration into cells and to a reduced solubility by changing the structure of cell walls and membranes.

Bacteria suspended in simple solutions were not inactivated with the same ease under exactly the same conditions due to the different constituents of the solutions (Hong and Pyun, 1999) and to the buffering capacity of the suspending medium (Erkmen, 2000a, 2000b, 2001a, 2001b; Erkmen and Karaman, 2001).

The water activity of the foods strongly affects microbial inactivation. Vegetative cells with low water content were poorly inactivated, and their resistance to inactivation increased with decreasing water content. In particular, Haas *et al.* (1989) suggested that a DPCD treatment would not successfully apply to dry substances. This was related to the direct result of an increased CO_2 solubility in wet microbial cells which were more prone to DPCD inactivation (Lin *et al.*, 1993; Dillow *et al.*, 1999).

Garcia-Gonzalez *et al.* (2009a) studied both the influence of different food components such as NaCl, oil, starch, whey protein, emulsifier, and food properties such as pH, fluid viscosity, and water activity on the DPCD inactivation of *P. fluorescens*. They found that the response of the microorganism to treatment was related to CO_2 characteristics in solubilization, cell wall penetration, and diffusivity. Sucrose and glycerol strongly protected the cells against inactivation.

Ferrentino *et al.* (2010c) demonstrated that the inactivation kinetics of *S. cerevisiae* cells was influenced by the concentration of sodium phosphate dissolved in the suspending medium. By progressively increasing the sodium phosphate concentration from zero (pure water) to 4.80 g in 100 g solution corresponding to a 0.4-molar solution, the inactivation level decreased at fixed conditions of pressure, temperature, and treatment time. The CO_2 solubility values obtained for these solutions (Ferrentino *et al.*, 2010b) show that the inactivation decrease is related to the lower CO_2 concentration in the suspending medium. Therefore, CO_2 solubility is a key parameter for a proper design of the DPCD process.

10.2.3.2 Experimental Methods to Measure CO_2 Solubility

Limited literature on CO_2 solubility in real liquid foods limits optimization of CO_2 use in DPCD treatments, since solubility of CO_2 at different temperatures, pressures, and food compositions is not known. This leads to excess CO_2 use to assure saturation of the liquid.

An experimental method was given by Dohrn *et al.* (1993), who measured CO_2 solubility in real liquid foods. They designed a high-pressure apparatus for obtaining reliable phase equilibrium data for the glucose–water–CO_2 and glucose–water–ethanol–CO_2 systems at temperatures up to 343 K and pressures to 30 MPa. The experimental apparatus had vapor-phase recirculation and a stainless steel autoclave used as a mixing and separating vessel. After the system reached equilibrium temperature, the evacuated gas sample vessel was carefully filled with gas by opening the valves connecting the two autoclaves. The vapor phase was allowed to circulate for 2 h. The samples were expanded in an evacuated vessel. The liquid part was collected by traps and the gaseous volume was measured by the pressure increase in a gas flask. A significant decrease in CO_2 solubility in solutions containing glucose was observed.

More recently, experimental set-ups were designed to evaluate CO_2 solubility in solutions emulating the compositions of fruit juice and in real juices (Calix *et al.*, 2008; Ferrentino *et al.*, 2009; Plaza *et al.*, 2009; Ramirez-Rodrigues *et al.*, 2009; Ferrentino *et al.*, 2010b). The system designed by Calix *et al.* (2008) had two superposed vessels immersed in a water bath and filled with the experimental solution. Gaseous CO_2 was bubbled upwards into the system from the bottom vessel while the liquid flowed down by gravity from the upper vessel into the lower vessel through a side-arm. After saturation of the liquid for 1 h, the bottom vessel was isolated while under pressure. The dissolved gas was drawn from the bottom vessel and the total volumetric flow was read by a gas meter. The liquid was collected at the end of the depressurization and weighted. The apparatus was used to measure CO_2 solubility in aqueous solutions such as water–sucrose–fructose–glucose–malic acid and water–sucrose–fructose–glucose–citric acid. Also, real fruit juices such as apple and orange juices, and later grapefruit juice, guava puree, and a *Hibiscus sabdariffa* beverage were used as experimental liquids. In all the tested solutions, CO_2 solubility decreased compared to that of pure water. There were no significant differences in CO_2 solubility between a real fruit juice and an aqueous solution emulating the composition of the juice (Calix *et al.*, 2008). These solubility experiments made it possible to use sufficient amounts of CO_2 to assure saturation, which resulted in a lower $\%CO_2$ to be used during the pasteurization of these liquids in the continuous DPCD equipment. The optimization and reduction of the excess of CO_2 use is of economic advantage in DPCD treatments.

10.2.3.3 Thermodynamic Methods to Predict Solubility

Experimental equilibrium data are important, even when thermodynamic models can calculate the phase behavior of a mixture. Models can reduce the number of experimental data needed for a special design problem but, at least some experimental data points are needed to evaluate binary or higher-order interaction parameters in thermodynamic models, or to test model predictions.

Modeling of the vapor–liquid phase equilibria of the CO_2–H_2O system has been extensively studied with different approaches based either on a single equation of state for all the phases (ϕ–ϕ models) or on an equation of state combined with an excess Gibbs free-energy model for the liquid phases ($\gamma-\phi$ models). However, few studies have modeled the equilibrium of more complex solutions. Recently the process simulation software Aspen Plus® was used to model the equilibrium of solutions emulating the composition of a fruit juice based on the experimental results obtained by Calix *et al.* (2008) and Ferrentino *et al.* (2010a, b). Three different thermodynamic models were tested: (1) the Peng–Robinson equation of state (EOS), with the Wong and Sandler mixing rules (PRWS) and the excess Gibbs free energy was calculated according to the UNIFAC method; (2) the Electrolytic Non-Random Two Liquids (ELECNRTL) with the Redlich–Kwong equation of state for aqueous and mixed solvent applications; (3) the completely predictive Soave–Redlich–Kwong (PSRK) equation of state. The thermodynamic model that predicted experimental data with good agreement was the Peng–Robinson EOS with Wong and Sandler mixing rules. This model also described the phase equilibrium of ternary systems (water, glucose, CO_2 and water, malic acid, CO_2), quaternary system (glucose, water, malic acid, CO_2) and more complex aqueous solutions emulating apple juice composition (water, citric acid, malic acid, sucrose, glucose, pectin, CO_2).

These findings suggest that suitable thermodynamic models, together with experimental data, can predict CO_2 solubility with reasonable accuracy over a wide range of experimental conditions, allowing prediction of the amount of CO_2 dissolved in the food for a better and more realistic correlation with microbial lethality.

10.2.4 Mixing Speed

Most laboratory-scale experimental set-ups designed for DPCD are batch systems equipped with a stirrer for agitation of the samples. By increasing the mixing speed, the transfer and solubilization rate of pressurized CO_2 in the liquid is increased, affecting the contact of CO_2 and microbial cells. The effect of mixing speed was studied by Lin *et al.* (1992b), who reported that a key factor for increasing the inactivation rate was agitation. Without agitation, the rates of yeast-cell inactivation decreased significantly at any experimental condition. Some authors (Lin *et al.*, 1992a; Hong *et al.*, 1997) found that, without agitation, microbial inactivation appeared to depend on the sample size.

Progressing from the laboratory to pilot scale, the design of the equipment changes from batch to plug-flow reactor. With this new configuration, the mixing speed can not be realized with a stirrer. Efficient contact between CO_2 and the sample is guaranteed by the static mixer at the beginning of the hold tube, and by the turbulent flow in the system (Balaban, 2003; Ferrentino *et al.*, 2009).

10.2.5 Depressurization Rate

Depressurization rate is frequently addressed in the literature on DPCD treatment. A fast rate is believed to burst cells and/or enhance mass transfer across cell membranes. Fraser (1951) proposed that the cells were mechanically ruptured like a "popped balloon" by the fast expansion of CO_2 within cells during flash discharge of pressure. Other researchers (Lin *et al.*, 1991, 1992a, 1992b; Nakamura *et al.*, 1994; Castor and Hong, 1992) claimed that rapid decompression enhanced the disruption of bacterial cells. More recently, Liu *et al.* (2005) concluded that the viability and the biological activity of *S. cerevisiae* (and also of spores of *Absidia coerulea*) were significantly influenced by CO_2 decompression rate (7.5 MPa, 30°C, 30 and 90 min). Although no direct evidence was presented by these authors regarding *S. cerevisiae* cells, this evidence was clearly presented for *A. coerulea* spores (which possess a thicker outer layer) using electron microscopy, which indirectly suggests that the decrease in *S. cerevisiae* cell viability was also partly correlated with the mechanical rupture of the cells.

To elucidate whether mechanical rupture due to explosive decompression was the predominating inactivation mechanism during a DPCD treatment, Enomoto *et al.* (1997) first exposed *S. cerevisiae* cells to pressurized CO_2 (4.0 MPa, 40°C, 240 min) and then decompressed to atmospheric pressure using two different rates (explosive *vs* slow decompression). They showed that a more rapid decompression would not always lead to increased inactivation, suggesting that the microbial cells may not be ruptured mainly due to explosive decompression. They concluded that most of the cells appeared to have been sterilized during pressurization, and not during the rapid discharge stage.

Other researchers (Arreola *et al.*, 1991a; Hong *et al.*, 1997; Kumagai *et al.*, 1996; Debs-Louka *et al.*, 1999) argued that the decompression rate (and the subsequent sudden pressure drop) had no significant effect on microbial reduction in their experimental set-up. It seems that under certain treatment conditions, cell rupture induced by a rapid release of the CO_2 pressure is possible. Secondary effects, such as intensive localized cooling due to the Joule–Thomson effect during expansion may also play some role in cell lysis during flash decompression. Although mechanical cell rupture may cause some cell death, other physiological mechanisms are predominantly involved in the lethal action of DPCD (Garcia–Gonzalez *et al.*, 2007).

10.2.6 Pressure Cycling

Pressure cycling involves repeated compression and release of CO_2. Two theories address increased deactivation: an enhanced cell rupture theory and an enhanced mass transfer theory. Fraser (1951) achieved 90% cell rupture only after two pressure cycles, a 15% increase from single pressurization. Lin *et al.* (1993, 1994) claimed a higher degree of microbial inactivation using pressure cycling, but no data were reported. Dillow *et al.* (1999) observed a significant increase in log reduction using pressure cycling. Inactivation increased from 3 log reductions with three pressure cycles to 9 log reductions with six pressure cycles, at 34°C, 20.5 MPa and total treatment time 0.6 h. Pressure cycling also enhanced the inactivation of bacterial spores. Spilimbergo *et al.* (2002) achieved 3.5-log reductions of *B. subtilis* spores by cycling pressure 30 times in 30 min, over a pressure range from 8 MPa to atmospheric at 36°C. At 60°C, Dillow *et al.* (1999) obtained 8 log-cycles inactivation of *B. cereus* spores with six pressure cycles during a 4-h operation at 20.5 MPa. The enhanced spore inactivation was due not only to a possibly enhanced rupture rate or an enhanced mass transfer rate, but also to the spore germination induced by high-pressure cycling. The literature on pressure cycling with CO_2 is not unanimous. Hong *et al.* (1997) treated *Lactobacillus* sp. suspended in MRS (de Man, Rogosa and Shape) broth with CO_2 pressure cycling, but observed the same inactivation level with static CO_2 pressure. Enomoto *et al.* (1997) reported a lower inactivation with a pressure-cycling process. With the same total treatment time of 4 h, the log reduction of baker's yeast decreased from 7 log-cycles without pressure cycling to 4 log-cycles with a frequency of 1 cycle/h. Ferrentino *et al.* (2008) reported inactivation of natural microflora in apple juice by six compression–decompression cycles. For each cycle, CO_2 contacted the juice at 13.0 MPa, 35°C and 10 min. For the same level of inactivation a treatment of 16.0 MPa, 60°C and 40 min was necessary.

10.3 MICROBIAL INACTIVATION MECHANISMS

For more than 100 years, it has been known that CO_2 can exert an inhibitory effect on microbial growth (Valley and Rettger, 1927). However, the specific mechanism of its bactericidal effect has not been elucidated. Daniels *et al.* (1985) thoroughly reviewed the major theories to explain the bactericidal action of CO_2. Most of these theories have also been used in more recent years by many different authors (Lin *et al.*, 1993; Hong and Pyun, 1999; Spilimbergo and Bertucco, 2003; Damar and Balaban, 2006) to explain the bacteriostatic action of pressurized CO_2. Figure 10.8 illustrates how pressurized CO_2 may exert lethal action on bacteria (Garcia Gonzalez *et al.*, 2007). The different steps in this simplified inactivation mechanism can be listed as follows:

- solubilization of pressurized CO_2 in the external liquid phase
- cell membrane modification
- intracellular pH decrease
- key enzyme inactivation/cellular metabolism inhibition due to internal pH lowering
- direct inhibitory effect of molecular CO_2 and HCO_3^- on metabolism
- disordering of the intracellular electrolyte balance
- extraction of vital constituents from cells and cell membranes
- physical disruption of cell membrane.

Most of these steps will not occur singly or consecutively, but take place simultaneously in a complex and interrelated manner.

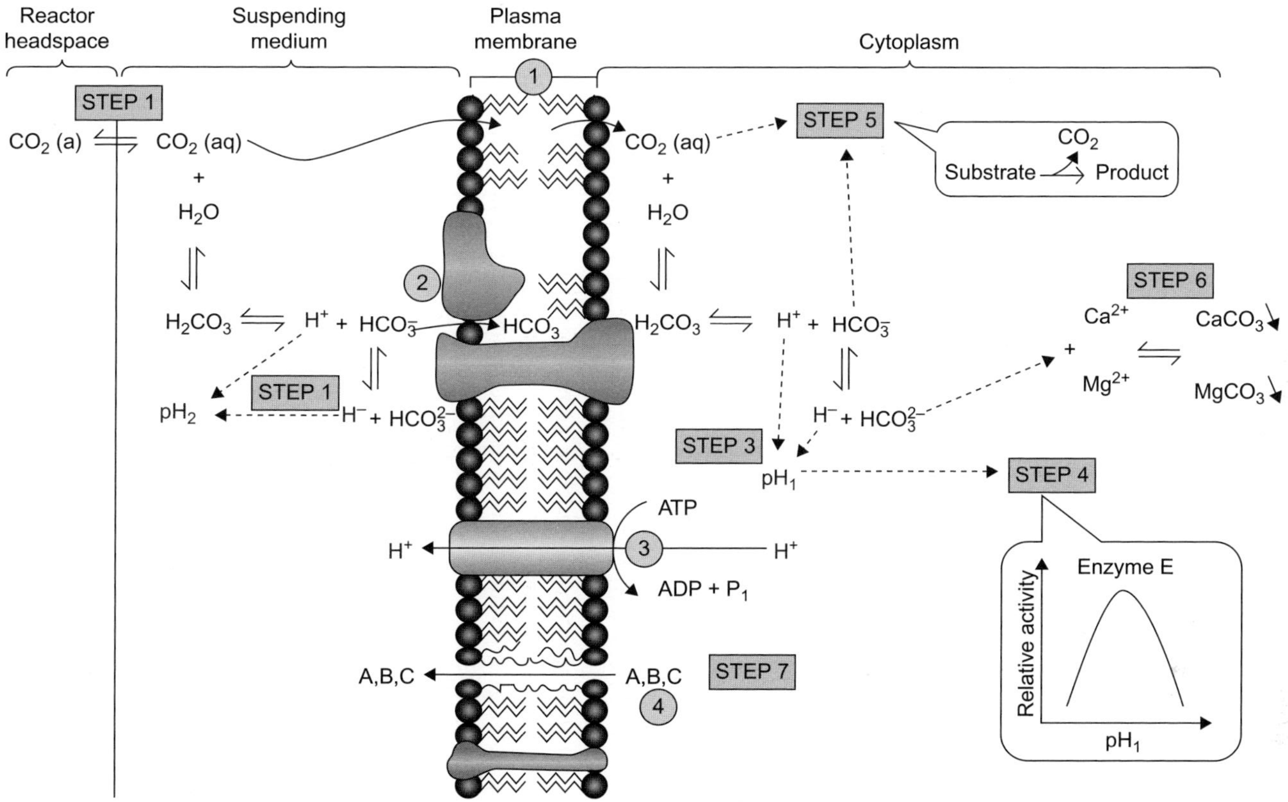

FIGURE 10.8

Schematic representation of the complex phenomena occurring to a microorganism during the high pressure carbon dioxide treatment (Garcia-Gonzalez *et al.*, 2007).

The microorganisms' membrane consists in great part of lipid components, thus the lipophilic CO_2 is likely to influence membrane properties (Isenschmid *et al.*, 1995). The high affinity between CO_2 and the plasma membrane was confirmed by Spilimbergo *et al.* (2002), who calculated that CO_2 can be dissolved in the phospholipids of a model cell membrane, mostly phosphotidietanol amines and phosphotiliglycerol to a very high extent. Due to the increased membrane permeability caused by the reaction of CO_2 with water, which lowers the extracellular pH, pressurized CO_2 may easily penetrate through the bacterial cell membrane and accumulate in the cytoplasmic interior of bacterial cells (Lin *et al.*, 1993, 1994). The accumulated CO_2 in the lipid phase may then structurally and functionally disrupt the cell membrane due to a loss of the order in the lipid chain (a process known as "anesthesia") which may increase the fluidity, and then the permeability of the membrane. The relative concentrations inside the cell of both aqueous CO_2 and HCO_3^- are, in first instance, controlled by the internal pH buffering as a result of pH homeostasis in order to maintain a constant cytoplasmic internal pH which is essential for optimal cell viability and cellular activity. Most organisms evolved mechanisms of controlling and regulating pH, such as (1) cytoplasmic buffering, (2) proton transfer systems, (3) production of acids or bases, and (4) proton pumps. However, although the cytoplasm of most microorganisms has a relatively high buffering capacity, it seems to be used in a limited way by the cells to counter variations in pH (Booth, 1985; Hutkins and Nannen, 1993). If too much dissolved CO_2 enters the cytoplasm, the cells may be unable to expel all the resulting protons and internal pH starts to decrease. If the internal pH is lowered too much, cell viability will be impaired. Large decreases in pH may inhibit cell metabolism or denature certain proteins and enzymes essential for metabolic and regulatory processes, such as glycolysis, amino acid and peptide transport, active transport of ions, and proton translocation (Hutkins and Nannen, 1993; Ballestra *et al.*, 1996; Hong and Pyun, 2001; Spilimbergo *et al.*, 2005).

Flow cytometric analysis showed that after high-pressure CO_2 treatment, a modification of cell membrane in terms of efflux pump activity and membrane integrity occurred (Kim *et al.*, 2009). The damage of the efflux pump induces microbial inactivation (Kim *et al.*, 2007a, b).

However, the reaction rate of the enzymatic reactions is not only a function of the pH but also of the intracellular concentrations of its substrates, products, and co-factors, which are primary elements in the regulation of enzymatic activity. For this reason, the concentration of HCO_3^- (which inside the cell is controlled due to the internal pH buffering) also appears to be central to the regulation of enzymatic activity and hence cellular metabolism with some enzymes is stimulated and some inhibited by this anion (Jones and Greenfield, 1982).

Another effect of the lower pH is linked to the conversion of HCO_3^- to CO_3^{2-} which can induce the precipitation of inorganic electrolytes such as calcium, magnesium, or similar ions from intracellular solutions and damage the functions of the proteins that bind these ions (Lin *et al.*, 1993). These inorganic electrolytes (apart from being important regulators of a large number of other cell activities) help in maintaining the osmotic relationships between cells and their surroundings. A change of this ionic equilibrium can have deleterious effects on the volume of cells (Gangola and Rosen, 1987).

Different authors suggest that accumulated CO_2, due to its relatively high solvent power, can extract vital constituents from the cells or cell membranes (Kamihira *et al.*, 1987; Lin *et al.*, 1992a, 1993). During this inactivation mechanism, CO_2 first penetrates into cells to build up to a critical level after which it removes intracellular constituents, such as phospholipids and hydrophobic compounds, disturbing or altering the structure of the biomembrane and the balance of the biological system, thus promoting inactivation (Lin *et al.*, 1992a, 1993). This removal process may be stimulated by a sudden release of pressure, leading to a rapid transfer of intracellular materials into the extracellular environment (Lin *et al.*, 1992a, 1993; Ballestra *et al.*, 1996). Pressurized CO_2 also has the ability to modify cell membranes as enlarging the perplasmic space between the walls and the cytoplasmic membranes induces the leakage of cytoplasmic membrane or cytoplasm (Hong *et al.*, 1999; Hong and Pyun, 2001; Bertoloni *et al.*, 2006).

Another suggested inactivation mechanism of microorganisms induced by pressurized CO_2 is the physical disruption of cells (Fraser 1951; Lin *et al.*, 1991; Hong and Pyun, 1999). This evidence is demonstrated through the scanning electron microscope (SEM) technique which shows that some cells completely burst whereas some have wrinkles or holes at their surface (Nakamura *et al.*, 1994; Ballestra *et al.*, 1996). The same results are reported by Dagan and Balaban (2006) for *Saccharomyces cerevisiae* cells in beer treated with a continuous DPCD system (27.5 MPa, 21°C, 10% CO_2/beer ratio, 5 min) demonstrating that cell rupture is possible by DPCD, but not necessary for cell inactivation. Cells may be completely inactivated even when they remain intact or show only signs of deformation (Hong and Pyun, 1999).

10.4 EFFECTS ON FLUID FOOD NUTRITIONAL AND QUALITY PARAMETERS

Most of the DPCD inactivation studies currently available in the scientific literature have been performed in inoculated or spoiled foods or in inoculated simple model systems and liquid substrates, such as physiological saline solution or culture media. In recent years, the number of published studies in which natural microorganisms in real foods are targeted and inactivated by DPCD has increased. Also, research on the effects of DPCD on the quality of foods, including some results published by companies offering commercial systems were reported.

In this section the effectiveness of the DPCD treatment on the microbial and enzyme inactivation and shelf-life will be addressed. The impact of the treatment on the physical, chemical, sensory and nutritional attributes will be described. In Table 10.3 a literature review on microbial inactivation in some liquid foods is reported.

Table 10.3 Summary of Some Studies on Microbial Inactivation in Liquid Foods by Dense-Phase CO_2

Solution	Inoculated Microorganism	Treatment	T (°C)	System	Maximum Reduction	Reference
Whole skim milk	*S. aureus*	9–15 MPa 2–5 h	25	Batch	7 log	Erkmen (1997)
Orange juice	*L. monocytogenes*	6 MPa 8 h	45	Batch	6 log	Erkmen (2000b)
Peach juice		6 MPa 4 h				
Carrot juice		6 MPa 12 h				
Whole milk	*E. coli*	10 MPa 6 h	30	Batch	6.4 log	Erkmen (2001b)
Skim milk					7.2 log	
Natural orange juice	Mold–Yeasts	30 MPa 15 min	28	Semi-cont. process	Total inact.	Spilimbergo *et al.* (2002)
Beer	Yeasts	26.5 MPa 4.77 min 9.6% CO_2	21	Continuous plant	7.3 log	Dagan and Balaban (2006)
Orange juice	*E. coli*	107 MPa 0.66 CO_2/juice 10 min	25	Continuous plant	5 log	Kincal *et al.* (2005)
	S. typhimurium	21 MPa 0.4 CO_2/juice 10 min			6 log	
	L. monocytogenes	38 MPa 0.20 CO_2/juice 10 min				
Mandarin juice	Aerobic plate count	4 MPa 9 min	35	Continuous plant	3.47 log	Yagiz *et al.* (2005)

Watermelon juice	Aerobic plate count	34.4 MPa 10% CO_2 5 min	40	Continuous plant	6.82 log	Lecky and Balaban (2005)
Apple juice	Mold–Yeasts	10 MPa 5 ml 1 min	36	Multi-batch system	Total inact.	Spilimbergo and Mantoan (2006)
Coconut water	Aerobic plate count	34.5 MPa 13% CO_2 6 min	40	Continuous plant	5.61 log	Damar and Balaban (2005)
Muscadine grape juice	Mold–Yeasts Total aerobic microorganisms	40 MPa 15.7% CO_2 6.25 min	30	Continuous plant	7 log 5.7 log	Del Pozo-Insfran *et al.* (2006a)
Pear juice Apple juice	*S. cerevisiae*	9 MPa 1.8 ml 15 min	32	Multi-batch system	5.5 log 4. 9 log	Parton *et al.* (2007)
Apple juice	*S. cerevisiae*	10 MPa 50 min	36	Multi-batch system	4.7 log	Spilimbergo *et al.*, 2007
Apple juice	Mold–Yeasts	16 MPa 40 min	60	Batch	5 log	Ferrentino *et al.* (2008)
Grapefruit juice	Mold–Yeasts Total aerobic microorganisms	34.5 MPa 5.7% CO_2 7 min	40	Continuous plant	5 log	Ferrentino *et al.* (2009)

10.4.1 Effect on Microbial Inactivation and Stability

Haas *et al.* (1989) reported that DPCD could kill bacteria, molds and yeasts in some real foods. The natural microflora in orange juice significantly decreased from 10^2–10^3 cfu/ml to less than 10 cfu/ml with a DPCD treatment at room temperature (Kincal *et al.*, 2005). Arreola *et al.* (1991a) also showed that 2 log-cycles reduction of natural aerobic bacteria in fresh-squeezed and quick-frozen single-strength orange juice was achieved by static DPCD equipment at 35°C for 60 min, 45°C for 45 min and 60°C for 15 min, respectively. The results highlighted that the treatment time of DPCD decreased with increasing the temperature from 35 to 60°C. Erkmen (1997) reported that DPCD led to a significant decrease in the natural aerobic bacteria in milk. The destruction of natural aerobic bacteria and yeasts and molds was also observed in a muscadine grape juice by Del Pozo-Insfran *et al.* (2006a).

Generally, it is more difficult to inactivate natural microorganisms in real foods than inoculated microorganisms in real foods or in buffers due to the complexity of natural microflora in real foods and their difference from inoculated pure strains in their susceptibility to DPCD (Garcia-Gonzalez *et al.*, 2007).

Supercritical CO_2 pasteurization was carried out on fresh apple juice by Spilimbergo and Mantoan (2006) achieving total microbial inactivation in the juice at 10.0 MPa, 36°C after 10 min in a multibatch system. Liao *et al.* (2010) carried out a more precise study on the same juice to evaluate the microbial inactivation and the stability of natural microorganisms in terms of aerobic bacteria and yeasts and molds. The experiments were performed at 20 MPa and 30 min with CO_2 concentration of 4.5–5.3% at different temperatures and compared to a mild heat treatment at atmospheric pressure. The DPCD process increased the susceptibility of the natural microorganisms to temperature and enhanced their microbial inactivation. Total aerobic bacteria inactivation was obtained at temperatures higher than 62°C for the mild heat treatment but higher than 57°C for DPCD. For the yeasts and molds a total inactivation occurred at temperatures higher than 57°C during the mild heat treatment but higher than 42°C during the DPCD.

Yeasts and molds were destroyed at significantly higher rates than aerobic bacteria under identical DPCD conditions (Del Pozo-Insfran *et al.*, 2006a; Liao *et al.*, 2010). These findings confirm that microbial inactivation is highly dependent on the type of microorganisms present in the food due to distinct microbial cell structure and the diffusion of CO_2 into the microbial cells.

Fresh-squeezed red blush (*Citrus paradisi*) grapefruit juice was treated with continuous DPCD equipment to inactivate yeasts and molds and total aerobic microorganisms (Ferrentino *et al.*, 2009). Five-log reductions for yeasts and molds and total aerobic microorganisms occurred at 34.5 MPa, 40°C, 7 min of treatment and 5.7% (w/w) CO_2/juice ratio.

The feasibility of DPCD processing as a non-thermal pasteurization technique was also investigated for liquid whole eggs (Garcia-Gonzalez *et al.*, 2009b), proving to be a promising non-thermal alternative to heat pasteurization.

The effect of DPCD treatment on microbial growth during refrigerated storage for different foods was also investigated. Park *et al.* (2002) evaluated the effect of a combined DPCD (0.1–4.90 MPa, 5 min, 25°C) and high hydrostatic pressure treatment (400 MPa, 5 min, 25°C) on the microbial quality of carrot juice during 4 weeks of storage at 4°C. During storage, no reactivation of aerobes was observed in the treated carrot juice, while the untreated sample showed 8.4 log-cycles after 4 weeks. Kincal *et al.* (2005) observed an increase in the bacterial numbers during storage at 1.7°C of orange juice treated with continuous DPCD equipment at 107 MPa and 1.03% (w/w) CO_2/juice ratio during 10 min, while immediately after DPCD treatment no culturable microorganisms were detected in the juice. By 6 weeks of storage, the microbial count in the DPCD-treated juice reached 6 log-cycles, considered unacceptable for juice quality.

Del Pozo-Insfran *et al.* (2006a) observed that the stability of the yeasts and molds in muscadine grape juice was similar between the DPCD and thermally pasteurized juices during the first 5 weeks of storage at 4°C, but significant differences were observed subsequently through 10 weeks. It was also observed that yeasts and molds counts for DPCD-treated juices continuously increased throughout subsequent storage whereas no changes were observed for thermally pasteurized juices. Whether this implies that injury and repair took place within the microorganisms in the treated juice during storage is not clear (Arreola *et al.*, 1991a). It was hypothesized that microorganisms exist in a viable but non-culturable state not detected by normal microbiological enumeration methods.

A shelf-life study was carried out on DPCD-treated liquid whole eggs for 35 days at 4°C. It was demonstrated that DPCD is comparable to the current shelf-life of heat-processed liquid whole eggs (Garcia-Gonzalez *et al.*, 2009b). Liao *et al.* (2010) showed that apple juice pasteurized at 20 MPa and 30 min with CO_2 concentration of 4.5–5.3% had a good stability during storage at 2 and 28°C for 40 days in terms of total aerobic bacteria for DPCD process temperature higher than 52°C and in terms of yeasts and molds for DPCD process temperature higher than 57°C.

10.4.2 Effect on Enzyme Inactivation

Inactivation of certain enzymes affecting food quality by DPCD has been shown by several researchers (Balaban *et al.*, 1991; Chen *et al.*, 1992, 1993; Park *et al.*, 2002). A summary of the literature including the enzymes, the activity reduction achieved and treatment conditions is given in Table 10.4.

DPCD can inactivate certain enzymes at temperatures where thermal inactivation is not effective (Balaban *et al.*, 1991). Among these enzymes, pectin methylesterase (PME) causes cloud loss in some fruit juices; polyphenol oxidase (PPO) causes undesirable browning in fruits, vegetables, juices, and some seafood; lypoxygenase (LOX) causes chlorophyll destruction and off–flavor development in frozen vegetables; peroxidase (POD) has an important role in the discoloration of foods and is used as an index of heat treatment efficiency in fruit and

Table 10.4 Summary of Some Studies on Enzymes Inactivated by Dense-Phase CO_2

Enzyme	Source	Treatment	T (°C)	System	Activity Loss (%)	Reference
Lipase	Commercial (62% to 68% water)	2 MPa 2 h	35	Batch	12 to 22	Kamihira *et al.* (1987)
α-Amylase	Commercial (62% to 68% water)	20 MPa 2 h	35	Batch	0	
Pectinerase	Orange juice	26.9 MPa 145 min	56	Batch	100	Balaban *et al.*, 1991
Polyphenol oxidase	Spiny lobster	5.8 MPa 1 min	43	Batch	98	Chen *et al.* (1992)
	Brown shrimp	5.8 MPa 1 min			78	
	Potato	5.8 MPa 30 min			91	
Polyphenol oxidase	Spiny lobster	0.1 MPa 30 min	33	Batch	98.5	Chen *et al.* (1993)
Lypoxygenase	Soybean	10.3 MPa 15 min	50	Batch	100	Tedjo *et al.* (2000)
Peroxidase	Horseradish	62.1 MPa 15 min	55		100	
Lypoxygenase	Soybean	62.1 MPa 15 min	35		95	

Polyphenol oxidase	Carrot juice	4.9 MPa 10 min	50	Batch	61	Park *et al.* (2002)
Lypoxygenase		2.94 MPa 10 min			>70	
Lipase	*C. rugosa*	0.1 MPa 200 ml/min CO_2 30 min	45	Semi-continuous system	88	Fadiloglu and Erkmen (2002)
α-Amylase	*B. subtilis*	10 MPa 15.6 g/min CO_2 14.4 min	45	Continuous system	83.9	Yoshimura *et al.* (2002)
Acid protease	*Aspergillus niger*	10 MPa 15.6 g/min CO_2 13.9 min	50		86.1	
Alkaline phosphatase	In buffer solution and raw milk	0.1 MPa 200 ml/min CO_2 30 min	50	Semi–continuous	81	Fadiloglu *et al.* (2004)
Polyphenol oxidase	Muscadine grape juice	27.6 MPa 7.5% CO_2 6.25 min	30	Continuous system	35	Del Pozo-Insfran *et al.* (2007)

vegetable processing. Although limited in number, studies on enzyme inactivation by CO_2 indicate its good potential, especially in fruit and vegetable juice processing where these enzymes cause quality deterioration if not inactivated. Enzyme inactivation by CO_2 could be due to many causes such as pH lowering, conformational changes in the enzyme, and inhibitory effect of molecular CO_2 on enzyme activity. Balaban *et al.* (1991) studied PME inactivation in orange juice demonstrating that the pH-lowering effect alone was not sufficient to explain inactivation. The results of Chen *et al.* (1992) supported this conclusion.

It has been shown that CO_2 under pressure can change isoelectric profiles and protein patterns of PPOs (Chen *et al.*, 1992), while CO_2 at atmospheric pressure does not affect these properties (Chen *et al.*, 1993). Chen *et al.* (1992) reported that DPCD caused noticeable alterations in the composition of α-helices and random coils in the secondary structures of lobster, brown shrimp, and potato PPO. Gui *et al.* (2006) showed that the α-helix relative content in the secondary structure of horseradish POD decreased with increasing DPCD pressures, and that after 7 days storage at 4°C the restoration of the structure was observed. Liao *et al.* (2009) also found that circular dichroism spectra of treated LOX were noticeably changed and the α-helix relative content decreased sharply to less than 10%. Ishikawa *et al.* (1995) reported that the residual activity of glucoamylase and acid protease treated by DPCD had a good linear relationship with the loss of the α-helix relative content. Modifications in the secondary and tertiary structures of PME were also observed by Zhou *et al.* (2009a) who reported an increase in the fluorescence intensity probably resulting from relocation of tryptophan residues in the three-dimensional structure of PME.

Further research should investigate whether the conformational changes are caused by pressure alone, and if not, elucidate the other causes. The extent of enzyme inactivation by DPCD is affected by the type and source of the enzyme, treatment conditions such as pressure, temperature, and time, and treatment-medium properties. A great number of studies reported differences in the inactivation level although the same enzyme was treated by DPCD. Balaban *et al.* (1991) observed that higher temperatures and pressures of CO_2 treatment result in higher percent PME inactivation and reported a complete inactivation of the enzyme. Zhi *et al.* (2008) reported that DPCD effectively inactivated PME in buffer and that pressure had a noticeable effect on the residual activity of apple PME. Other studies demonstrated that, despite high DPCD process conditions, a minimal residual activity remained after the process (Kincal *et al.*, 2006; Zhou *et al.*, 2009a). A combined process was used by Park *et al.* (2002) who found that the minimum residual activity of carrot PME was 35.1% after a combined process of DPCD at 4.90 MPa and high hydrostatic pressure at 600 MPa. Corwin and Shellhammer (2002) found that CO_2 was a significant factor in further inactivating PME in orange juice beyond that which pressure would achieve alone at 25°C and at 50°C.

An enzyme isolated from different sources has different resistance to treatment. For example, potato PPO is more resistant to inactivation by CO_2 compared with spiny lobster and shrimp PPOs (Chen *et al.*, 1992). Liu *et al.* (2010) reported the inactivation of POD and PPO in red beet extract, at 7.5 MPa, 55°C and 30 min, with continuous DPCD and demonstrated that both enzymes achieved more retention compared to the thermal treatment.

The presence of other soluble compounds in the medium may have a protective effect against DPCD treatment. Tedjo *et al.* (2000) showed that residual per cent LOX and per cent POD activity increased by increasing the sucrose concentration up to 40%. This could be explained by the decrease of CO_2 solubility as sucrose concentration increases.

Water content is a key parameter in enzyme inactivation. Fricks *et al.* (2006) performed experiments on radish POD extracts subjected to lyophilization to change the water content of the samples. Results indicated a decrease in the enzyme activity for samples treated at 7.1 MPa, 30°C for 1 h. Samples lyophilized by 24 h and 12 h had a residual activity of 15% and 42%, respectively. This demonstrated the importance of water in POD stability under DPCD process.

10.4.3 Effect on the Physical, Chemical, Sensory and Nutritional Attributes

The physical, chemical, sensory, and nutritional attributes of DPCD-treated juices have been studied and some improvements have been observed in the processed juice. It is important to carry out taste panel tests or consumer acceptance of food products processed by DPCD. The extensive use and commercialization of this technology depends on the positive results of these tests.

Whole fruits such as strawberry, honeydew melon, and cucumber were DPCD-treated to inhibit mold growth. However, it was demonstrated that DPCD may cause severe tissue damage in some fruits even at low pressures (Haas *et al.*, 1989). Arreola *et al.* (1991a) observed significantly higher ascorbic acid retention of DPCD-treated orange juice than that of the temperature controls. Del Pozo-Insfran *et al.* (2006b) observed no significant changes in total anthocyanins, total soluble phenolics, and antioxidant capacity of DPCD-treated muscadine grape juice compared to control, whereas heat-treated samples caused a decrease of 16%, 26%, and 10%, respectively. DPCD-treated juices retained higher total anthocyanin, total soluble phenolics and antioxidant capacity content than thermally pasteurized juices after 10 weeks of storage at 4°C.

Dagan and Balaban (2006) used a continuous CO_2 system to pasteurize beer and compared physical and sensory attributes with that of fresh and heat pasteurized beer. Aroma and flavor of DPCD-treated beer was not significantly different from fresh beer after being stored for 1 month at 1.67°C, but heat-treated beer was significantly different in taste and aroma. DPCD-treated beer had significantly less foam capacity and stability compared to heat pasteurized, but not at levels detrimental to product quality.

Yagiz *et al.* (2005) treated mandarin juice with CO_2 using a continuous system and measured the pH, °Brix, titratable acidity, cloud, and color after treatment

(13.8–41.4 MPa, 25°C to 45°C, 7–9 min). DPCD enhanced cloud up to 38.4%, increased lightness and yellowness, and decreased redness of mandarin juice. The treated samples had higher titratable acidity than the untreated samples, but pH and °Brix did not change after treatment.

The influence of thermal (90°C for 60 s) and DPCD (55°C, 60 min, and 35 MPa) pasteurization on physicochemical properties and flavor compounds in Hami melon juice was also investigated. DPCD induced microbial inactivation as well as preserving some quality attributes, including most of the aroma compounds in the melon juice. DPCD-treated melon juice retained more volatile compounds, ascorbic acid, and β-carotene. The same juice treated with conventional thermal treatment had significant losses in ascorbic acid, β-carotene, six-and nine-carbon alcohol, and aldehydes, while some new compounds were produced with cooked-off odor (Chen *et al.*, 2009). DPCD treatment has also been applied to more sensitive juices such as carrot juice which is a natural source of provitamin A and has a higher risk of bacterial contamination since it is a low-acid drink. The treatment affected the degree of browning and decreased the pH compared to the untreated sample while the cloud and titratable acidity increased significantly. PPO was inactivated, the total soluble solid content and carotenoids were stable but the carrot juice viscosity increased (Zhou *et al.*, 2009b).

In some studies (Arreola *et al.*, 1991b; Balaban *et al.*, 2001; Dagan and Balaban 2006; Damar and Balaban 2006; Gunes *et al.*, 2005; Lecky and Balaban 2005; Del Pozo-Insfran *et al.*, 2006b), sensory evaluation was conducted by panelists to better understand consumer perspective and acceptance of DPCD-treated foods. Sensory analyses were performed on fresh control (i.e. untreated) and DPCD-processed juices using difference from control and ranking tests. Additionally, three studies (Dagan and Balaban 2006; Damar and Balaban, 2006; Del Pozo-Insfran *et al.*, 2006b) also conducted sensory analyses on heat-pasteurized samples. Results indicated that the DPCD-treated juices were indistinguishable from the untreated sample (Balaban *et al.*, 2001; Gunes *et al.*, 2005; Lecky and Balaban 2005). There was no significant difference in flavor, aroma, or overall acceptability of these samples (Arreola *et al.*, 1991b; Dagan and Balaban, 2006; Damar and Balaban, 2005; Del Pozo-Insfran *et al.*, 2006a). DPCD-treated juices were ranked higher (more likeable) by the sensory panel than the heat-pasteurized samples (Damar and Balaban, 2005; Del Pozo-Insfran *et al.*, 2006a) or the overall likeability was not significantly different from fresh control (Damar and Balaban, 2005).

A triangle sensory test was performed on the DPCD apple juice, indicating a moderate effect of the treatment and an effect on the perceptible juice quality. Instrumental analysis showed no significant differences regarding sugar and acid content and polyphenol concentration but more specific analysis highlighted a reduction in the concentration of many volatile compounds such as acetates and aldehydes responsible for the odor and flavor of juices (Gasperi *et al.*, 2009).

10.5 CURRENT STATUS AND FUTURE TRENDS

10.5.1 Commercialization Efforts

DPCD pasteurization is operational and almost ready to be used on a commercial scale. Most of the commercialization efforts so far have been performed by Praxair Inc. (Burr Ridge, IL, US). Based on the technology licensed from the University of Florida (Balaban *et al.*, 1995; Balaban, 2004), Praxair developed a continuous process which utilizes DPCD as a non-thermal alternative to thermal pasteurization (Connery *et al.*, 2005). This system was given the Trade Mark "Better Than Fresh (BTF)". Praxair has constructed four mobile BTF demonstration units for processing about 1.5 l/min. In addition, a commercial-scale unit of 150 l/min has also been constructed (Connery *et al.*, 2005) and tested at an orange juice processing plant in Florida. For the continuous treatment of liquid foods, pilot-scale equipment was also manufactured by Mitsubishi Kakoki Co. (Tokyo, Japan) on behalf of and according to the patents owned by Shimadzu Co. (Kyoto, Japan) (Osajima *et al.*, 1997, 1999). This equipment consisted of a treatment vessel of 5.8 l through which CO_2 and liquid foodstuff were simultaneously pumped at maximum flow rates of 3.0 kg/h and 20 kg/h, respectively. In 2003, the instrument was only made available to research laboratories in Japan (private communication, Shimadzu Belgium). At the moment, we have no information on further commercialization efforts of Shimadzu in HPCD processing for liquid foods. Based on their patent (Sims, 2000), PoroCrit LLC (Berkeley, CA, US) also made a commercial membrane contactor consisting of several hollow-fiber membranes for the continuous HPCD pasteurization of liquid foods mainly beverages, juices, milk and wine (Garcia-Gonzalez *et al.*, 2007).

10.5.1.1 Potential Advantages

DPCD has progressed, both in terms of theoretical and predictive capabilities, and in the accumulation of experience in its applications to many juices and beverages. Recent findings regarding the expression of the rate of inactivation of microorganisms based on the solubility of the CO_2 in the liquid allow theoretical prediction of CO_2 solubility in a liquid to be used in actual prediction of the microbial inactivation rates (Ferrentino *et al.*, 2010a). This still requires extensive testing and generation of data to satisfy regulatory requirements.

Many examples of the applications of the DPCD to juices demonstrated the protective nature of the process to antioxidants, phytochemicals (Del Pozo-Insfran *et al.*, 2006a, 2006b), organoleptic attributes (Arreola *et al.*, 1991b, Lim *et al.*, 2006) such as taste, color, and appearance (Kincal *et al.*, 2006, Damar *et al.*, 2009). The relatively low process temperature, the lack of oxygen in the environment, and for some nutrients, the lower pH, protect the vitamins such as vitamin C (Arreola *et al.*, 1991b). Since the process can be made continuous, its control is easy.

10.5.1.2 Potential Difficulties

The lack of the first commercially successful DPCD operation is making new entries into the field difficult. Although during commercial testing with orange juice most of the technical issues have been resolved, the added cost of the operation is higher than, for example, heating for pasteurization.

Another issue is the use of CO_2, which is a greenhouse gas. With environmental regulations getting tighter regarding the release of CO_2 into the atmosphere, both total capture and recycling of the gas needs to be designed into new systems, or a carbon-neutral source of CO_2 needs to be used.

Despite the long list of applications and research in the area, there is barely enough data to satisfy the regulatory requirements, and new beverages and new treatment systems need to go through a process of proof of safety of the final product. There are niche areas where this technology will fit perfectly. Our challenge is to find these areas, prove successful commercial operation, and then expand to other applications.

References

Arreola, A.G., Balaban, M.O., Wei, C.I., Peplow, A., Marshall, M.R., 1991a. Effect of supercritical CO_2 on microbial populations in single strength orange juice. J. Food Qual. 14, 275–284.

Arreola, A.G., Balaban, M.O., Marshall, M.R., Peplow, A.J., Wei, C.I., Cornell, J.A., 1991b. Supercritical CO_2 effects on some quality attributes of single strength orange juice. J. Food Sci. 56, 1030–1033.

Balaban, M.O., Arreola, A.G., Marshall, M.R., Peplow, A., Wei, C.I., Cornell, J., 1991. Inactivation of pectinesterase in orange juice by supercritical CO_2. J. Food Sci. 56, 743–750.

Balaban, M.O., Marshall, M.R., Wicker, L., 1995. Inactivation of enzymes in foods with pressurized CO_2. US Patent 5, 393, 547.

Balaban, M.O., Kincal, D., Hill, S., Marshall, M.R., Wildasin, R., 2001. The synergistic use of carbon dioxide and pressure in nonthermal processing of juices. In: IFT Annual Meeting Book of Abstracts, June 23–27, 2001, New Orleans, La. Chicago Ill.

Balaban, M.O., 2003. Method and apparatus for continuous flow reduction of microbial and/or enzymatic activity in a liquid beer product using carbon dioxide. US Patent Application 2004/0234661 A1 (Nov. 25, 2004) and WO Patent 2005/034655 A1 (Apr. 21, 2005).

Balaban, M.O., 2004. Method and apparatus for continuous flow reduction of microbial and/or enzymatic activity in a liquid product using carbon dioxide. US Patent 6,723,365 B2 and US Patent Application 2004/0131739 A1.

Ballestra, P., Da Silva, A.A., Cuq, J.L., 1996. Inactivation of Escherichia coli by carbon dioxide under pressure. J. Food Sci. 61, 829–836.

Ballestra, P., Cuq, J.L., 1998. Influence of pressurized CO_2 on the thermal inactivation of bacterial and fungal spores. Lebensm. Wiss. Technol. 31, 84–88.

Battino, R., Clever, H.L., 1965. The solubility of gases in liquids. Chem. Rev. 66, 395–463.

Bertoloni, G., Bertucco, A., De Cian, V., Parton, T., 2006. A study on the inactivation of micro organisms and enzymes by high pressure CO_2. J. Biotechnol. Bioeng. 95, 155–160.

Booth, I.R., 1985. Regulation of cytoplasmic pH in bacteria. Microbiol. Rev. 49, 359–378.

Brown, H.D., Fitzgerald, C., Neuman, F., 1939. Preserving cider by carbonation. Am. Soc. of Hortic. Sci. 36, 371–373.

Calix, T.F., Ferrentino, G., Balaban, M.O., 2008. Measurement of high-pressure carbon dioxide solubility in orange juice, apple juice and model liquid solutions. J. Food Sci. 73, E439–E445.

Castor, T.P., Hong, G.T., 1992. Supercritical fluid disruption of and extraction from microbial cells. US Patent 5, 380, 826.

Chen, J.S., Balaban, M.O., Wei, C.I., Marshall, M.R., Hsu, W.Y., 1992. Inactivation of polyphenol oxidase by high pressure CO_2. J. Agric. Food Chem. 40, 2345–2349.

Chen, J.S., Balaban, M.O., Wei, C.I., Gleeson, R.A., Marshall, M.R., 1993. Effect of CO_2 on the inactivation of Florida spiny lobster polyphenol oxidase. J. Sci. Food Agric. 61, 253–259.

Chen, J., Zhang, J., Feng, Z., Song, L., Wu, J., Hu, X., 2009. Influence of thermal and dense-phase carbon dioxide pasteurization on physicochemical properties and flavor compounds in Hami melon juice. J. Agric. Food Chem. 57, 5805–5808.

Connery, K.A., Shah, P., Coleman, L., Hunek, B., 2005. Commercialization of Better Than Fresh™ dense phase carbon dioxide processing for liquid food. In: ISSF 2005, Orlando (US).

Corwin, H., Shellhammer, T.H., 2002. Combined carbon dioxide and high pressure inactivation on pectin methylesterase, polyphenol oxidase, Lactobacillus plantarum and Escherichia coli. Food Eng. Phys. Prop. 67, 697–701.

Dagan, G.F., Balaban, M.O., 2006. Pasteurization of beer by a continuous dense phase CO_2 system. J. Food Sci. 71, E164–169.

Damar, S., Balaban, M.O., 2005. Cold pasteurization of coconut water with a dense phase CO_2 system. In: IFT Annual Meeting Book of Abstracts; 2005 July 15–20; New Orleans, La. Chicago Ill.

Damar, S., Balaban, M.O., Sims, C.A., 2009. Dense Phase CO_2 Processing of Coconut Water. Int. J. Food Sci. Techn. 44, 666–673.

Damar, S., Balaban, M.O., 2006. Review of dense phase CO_2 technology: Microbial and enzyme inactivation, and effects on food quality. J. Food Sci. 71, R1–R11.

Daniels, J.A., Krishnamurthi, R., Rizvi, S.S.H., 1985. A review of effects of carbon dioxide on microbial growth and food quality. J. Food Prot. 48, 532–537.

Debs-Louka, E., Louka, N., Abraham, G., Chabot, V., Allaf, K., 1999. Effect of compressed carbon dioxide on microbial cell viability. Appl. Environ. Microbiol. 65, 626–631.

Del Pozo-Insfran, D., Balaban, M.O., Talcott, S.T., 2006a. Microbial stability, phytochemical retention and organoleptic attributes of dense phase CO_2 processed Muscadine grape juice. J. Agric. Food Chem. 54, 5468–5473.

Del Pozo-Insfran, D., Balaban, M.O., Talcott, S.T., 2006b. Enhancing the retention of phytochemicals and organoleptic attributes in muscadine grape juice by dense phase-CO_2 processing and copigmentation. J. Agric. Food Chem. 54, 6705–6712.

Del Pozo-Insfran, D., Balaban, M.O., Talcott, S.T., 2007. Inactivation of polyphenol oxidase in muscadine grape juice by dense phase-CO_2 processing. Food Res. Int. 40, 894–899.

Descoins, C., Mathlouthi, M., Moual, M., Hennequin, J., 2006. Carbonation monitoring of beverage in a laboratory scale unit with on-line measurement of dissolved CO_2. J. Food Chem. 95, 541–553.

Dillow, A.K., Dehghani, F., Hrkach, J.S., Foster, N.R., Langer, R., 1999. Bacterial inactivation by using near- and supercritical CO_2. Proc. Natl. Acad. Sci. USA 96, 10344–10348.

Dodds, W.S., Stutzman, L.F., Sollami, B.J., 1956. CO_2 solubility in water. Ind. Eng. Chem. 1, 92–95.

Dohrn, R., Bunz, A.P., Devlieghere, F., Thelen, D., 1993. Experimental measurements of phase equilibria for ternary and quaternary systems of glucose, water, CO_2 and ethanol with a novel apparatus. Fluid Phase Equilib. 83, 149–158.

Enomoto, A., Nakamura, K., Hakoda, M., Amaya, N., 1997. Lethal effect of high-pressure carbon dioxide on a bacterial spore. J. Ferment. Bioeng. 83, 305–307.

Erkmen, O., 1997. Antimicrobial effect of pressurized CO_2 on Staphylococcus aureus in broth and milk. Technol. Lebensm. Wissen. 30, 826–829.

Erkmen, O., 2000a. Antimicrobial effects of pressurized carbon dioxide on Brochothrix thermosphacta in broth and foods. J. Sci. Food Agr. 80, 1365–1370.

Erkmen, O., 2000b. Effect of carbon dioxide on Listeria monocytogenes in physiological saline and foods. Food Microbiol. 17, 589–596.

Erkmen, O., 2000c. Predictive modelling of Listeria monocytogenes inactivation under high pressure carbon dioxide. Lebensm. Wiss. Technol. 33, 514–519.

Erkmen, O., 2001a. Kinetics analysis of Listeria monocytogenes inactivation by high pressure carbon dioxide. J. Food Eng. 47, 7–10.

Erkmen, O., 2001b. Mathematical modelling of Escherichia coli inactivation under high pressure carbon dioxide. J. Biosci. Bioeng. 92, 39–43.

Erkmen, O., Karaman, H., 2001. Kinetic studies on the high pressure carbon dioxide inactivation of Salmonella typhimurium. J. Food Eng. 50, 25–28.

Erkmen, O., 2003. Mathematical modeling of Saccharomyces cerevisiae inactivation under high-pressure carbon dioxide. Nahrung/Food 47, 176–180.

Fadiloglu, S., Erkmen, O., 2002. Inactivation of lipase by carbon dioxide under atmospheric pressure. J. Food Eng. 52, 331–335.

Fadiloglu, S., Erkmen, O., Sekeroglu, G., 2004. Thermal and carbon dioxide inactivation of alkaline phosphatase in buffer and milk. Food Technol. Biotechnol. 42, 27–32.

Ferrentino, G., Bruno, M.C., Ferrari, G., Poletto, M., Balaban, M.O., 2008. Microbial inactivation and shelf life of Apple juice treated with high pressure carbon dioxide. In: Proceedings of the 1st European Food Congress. 4–9 November, Ljubliana Slovenia.

Ferrentino, G., Plaza, M.L., Ramirez-Rodrigues, M., Ferrari, G., Balaban, M.O., 2009. Effects of dense phase carbon dioxide pasteurization on the physical and quality attributes of a red grapefruit juice. J. Food Sci. 74, E333–E341.

Ferrentino, G., Barletta, D., Balaban, M.O., Ferrari, G., Poletto, M., 2010a. Measurement and prediction of CO_2 solubility in sodium phosphate monobasic solutions for food treatment with high pressure carbon dioxide. J. Supercrit. fluid. 52, 142–150.

Ferrentino, G., Balaban, M.O., Ferrari, G., Poletto, M., 2010b. Food treatment with high pressure carbon dioxide: S. cerevisiae inactivation kinetics expressed as a function of CO_2 solubility. J. Supercrit. fluid. 52, 151–160.

Ferrentino, G., Barletta, D., Donsi, F., Ferrari, G., Poletto, M., 2010c. Experimental measurements and thermodynamic modeling of CO_2 solubility at high pressure in model apple juices. Ind. Eng. Chem. Res. 49 (6), 2992–3000.

Fraser, D., 1951. Bursting bacteria by releasing gas pressure. Nature 167, 33–34.

Fricks, A.T., Souza, D.P.B., Oestreicher, E.G., Antunes, O.A.C., Girardi, J.S., Oliveira, D., et al., 2006. Evaluation of radish (Raphanus sativus L.) peroxidase activity after high-pressure treatment with carbon dioxide. J. Supercrit. fluid. 38, 347–353.

Furukawa, S., Watanabe, T., Tai, T., Hirata, J., Narisawa, N., Kawarai, T., et al., 2004. Effect of high pressure CO_2 on the germination of bacterial spores. Int. J. Food Microbiol. 91, 209–213.

Gangola, P., Rosen, B.P., 1987. Maintenance of intracellular calcium in Escherichia coli. J. Biol. Chem. 262, 12570–12574.

Garcia-Gonzalez, L., Geeraerd, A.H., Spilimbego, S., Elst, K., van Ginneken, L.V., Debevere, J., et al., 2007. High pressure carbon dioxide inactivation of microorganisms in foods: the past, the present and the future. Int. J. Microbiol. 117, 1–28.

Garcia-Gonzalez, L., Geeraerd, A.H., Elst, K., van Ginneken, L.V., van Impe, J.F., Devlieghere, F., 2009a. Influence of type of microorganism, food ingredients and food properties on high-pressure carbon dioxide inactivation of microorganisms. Int. J. Food Microbiol. 129, 253–263.

Garcia–Gonzalez, L., Geeraerd, A.H., Elst, K., van Ginneken, L.V., van Impe, J.F., Devlieghere, F., 2009b. Inactivation of naturally occurring microorganisms in liquid whole eggs using high pressure carbon dioxide processing as an alternative to heat pasteurization. J. Supercrit. Fluid. 51, 74–82.

Gasperi, F., Aprea, E., Biasioli, F., Carlin, S., Endrizzi, I., Pirretti, G., et al., 2009. Effects of supercritical CO_2 and N_2O pasteurization on the quality of fresh apple juice. Food Chem. 115, 129–136.

Gui, F., Chen, F., Wu, J., Wang, Z., Liao, X., Hu, X., 2006. Inactivation and structural change of horseradish peroxidase treated with supercritical carbon dioxide. Food Chem. 97, 480–489.

Gunes, G., Blum, L.K., Hotchkiss, J.H., 2005. Inactivation of yeasts in grape juice using a continuous dense phase carbon dioxide processing system. J. Sci. Food Agric. 85, 2362–2368.

Haas, G.J., Prescott, H.E., Dudley, E., Dik, R., Hintlian, C., Keane, L., 1989. Inactivation of microorganisms by carbon dioxide under pressure. J. Food Saf. 9, 253–265.

Hong, S.I., Park, W.S., Pyun, Y.R., 1997. Inactivation of Lactobacillus sp. from kimchi by high pressure carbon dioxide. Lebensm. Wiss. Technol. 30, 681–685.

Hong, S.I., Pyun, Y.R., 1999. Inactivation kinetics of Lactobacillus plantarum by high pressure carbon dioxide. J. Food Sci. 64, 728–733.

Hong, S.I., Park, W.S., Pyun, Y.R., 1999. Non thermal inactivation of Lactobacillus plantarum as influenced by pressure and temperature of pressured carbon dioxide. Int. J. Food Sci. Technol. 34, 125–130.

Hong, S.I., Pyun, Y.R., 2001. Membrane damage and enzyme inactivation of Lactobacillus Plantarum by high pressure CO_2 treatment. Int. J. Food Microbiol. 63, 19–28.

Hutkins, R.W., Nannen, N.L., 1993. pH homeostasis in lactic-acid bacteria. J. Dairy Sci. 76, 2354–2365.

Ikeda, T., Mae, S., 1998. Effect of guest–host interaction on Raman spectrum of a CO_2 clathrate hydrate single crystal. J. Chem. Phys. 108, 1352–1359.

Insalata, N., 1952. CO_2 versus beverage bacteria. Food Eng. 190, 84–85.

Isenschmid, A., Marison, W.I., Stockar, V.U., 1995. The influence of pressure and temperature of compressed CO_2 on the survival of yeast cells. J. Biotechnol. 39, 229–237.

Ishikawa, H., Shimoda, M., Kawano, T., Osajima, Y., 1995. Inactivation of enzymes in an aqueous solution by micro-bubbles of supercritical CO_2. Biosci. Biotechnol. Biochem. 59, 628–631.

Ishikawa, H., Shimoda, M., Tamaya, K., Yonekura, A., Kawano, T., Osajima, Y., 1997. Inactivation of Bacillus spores by the supercritical carbon dioxide micro-bubble method. Biosci. Biotechnol. Biochem. 61, 1022–1023.

Jones, R.P., Greenfield, P.F., 1982. Effect of carbon dioxide on yeast growth and fermentation. Enzym. Microb. Tech. 4, 210–223.

Kamihira, M., Taniguchi, M., Kobayashi, T., 1987. Sterilization of microorganisms with supercritical carbon dioxide. Agr. Biol. Chem. 51, 407–412.

Karukstis, K.K., Van Hecke, G.R., 2000. Chemistry Connections: The Chemical Basis of Everyday Phenomena. Academic Press, San Diego, CA, pp. 11–28

Kim, S.R., Rhee, M.S., Kim, B.C., Kim, K.H., 2007a. Modeling the inactivation of Escherichia coli O157:H7 and generic Escherichia coli by supercritical carbon dioxide. Int. J. Food Microbiol. 118, 52–61.

Kim, S.R., Rhee, M.S., Kim, B.C., Lee, H., Kim, K.H., 2007b. Modeling of the inactivation of Salmonella typhimurium by supercritical carbon dioxide in physiological saline and phosphate-buffered saline. J. Microbiol. Method. 70, 132–141.

Kim, H.T., Choi, H.J., Kim, K.H., 2009. Flow cytometric analysis of Salmonella enterica serotype Typhimurium inactivated with supercritical carbon dioxide. J. Microbiol. Method. 78, 155–160.

Kincal, D., Hill, W.S., Balaban, M.O., Portier, K.M., Wei, C.I., Marshall, M.R., 2005. A continuous high pressure carbon dioxide system for microbial reduction in orange juice. J. Food Sci. 70, M249–M254.

Kincal, D., Hill, S., Balaban, M.O., Marshall, M.R., Wei, C., 2006. A Continuous high pressure carbon dioxide system for cloud and quality retention in orange juice. J. Food Sci. 71, C338–C344.

King, M.B., Mubarak, A., Kim, J.D., Bott, T.R., 1992. The mutual solubilities of water with supercritical and liquid carbon dioxide. J. Supercrit. Fluids 5, 296–302.

Kumagai, H., Hata, C., Nakamura, K., 1996. CO_2 Sorption by microbial cells and sterilization by high-pressure CO_2. Biosci. Biotech. Biochem. 61, 931–935.

Lecky, M., 2005. Continuous High Pressure Carbon Dioxide Processing of Watermelon Juice. University of Florida, Gainesville (FL) (MSc. Thesis)

Lecky, M., Balaban, M.O., 2005. Shelf life evaluation of watermelon juice after processing with a continuous high pressure CO_2 system. In: IFT annual meeting book of abstracts; July 15–20; New Orleans La. Chicago Ill.

Liao, X., Zhang, Y., Bei, J., Hu, X., Wu, J., 2009. Alterations of molecular properties of lipoxygenase induced by dense phase carbon dioxide. Innovat. Food Sci. Emerg. Tech. 10, 47–53.

Liao, H., Zhang, L., Hu, X., Liao, X., 2010. Effect of high pressure CO_2 and mild heat processing on natural microorganisms in apple juice. Int. J. Food Microbiol. 137, 81–87.

Lim, S., Yagiz, Y., Balaban, M.O., 2006. Continuous high pressure carbon dioxide processing of mandarin juice. Food Sci. Biotechnol. 15, 13–18.

Lin, H.M., Chan, E.C., Chen, C., Chen, L.F., 1991. Disintegration of yeast cells by pressurized CO_2. Biotechnol. Prog. 7, 201–204.

Lin, H.M., Yang, Z., Chen, L.F., 1992a. Inactivation of saccharomyces cerevisiae by supercritical and subcritical carbon dioxide. Biotechnol. Prog. 8, 458–461.

Lin, H.M., Yang, Z.Y., Chen, L.F., 1992b. An improved method for disruption of microbial cells with pressurized carbon dioxide. Biotechnol. Prog. 8, 165–166.

Lin, H.M., Yang, Z.Y., Chen, L.F., 1993. Inactivation of Leuconostoc dextranicum with carbon dioxide under pressure. Chem. Eng. J. Biochem. Eng. J. 52, B29–B34.

Lin, H.M., Cao, N., Chen, L.F., 1994. Antimicrobial effect of pressurized carbon dioxide on Listeria monocytogenes. J. Food Sci. 59, 657–659.

Liu, X.F., Zhang, B.Q., Li, T.J., 2005. Effects of CO_2 compression and decompression rates on the physiology of microorganisms. Chinese J. Chem. Eng. 13, 140–143.

Liu, X., Gao, Y., Xu, H., Hao, Q., Liu, G., Wang, Q., 2010. Inactivation of peroxidase and polyphenol oxidase in red beet (Beta vulgaris L.) extract with continuous high pressure carbon dioxide. Food Chem. 119, 108–113.

Lucien, F.P., Foster, N.R., 1999. Phase behavior and solubility. In: Jessop, P.G., Leitner, W. (Eds.), Chemical Synthesis using Supercritical Fluids. Wiley-VCH Press, Weinheim (G), pp. 37–53.

Meyssami, B., Balaban, M.O., Teixeira, A.A., 1992. Prediction of pH in model systems pressurized with carbon dioxide. Biotechnol. Progr. 8, 149–154.

Molin, G., 1983. The resistance to carbon dioxide of some food related bacteria. European J. Appl. Microbiol. Biotechnol. 18, 214–217.

Nakamura, K., Enomoto, A., Fukushima, H., Nagai, K., Hakoda, M., 1994. Disruption of microbial cells by the flash discharge of high-pressure carbon dioxide. Biosc. Biotechnol. Biochem. 58, 1297–1301.

Osajima, Y., Shimoda, M., Kawano, T., 1997. Method for modifying the quality of liquid foodstuff. US Patent 5, 667, 835.

Osajima, Y., Shimoda, M., Kawano, T., Okubo, K., 1999. System for processing liquid foodstuff or liquid medicine with a supercritical fluid of carbon dioxide. US Patent 5, 869, 123, and EP 0 786 513 B1.

Oule, M.K., Tano, K., Bernier, A.M., Arul, J., 2006. Escherichia coli inactivation mechanism by pressurized CO_2. Canadian J. Microbiol. 52, 1208–1217.

Park, S.J., Lee, J.I., Park, J., 2002. Effects of a combined process of high-pressure carbon dioxide and high hydrostatic pressure on the quality of carrot juice. J. Food Sci. 67, 1827–1834.

Parton, T., Bertucco, A., Elvassore, N., Grimolizzi, L., 2007. A continuous plant for food preservation by high pressure CO_2. J. Food Eng. 79, 1410–1417.

Plaza, M.L., Ramirez-Rodrigues, M., Ferrentino, G., Balaban, M.O., 2009. Optimization of dense phase carbon dioxide pasteurization of guava puree. In: IFT Annual Meeting Book of Abstracts, 6–10 June. Anaheim, California, USA.

Ramirez-Rodrigues, M., Plaza, M.L., Ferrentino, G., Balaban, M.O., 2009. Dense phase CO_2 process optimization of a Hibiscus sabdariffa beverage. In: IFT Annual Meeting Book of Abstracts, 6–10 June. Anaheim, California, USA.

Rizvi, S.S.H., Benado, A.L., Zollweg, J.A., Daniels, J.A., 1986. Supercritical fluid extraction process. Part I. Fundamental principles and modelling methods. Food Technol. 40, 55–65.

Shimoda, M., Cocunubo-Castellanos, J., Kago, H., Miyake, M., Osajima, Y., Hayakawa, I., 2001. The influence of dissolved CO_2 concentration on the death kinetics of Saccharomyces cerevisiae. J. Appl. Microbiol. 91, 306–311.

Shimoda, M., Kago, H., Kojima, N., Miyake, M., Osajima, Y., Hayakawa, I., 2002. Accelerated death kinetics of Aspergillus niger spores under high pressure carbonation. Appl. Environ. Microbiol. 68, 4162–4167.

Sims, M., 2000. Method and membrane system for sterilizing and preserving liquids using carbon dioxide. US Patent 6,331,272 B1 and WO Patent 00/41805.

Sims, M., Estigarribia, E., 2003. Membrane carbon dioxide sterilization of liquid foods: scale up of a commercial continuous process. In: Proceedings of the 6th International Symposium on Supercritical Fluids, 28–30 April, Versailles, France, pp. 1457–1460.

Spilimbergo, S., Elvassore, N., Bertucco, A., 2002. Microbial inactivation by high-pressure. J. Supercrit. Fluids 22, 55–63.

Spilimbergo, S., Bertucco, A., 2003. Non-thermal bacteria inactivation with dense CO_2. Biotechnol. Bioeng. 84, 627–638.

Spilimbergo, S., Elvassore, N., Bertucco, A., 2003. Inactivation of microorganisms by supercritical CO_2 in a semi-continuous process. Italian J. Food Sci. 1, 115–124.

Spilimbergo, S., Mantoan, D., 2005. Stochastic modeling of S. cerevisiae inactivation by supercritical CO_2. Biotechnol. Prog. 21, 1461–1465.

Spilimbergo, S., Bertucco, A., Basso, G., Bertoloni, G., 2005. Determination of extracellular and intracellular pH of Bacillus subtilis suspension under CO_2 treatment. Biotechnol. Bioeng. 92, 447–451.

Spilimbergo, S., Mantoan, D., 2006. Kinetic analysis of microorganisms inactivation in apple juice by high pressure carbon dioxide. Int. J. Food Eng. 2, 1–9.

Spilimbergo, S., Mantoan, D., Dalser, A., 2007. Supercritical gases pasteurization of apple juice. J. Supercrit. Fluids 40, 485–489.

Tedjo, W., Eshtiaghi, M.N., Knorr, D., 2000. IMPact of supercritical CO_2 and high pressure on lypoxygenase and peroxidase activity. J. Food Sci. 65, 1284–1287.

Valley, G., Rettger, L.F., 1927. The influence of carbon dioxide on bacteria. J. Bacteriol. 14, 101–137.

Watanabe, T., Furukawa, S., Hirata, J., Koyama, T., Ogihara, H., Yamasaki, M., 2003. Inactivation of *Geobacillus stearothermophilus* spores by high-pressure carbon dioxide treatment. Appl. Environ. Microbiol. 69, 7124–7129.

Watanabe, T., Furukawa, S., Kawarai, T., Wachi, M., Ogihara, H., Yamasaki, M., 2007. Cytoplasmic acidification may occur in high-pressure carbon dioxide-treated *Escherichia coli* K12. Bioschi. Biotechnol. Biochem. 71, 2522–2526.

Wei, C.I., Balaban, M.O., Fernando, S.Y., Peplow, A.J., 1991. Bacterial effect of high pressure CO_2 treatment on foods spiked with Listeria and Salmonella. J. Food Protect. 54, 189–193.

Yagiz, Y., Lim, S.L., Balaban, M.O., 2005. Continuous high pressure CO_2 processing of mandarin juice. In: IFT Annual Meeting Book of Abstracts, July 15–20, New Orleans La. Chicago Ill.

Yoshimura, T., Futurera, M., Shimoda, M., Ishikawa, H., Miyake, M., Matsumoto, K., et al., 2002. Inactivation efficiency of enzymes in buffered system by continuous method with microbubbles of supercritical carbon dioxide. Food Chem. Toxicol. 9, 3227–3231.

Zhang, J., Burrows, S., Gleason, C., Matthews, M.A., Drews, M.J., LaBerge, M., et al., 2006. Sterilizing Bacillus pumilus spores using supercritical carbon dioxide. J. Microbiol. Method. 66, 479–485.

Zhi, X., Zhang, Y., Hu, X., Wu, J., Liao, X., 2008. Inactivation of apple methylesterase induced by dense phase carbon dioxide. J. Agr. Food Chem. 56, 5394–5400.

Zhou, L., Wu, J., Hu, X., Zhi, X., Liao, X., 2009a. Alterations in the Activity and Structure of Pectin Methylesterase treated by high pressure carbon dioxide. J. Agr. Food Chem. 57, 1890–1895.

Zhou, L., Wang, Y., Hu, X., Wu, J., Liao, X., 2009b. Effect of high pressure carbon dioxide on the quality of carrot juice. Innov. Food Sci. Emerg. Tech. 10, 321–327.

CHAPTER

Ohmic Heating of Fluid Foods

11

Filiz Icier

Food Engineering Department, Engineering Faculty, Ege University, Bornova, Izmir, Turkey

11.1 INTRODUCTION

Conventional heating processes essentially consist of heat-transfer mechanisms of conduction, convection, and radiation, in both steady- and unsteady-state operations (Goullieux and Pain, 2005). The internal resistance by conduction results in very heterogenous treatment and the notable loss of product quality. To overcome these problems, alternative technologies utilizing electrical energy directly in the food processing have attracted interest in the food industry in recent decades. Some of those are now being used for processing of a broad range of food products, especially those containing particulates, on a commercial scale. Research in this area provides the food processor with the opportunity to produce new and value-added food products with enhanced quality attributes preferred by consumers.

Ohmic heating is an electroheating technique based on the passage of electrical current through a food product having electrical resistance (Reznick, 1996; Sastry and Salengke, 1998; Icier, 2003) (Fig. 11.1). Heat is generated instantly inside the food, and its amount is directly related to the voltage gradient, and the electrical conductivity (Sastry and Li, 1996). The uniform heat generation results to uniform temperature distribution.

The obvious advantage of ohmic treatments over conventional methods is the lack of high wall temperatures and limiting heat transfer coefficients requirements. Its other advantages compared to conventional heating include maintaining the color and nutritional value of food, short processing time, and higher yield (Wang and Sastry, 2002; Castro *et al.*, 2004a; Icier and Ilicali, 2005a; Leizerson and Shimoni, 2005a, b; Vikram *et al.*, 2005). Ohmic treatment is used in a wide range of applications such as preheating, blanching, pasteurization, sterilization, extraction of food products (Mizrahi, 1996; Lima and Sastry, 1999; Leizerson and

Novel Thermal and Non-Thermal Technologies for Fluid Foods. DOI: 10.1016/B978-0-12-381470-8.00011-6

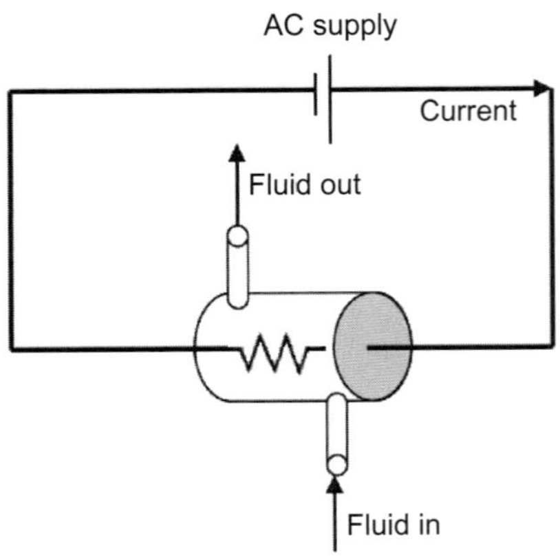

FIGURE 11.1

Ohmic heating of fluid food.

Shimoni, 2005a, b). Since USDA and FDA suggested the usage of ohmic technology for pumpable foods, it is currently being used commercially throughout the world (USA, Japan, UK, and other several European countries) for the pasteurization of pumpable foods (syrups including whole fruits, fruit juices, egg, milk, etc.) and aseptic packaging (Anonymous, 2000; Icier, 2003; Ramaswamy *et al.*, 2005; Zell *et al.*, 2009a; Icier and Bozkurt, 2010).

This chapter will focus on the application of the novel thermal technology of ohmic heating. The liquid food conductivity as a controlling process parameter will be discussed. The effects of this electric resistance heating on the microbial and chemical properties of different studied liquid food products will be reviewed, and the main industrial applications as well as future trends of this thermal treatment will be tackled.

11.2 PRINCIPLES

Ohmic heating is a thermal–electrical method where food is in contact with the electrodes, also known as Joule heating, electroconductive heating, electrical resistance heating, direct electrical resistance heating, and electroheating in the literature. Ohmic heating is very often used in pasteurization/sterilization of fluid food products, in which the contact with the electrodes is not a serious problem, resulting in excellent quality.

Ohmic technology was firstly used in the 19th century (Anderson and Finkelsten, 1919; Prescott, 1927) for milk heating, and was then investigated regularly in the earlier part of the last century. The electro-process used for pasteurization of milk had been named as "Electropure". Unfortunately, such a technology did not succeed at that time due to higher electricity prices and electrolysis-related effects, process regulations, and other technical limitations (de Alwis and Fryer, 1990). The emphasis on the ohmic heating research slowed down during 1930s to 1960s. In the 1980s, Electricity Research and Development Centre (UK) revised this technology and improved the design procedures of ohmic heating systems. APV Baker Ltd holds the patent for industrial use of this technology (Biss *et al.*, 1989). Over the past 20 years, new, improved materials and equipment design for ohmic heating have become available (Ayadi *et al.*, 2004a). The technology has found application in materials manufacturing, consumer products and food processing, and potentially in food and beverage vending and dispensing applications (Herrick *et al.*, 2000). Further, with the development of solid-state power supply technology, it is now possible to use ohmic heating in pulse mode, to economically control electrolytic effects to innocuous levels. Ohmic systems are now better-engineered, more sophisticated, and far less expensive than their predecessors, and four manufacturers currently produce ohmic heating equipment (Sastry, 2008; Anderson, 2008).

Ohmic heating takes its name from Ohm's law, which is known as the relation between current, voltage, and resistance (Equation 11.1). The food material switched between electrodes has a role of resistance in the circuit (Figs 11.1, 11.2).

$$I = \frac{V}{R} \tag{11.1}$$

The resistance of food material to electrical current passing causes heat generation within the food. In other words, electrical energy is converted to heat energy (Sastry, 1992).

The voltage distribution within the ohmic heater can be developed from Maxwell's equations, or by combining Ohm's law and continuity equation for electric current (Sastry and Palaniappan, 1992; Sastry and Salengke, 1998):

$$\nabla(\sigma \nabla V) + \frac{\partial \rho_c}{\partial t} = 0 \tag{11.2}$$

for the steady state case which is typical:

$$\nabla(\sigma \nabla V) = 0 \tag{11.3}$$

which must be solved over the domain of the sample within the ohmic heater. σ is the electrical conductivity of the food material (S/m), V is the electrical potential difference (V), ρ_c is the current density (A/m^3), and t is the time (s).

Once the voltage distribution is available the heat generation is calculated. For a constant voltage process, the volumetric heat generation rate ($\dot{u}$, W/m^3) is (Sastry and Palaniappan, 1992):

$$\dot{u} = |\nabla V|^2 \sigma \tag{11.4}$$

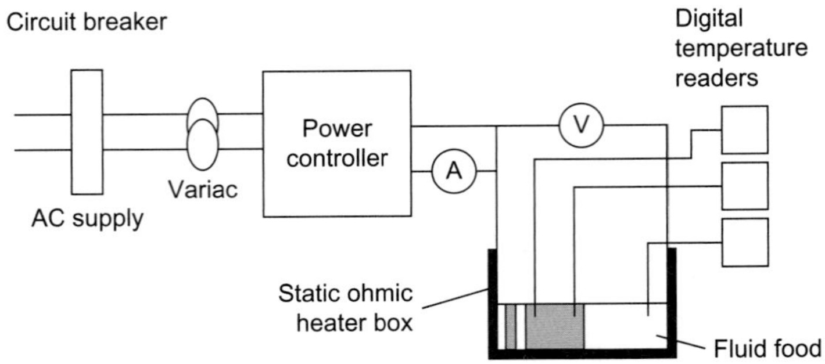

FIGURE 11.2

Schematic illustration of laboratory scale ohmic heating system.
[Modified, with permission from Davies et al. *(1999).* J. Food Eng. *40, 245–258.]*

The heat transfer occurring during ohmic processing of a liquid food as a single phase is described by the unsteady heat conduction equation with a internal heat generation as given below (Marra *et al.*, 2009):

$$\nabla(k\nabla T) + \dot{u} = \rho C_p \frac{\partial T}{\partial t} \tag{11.5}$$

where k, ρ and Cp are the temperature-dependent thermophysical properties of liquid food: the thermal conductivity (W/mK), the density (kg/m^3) and the specific heat capacity (J/kgK), respectively. The convective heat transfer within the liquid can be taken account by adding the convection term directly to Equation 11.5. The uniform heat generation ($\dot{u}$) results in remarkably rapid and relatively uniform heating compared to other heating methods, especially in liquid foods. Therefore, it is suitable for continuous processing of fluid foods.

Ohmic heating system basically consists of an AC supply to give electrical energy to system, a variac to apply the desired voltage, the current and voltage measurement units, an ohmic heater test unit including sample cell and electrodes, the temperature measurement system, and the microcomputer system to record the data (Fig. 11.2).

Large-scale process can be carried out in heavy-duty ohmic cookers or batch ohmic heaters (Fellows, 2000). A wide variety of potential designs exists for ohmic heaters, depending on the application. The design aspects are only now being understood. Ohmic heaters are also being developed in batch-mode (Fig. 11.2). A NASA project has involved the development of a special pouch with ohmic heating capability for reheating of foods for space missions, as well as sterilization of waste (Jun and Sastry, 2005; Sastry, 2008).

Continuous ohmic heating systems made for commercial applications can vary greatly. However, they include the flow system and cooling parts as well as the main parts of the ohmic heating system. They have several ohmic heater columns,

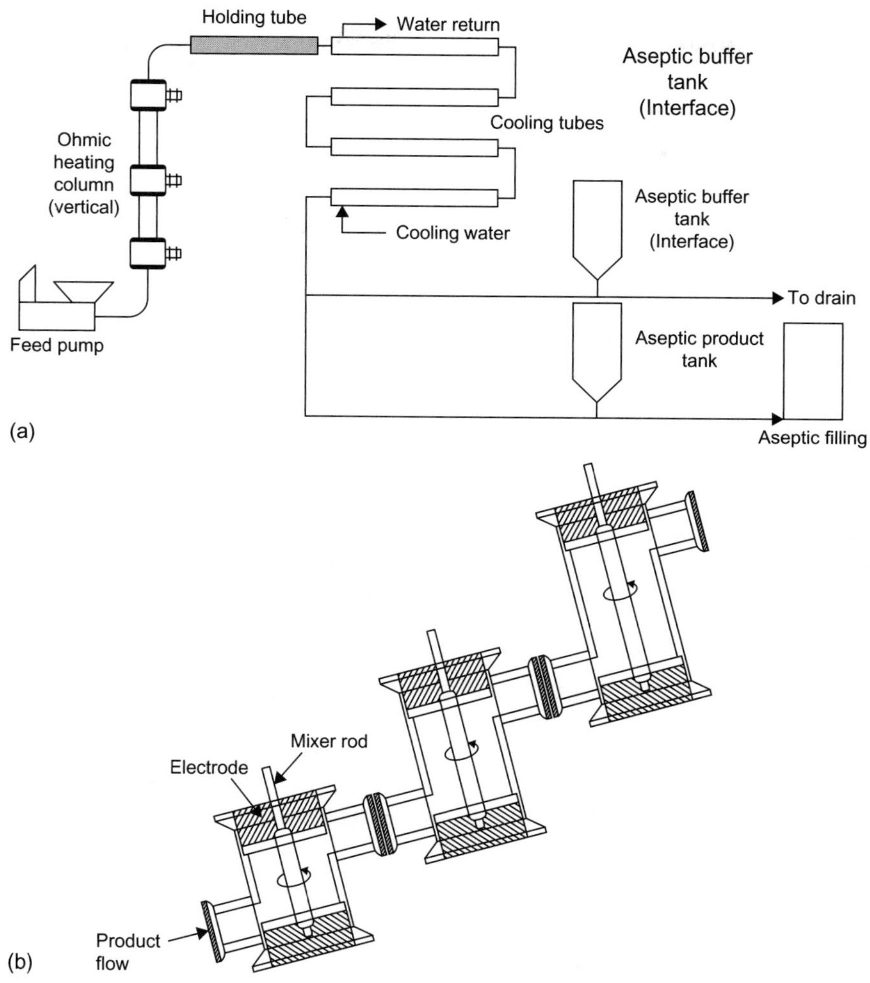

FIGURE 11.3

(a) Schematic illustration of continuous ohmic heating system.

[Reproduced with permission from Eliot-Godereaux et al. *(2001).* Innov. Food Sci. Emerging Technol. *2, 279–281.]*

(b) Ohmic heating columns in series.

[Reproduced with permission from Tulsiyan et al. *(2009).* J. Food Eng. *93(3), 313–317.]*

each comprised of insulating coverings (such as polytetrafluoroethylene, PTFE) and a single cantilever electrode (Fig. 11.3). These columns are structured vertically or inclined to allow for an upward flow of product, and are connected with insulated tubes (Fig.11.3b). Due to the increased electrical conductivity of products as heat increases, the connecting tubes increase in length throughout the system to maintain the same electrical impedance. A process control system constantly monitors the temperature, flow rate, heat capacity, and specific heat of a product to calculate the electrical power needed for the system (Tempest, 1992; Anderson, 2008).

Several experimental studies have been conducted on the application of ohmic heating to fluid foods and effects on their quality. Thus, the main effects of process parameters on ohmic heating rate characteristics have been found (Palaniappan and Sastry, 1991; de Alwis and Fryer, 1990; Castro *et al.*, 2003; Icier and Ilicali, 2005a; Assiry *et al.*, 2006; Salengke and Sastry, 2007a). Studies including mathematical modeling and sensitivity analysis provided the opportunity to understand the critical factors affecting the ohmic heating process and its lethality (Fryer *et al.*, 1993; Salengke and Sastry, 2007b; Chen *et al.*, 2010).

Ohmic heating has some advantages over conventional heating methods. The electric current passing through food material causes rapid heat generation, and thus faster heating of food. Since the heating takes place volumetrically, temperature increase is uniform. If the liquid food contains food particles, they can be heated uniformly at the same time in the case of similar electrical conductivities. This reduces the cold point possibility, thermal damage, and nutritional losses, and increases the overall lethality in the mixture. It is possible to obtain a product having acceptable textural properties, minimum aroma loss, and high sensory quality (Tempest, 1992).

Ohmic heating systems can be adapted to aseptic food-processing lines (Kim *et al.*, 1996). The temperature required for UHT processing can be achieved. The shelf-life of such products is prolonged and there is no need to transport these products in cold chain (Biss *et al.*, 1989). It requires precise process control. The prescribed temperature can be reached faster than in other conventional heating methods. The short processing time decreases the adverse effects of heating on the quality characteristics. Since it has the ability to heat materials rapidly and uniformly leading to a less aggressive thermal treatment, it can pasteurize the proteinous foods, such as liquid egg and whey, without coagulation (Icier and Bozkurt, 2010; Icier, 2010). Since the degradation of proteins and fouling in the equipment surfaces during ohmic heating is less compared to conventional heating methods, the clean-up and maintenance costs are lower (Tempest, 1992; Reznick, 1996). There is no need for the mixing of fluid foods for homogenous heating, which is important for foods sensitive to mechanical damage. Power supply units are not complex when performed at low frequencies (Tempest, 1992; Reznick, 1996). Energy conversion efficiencies are very high, systems have small equipment footprints, and are generally quiet. Since instant on/off can be obtained, accurate temperature control is possible.

However, there are still problems that arise in ohmic heating systems. The system needs proper electrical insulation, accurate process control systems, and well-trained personnel. The industrial applications are limited due to consumer constraints to electrically processed products; hence there is no available data on the effects of ohmic heating on formation of some toxicological and mutagenic substances, and precise information on assurance of its lethality in various kinds of foods. Furthermore, the costs of commercial ohmic heating systems, including installation, can be in excess of $9,000,000 USD, which is a huge investment for a manufacturing facility (Anderson, 2008). However, its process costs are comparable to commercial conventional systems.

A disadvantage related to the type of food that can be processed occurs in the presence of non-conductive parts or some components such as fat globules (Salengke and Sastry, 2007a). If these globules are present in a highly electrical-conductive region, the current flow can bypass them. A uniform temperature distribution may not be achieved within such products. Any pathogenic bacteria that may be present in these globules may receive less heat treatment than the rest of the substance (Sastry, 1992). However, in the case of highly conductive fluids, the electrical conductivity also increases as the temperature increases. If the temperature increase is highly rapid and uncontrollable, it creates the possibility of "runaway" heating (Anonymous, 2000).

The full commercialization of ohmic heating technology depends in part on the development of adequate safety and quality-assurance protocols in order to obtain an approved filing of the process with the FDA for all possible food materials (Ye *et al.*, 2003).

11.3 PROCESS CONTROL PARAMETERS

Ohmic heating is an alternative heating system for pumpable foods. It can be used as a continuous in-line heater for cooking and sterilization of viscous and liquid food products. Therefore, the process control in the ohmic heating system is related to electrical system parameters, flow properties in the pumping system, food properties, and other design criteria. In industrial-scale production, the voltage gradient, frequency, temperature range, electrical conductivity differences, concentration and type of food, electrode type used, and flow properties are all taken as critical design parameters in general (Icier and Ilicali, 2005a, b).

Data on electrical properties of the liquid foods are especially important in their ohmic processing, whether as a whole product or as fluid medium in the mixture. The most crucial process-control parameters to be taken into account in the design of ohmic heating systems for fluid foods are discussed in detail in the following subsections.

11.3.1 Voltage, Current and Power Applied

The power needed is determined depending on the resistance of the food to the voltage gradient applied and the current passing through the food. The voltage applied can be arranged using a variac. The critical current limit and maximum power capacity must be known to design the replacement of electrodes in the ohmic treatment unit. At low voltages, the current reaches very high values to obtain the power required. For this purpose, transformers, are placed within the system to facilitate processing under these conditions (Roberts *et al.*, 1998).

One of the most critical parameters is current density. This is the ratio of the current to the electrode surface area (Reznick, 1996). The arc discharge occurs in the system when the critical current density is reached. The maximum current limit and critical current density are used in the design of the electrode dimension. The maximum and minimum electrode gap values are assessed depending on electrode surface area and the resistance of the food (Roberts *et al.*, 1998). The overall system must be electrically isolated.

Ohmic heating times are dependent on the voltage gradient used. As the voltage gradient increases, the heat generation per unit time increases, and hence the heating time necessary to reach the prescribed temperature decreases. A time scale can be arranged by choosing the correct voltage gradient parameter (Icier and Ilicali, 2005b). Ohmic heating rates depend on the type and composition of the food material, relating with its electrical conductivity, and the voltage gradient applied.

The sharp decrease of the ohmic heating time with increasing voltage gradient was obvious in milk and reconstituted whey solutions having various concentrations (Icier, 2004, 2009) Similarly, as the voltage gradient increased from 10 V/cm to 60 V/cm, the ohmic heating times required to heat up the ice-cream mixes to 80°C from 4°C decreased approximately 50 and 31 times for Maras-type and standard type ice-cream mixes, respectively (Icier and Tavman, 2006). The higher fat content of standard type ice-cream mix results in lower electrical conductivity and slower heating compared to Maras-type ice-cream mixes at all voltage gradients. The liquid whole egg can be heated up to 60°C from 10°C in 105 s by applying a voltage gradient of 20 V/cm (Icier and Bozkurt, 2010). Ohmic heating is a very fast heating method for liquid whole egg.

Fruit and vegetable products (juices, purees, and pulps) are also successfully ohmically heated. Their high heating rates, which are dependent on the voltage gradient applied, provide the opportunity to blanch or pasteurize them quickly and uniformly (Palaniappan and Sastry, 1991; Lima *et al.*, 1999; Icier and Ilicali, 2004, 2005a, b; Assiry *et al.*, 2003; Castro *et al.*, 2004b; Leizerson and Shimoni, 2005a; Icier *et al.*, 2006, 2008; Allali *et al.*, 2008; Yildiz *et al.*, 2009, 2010).

The main property related to the food composition is the electrical conductivity. It determines the resistance of the food to the current passing through it for any power applied.

11.3.2 Electrical Conductivity

The most important parameter in ohmic heating of liquid food product is its electrical conductivity behavior. It depends on the temperature, applied voltage gradient, frequency, concentration of the electrolytes (Sastry and Palaniappan, 1992; Icier and Ilicali, 2005c).

The instantaneous values of current and voltage recorded during ohmic heating leads to the determination of electrical conductivities at various temperatures, provided the cell dimensions are known (de Alwis and Fryer, 1992; Qihua *et al.*, 1993; Reznick, 1996; Lima *et al.*, 1999).

$$\sigma = \frac{L}{A\,R} \tag{11.6}$$

Electrical conductivity (σ) increases linearly with temperature when subjected to ohmic heating at representative voltage gradients (Equation 11.7) while the soluble and insoluble solid concentrations affect it non-linearly (Equation 11.8) (Palaniappan and Sastry, 1991; Icier and Ilicali, 2004,2005b, c);

$$\sigma = B_1 T + C_1 \tag{11.7}$$

$$\sigma = E_2 (Conc\%)^{N_2} + B_2 T + C_2. \tag{11.8}$$

Several studies have been conducted on electrical conductivity changes of fluid foods. The relationships obtained are given in Table 11.1. The temperature dependency of the electrical conductivity liquid products follows linear or quadratic relations, depending on the product type tested (Castro *et al.*, 2003). The introduction of the voltage gradient term to the electrical conductivity equation (Equation 11.8) results in negligible improvement in the model predictions (Icier and Ilicali, 2005b).

The electrical conductivity depends on the ionic breakup and microstructure of the food material undergoing heating (Parrott, 1992; Marcotte *et al.*, 2000a). Some components such as sugar and fat globules may influence the electrical conductivity of the sample, depending on their electrolytic characteristics. In fruit juices, the main solute component is the sugar, having non-electrolytic behavior. Icier (2003) reported that, as the sugar content increased, the electrical conductivities of the liquid solutions decreased while the acidity of the juices enhanced their electrical conductivities. The variation in sugar content and the nature of the other components may cause different electrical conductivities for juices having similar concentrations. Furthermore, since the ohmic heating rates of the fluid foods are dependent on the electrical conductivity values for the same voltage gradient, acidity, and pulp content, differences between different fruit juices and purees directly affect their ohmic heating rates.

Palaniappan and Sastry (1991) re-suspended a known amount of insoluble solids previously separated by centrifugation into juice serums (tomato and orange juices), and studied the effects of insoluble solids (0–16.7%) and applied voltage on electrical conductivity of the pre-pasteurized commercial juices during ohmic heating. They reported that the electrical conductivities of fruit and vegetable juices increased linearly by decreasing insoluble solid contents.

Typical electrical conductivity curves of apple and sourcherry juices having 30% soluble solid concentration are shown in Figure 11.4. Icier and Ilicali (2004)

Table 11.1 Electrical Conductivity Ranges for Selected Fluid Foods During Ohmic Heating

Fluid Food	Temperature Range (°C)	Voltage Gradients Applied (V/cm)	Concentration Range	Electrical Conductivity (S/m)	Reference
Salt solutions	25–80	20	0.2–0.8% w/v NaCl	0.40–3.20	Icier *et al.* (2006)
Sugar solutions	30–60	20–70	13–14% w/v	0.10–0.20	Icier (2003)
Apple juice	30–80	20–60	20–60% v/v soluble solids	0.30–1.00	Icier and Ilicali (2004)
Sourcherry juice	30–75	20–60	20–60% v/v soluble solids	0.70–1.60	Icier and Ilicali (2004)
Orange juice	25–80 30–60	42.42 20–60	0–21% solid 0.2–0.6 mass fraction	0.50–1.25 0.15–1.15	Palaniappan and Sastry (1991) Icier and Ilicali (2005b)
Grape juice	20–80	20–40	Fresh, not reported	0.38–0.78	Icier *et al.* (2008)
Strawberry pulp	20–80				Castro *et al.* (2004a)
Pulp 1		32–70	Brix 14.5°, 2.5% w/w starch	0.20–0.50	
Pulp 2		40–80	Brix 26.5°, no starch	0.10–0.35	
Strawberry topping	20–100	55–100	Not reported	0.005–0.06	Castro *et al.* (2004a)
Strawberry filling	20–100	55–100	Not reported	0.03–0.25	Castro *et al.* (2004a)
Strawberry–apple sauce	20–100	25–70	Not reported	0.11–0.34	Castro *et al.* (2004a)

Apricot puree	20–60	20–70	12.79% soluble solid	0.61–1.25	Icier and Ilicali (2005a)
Peach puree	25–60	20–70	9.90% soluble solid	0.51–0.91	Icier and Ilicali (2005a)
Carrot puree–starch mixture	20–80	12.5	30–50% puree in starch solution (4% starch, 0.5% salt)	0.70–1.80	Zareifard *et al.* (2003)
Tomato juice	25–80	42.42	0–16.7% solid	0.70–1.65	Palaniappan and Sastry (1991)
Tomato soup	20–120, under pressure of 30 psi.	0–20.5	Dilution of commercial condensed tomato soup in distilled water (1:1 v/v)	1.50–4.80	Jun and Sastry (2007)
Tomato sauce (model)	25–80	12.5	75.3% w/w water, 0.60% w/w xanthan gum, 22.6% w/w sugar, 1.5% w/w NaCl	1.25–3.20	Legrand *et al.* (2007)
Milk	10–70	15–45	3% fat	0.30–1.00	Icier (2004)
Ice-cream mix	4–80	10–60			Icier and Tavman (2006)
Standard mix			9.8% fat	0.15–1.00	
Maras-type mix			3.3% fat	0.18–1.00	
Whole liquid egg	4–60	20	Fresh, not reported	0.32–0.83	Icier and Bozkurt (2010)
Reconstituted whey solutions	20–80	20–40	8–24% w/v solute	0.98–3.90	Icier (2009)
Aqueous solution of native whey protein	10–100	Not reported	1% w/v	0.10–0.37	Ayadi *et al.* (2004b)
Aqueous mixed solution of native whey protein and xanthan gum	10–100	Not reported	1% w/v protein, 0.2% w/v xanthan	0.10–0.42	Ayadi *et al.* (2004b)

(*Continued*)

Table 11.1 (Continued)

Fluid Food	Temperature Range (°C)	Voltage Gradients Applied (V/cm)	Concentration Range	Electrical Conductivity (S/m)	Reference
Corn starch suspension	30–90	20	20% w/w starch, 1% salt	1.50–3.10	Wang and Sastry (1997)
	30–65	10	33.3% starch, 0.4% salt	0.50–0.88	Li *et al.* (2004)
Potato starch suspension	30–90	20	20% w/w starch, 1% salt	1.65–2.90	Wang and Sastry (1997)
	30–60	10	33.3% starch, 0.4% salt	0.50–0.84	Li *et al.* (2004)
Mugbean starch suspension	30–70	10	33.3% starch, 0.4% salt	0.52–1.00	Li *et al.* (2004)
Hydrocolloids	25–100	7.25			Marcotte *et al.* (1998)
Pectin			1–5% w/w	0.06–0.37	
Carrageenan			1–3% w/w	0.20–1.40	
Xanthan			1–3% w/w	0.14–1.10	
Gelatin			2–4% w/w	0.06–0.37	
Starch			4–6% w/w	0.05–0.21	
Surimi paste	20–70	8	79% moisture,	DC conductivity	Wu *et al.* (1998)
			1% salt	1.20–3.10	
			2% salt	2.10–6.80	
			3% salt	3.50–7.90	

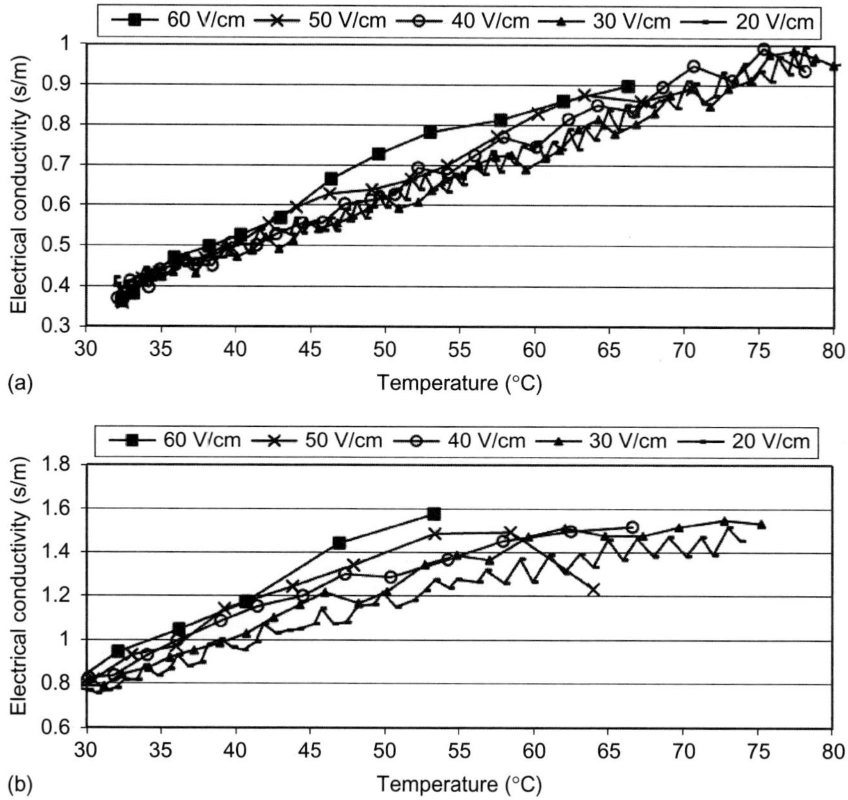

FIGURE 11.4

The electrical conductivity curves of fruit juices having 30% soluble solid concentration with five different voltage gradients applied: a, apple juice; b, sourcherry juice.
[Reproduced with permission from Icier and Ilicali (2004). J. Food Process Eng. *27, 159–180.]*

reported that the electrical conductivity values of apple and sourcherry juices were in the range of 0.1–1.6 S/m having a decreasing trend with increasing soluble solid concentration (20, 30, 40, 50, and 60% w/v) (Table 11.1). For the same voltage gradient (Fig. 11.5), the electrical conductivity values of fruit juice measured at the same temperature decreased as the concentration increased (Icier and Ilicali, 2005b). The concentration dependency of the electrical conductivity of the juices has been explained by the increased drag for the movement of ions with increasing concentration. Castro *et al.* (2003) suggested that electrical conductivity decreases with increase in solids and sugar content of strawberry-based products. They mentioned that, for formulations of products having solid content over 20% w/w and over 40 °Brix, a different design of ohmic heater may be necessary because of the low values of electrical conductivity.

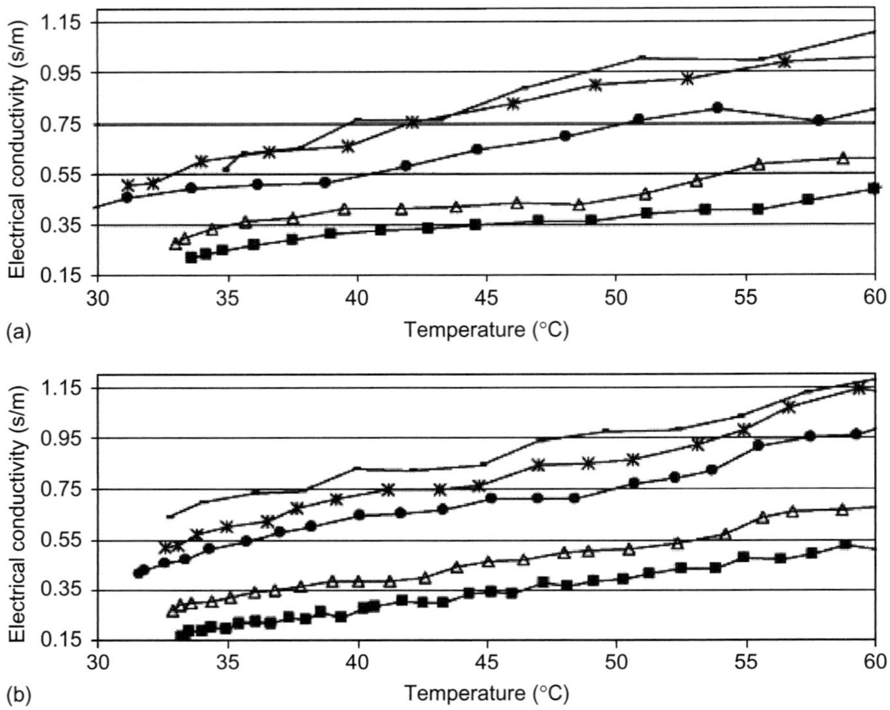

FIGURE 11.5

The electrical conductivity curves of orange juice with five different concentrations (mass fraction) at two different voltage gradients: a, 60 V/cm; b, 40 V/cm; ■ 0.6, Δ 0.5, ● 0.4, × 0.3, – 0.2.

[Reproduced with permission from Icier and Ilicali (2005). Eur. Food Res. Technol. *220(3–4), 406–414.]*

The frequency and waveform of applied voltage affects the electrical conductivity values and the process of heating the food samples (Lima *et al.*, 1999). The electrical conductivity observed by using sine and saw-tooth waves at low frequency (4 Hz) values are higher than those observed by using square waves (Lima *et al.*, 2001). Castro *et al.* (2004a) reported that an increase in electrical conductivity with field strength (12.5–50 V/cm) was clear for strawberry pulp and strawberry filling, but not evident for strawberry topping and strawberry–apple sauce. It was explained by differences in the membrane destruction and fluid motion through the capillaries between the samples at higher electrical field strengths. For strawberry topping and sauce, their previous processing might already have led to membrane destruction; so there were no further structural changes and variations in the mobility of the fluid and of the ionic components present.

The common problem observed during ohmic heating of fruit juices is bubbling. It is observed above 50°C at especially high voltage gradients, and the sharp decrease occurs in electrical conductivities after bubbling starts (Icier and Ilicali, 2005a). The temperatures of apricot puree and peach puree were reported not to be able to reach 60°C and 65°C, respectively, at 70 V/cm due to excessive bubbling. Fruit juices are acidic, resulting in the potential for electrolytic hydrogen bubble formation (Palaniappan and Sastry, 1991). Effects of electrolytic reactions will be discussed in the following sections.

Furthermore, when solid particles are suspended in a fluid medium having similar electrical conductivities, the component among them having lower heat capacity will have the tendency to heat faster. Electrical conductivity of the mixture and the amount of voltage applied significantly affects the heating rate of the solid–liquid mixtures during ohmic heating. The simultaneous and uniform heating of solid and liquid phases can be achieved, thus reducing the danger of under-processing as well as nutritional loss (Bhale, 2004; Salengke and Sastry, 2007a).

Although proteins increase the overall electrical conductivity of milk products, the increase in fat and lactose content decreases it. For temperatures over 35°C, the electrical conductivity of the aqueous protein–xanthan mixture is higher than that of the protein solution. This electrical conductivity behavior could be explained by the interaction between xanthan gum and β-lactoglobulin molecules. The quantity of free native β-lactoglobulin in the protein–xanthan mixture ($3.784 + 0.05\ g\ l^{-1}$) has been established as less than that in the protein solution ($7.690 + 0.05\ g\ l^{-1}$) (Ayadi *et al.*, 2004b). The presence of such a protein–xanthan complex does not significantly change the electrical conductivity of the aqueous mixture in the lower temperature region (5–35°C), but as the temperature increases from 35 to 100°C, the electric conductivity of the aqueous protein–xanthan mixture become greater. Ayadi *et al.* (2004b) investigated the fouling ability of ohmic heating cells and the electrical conductivity of deposit layers on electrode surfaces. The electrical conductivity values of the deposits were markedly lower than those of the fluids. The low-deposit electrical conductivity has a practical significance for the linking of fouling layers and thermoelectric phenomena during the fouling build-up. Indeed, when electricity is applied to a series of materials (electrode–deposit fluid–deposit–electrode), the Joule effect is strongest in the material which has the highest conductivity (fluid).

Similar linear band shape electrical conductivity changes were observed for both of the ice-cream mixes (Icier and Tavman, 2006). At the voltage gradient range of 10–40 V/cm, the electrical conductivity of Maras-type ice-cream mix was higher than that of the standard type ice-cream mix having lower fat content, at the same temperature.

Ohmic heating is longer for lower solute concentrations of reconstituted whey (8–24% w/v solute concentrations) (Icier, 2009). It is explained by the decreasing amount of free ions/charge carriers, which decrease the electrical conductivity, at lower concentrations. In that, the fast heating occurs in whey solutions having higher electrical conductivity. The application of ohmic heating at high voltage gradients and at higher solute concentrations could be advantageous to obtain faster heating at the industrial processing of whey solutions.

The electrical conductivity measurement can be used as the indicator for the occurrence of starch gelatinization in liquid food. Electrical conductivity of starch increased with temperature, but decreased with degree of starch gelatinization (Wang and Sastry, 1997). Li *et al.* (2004) found a decrease of the electrical conductivity in the gelatinization range due to the reduction in the area for starch particle movement and the increase in the resistance to motion of the swollen particles caused by starch granule swelling and viscosity increases.

The specific electrical resistance can also be expressed as the inverse of the electrical conductivity. It is different from electrical resistance since this specific resistance decreases with a rise in temperature (Bhale, 2004). The temperature dependency of the specific electrical resistance of the food and the dimensions of the ohmic application unit determines the power required to be applied (Reznick, 1996).

The precise measurement and the accurate control of the temperature during ohmic heating are crucial in the design procedure of ohmic heating systems.

11.3.3 Temperature

The change in the temperature during ohmic heating is very fast and could result in runaway heating in the fluid food if the control is not properly designed. In continuous systems, the power applied during heating can be adjusted by using feedback or feed-forward control by measurements of temperature, mass flow rate, and specific heat-capacity changes. In batch systems, the temperature measurements taken from different points of the food help the observation of the maximum temperature limit and necessitate on/off of the power supply.

The electrical insulation of the temperature probe used is important in avoiding undesired signal disturbances in the measurement system. Since the ohmic heating takes place in seconds for high-acidity fluid foods, the sensitivity of the temperature probes should be high and the response times should be low enough. The electrical insulation layer or shielding for the temperature probe is selected by taking account of its effects on the measurement response and accuracy. In the literature, specially coated T-type thermocouples have been used in several studies for the purpose of temperature measurements during ohmic heating of fluid foods (Sastry, 1992; Qihua *et al.*, 1993; Fryer *et al.*, 1993; Sastry and Salengke, 1998; Icier *et al.*, 2006). Zell *et al.* (2009b) recently developed the rapid-response thermocouple probes for use in ohmic heating systems.

Furthermore, liquid crystal tablets or layers can be used to assess the temperature distribution in continuous ohmic heating systems (Sastry and Li, 1996). However, the adaptation of this method to automatic control systems is problematic. Some non-invasive temperature mapping techniques (such as magnetic resonance imaging, etc.) enabling fast spatial temperature monitoring during ohmic heating process are used (Speller *et al.*, 1999; Ruan *et al.*, 1999; Ye *et al.*, 2003,

2004). This technique allows the production of two- or three-dimensional temperature maps of ohmically heated food materials without disconnecting the electrical heating power and also allows the estimation of the fluid-particulate heat transfer coefficient for food mixtures in the ohmic heater during the holding (Ye *et al.*, 2003).

The mathematical models are also useful in estimation of temperature distribution and assessing worst-case scenarios in ohmic heater cells (Zhang *et al.*, 1992; Fryer *et al.*, 1993; Zhang and Fryer, 1995; Sastry and Salengke, 1998; Salengke and Sastry, 2007b; Marra *et al.*, 2009)

11.3.4 Electrode Types and Electrochemical Reactions

The most important electrolytic problems are the contamination of food with metal ions migrated from electrodes and the resulting electrochemical reaction products. This contamination could be either toxic (carcinogenic) or bring undesired taste and coloring to processed food (Amatore *et al.*, 1998).

The migration of electrode materials into foods needs to be minimized by choosing better electrode materials or by operation at higher frequencies (Zhao and Kolbe, 1999; Sastry, 2008). The use of surface materials with lower corrosion tendencies (titanium, gold, etc.) or the coating of the existing surfaces with less corroding materials may be an option (platinized titanium, etc.) (Ibrahim, 1999; Samaranayake and Sastry, 2005a). Pulsed ohmic heating is another option for minimizing the electrochemical reactions and the formation of gas bubbles during ohmic heating (Samaranayake and Sastry, 2005b; Jun *et al.*, 2007).

The gas bubbles are the results of either water boiling due to localized high current densities or the by-products of various oxidation/reduction reactions (e.g., H_2 or O_2 gas) (Zhao and Kolbe, 1999; Jun *et al.*, 2007). Castro *et al.* (2004a) explained that in an unpressurized heater, if air is occluded in the sample, the air bubbles expanded with temperature, with the pressure remaining constant.

Noticeable gas liberation at the stainless steel electrodes during ohmic heating of orange juice was reported by Lima *et al.* (1999). However, there was no visible gas liberation, electrode cavitation, metal dissolution, or rust tinge in the samples heated in the case of using specially coated titanium electrodes. Similarly, Assiry *et al.* (2003) concluded that electrolytic reactions were not evident with specially coated electrodes or in the use of high-frequency power.

Faradaic current increases with the amplitude of the applied voltage signal. Since the interfacial potential achieved across the electrode is proportional to the peak voltage amplitude, the faradaic current and thereby rates of faradaic reactions increase with increasing voltage (Assiry *et al.*, 2006);

Anodic half-reaction:

$$2H_2O_{(l)} \leftrightarrow O_{2(g)} + 4H^+_{(aq)} + 4e \qquad (11.9)$$

Cathodic half-reactions:

$$2H^+_{(aq)} + 2e \leftrightarrow H_{2(g)} \tag{11.10}$$

$$2H_2O_{(l)} + 2e \leftrightarrow H_{2(g)} + 2OH^-_{(aq)} \tag{11.11}$$

Overall reaction:

$$2H_2O_{(l)} + 2e \leftrightarrow 2H_{2(g)} + O_{2(g)} \tag{11.12}$$

pH is an important factor in electrode corrosion and electrolytic reactions, influencing the degradation of ascorbic acid in the fluid food (Samaranayake and Sastry, 2005a). At high NaCl concentrations, the following anodic half-reaction (Equation 11.13) hinders oxygen generation (Equation 11.9) resulting in a lower degradation rate (Assiry *et al.*, 2003).

$$2Cl^-_{(aq)} \leftrightarrow Cl_{2(g)} + 2e \tag{11.13}$$

Since the capacitance is directly proportional to the surface area, an electrode having a larger microscopic surface area possesses a higher electrical double-layer capacitance per unit apparent geometric surface area (Samaranayake and Sastry, 2005a). The electrical double layer of such an electrode is capable of holding more charge before the double-layer capacitor becomes "leaky", inhibiting faradaic-type reactions at the electrode/solution interface. The use of electrodes having a large microscopic surface area is beneficial in terms of inhibiting the faradaic processes at the interfaces, as well as in achieving more capacitive current for the heat generation. According to the SEM analysis (Fig. 11.6), variation of the double-layer capacitance of the electrodes can be represented as: titanium $<$ stainless steel $\leq$ platinized-titanium $\leq$ graphite (Samaranayake and Sastry, 2005a).

The Insulated Gate Bipolar Transistor (IGBT), a member of a broad-power semiconductor family, is basically a rapid switching device that enables the application of current and voltage as high frequency short duration pulses. The use of IGBT switching devices for ohmic heater circuitry is a relatively inexpensive alternative route of moving into high frequencies (Samaranayake and Sastry, 2005b). Pulse waveforms derived from an IGBT can be independently manipulated by adjusting various pulse parameters including frequency, pulse width, and delay time (off-time between adjacent pulses) and, therefore, differ from the waveforms typically generated by means of high-frequency generators. Samaranayake and Sastry (2005b) significantly reduced the electrochemical reactions during ohmic heating with stainless steel, titanium, and platinized-titanium electrodes by using this generator. They concluded that pulsed ohmic heating at higher frequencies and shorter pulse widths yielded the lowest rates of electrochemical reactions for stainless steel electrodes. However, the pulsed ohmic heating at lower frequencies and longer pulse widths was more effective in suppressing the electrochemical reactions of titanium and platinized-titanium electrodes. Delay time was found to be a critical factor in pulsed ohmic heating.

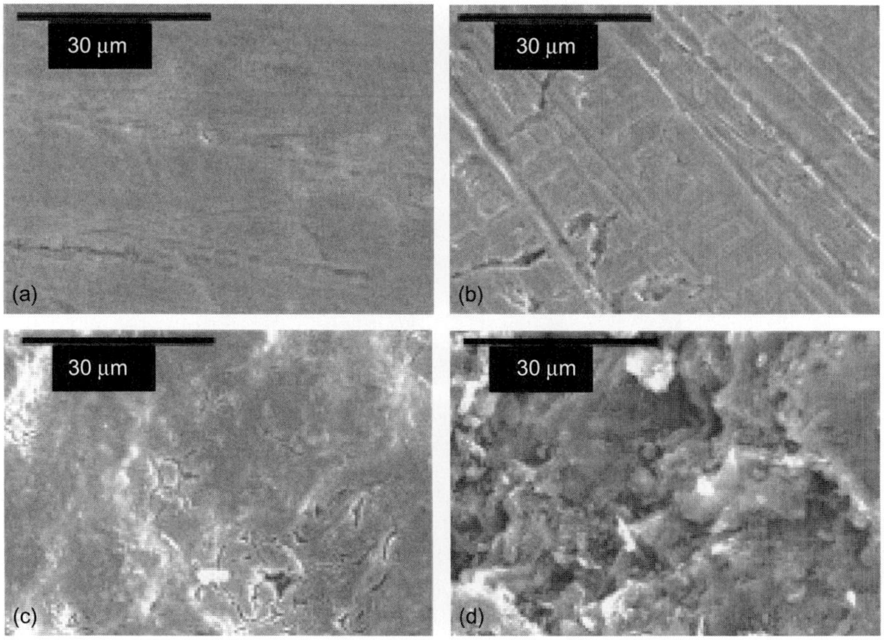

FIGURE 11.6

Typical SEM micrographs of the electrode materials: a, titanium; b, stainless steel; c, platinized-titanium; d, graphite.

[Reproduced with permission from Samaranayake and Sastry (2005a). J. Electroanal Chem. *577, 125–135.]*

The effects of corrosion begin to emerge over a period of ohmic heating time at the lower frequency (50 Hz) and the electrode surface starts experiencing severe pit corrosion. Conversely, the electrode surface remains unaffected at the higher frequency (10 kHz). The effect of the frequency is clearly visible in Figure 11.7 where the photographs of the electrodes are shown after the fouling deposits are removed (Bansal and Chen, 2006a). Increasing the frequency reduces the cycle time which apparently controls the electrochemical reactions. The use of a high-frequency power supply can be a useful tool in controlling the corrosion of electrode surfaces during ohmic heating.

The fouling layer formed may represent an inner overheating during ohmic heating, causing overburning in electrode surfaces. In a recent study, Stanel and Zitny (2010) have shown the risk of overburning in skim milk, especially at thin deposit layers, depending on the specific electrical conductivities of the deposits.

The effects of different types of electrodes on energy loss in ohmic heaters and the risks brought up by the reactions occurred by using them must be further studied (Sastry and Salengke, 1998; Icier and Ilicali, 2005b).

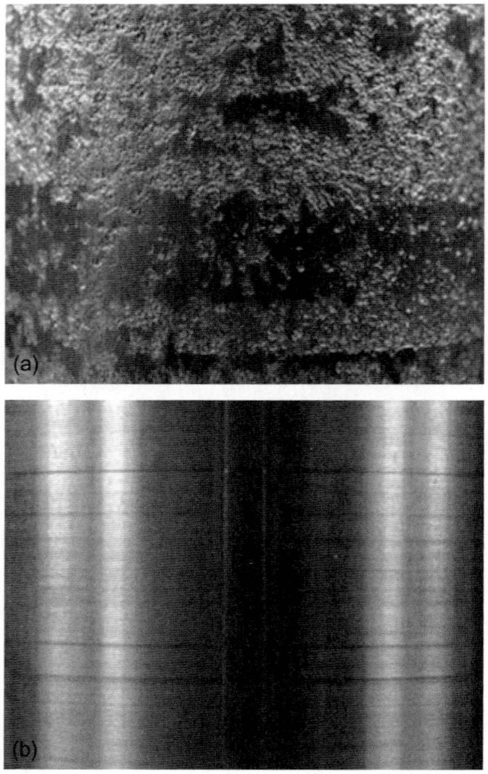

FIGURE 11.7

Photographs of electrode surfaces: a, 50 Hz; b,10 kHz.

[Reproduced with permission from Bansal and Chen (2006a). Food Bioprod. Process. *84(4), 286–291.]*

11.3.5 Food Properties

Acidity, composition, total solid content, and viscosity of liquid food can affect the ohmic heating rate. If fluid food is in the form of a solid–liquid mixture, particle dimension, orientation, density, and electrical conductivity ratios of liquid to solid are critical in assessing overall ohmic heating lethality. The difference between electrical conductivity and specific heats of solid and liquid phases in the mixture will extensively affect the temperature distribution in the ohmic column (Zoltai and Swearingen, 1996; Sastry and Li, 1996; Larkin and Spinak, 1996). In some cases, overheating or cold-point situations occur. The physical and electrical properties of food should be known and taken into account in the design procedures.

The efficiency of thermal processes applied to the non-Newtonian fluid food mixtures and purees depends on composition and rheological properties (Cogné *et al.*, 2003; Ditchfield *et al.*, 2004). It is difficult to obtain uniform heating in fluid foods having high consistency. It requires the special design of heat transfer equipment and affects the operational costs. The high consistency causes heat transfer problems during pasteurization. Uniform heating could not be achieved as a result of velocity differences within the ohmic column. The design of ohmic heater having proper process control equipment, in which uniform heating could be obtained, is important in ensuring the sufficient pasteurization of non-Newtonian fluids (Icier and Tavman, 2006).

When ohmic heating is applied to Newtonian fluids, the electrode surfaces are clean and the flow regime is close to the turbulent regime (Re = 1900), so that no temperature gradient is observed. When non-Newtonian fluid is switched to the ohmic system, viscosity changes and the flow regime become close to a laminar regime (Re = 65). Therefore, a low velocity close to the wall is observed, which then leads to a slight overheating (Ayadi *et al.*, 2004a).

There are various studies which deal with the effects of fluid viscosity on ohmic heating rate; some of them report that higher-viscosity fluids tend to result in faster ohmic heating than lower viscosity fluids, and others report *vice versa*. The conflict in these results is probably due to different reactions occurring during ohmic heating of different foods, depending on their composition or other process parameters applied.

Since Maras-type ice-cream mixes have higher consistencies than standard-type ice-cream mixes, their heating via conventional methods involving conductive and/or convective heat transfer mechanisms is problematic. However, their lower fat content results in higher electrical conductivity values, and thus they can be ohmically heated faster than standard-type ice-cream mixes (Icier and Tavman, 2006). The temperature control is more critical for Maras-type ice-cream mixes during ohmic heating.

The change in the composition of the food during ohmic heating must be accurately determined, and its effects on ohmic heating rate and temperature distribution should be taken into account in the process control procedure.

11.3.6 Flow Properties

The regulation of the mass flow rate depending on the change of the ohmic heating rate is crucial in continuous ohmic heating systems. It is an especially critical factor during heating of proteinous foods, in which the coagulation is not desirable. Even a slight change in the flow rate of the food can result in substantial temperature differences within it. Therefore, the characteristics of pumps used and the piping system are important in rapidly adjusting the residence time of the liquid within the column.

The determinations of residence time distribution and/or velocity distribution and its control during ohmic heating are crucial in ensuring the lethality of fluid foods. There are several methods of determining the residence time distribution of particles during ohmic heating of particle–liquid mixtures, such as visual

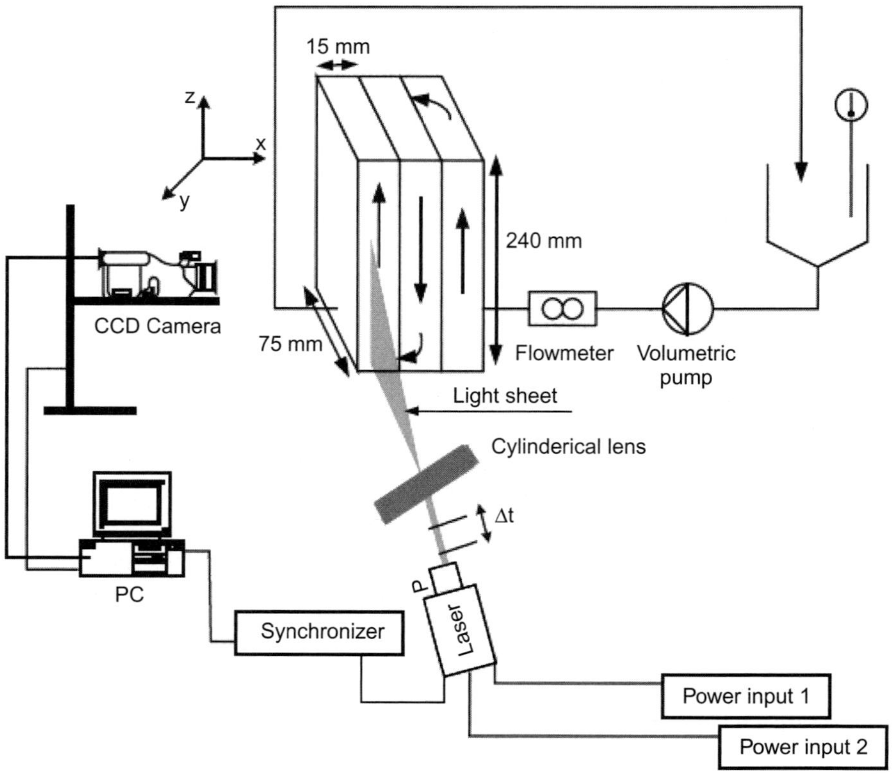

FIGURE 11.8

PIV measurement: a pilot plant.

[Reproduced with permission from Ayadi et al. *(2005).* J. Food Eng. *70, 489–498.]*

observation, laser-beam detection, chemical markers and thermal memory cells, radioactive tracers, salt tracers, ultrasound, magnetic response, and photosensor methodology (Tulsiyan et al, 2009). Marcotte *et al.* (2000b) suggested the use of ultrasonic methods for measuring the residence time of particulate liquid mixtures in the ohmic heating column.

For non-Newtonian fluids the flow regime becomes close to a laminar regime in the ohmic heating column (Ayadi *et al.*, 2004a). Therefore, low-velocity regions are exposed to a slight overheating. Ayadi *et al.* (2005) carried out a hydrodynamic study with a Newtonian model fluid under isothermal conditions, using a flow visualization technique (colored tracer) and the velocity fields were measured using particle image velocimetry (PIV) (Fig. 11.8).

They injected a colored tracer just before the cell inlet and its distribution was recorded in the form of video sequences through the transparent surface. Velocity fields in the ohmic cell were measured using the PIV technique. A light source including a crystal harmonic generator to produce the double-frequency green light was added to the experimental set-up used for flow visualization. They directly linked the presence and intensity of the deposit in the cells to the uniformity and non-uniformity of the velocities. The results of their study show that the quantity of deposit is greater in the zone where the temperature is lowest (entrance zone) and the velocity is non-uniform. Even the slightest hydrodynamic disturbance (re-circulation, poor filling, singularity, etc.) results in a thermal and electric disturbance and thereby creates zones, which are subject to fouling in continuous ohmic heating.

For liquid–particle mixtures, residence time determination of the fastest particle is important in calculating the minimum lethality in the ohmic heater. However, the tracking of multiple particles gives reliable information on the residence time distribution in the column. Tulsiyan *et al.* (2009) used radio-frequency identification to measure the residence time distribution of analog chicken particles in chicken chowmein during ohmic heating. They recommended the use of this measurement technique as an innovative and time-saving method which is suitable for aseptic conditions.

In the following sections, the effects of ohmic heating on microorganisms and nutritional and quality characteristics are discussed.

11.4 MICROBIAL INACTIVATION MECHANISM

Ohmic heating is a continuous high-temperature, short-time (HTST) sterilization process (de Alwis and Fryer, 1990). The mild electroporation during ohmic heating may contribute to cell inactivation owing to the fact that the presence of mild electroporation can improve transfer of substrates at the early stages of fermentation (Bhale, 2004). Ohmic heating can also be used for ultra-high temperature (UHT) sterilization of foods, and especially those that contain large particles (up to 2.5 cm) that are difficult to sterilize by other means. Therefore, heat processing by electricity can also be termed as aseptic processing; in which a sterile product is obtained by applying a sterile packaging in order to preserve the perishable food material for a long duration of time (Parrott, 1992). A current NASA project has involved the development of a specially designed pouch with ohmic heating capability for reheating of foods for space missions, as well as sterilization of waste (Jun and Sastry, 2005).

Microbial inactivation in relation to ohmic heating is primarily due to thermal effects in nature. Although some findings show the additional electrical effects, some results are not convincing since it is necessary to exactly match the thermal histories of ohmic and conventional heating experiments for the purpose of comparison the effects. The difference in slopes in the microbial inactivation curves of ohmic heating and conventional heating, which have similar thermal history, can most likely be explained by the presence of the electric field (Anonymous,

2000). However, new methods are needed for efficient bio-validation of ohmic heating in order to realize its full potential in microbial inactivation (Somawat *et al.*, 2009).

The low frequency used in ohmic heating (50–60 Hz) allows cell walls to build up charges and form pores, which is not the case with high-frequency methods such as microwave heating, where the electric field is reversed before sufficient charge build-up occurs at the cell walls (Bhale, 2004). However, Sitzmann (1995) hypothesized that inactivation of spores may result from an indirect bactericidal effect of electrolysis products formed during the electrical treatment.

Park *et al.* (2003) reviewed the effective mechanisms of electrical current on microorganisms. The mechanism of electric current activity may include disruption of bacterial membrane integrity or electrolysis of molecules on the cell surface (Liu *et al.*, 1997). When a voltage is applied, it increases the energy of the membrane such that an increase in membrane pore size takes place up to a transition to hydrophilic pores, where free diffusion may occur (Sugar and Neumann, 1984). Another hypothesis is called dielectric breakdown (Zimmermann *et al.*, 1974). Because of the attraction of opposite charges induced on the inner and outer surfaces of the cell membrane, compression pressure occurs, resulting in a decrease in membrane thickness, an irreversible rupture may take place. The most widely accepted model is that of severe electroporation (Park *et al.*, 2003). Electroporation is the formation of holes in a cell membrane due to individual ion pressure, which cause change in permeability of the cell membrane, due to the varying the electric field (Weaver and Chizmadzhev, 1996). If critical electrical field strength is exceeded, the membrane is permeabilized by pore formation. This permeabilization can be reversible or irreversible, depending on the electrical field strength, treatment time, cell size, membrane surface charge, cytoplasm, and suspending liquid medium (Lojewska *et al.*, 1989). The dielectric strength of a cell membrane is related to the amount of lipids (acting as an insulator) present in the membrane itself. The pores formed can vary in size depending on the strength of the electric field, and can reseal after a short period of time. It is generally thought that the critical membrane potential induced by electric fields causes microbial inactivation and is about 1 V. At this level, it is thought that the permeability of the membrane increases such that cell death occurs (Rowan *et al.*, 2000).

At low frequencies (50–60 Hz) and high field strengths (>100 V/cm) most commonly associated with ohmic heating, the naturally porous cell walls can allow the cell membrane to build up charges, forming disruptive pores (Cho *et al.*, 1996). Excessive exposure causes cell death due to the leakage of intracellular components through the pores (Lee and Yoon, 1999). If electricity is specifically applied during the microbial growth cycle, the enhanced transport of inhibitory substances across the cell membranes is possible (Bhale, 2004).

Corresponding electroporation also ruptures the structure of the food sample during ohmic heating and moderate electric field. It is sometimes desirable and sometimes unacceptable. Moderate electric field (MEF) processes involve the application of electric fields typically under 1000 V/cm, with or without heating, to accomplish specific objectives (Sastry, 2008). There is an obvious effect of temperature on efficiency of damage under AC treatment of plant tissue. At MEF with an electric field strength E under 100 V/cm, the ohmic heating to temperatures not exceeding 50°C results in a high degree of tissue damage (Lebovka *et al.*, 2005). They discussed that, with the same conditions of AC treatment, the tissue damage observed for apples is higher than for potatoes. The observed effects show the importance of the electroporation mechanism in plant tissue damage induced by ohmic heating, which is controlled by temperature-induced changes in the cell membrane structure.

Studies in the area of electrical treatment of foods dealt with inactivation of vegetative cells and the lethal effect of electricity on bacterial spores is currently limited (Palaniappan *et al.*, 1990, 1992; Cho *et al.*, 1996; Qin *et al.*, 1995; Leizerson and Shimoni, 2005a, b; Baysal and Icier, 2007, 2010; Anderson, 2008).

A shelf-life study by Raztek compared conventional heating to ohmic heating using pasteurized liquid eggs. Although the initial plate counts were similar, as time progressed, conventionally heated samples reached plate counts of 10,000 cfu/ml compared with ohmically heated samples with plate counts of <10 cfu/ml after 12 weeks (Reznick, 1996). This result was explained by the injury effects of the ohmic heating to the cells due to possible electroporation.

Fruit juices in general are characterized by high acidity conditions, which lead to the growth of yeast and mold ,in addition to a few types of low-acid-tolerant bacteria (Splittstoesser, 1996). To avoid microbial spoilage, it is necessary to cause inactivation by applying heat. In the study of Palaniappan *et al.* (1992) suspensions of yeast cells (*Zygo Saccharomyces bailii*) and cells of *Escherichia coli* were subjected to conventional and ohmic heating. They found that any lethal effect caused by electricity was insignificant when compared to that produced by heat, at thermally lethal conditions. However, the ohmic heating enhanced the leakage of intracellular components of *Saccharomyces cerevisiae*, indicating irreversible damage to cell walls, as compared with the conventional heating (Lee and Yoon, 1999; Yoon *et al.*, 2002). The application of ohmic heating (10–20 V/cm) to 10% yeast suspension in phosphate buffer gave higher destruction rates (in the range of 70–90°C) than conventional heating (using a water bath) (Yoon *et al.*, 2002). They concluded that the inactivation mechanism of microorganisms was due to electroporation created by ohmic heating.

Furthermore, Cho *et al.* (1999) conducted conventional or ohmic heating in *Bacillus subtilis* spores suspended in 0.1% NaCl solution (ca. 10^7 CFU/ml), treated under identical temperatures histories. They reported that spores heated at 92.3°C had significantly smaller D values when heated by ohmically (8.55 ± 0.23 min) than by the conventional method (9.87 ± 0.08 min) (Fig. 11.9). z-Values for the temperature range of 88–97°C were determined as 9.16 ± 0.28°C during ohmic heating and 8.74 ± 0.39°C during conventional heating. They concluded that the spore inactivation during ohmic heating was primarily due to the thermal effect, but there was an additional killing effect caused by

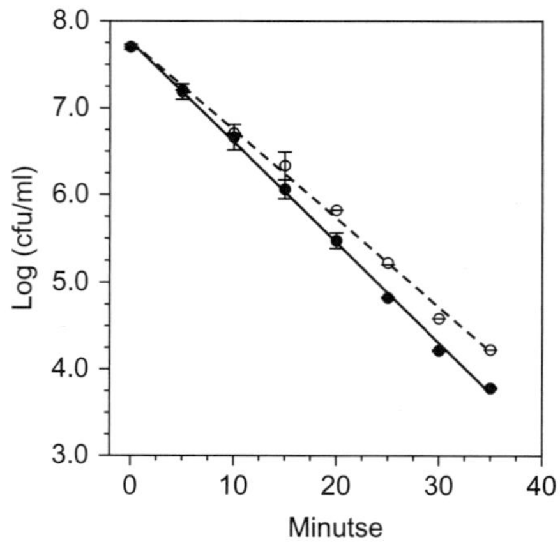

FIGURE 11.9

Survivor plot for *B. subtilis* spores treated by a single-stage conventional or ohmic heating at 92.3°C. Each data-point represents the survivor's average ± standard deviation: (- - -) conventional heating; (—) ohmic heating.

[Reproduced with permission from Cho et al. *(1999).* Biotechnol. Bioeng. *62 (3), 368–372.]*

the electric current, which caused injury to spores and thus made them susceptible to the associated ohmic heating. In addition, they concluded that electricity affected the death rate but did not affect the temperature dependency of the spore-inactivation process since z and activation-energy values for ohmic and conventional heating were not different.

The increase in lethality caused by the double heating is known as the Tyndallization effect. Ohmic heating has a greater Tyndallization effect, when compared with conventional heating. The enhanced spore inactivation by ohmic, compared with conventional heating, resulted from a greater rate of spore death during the first stage of heating and a greater decrease in the count of viable spores immediately after the incubation period that intervened in the heating process (Cho *et al.*, 1999). They explained the Tyndallization effect under ohmic heating conditions by the additional effects of the electrical current on the spore germination and cell injury. Baysal and Icier (2010) similarly found additional electrical effects on inactivation of Alicyclobacillus acidoterrestris spores during ohmic heating of orange juice. They matched thermal history of ohmic and conventional heating, and reported D-values for ohmic heating which were significantly lower than those for conventional heating.

Leizerson and Shimoni (2005a) applied ohmic heating as an alternative pasteurization method. They treated orange juice at temperatures of 90, 120, and 150°C for 1.13, 0.85, and 0.68 s in an ohmic heating system. Both ohmic and conventional thermal treatments reduced microbial counts by at least two to three orders of magnitude compared to their number in fresh orange juice. Their other

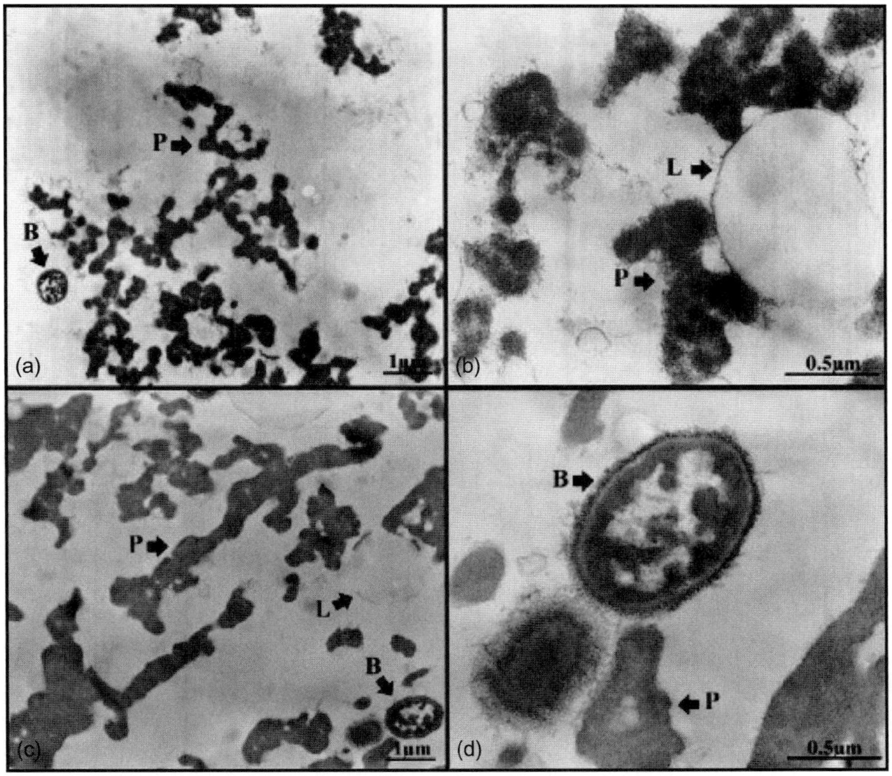

FIGURE 11.10

Transmission electron micrographs of kefir's microstructure following two different heat treatments. a, b, Ohmic heating treatment at 72°C for 10 s. c, d, Treatment by autoclaving at 110°C for 3 min. B, bacteria; L, lipids; P, proteins.

[Reproduced with permission from Mainvilee et al. *(2001).* Int. Dairy J. *11, 45–49.]*

study (Leizerson and Shimoni, 2005b) showed that, although both thermal treatments prevented the growth of microorganisms for 105 days, the sensory shelf-life of ohmic-treated orange juice was >100 days and was almost two-times longer than that of conventionally pasteurized juice.

Heat treatments (autoclaving, at 110°C for 3 min and ohmic heating up to internal temperature 50–72°C) deactivated the bacteria and yeast in kefir (Mainville *et al.*, 2001). Untreated kefir had 8.58 log cfu g^{-1} lactococci, 8.60 log cfu g^{-1} lactobacilli and 5.09 log cfu g^{-1} total yeasts. Deactivated kefir had microflora counts of 0.59 log cfu g^{-1} or less. Both heat treatments applied to kefir caused changes of the protein structure which were extensive in the case of the autoclaved kefir. In both cases transmission electron microscopy (TEM) showed that, the lipid globules were affected and the inner structure of the bacteria was disrupted (Fig. 11.10). The clumping was noted in samples treated at 60°C and 72°C due to thermal effects on milk proteins.

Pereira *et al.* (2007) studied inactivation kinetics of *E. coli* in goat's milk, and *Bacillus licheniformis* ascospores in cloudberry jam during ohmic and conventional heating under identical temperature histories. They found that the thermal death times under ohmic heating were shortened. Sun *et al.* (2008) similarly found that the ohmic heating reduced the decimal reduction times of viable aerobes and *Streptococcus thermophilus* 2646 in milk compared to conventional heating having matched their thermal history.

Anderson (2008) applied ohmic heating (18 V) to a solution containing *L. acidophilus*. The ohmic heating of the solution to 90°C at approximately 8 min caused 2 log reductions. However, there was not any comparison with the conventional heating at the identical heating conditions.

Somawat *et al.* (2009) recently developed a precise universal capillary cell (UCC) ohmic heating device to quantify additional non-thermal lethal effects of electricity during ohmic heating on *Geobacillus stearothermophilus* spores. They applied ohmic heating at frequencies of 60 Hz and 10 kHz to tomato soup inoculated with 10^8 spore cfu/ml and compared its results with conventional heating at 121°C, 125°C, and 130°C for four different holding times. Ohmic heating at 10 kHz showed more inactivation of the bacterial spores than conventional heating. They explained it by the release of polar ionic materials like dipropylamine (DPA) from the spore core together with the spore coat proteins at high temperature, which assisted in the increased lethal effect of high-frequency electricity.

The sterilizing effect of low-amperage electric treatment for various times has been also studied with natural seawater and seawater inoculated with *V. Parahaemolyticus* (Park *et al.*, 2003). In both cases, bacteria including *V. parahaemolyticus* in seawater were completely eliminated in 100 ms by a 0.5-A, 12-V direct current. They explained the high lethal effects by electropermebilization of cells. Electron-microscopic investigation of the electrically treated bacteria revealed substantial structural damage at the cellular level. They recommended the low-amperage electric treatment for rapid inactivation of microorganisms in seawater. Guillou and Murr (2002) concluded that electric current and temperature seem to act synergistically for the inactivation of *Saccharomyces cerevisiae* and this interesting phenomenon can be applied in the sterilization of thermosensitive food products by electrolysis associated with moderate heat treatment.

Sublethal electrical treatment, however, reduces the subsequent thermal requirement for inactivation of microorganisms (Palaniappan *et al.*, 1992). An ohmic heating system may be particularly useful in this important process of food manufacturing, particularly in the dairy industry, where fermentation by *Lactobacillus acidophilus* is necessary for the production of cheese and yoghurt (Anderson, 2008). Cho *et al.* (1996) subjected the inoculated solution of *L. acidophilus* OSU133 in MRS broth to fermentation temperatures (30, 35 or 40°C) attaining by ohmic heating (15 and 40 V) or conventional heating. They showed that ohmic heating reduced the lag period of the bacteria in the early stages of growth, but tended to hinder the later stages. At low fermentation temperature, the lag period in conventional heating was 6.09 h while ohmic heating reduced it to 0.34 h (15 V). Ohmic heating at 35°C had minimal effect on glucose utilization and lactic acid production by *L. acidophilus*. Cho *et al.* (1996) hypothesized that the lag period in the early stages of the fermentation of *L. acidophilus* was appreciably affected by the ohmic heating treatment due to the oscillating electric field dislodging the polar antimicrobials (found in MRS broth) and the macromolecules (produced by the *L. acidophilus*) which were adhered to the cell walls and membranes. This activity might improve the absorption of nutrients and minimize the inhibitory action of fresh MRS broth which in turn can shorten the lag period. A dairy manufacturer utilizing the ohmic heating process in the early stages of fermentation may shorten the total processing time of a dairy product. This speedier process would save untold amounts of costs in overheads and labor time (Anderson, 2008).

Loghavi *et al.* (2007) applied MEF to an inoculated solution containing *L. acidophilus* to determine whether the permeability of the cell membranes is affected during fermentation. The MEF (1 V/cm for 5 h and then turned off for the remaining 35 h) treatment affected the bacteriocin activity the greatest, even though this activity was associated with the growth of bacteria. It was explained by the extra stress placed on the bacteria; an increase in production of defensive molecules such as bacteriocin or effects on cell membranes; such as causing temporary pores (electroporation) and an increase in transmembrane conductivity and diffusive permeability of nutrients, surfactants, autoinducers, and bacteriocin.

Furthermore, the fermentation process is accelerated by applying pure sinusoidal moderate electrical field at the early stage of growth. However, a significant increase in the bacteriocin production occurs when a sinusoidal field at 60 Hz with harmonics is applied at the early stage of the growth of *L. acidophilus* OSU 133 (Loghavi *et al.*, 2008). The application of MEF during fermentation of *L. acidophilus* is useful for increased production of bacteriocin, which acts as a food and feed preservative due to its antimicrobial effects. This knowledge is invaluable as a natural way to preserve food and feed, increasing its shelf-life and attractiveness to consumers who are interested in using organic products or products without the addition of chemical preservatives (Anderson, 2008).

Whether ohmic heating has electrical rather than thermal effects or not, ohmic heating is presumed to produce lethal effects; it may be advantageous in reducing process times for heat-sensitive products in the future (Sastry, 2008).

The effects of ohmic heating on nutritional and quality characteristics of fluid foods are discussed in the following section.

11.5 EFFECTS ON NUTRITIONAL AND QUALITY CHARACTERISTICS OF FLUID FOOD

The quality parameters concerning liquid food products, such as inactivation of microorganisms and enzymes, heat-sensitive compounds, and physical characteristics, play an important role in the industry. Various bioactive components can degrade differently by different thermal methods of processing.

First-order degradation/inactivation rate kinetics of several nutrients and enzymes, respectively, can be written as follows (Castro *et al.*, 2004b):

$$\frac{logC_A - logC_{A0}}{t} = \frac{1}{D} \tag{11.14}$$

$$\frac{logD_2 - logD_2}{T_2 - T_1} = \frac{1}{z} \tag{11.15}$$

$$k = k_0 e \frac{E_a}{RT} \tag{11.16}$$

where C_A is the activity or concentration, C_{A0} is the initial activity or initial concentration, D is the decimal reduction time (min), t is the processing time (min), T is the processing temperature, z is the temperature sensitivity indicator (°C), E_a is the activation energy (kJ•mol^{-1}), k is the frequency constant which is equal to 2.303/D (s^{-1}), and k_O is the pre-exponential factor (s^{-1}).

If the temperature dependency of the kinetic constant is found by taking account of the reference temperature (T_{ref}), Equation 11.16 takes the form of Equation 11.17 (Icier *et al.*, 2008):

$$k = k_{0,ref} e^{\left(\frac{-E_a}{R}\left(\frac{1}{T} - \frac{1}{T_{ref}}\right)\right)} \tag{11.17}$$

where $k_{0,ref}$ is the pre-exponential term of the Arrhenius equation at reference temperature (s^{-1}).

Eliot-Godereaux *et al.* (2001) demonstrated the feasibility of processing cauliflower florets in starch solution (25% w/w cauliflower) by continuous ohmic heating. Their experiments performed in a 10 kW APV ohmic heating pilot resulted in a final product with interesting firmness properties and stabilized at 25°C and 37°C. They recommended the use of this electrical technology to process brittle products such as ready meals containing cauliflower.

In the following subsections, as well as changes in composition, proteins, vitamins, phenolic contents, effects on color, enzyme inactivation kinetics, rheological and sensory properties during ohmic heating are discussed in detail.

11.5.1 Soluble Solid Content and Acidity

Ohmic heating generally does not cause any differences in acidity and °Brix values of fluid foods (Icier and Ilicali, 2004, 2005a, b; Icier *et al.*, 2008; Bozkurt and Icier, 2009, 2010; Yildiz *et al.*, 2009, 2010).

Ohmic treatment results in the effective damage of cells by a combination of electrical and thermal effects, resulting in the important diffusion of soluble nutrients during ohmic heating. If electropermeabilization occurs in vegetable/fruit pulps or purees, pores formed in the cell membranes upon electric field exposure cause a drop in resistance as ions are allowed to pass through the membrane (Kulshrestha and Sastry, 2006). Thus, a slight increase in pH and soluble solid contents of fluid foods during ohmic heating is possible.

However, the possible fouling during ohmic heating of proteinous fluids can change their concentration and pH. If a deposit layer starts to be removed due to boiling at high temperatures over long process times, the degraded or aggregated proteins may also be absorbed back into solution.

11.5.2 Proteins

The hypothesis of the additional effect of the electric field being due to the separation of the subunits or breakage of the protein structures (enzyme molecule) was raised by Castro *et al.* (2004b). They checked this hypothesis by polyacrylamide gel electrophoresis (PAGE). When they run a PAGE system with proteins under their native form (native PAGE), these proteins move through the gel when an electric field is applied and each band on the gel represents a different protein (or a protein subunit). If there had been a subunit separation due to heating, different bands would have appeared in the gel, corresponding to different subunits. Nevertheless, both the untreated enzyme and the heat-treated enzyme (in this latter case, enzymes used were both conventionally and ohmically treated) presented the same number and position of bands in the gel. This leads to the conclusion that the enzyme had the same molecular weight and no subunit separation occurred before or after the thermal treatments (conventional or ohmic).

When the material used is a proteinous liquid, the fouling at the electrode surfaces during ohmic heating especially for high solute concentrations at high temperatures is observed. The design evaluation of continuous ohmic systems is generally carried out by minimizing their fouling generation. Ohmic heating of dairy products in particular generates fouling on electrode surfaces. The presence of this motionless deposit causes an increase in electricity consumption and leads to an increase in the temperature of the electrode surfaces (Ayadi *et al.*, 2004a). The rapid increase in the pressure drop can be observed when channels are blocked by deposit and an overall heat-transfer coefficient decreases as the fouling layer formed on heat transfer surfaces.

Ayadi *et al.* (2004a) used a whey protein–xanthan gum mixture chosen as a model fouling fluid. They heated this fluid ohmically from 75 to 100°C by fixing the flow rate at 300 l/h (Re = 63) in the rectangular continuous ohmic system and monitored the fouling in ohmic cells for durations of 1–6 h (Fig. 11.11). The deposits on the surfaces increased as the temperature of model fluid increased from cell 1 to cell 5. They observed that, after 3 h of processing, the surfaces of the electrodes (cell 4) were almost entirely covered by a fouling deposit, which was voluminous, spongy, and of a whitish color. The photograph of the electrode surface after a 4h fouling test shows a change in the deposit aspect (appearance of

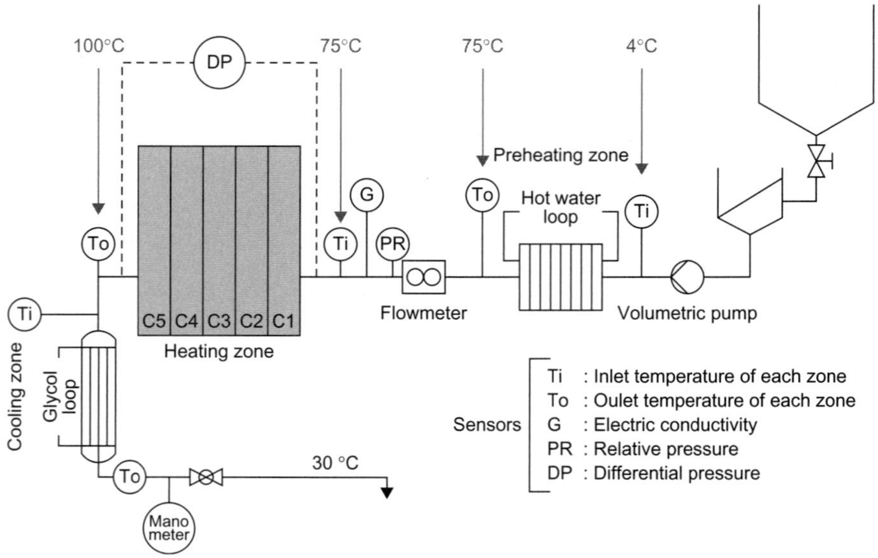

FIGURE 11.11

Fouling pilot plant test rig.

[Reproduced with permission from Ayadi et al. *(2004a).* Innov. Food Sci. Emerging Technol. *5, 465–473.]*

a local over-heated zone and the trace of tiny bubbles). It was concluded that, since the fluid temperature was approximately 100°C and the residence time in each ohmic cell (cell no.s 2, 3, 4, and 5) were sufficiently long that β-lactoglobulin denatured and adhered on to the electrode surfaces. The photograph of the electrode surface after a 6h fouling test shows the presence of an over-heated deposit and some deposit removal due to boiling started (Fig. 11.12). However, considering the temperature (no heating, fouling fluid 75°C) and the residence time in this cell, no h-lactoglobulin adhesion occurred in cell 1. and, thus, no fouling took place (Ayadi *et al.*, 2004a).

The fouling build-up causes an increase in the electric power required and the deposit on the electrode surfaces acts as an additional electrical resistance. Given the fact that motionless deposit is continually subject to the Joule effect, the electrode surface temperature increases during the experiment (Ayadi *et al.*, 2004a). The deposit layer and the fouling ability in laminar regime can be minimized by changing the inlet design and optimizing the flow field in the channels of ohmic cells (Ayadi *et al.*, 2003).

Ayadi *et al.* (2004b) used model fluid solutions of whey protein and xanthan gum based on the fact that the heat denaturation of β-lactoglobulin protein governs milk deposit formation when the temperature exceeds 75°C, and that xanthan gum is added to modify the viscosity of the model fluid as a dairy product

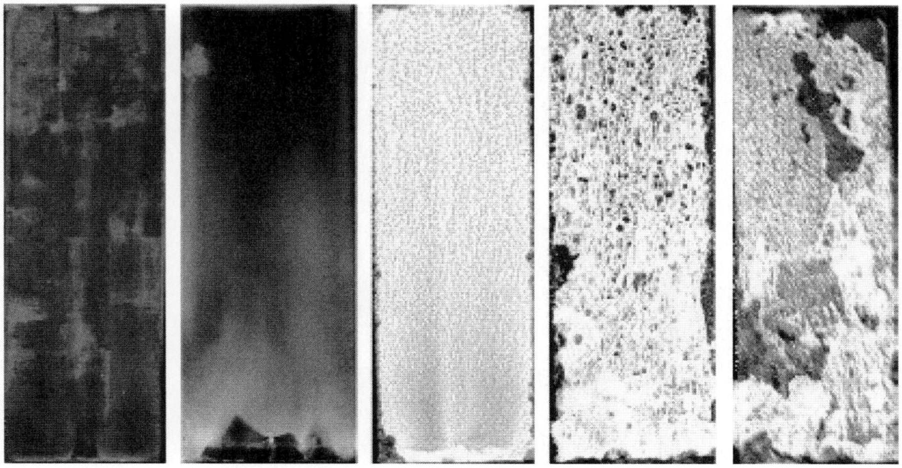

FIGURE 11.12

Photographs of cell no. 4 electrode surface after 1, 2, 3, 4 and 6h fouling experiments.
[Reproduced with permission from Ayadi et al. *(2004a).* Innov. Food Sci. Emerging Technol. *5, 465–473.]*

thickener. They showed the microstructure of the deposit layers, revealing that the deposit generated by heating the protein solution was less fibrous than that generated by heating the protein–xanthan solution (Fig. 11.13). The structural difference between the two deposit layers was used to explain the difference between the evolution and the values of the electrical conductivity. When a protein solution is heated, protein molecules partially unfold and then aggregate to form a three-dimensional network that entrap water through capillary forces (Mulvihill and Kinsella, 1987). The presence of xanthan gum reinforces this three-dimensional structure and makes it more porous and more water will be entrapped by capillary action and consequently a higher electrical conductivity is observed (Ayadi *et al.*, 2004b)

Similarly, the fouling of the reconstituted milk solution (5% solute concentration) during ohmic heating was also observed in a cylindrical continuous ohmic heating system consisting of two concentric cylinders (Fig. 11.14) (Bansal and Chen, 2006b). They reported that as soon as heating was commenced inside the heater, both protein denaturation and aggregation reactions took place and as a result, deposits started settling on the electrode surfaces. The additional electrical resistance caused less current to flow through milk and as a result the bulk temperature decreased. The changing temperatures led to changes in the electrical conductivities of milk and deposits. As milk was heated inside the heater, an increasing amount of the aggregated protein was formed. Less fouling (i.e. lower

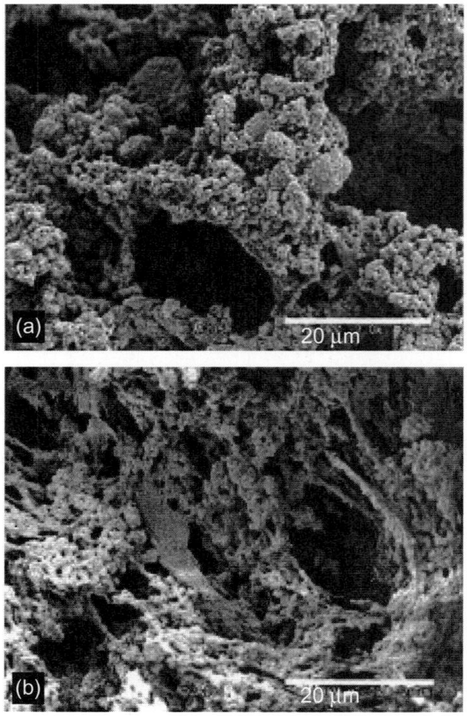

FIGURE 11.13

SEM observations of deposit generated by ohmic heating of: a, protein solution; b, protein–xanthan mixture solution.

[Reproduced with permission from Ayadi et al. *(2004b).* Food Bioprod. Process. *82(4), 320–325.]*

reduction in the current value) was observed at higher temperatures and this was explained by the domination of the fouling rate due to aggregated proteins. The aggregated protein is considered to cause less fouling compared with denatured protein. There may be two reasons for this: firstly its larger size restricts its transport from the bulk to the electrode surface; and secondly its comparatively compact shape lowers its ability to stick to the surface (Bansal and Chen, 2006b).

Corrosion of heat transfer surfaces is an issue which needs to be dealt with during fouling due to the electrochemical degradation of the electrode surfaces. The electrical power supplied in an ohmic heater can also result in electrochemical reactions at the electrode/liquid interface which are generally not desirable. Stainless steel, a widely used material of construction in the dairy industry, shows significant resistance against corrosion but the overall performance is found to be several orders of magnitude inferior to titanium (Samaranayake and Sastry, 2005a). Increasing the frequency of the power supply may be a method of controling the corrosion process (Bansal and Chen, 2006a).

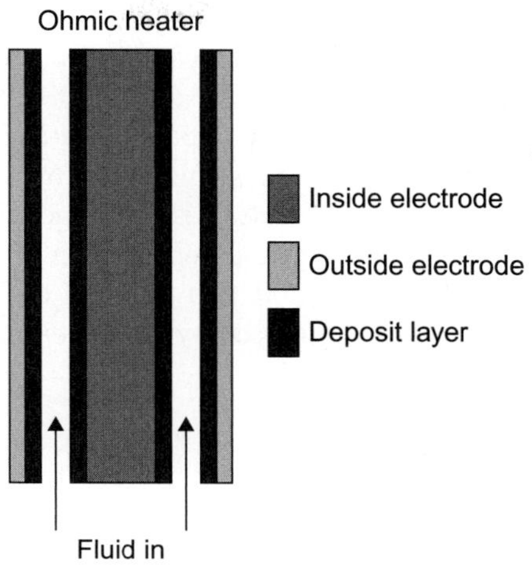

FIGURE 11.14

Concentric-cylinders type ohmic heater.

[Reproduced with permission from Bansal and Chen (2006). Food Bioprod. Process. *84(4), 286–291.]*

Conversely, degradation of protein via ohmic heating can be the main purpose of a process. The separation of protein bulk from the fish waste water after coagulation by ohmic heating is used as an industrial process. Huang *et al.* (1997) studied the effect of heating on coagulation of protein from fish mince wash water, which was obtained from frozen pacific whiting, using batch ohmic heating. They suggested that the highest heating temperature for effective coagulation of protein was 70°C. Similarly, Kanjanapongkul *et al.* (2009) developed a continuous ohmic heating system to coagulate protein from surimi wastewater and to reduce the biological oxygen demand of the wastewater. They heated the waste samples under different conditions (electric field strength of 20, 25, and 30 V/cm; flow rates of 100, 200, and 300 cc/min). The lab-scale ohmic heating system possessed good performance to coagulate protein (60%) from surimi wastewater.

Ohmic heating is a new method in the production of protein–lipid film and tofu from soymilk (Lei *et al.*, 2007; Wang *et al.*, 2007). Lei *et al.* (2007) maintained the temperature of the precooked soymilk (6.5% soluble solid concentration) at 85°C ohmically to form a protein–lipid film. They concluded that protein–lipid films from soymilk of higher quality were obtained by ohmic heating. The yield and protein incorporation efficiency by ohmic heating were higher than those by water-bath heating due to possible electroporation of soymilk and increasing diffusion of the components. Increased film formation rate by ohmic heating was explained. According to the conclusion of Lei *et al.* (2007), heat rapidly penetrates into the whole soybean milk rapidly and accelerates colliding between inner molecules during the ohmic heating process. Thus, it accelerates the surface of soybean milk dehydration and the three-dimensional structure of proteins exposing sulphydryl groups and hydrophobic side-chains alternately. Meanwhile, lipids act as active-surface agents; they go to air interfaces quickly and interact with proteins with hydrophobic bonding. In addition, the rehydration capacity of protein–lipid film improved significantly by ohmic heating due to less thermal damage, while the whiteness showed no significant difference. Similarly, the two-stage ohmic heating (the combination of 70°C for 10 min and 100°C for 5 min) not only improved soft tofu physical quality significantly, but also increased soft-tofu yield and solid recovery (Wang *et al.*, 2007).

11.5.3 Vitamins

The degradation of ascorbic acid during ohmic heating follows first-order kinetics. The degradation rate constants obtained in various studies are given in Table 11.2.

Lima *et al.* (1999) reported that the difference in the per cent degradation of ascorbic acid during ohmic and conventional heating was not statistically different. The activation value (52.8 kJ/mol) for ascorbic acid degradation in the pre-pasteurized orange juice was similar during conventional and ohmic heating, in the temperature range of 65–90°C. Similarly, the presence of an electric field (<20 V/cm, temperature range of 60–97°C) did not affect ascorbic acid degradation of strawberry pulp (Castro *et al.*, 2004a). Even if the temperature was higher (90–150°C) and treatment times were lower (0.68–1.13 s), the results were similar for both ohmic and conventional pasteurization, reducing the Vitamin C content of orange juice by approximately 15% compared to fresh juice (Leizerson and Shimoni, 2005a). However, Vikram *et al.* (2005) reported that ohmic heating facilitated higher nutrient retention in orange juice at all temperatures (50–90°C) (Fig. 11.15), compared to other methods (infrared, microwave and conventional heating). The higher value of activation energy for vitamin C destruction during ohmic heating (47.27 ± 0.80 kJ/mK) than during conventional heating (39.84 ± 0.61 kJ/mK) was explained by the higher dielectric properties of the juice and instantaneous heat generation due to passage of electric current. The z-value observed in the same study was 20.70°C and 24.39°C for ohmic heating and conventional heating, respectively.

The variation of the salt concentration, buffer pH, and applied electrical power during ohmic heating may alter the reaction kinetics of ascorbic acid (Assiry *et al.*, 2006). The pH level of the reaction environment is considered to be one of the most important factors that influence the degradation rate. However, pH is an important factor in electrode corrosion and electrolytic reactions. Assiry *et al.* (2003) described the influence of corrosion and electrolysis on the degradation of ascorbic acid at pH 3.5. The increase in the voltage and/or the NaCl concentration resulted in higher pH. This increase in the pH can be related to loss of buffering capacity due to metal–citrate complexation, as described by Assiry *et al.* (2003).

Table 11.2 Kinetic Constants for Ascorbic Acid Degradation During Ohmic Heating

Food	Heating Method	Temperature (°C)	Voltage gradient (V/cm)	% loss	D (min)	z (°C)	k_o (s^{-1})	E_a (kJ/mol)	Reference
Orange juice	Ohmic	65	18.2	13.1 ± 2.5			3.27×10^4	2.99	Lima *et al.* (1999)
		75		11.3 ± 1.5					
		80		12.9 ± 3.2					
		90		21.2 ± 3.0					
	Conventional	65		10.0 ± 2.5			3.33×10^4	3.01	
		75		13.7 ± 1.9					
		80		13.7 ± 2.6					
		90		23.3 ± 3.8					
Orange juice	Ohmic	50	42		95.96				Vikram *et al.* (2005)
		60			58.55				
		75			23.72				
		90			14.66				
	Conventional	50			65.67				
		60			49.81				
		75			27.02				
		90			12.91				
Strawberry pulp	Ohmic	60	32–80		256	46.73	0.14	21.05	Castro *et al.* (2004a)
		70			192				
		75			169				
		80			154				
		90			130				
		97			114				

(Continued)

Table 11.2 (Continued)

Food	Heating Method	Temperature (°C)	Voltage gradient (V/cm)	% loss	D (min)	z (°C)	k_o (s^{-1})	E_a (kJ/mol)	Reference
	Conventional	60			294	46.73	0.15	21.36	
		70			196				
		75			175				
		80			152				
		90			123				
		97			111				
Artichoke by-product	Ohmic	85	25	63.13 ± 0.65					Icier (2010)
			30	65.80 ± 0.53					
			40	48.02 ± 0.67					
	Conventional	85		71.19 ± 0.41					
		100		47.39 ± 0.36					
NaCl sol., (0.5–1%) pH 5.7	Ohmic 100 W	40–80					3.63×10^{-4} – 9.43×10^{-9}	(−26.92) – (2.78)	Assiry *et al.* (2006)
	150 W						1.08×10^{-7} – 6.50×10^{-9}	(−27.89) – (−32.39)	
	Conventional	40–80					2.22×10^{-5} – 6.62×10^{-4}	3.97 – 8.45	

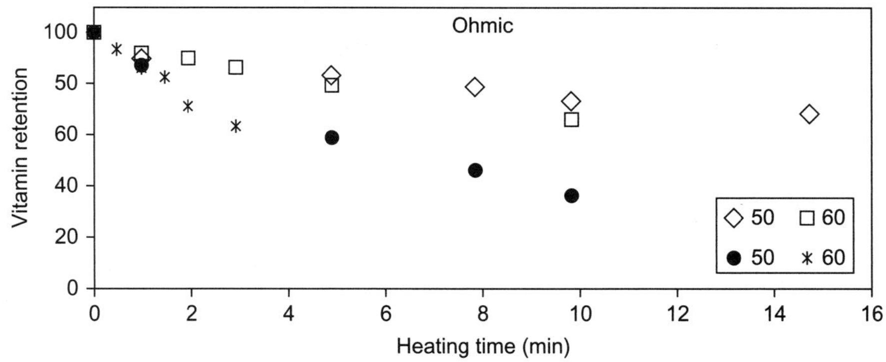

FIGURE 11.15

Vitamin C retention during ohmic heating at different temperatures.

[Reproduced with permission from Vikram et al. *(2005).* J. Food Eng. *69, 31–40.]*

Assiry *et al.* (2006) described the ascorbic acid degradation in a pH 5.7 buffer system by a first-order model during conventional and ohmic heating treatments. The effect of power and temperature on the rate of ascorbic acid degradation was interdependent. For the same power level, the reaction rate increased with decreasing temperature or increasing voltage gradient. Since the dissolved oxygen in the system increased due to electrolysis which was enhanced at high power (300 W) and low NaCl (0.25%), it apparently caused greater degradation especially at lower temperatures (Assiry *et al.*, 2006). They concluded that the enhanced degradation at pH 5.7 might be attributed to the increasing dissociation of ascorbic acid with increasing pH. The dissociated (anionic) form of ascorbic acid is more susceptible to degradation than the undissociated form.

Leizerson and Shimoni (2005b) studied the destruction of Vitamin C in ohmically pasteurized orange juice during storage. Degradation curves of ascorbic acid followed a linear decrease in both ohmic-heated and conventionally pasteurized orange juices during storage at 4°C. They discussed that neither electrical heating nor conventional pasteurization had any influence on the degradation rate of ascorbic acid.

Vegetable purees can be heated ohmically, resulting in high retention of vitamins (Table 11.2). For artichoke by-product puree, vitamin C loss as a result of ohmic and water blanching at 85°C has been observed in the range of 48.02–65.80% and 71.19%, respectively (Icier, 2010). Ohmic heating (in the range of 10–40 V/cm) has resulted in high retention of β-carotene in spinach puree, in the study of Yildiz *et al.* (2010). They determined that the effect of voltage gradient on the β-carotene content was not statistically significant ($p < 0.05$). However, the holding time (600 s) at constant temperature resulted in the increase of β-carotene contents in the range of 6.8–28.4%, depending on voltage gradient applied ($p < 0.05$). This increase has been explained by a possible enhancing effect of ohmic heating on β-carotene biosynthesis.

11.5.4 Phenolic Compounds and Flavor Characteristics

Thermal treatment can negatively affect the flavor of the fluid foods. Irreversible damage to fruit juice flavor results from chemical reactions accelerated during the heating process (Leizerson and Shimoni, 2005a). Flavor compounds are a major contributor to the unique aroma of citrus juices. These characteristic compounds do not exhibit high stability in the presence of microorganisms and exposure to high temperatures (Leizerson and Shimoni, 2005b).

There are interesting observations reported on the retention of flavor compounds during ohmic treatments. Leizerson and Shimoni (2005a) observed the relative concentrations (limonene, pinene, myrcene, octanal, and decanal) during 90 and 120°C ohmic-heating treatments as above 100%. After ohmic-heating treatment at 150°C, flavor compound concentrations were <100%, but still higher than conventionally pasteurized ones. These results were explained by two corresponding phenomena, that thermal treatment may cause a release of bonded components from the medium and/or that the released flavor compounds are not as quickly degraded as in conventional pasteurization due to the short residence time of ohmic heating. However, the retention of flavor compounds (limonene, myrcene, octanal, and decanal) was significantly higher in ohmic-heated orange juice than in conventionally pasteurized juice during storage at 4°C (Leizerson and Shimoni, 2005b).

Total phenolic content of fruit juice and vegetable purees can be affected during ohmic heating. Phenolic compounds have medicinal and industrial value since they have antioxidant and anti-inflammatory activity. However, the increase in some phenolic compounds of some fruit juices negatively affects the taste. Yildiz *et al.* (2009) heated fresh pomegranate juice, which is well known as a fruit juice having high phenolic content, from 20°C to 90°C and holding at 90°C for different treatment times (0, 3, 6, 9, 12 min) by matching the thermal history of ohmic treatment (10–40 V/cm) with the conventional method. They determined that total phenolic content (TPC) values changed at the initial heating up period and there were no significant changes during the holding period. Interestingly, similar to flavor results of Leizerson and Shimoni (2005a), the relative total phenolic compounds contents (compared to fresh juice) were greater than 1.0, i.e. heating processes either ohmic or conventional caused an increase in the total amount of phenolics. Ohmic heating did not cause any different decrease in TPC when compared with conventional heating.

Similarly, Icier (2010) obtained total phenolic content retentions in artichoke by-product puree as in the similar range during ohmic blanching at 25 V/cm (62.94 ± 1.21%) when compared with water blanching (64.97 ± 0.92%), at the same blanching temperature. However, they reported that ohmic blanching at 40 V/cm caused higher total phenolic content retention (89.14 ± 1.14%), in addition to similar vitamin C loss and faster enzyme inactivation at the lower temperature, than water blanching.

11.5.5 Color Properties

Browning in juice and purees during manufacture and storage is of vital interest to the industry (Garza *et al.*, 1999). Changes in food color can be associated with its previous heat-treatment history. Various reactions such as pigment destruction and non-enzymatic browning reactions can occur during heating of fruits and vegetables and, therefore, affect their color. Color can be used as a quality indicator in evaluating the extent of deterioration due to thermal processing (Avila and Silva, 1999).

Nonenzymatic browning may result in the formation of off-flavor, a decrease in nutrient content, a loss of color, and, above all, the appearance of brown pigments (Leizerson and Shimoni, 2005a). The change in the visual color properties or the browning index can be taken as critical tools to assess the overall color changes in the samples during any treatment applied.

Ascorbic acid degradation is considered to be a major chemical reaction responsible for browning in citrus juices. The increase in absorbance (at 420 nm) indicates initiation of browning reactions due to exposure to high temperatures during thermal treatments. Leizerson and Shimoni (2005a) determined similar browning-index values for both ohmic and conventional pasteurization of orange juice at an equivalent F-value of thermal treatments. The decrease in lightness of ohmic-heated juices was correlated with the increase in their browning levels. However, in a study by Leizerson and Shimoni (2005b), the browning measurements showed significant increases in the browning values for ohmic-heated orange juice compared to those for conventionally pasteurized juice during storage. In ohmic-heated orange juice, absorbance values were stable until day 35. From that day on, there was a continuous increase in browning index up to a level of 0.367, which was invisible to the human eye. During the initial lag period, colorless compounds were probably formed, which did not contribute to the increase in absorbance.

Ohmic treatment can be used successfully as an alternative blanching method for vegetable purees, resulting in a high retention of color attributes. The browning of leafy vegetable puree is measured with increasing hue angle (Equation 11.18) and lower tristimulus-combination values ("Lb/a," "Lab") (Icier *et al.*, 2006).

$$\text{Hue angle} = \tan^{-1}\frac{b}{a} \tag{11.18}$$

Pea puree samples heated up to 100°C and held at this temperature by ohmic blanching at voltage gradients of 20–50 V/cm were compared to water-blanched samples (Icier *et al.*, 2006). Since the peroxidase enzyme in the pea puree was inactivated at lower critical inactivation times for ohmic blanching, the water-blanched pea puree had higher browning than that blanched ohmically at voltage gradients above 30 V/cm (Table 11.3). The effect of voltage gradients on the color of pea puree has been found to be significant ($p < 0.01$). When the treatment times were close to each other for water and ohmic blanching, the browning in the water- and ohmic-blanched samples were similar. For the same holding times, ohmic heating was found to cause browning of pre-blanched spinach puree more than conventional water heating for the same temperature range (60–90°C) (Yildiz *et al.*, 2010) (Table 11.3).

Dimensionless "Hue angle ratio" can be used to explain the change in color of fruit juices during heating treatment. This value is the hue angle value with respect to that of raw juice, thus it is equal to unity in raw juices. Lower hue angle ratio values correspond to a lower amount of browning. In this respect, the hue angle ratio values of ohmically heated pomegranate juice have been found to be lower than conventionally heated samples (Yildiz *et al.*, 2009).

Vikram *et al.* (2005) evaluated the degradation of visual color by combination values (a × b), which followed first-order kinetics (Table 11.3). The activation energy and thermal resistance z-values for degradation of color during ohmic heating of orange juice were 79.92 ± 0.68 kJ/mK and 12.32°C, respectively. They discussed the fact that microwave heating caused lower color degradation than ohmic heating.

It is not always possible to apply kinetics as simple as first order or zero order to describe the color changes in fruit and vegetable purees since these changes may not be only due to the Maillard reaction, but also due to the thermal destruction of pigments present in the samples (Ibarz *et al.*, 1999). Hence, a combined kinetics has been developed: a first stage of colored polymeric compound formation following a zero-order kinetics, the second stage supposes decomposition of the colored polymers into non-colored compounds following a first-order kinetics (Garza *et al.*, 1999):

$$\text{sugar} + \text{aminoacids} \xrightarrow{k_0} \text{colored polymers} \tag{11.19}$$

$$\text{pigments} \xrightarrow{k_1} \text{non-colored pigments} \tag{11.20}$$

The modified two stage combining process can be written as (Icier *et al.*, 2006):

$$C = \frac{k_0}{k_1} - \left(\frac{k_0}{k_1} - C_0\right)e^{-k_1 t} \tag{11.21}$$

The kinetic constants of this relation for different voltage gradients during ohmic blanching of pea puree are given in Table 11.3. The negative sign of the reaction constant is the indication of a decreasing trend. Hue angle was the most appropriate combination, which closely described the first-order reaction kinetics of total color changes of pea puree for ohmic blanching at 20 V/cm

Table 11.3 Color Changes During Ohmic Heating Compared with Other Heating Methods

Food	Heating Method	Temperature (°C)	Voltage Gradient (V/cm)	Color Parameter	Value	k (min^{-1}), Equation 11.14	k_O and k_1 (min^{-1}), Equation 11.21	Reference
Orange juice	Ohmic	50	42	$a \times b$		0.0183		Vikram *et al.* (2005)
		60				0.1103		
		75				0.2885		
		90				0.5697		
	Infrared, 250 W	50		$a \times b$		0.0706		
		60				0.1294		
		75				0.3256		
		90				0.6108		
Pea puree	Ohmic	100	20	Hue angle	68.07 ± 0.20	0.0088	1.2692, 0.0088	Icier *et al.* (2006)
			30		65.13 ± 0.39			
			40		63.87 ± 0.53			
			50		63.89 ± 0.16			
	Conventional	100		Hue angle	65.85 ± 0.41			
Pomegranate juice	Ohmic	90	10–40	Hue angle ratio	0.25–1.15			Yildiz *et al.* (2009)
	Conventional	90		Hue angle ratio	1.20–2.00			
Spinach puree	Ohmic	60	30	Hue angle	47.2 ± 0.9			Yildiz *et al.* (2010)
		70			46.2 ± 0.7			
		80			47.1 ± 0.7			
		90			46.7 ± 0.4			
	Conventional	60		Hue angle	46.9 ± 0.3			
		70			47.7 ± 0.7			
		80			47.2 ± 0.1			
		90			49.6 ± 0.3			

(Icier *et al.*, 2006). As the zero-order reaction rate constant of Equation 11.19 was higher than the first-order reaction constant Equation 11.20, it was concluded that color formation could be higher than color destruction during ohmic blanching (Table 11.3)

The change in chlorophyll content of spinach puree during ohmic heating has been investigated by Yildiz *et al.* (2010). They reported that the effect of voltage gradient applied was not found statistically significant on chlorophyll content (total, *a* and *b*) ($p < 0.05$). However, temperature (60, 70, 80, or 90°C) at constant voltage gradient (30 V/cm) affected color values of spinach puree ($p < 0.05$), whereas holding time (0 and 600 s) at constant temperature (70°C) increased chlorophyll-*a* content in the range of 12–40% ($p < 0.05$). Ohmically heated puree samples at 90°C for 10 min contained the highest chlorophyll *a*, *b* and total chlorophyll. This increase could be caused by a possible enhancing effect of ohmic heating on formation of chlorophyll derivatives. Chlorophyll is more susceptible to degradation by wet heating methods than by electrical methods (Teng and Chen, 1999). Pyropheophytins could be formed from pheophytins through decarboxylation or from pyrochlorophylls through elimination of magnesium ions during steaming or microwave cooking. Although there is not enough information about this formation during ohmic heating, possible formations could be observed as slight increases in chlorophyll contents (Yildiz *et al.*, 2010). Further research is necessary to explore the electrical effects on the formation of secondary substances and biosynthesis during ohmic heating.

11.5.6 Enzyme Inactivation

Several enzymes are used in the food industry for improving food quality, for the recovery of by-products and for achieving higher juice yields. However, some have negative effects on food quality. Thus, the control of enzymatic activity is required in many food-processing steps to promote/inhibit enzymatic activity during processing (Castro *et al.*, 2004b).

Most enzymes are used as time–temperature indicators in food technology (Tucker *et al.*, 2002). The design aspects of thermal treatments of fluid foods take into account the inactivation characteristics of several enzymes depending on the type of fluid food.

Castro *et al.* (2004b) applied equal thermal histories of samples (conventional and ohmically processed) to determine if there was an additional inactivation caused by the presence of an electric field, thus eliminating temperature as a variable. For this purpose, they matched the thermal histories of both methods by arranging the voltage gradient in the range of 50–90 V/cm during heating periods and lower than 20 V/cm during holding periods. They studied several enzymes: lipoxigenase in Tris HCl buffer, pectinase in citrate buffer, polyphenoloxidase in phosphate buffer, alkaline phosphotase in milk and β-galactosidase in fermentation broth. All the enzymes followed first-order inactivation kinetics for both conventional and ohmic heating treatments. The presence of an electric field does not

Table 11.4 *z*-Values Determined Under Ohmic and Conventional Heating Conditions for Different Enzymes[1]

Enzyme	Prosthetic Group	$z_{conventional}$	z_{ohmic}
Lipoxygenase	Fe	11.10 ± 0.34	15.04 ± 0.38
Pectinase	–	8.12 ± 0.21	7.72 ± 0.34
Polyphenoloxidase	Cu	12.20 ± 0.08	12.80 ± 0.04
Alkaline phosphatase	Zn, Mg	9.05 ± 0.13	9.30 ± 0.05
β-galactosidase	–	5.12 ± 0.08	5.08 ± 0.17

[1]*Reproduced with permission from Castro* et al. *(2004b).* J. Food Sci. *69(9), C696–C701.*

cause an enhanced inactivation of alkaline phosphatase, pectinase, and β- galactosidase (Table 11.4). However, lipoxygenase (LOX) and polyphenoloxidase(PPO) kinetics are significantly affected by the electric field, reducing the time needed for inactivation. The presence of an electric field can influence biochemical reactions by changing molecular spacing and increasing interchain reactions; such as removing the metallic prosthetic groups present in the enzymes LOX and PPO, thus causing the enhancement of activity loss (Castro *et al.*, 2004b).

Since both conventional and ohmic inactivation mechanisms of alkaline phosphatase were reported as similar, even if milk is ohmically pasteurized, this enzyme may continue to be used as a time–temperature indicator (TTI) by the dairy industry (Castro *et al.*, 2004b).

The inactivation of PPO by ohmic treatment has been investigated also by Icier *et al.* (2008). Fresh grape juice was ohmically heated at different voltage gradients (20, 30, and 40 V/cm) from 20°C to temperatures of 60, 70, 80, or 90°C. The critical deactivation temperatures were found to be 60°C for 40 V/cm, and 70°C for 20 and 30 V/cm probably because of the faster increase in electrical conductivity at higher voltage gradients causing higher deactivation in PPO (Fig. 11.16). At a constant voltage gradient of 30 V/cm, a small increase in the activity with holding time was observed at 60°C until the deactivation started after 15 min. The increase in the enzyme activity at constant temperature was explained by the change of conformation of the enzyme to give higher enzyme–substrate interaction and consequently to an optimal consumption of the substrate and/or the possible influence of the biochemical reactions by changing molecular spacing and increasing interchain reactions in the presence of an electric field. For the same holding time at higher temperatures, it was reported that as the ohmic holding temperature increased, the deactivation of the PPO enzyme increased (Fig. 11.17). The simplest kinetic model involving one-step first-order deactivation of PPO (at 30 V/cm) was better than more complex models (Table 11.5). The activation energy of the PPO deactivation for the temperature range of 70–90°C was found to be 83.5 kJ/mol.

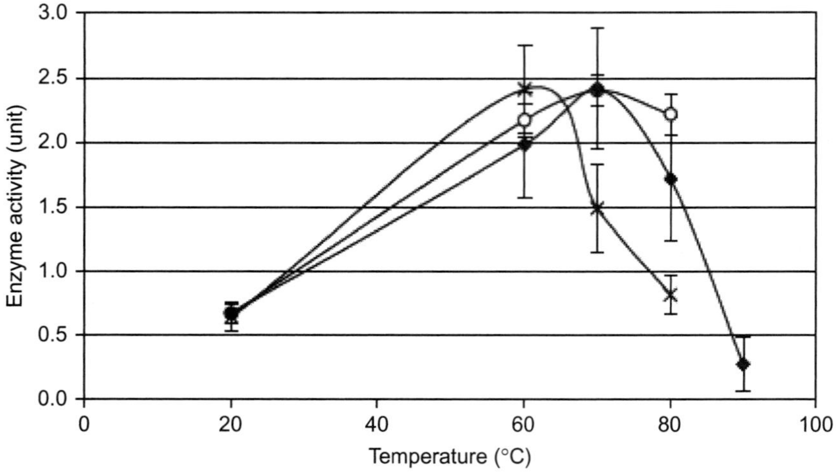

FIGURE 11.16

The changes in PPO activity of grape juices heated to different temperatures at different voltage gradients. (-o-) 20 V/cm; (-●-) 30 V/cm; and (-×-) 40 V/cm.

[Reproduced with permission from Icier et al. *(2008).* J. Food Eng. *85, 410–417.]*

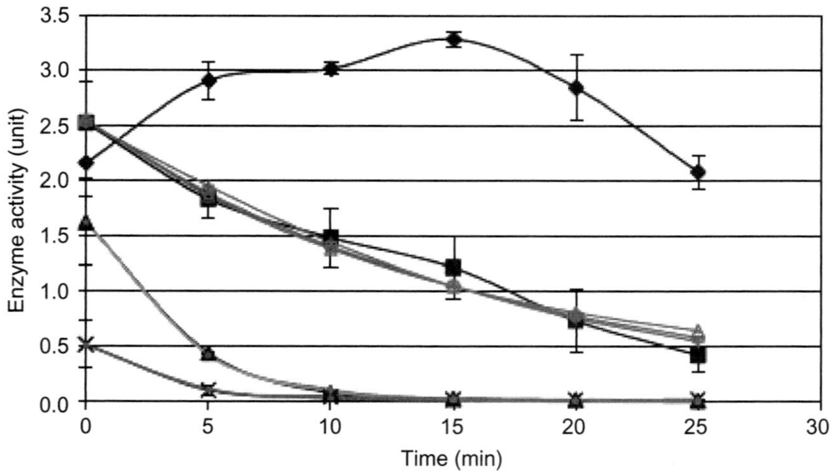

FIGURE 11.17

The inactivation curves of PPO at different ohmic holding temperatures at 30 V/cm (-♦-) 60°C; (-■-) 70°C; (-▲-) 80°C; and (-×-) 90°C. (—) model 1, (-o-) model 2, (-Δ-) model 3 and (÷) model 4.

[Reproduced with permission from Icier et al. *(2008).* J. Food Eng. *85, 410–417.]*

Table 11.5 The Temperature Dependency of Rate Constant k_1 Obtained by Different Kinetic Models[1,2]

Model	E_a, kJ/mol (std. error)	$k_{0,ref}$ (standard error)	R^2	Adj.R^2	S.E.E.	F	R.M.S.
$\frac{C}{C_0} = e^{-k_1 t}$	83.45 (41.48)	0.169 (0.089)	0.802	0.604	0.567	4.047	0.321
$C = C_{01}e^{-k_1 t} + C_{02}e^{-k_2 t}$	91.10 (36.71)	0.178 (0.076)	0,860	0.721	0.502	6.157	0.252
$\frac{C}{C_0} = (1-\beta)e^{-k_1 t} + \beta$	80.98 (31.84)	0.190 (0.080)	0.866	0.732	0.435	6.470	0.189
$\frac{C}{C_0} = \left(1 + \beta_1 \frac{k_1}{k_2 - k_1} - \beta_2 \frac{k_2}{k_2 - k_1}\right) e^{-k_1 t} - \frac{k_1}{k_2 - k_1}(\beta_1 - \beta_2)e^{-k_2 t} + \beta_2$	44.83 (2.78)	0.332 (0.022)	0.996	0.992	0.038	260.546	0.001

Confidence level for all parameters: 95%.
[1] *With permission from Icier* et al. *(2008).* J Food Eng. *85, 410–417.*
[2] *For polyphenoloxidase enzyme during ohmic heating of grape juice.*

The design for thermal pasteurization of orange juice is based on pectin methyl esterase, which is more thermally stable than many vegetative microorganisms. As the impact of thermal treatments during ohmic heating increases, the residual pectin esterase (PE) activity decreases. During ohmic heating, PE activity showed a reduction of 90–98% compared to its activity in fresh orange juice while there was no significant difference in inactivation effects of conventional and ohmic pasteurization (Leizerson and Shimoni, 2005a). No regeneration or retardation of PE activity was found for the orange juices (Leizerson and Shimoni, 2005b).

Since a less aggressive thermal treatment is needed, the thermal destruction of nutrients (for example, vitamins and pigments) and texture (a quality parameter) will obviously be lower, thus increasing the final quality of the product (Castro *et al.*, 2004b).

Particle size analysis can be performed to evaluate the effect of the residual enzyme activity on the stability of the juice cloud. Particle size can increase during storage depending on residual enzyme activity. The upper limit is 2 μm in the normal distribution of particle size in citrus juices, and above this value there is an indication of separation (Corredig *et al.*, 2001). High levels of surface area mean diameters and volume-weighted mean diameters indicate a higher potential for interaction between particles in pasteurized orange juice, thus leading more easily to aggregation. Leizerson and Shimoni (2005b) obtained higher mean diameters for particles in conventionally pasteurized orange juice than those in ohmically heated ones.

Peroxidases are known to be the most heat-stable enzymes in vegetables, and their inactivation is usually used to indicate the adequacy of blanching (Akyol *et al.*, 2004).

Peroxidase enzyme has been used as the indicator for testing the efficiency of ohmic and water blanching of pea puree. Icier *et al.* (2006) reported that the critical inactivation time decreased as the voltage gradient increased. Further, critical inactivation time for water blanching was longer than that for ohmic blanching (Table 11.6). The temperature distribution during ohmic blanching of pea puree has been obtained as more uniform, and the electrical current passing through the samples was higher at high voltage gradients, in that, the uniform and faster peroxidase inactivation has been achieved ohmically compared with water blanching.

Furthermore, it could be used as an alternative low-temperature blanching method, due to faster enzymatic inactivation, higher vitamin C and total phenolic content retention compared with water blanching and without causing any blanching wastewater (Icier, 2010). Similar inactivation periods have been observed for blanching in boiling water (100°C) and ohmic blanching (40 V/cm) at 85°C by Icier (2010). It was reported that total peroxidase inactivation time for artichoke by-product decreased to 310 s from 470 s when the voltage gradient was increased from 25 V/cm to 40 V/cm.

Table 11.6 The Critical Inactivation Times of Peroxidase in Pea Purees Blanched by Ohmically and Conventionally[1]

Blanching Method	Voltage Gradient (V/cm)	Heating Time (s) (30°C–100°C)	Holding Time at 100°C (s)	Critical Inactivation Time (s)
Ohmic blanching	20	516	60	576
	30	141	60	201
	40	74	30	104
	50	54	0	54
Water blanching	–	–	300	300

[1]*Reproduced with permission from Icier* et al. *(2006).* J. Food Eng. *74, 424–429.*

11.5.7 Rheological Properties

The determination of changes in the rheological properties of liquid foods during ohmic heating is important in the design of continuous ohmic heating units.

An and King (2006) reported that ohmic heating made commercial rice starch swell faster and resulted in different pasting characteristics for rice starch and rice flours than conventional heating. They observed that white rice flour heated ohmically at 70 V/cm had better cooking stability, less retrogradation potential, and lower pasting temperature than the conventionally heated sample. They discussed that ohmically heated starch samples at high voltage might not have had enough time for starch granules to be disrupted or for amylose to leach out; therefore, more starch granules remained intact for swelling and less amylose was available for re-aggregation upon cooling.

Ohmic heating results in more pregelatinization than conventional heating. The effect of ohmic heating varies depending on the type of starches with different constituents of fats, proteins, and amylose contents, while conventional heating simply showed thermal resistance and narrower gelatinization temperature ranges regardless of starch source (An and King, 2007). An and King (2007) explained the greater effect on gelatinization at a low voltage by the slower heating rate, resulting in a slower rate of swelling and solubilization. Longer cooking causes greater starch gelatinization and resistance to heat, and hence less energy was required for the enthalpy transition. An and King (2007) reported that at higher electric field strength, ohmically heated starch samples showed higher enthalpy (2.30 kJ/kg). Conventionally heated starches averaged overall for each time period showed different results from ohmically heated samples, in which 76 s cooked

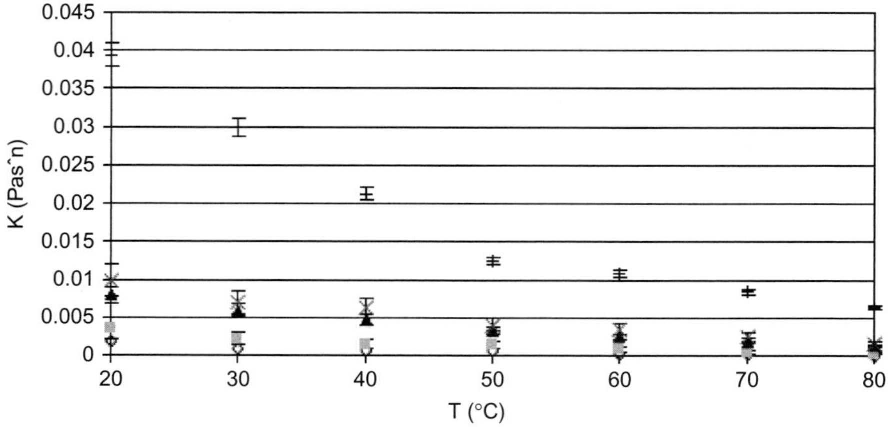

FIGURE 11.18

Change of consistency coefficient of whey solutions during ohmic heating at 40 V/cm. (Δ) 8%, (■) 12%, (▲) 16%, (×) 20% and (+) 24%.

[Reproduced with permission from Icier (2009). Food Bioprod. Process. *87, 308–316.]*

starch samples presented the highest onset gelatinization temperature and the lowest enthalpy (0.305 kJ/kg). Ohmic heating at low voltage can produce a starch that requires less energy to cook and that is stabilized. Ohmic treatment may cause hydrolysis of starch resulting from electroporation, in which disruption occurs in a food product followed by formation of permeable pores (An and King, 2007).

Icier (2009) investigated the rheological changes of reconstituted whey solutions during ohmic heating. It was observed that the consistency coefficient of whey solutions decreased as the temperature increased during both ohmic and water-bath heating methods. The solute concentration and its interaction with temperature were effective on the consistency coefficient statistically while the voltage gradient was not important ($p < 0.01$). Consistency coefficient decreased as the temperature increased (Fig. 11.18). Flow behavior index "n" increased with increasing temperature, and decreased with increasing concentration. Rheological characteristics of whey solutions during ohmic heating at the temperature range of 20–80°C was found to be time independent. For the solute concentration of 24%, solutions had yield stress values. To determine the possible electrical effect of ohmic heating on rheological properties, thermal history of conventional heating was matched to ohmic heating at 20 V/cm. Although flow behavior indexes were not different for both methods, the effect of heating method on consistency coefficient was statistically significant. Arrhenius-type relation constants of temperature dependency of the consistency coefficients for different solute concentrations are shown in Table 11.7. Activation energies for the temperature

Table 11.7 Temperature Dependency of Consistency Coefficient for Different Solute Concentrations of Reconstituted Whey Solutions[1]

Heating Method	Solute Concentration (%w/v)	k_0(Pa•s^n) × 10^3	E_a (kJmol^{-1})	R^2
Ohmic[2]	8	10.6 ± 1.2	45.57 ± 0.22	0.938
	12	14.0 ± 1.4	39.95 ± 0.16	0.974
	16	16.5 ± 0.9	26.87 ± 0.18	0.967
	20	27.6 ± 2.3	26.37 ± 0.03	0.931
	24	99.4 ± 8.7	27.46 ± 0.03	0.976
Conventional heating	8	10.6 ± 1.2	49.93 ± 0.11	0.967
	12	14.0 ± 1.4	44.85 ± 0.03	0.963
	16	16.5 ± 0.9	31.92 ± 0.08	0.976
	20	27.6 ± 2.3	26.96 ± 0.04	0.931
	24	99.4 ± 8.7	26.73 ± 0.03	0.967

[1]*Reproduced with permission from Icier (2009).* Food Bioprod. Process. *87, 308–316.*
[2]*At voltage gradient of 20 V/cm.*

dependency of the consistency coefficient were in the range of 26.34–45.79 kJ/mol for ohmic heating, and 26.70–50.04 kJ/mol for conventional heating. Activation energies for lower solute concentrations during ohmic heating were lower than conventional heating. This means reconstituted whey solutions are less sensitive to temperature changes during ohmic heating. It is explained by the possible electrical effects of ohmic heating causing decreasing effects in consistency and thermal effects increasing the consistency. It was discussed that electricity could have delayed some processes such as gelation and degradation processes due to fast heating, and/or enhanced the occurrence of reaction products resulting in rupture of some chemical bonds and causing the decrease in consistency. However, possible electrochemical reactions and the fouling which occurred on the electrode surfaces of the ohmic heater could have affected the decrease in consistency of the bulk fluid. This hypothesis should be proved by micro-scale investigations of structural changes of milk products as a result of electrical effects during ohmic heating.

Icier (2009) reported attractive results that both electrical conductivity and consistency coefficients of reconstituted whey solutions were dependent on concentration while independent of the voltage gradient. The correlation between the electrical conductivity and the consistency coefficient was examined, with correlation coefficients for all voltage gradients and solute concentrations in the range of −0.910 and −0.991 (Fig. 11.19).

Characterization of the flow characteristics of liquid whole egg are crucial for modeling, design and scale up of the continuous ohmic heating systems for

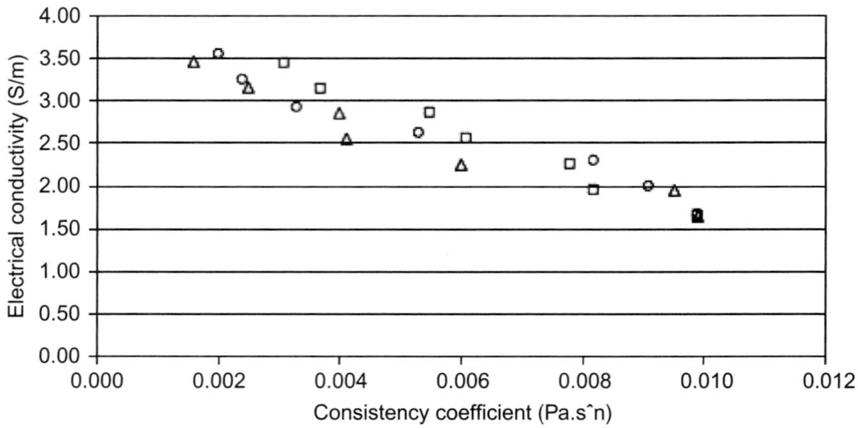

FIGURE 11.19

Correlation between consistency coefficient and electrical conductivity value of reconstituted whey solution having 20% solute concentration. (o) 20 V/cm, (□) 30 V/cm and (Δ) 40 V/cm.

[Reproduced with permission from Icier (2009). Food Bioprod. Process. *87, 308–316.]*

pasteurization of liquid egg products. The thermal pasteurization of liquid egg is problematic because of its great instability in the range of effective pasteurization temperatures (Geveke, 2008).

The change of rheological behavior of the ohmically heated liquid whole egg, across a temperature range of 4–60°C was determined by Icier and Bozkurt (2010). They observed that the liquid whole egg represented power law characteristics during ohmic heating since either regression coefficients of power law equation were comparatively higher or both χ^2 and RMSE (root mean square errors) values were comparatively lower than Newtonian type equation for all temperatures studied. The flow behavior index (n) of liquid whole egg was in the range of 0.86–0.90 depending on temperature. Liquid whole egg heated at 60°C ohmically exhibited a higher degree of thixotropic index (38.53 Pas −1) indicating the occurrence of protein denaturation. The approach of the ascending and descending curves in the thixotropic cycle indicated that less denaturation occurred depending on the temperature in the direction of shear negligible orientation (Fig. 11.20a). Thixotropic index decreased as the temperature decreased. It was 5.44 and 14.89 Pas −1 for 4 and 10°C, respectively. Hence, it can be concluded that less denaturation occurs below the pasteurization temperature. The change in time dependency with the increasing temperature showed the temperature sensitivity of the egg protein during thermal treatment (Fig. 11.20b). The effect of the temperature on the time-dependency value was similar to that on the thixotropy index value. Thus, it could be concluded that temperature has a significant role during ohmic treatment of liquid whole egg, which is a major factor for the occurrence

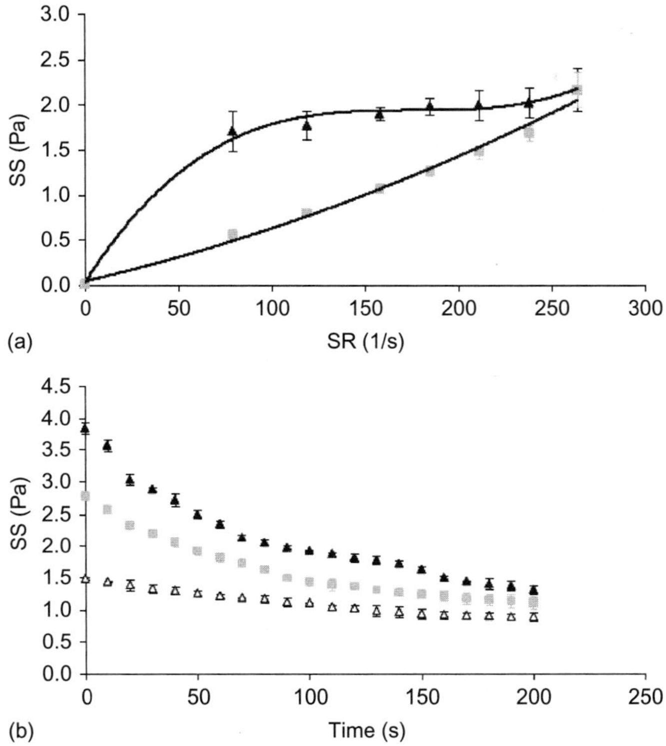

FIGURE 11.20

a, Thixotropic behaviour of liquid whole egg at 60°C during ohmic heating; ▲ up term, ■ down term. b, Time-dependency behaviour of liquid whole egg at 60°C; Δ 60 rpm, ■ 140 rpm, ▲200 rpm.

[Reproduced with permission from Icier and Bozkurt (2010). Food Bioprocess Technol. *DOI 10.1007/s11947–009–0229–4.]*

of the protein denaturation and the change in the rheological behavior. The flow behavior of liquid whole egg in the continuous ohmic heating system was predicted as laminar (Generalized Reynolds Number range of 87.59–538.87) for the mass flow rate range of 0.0056–0.0166 kg/s.

Bozkurt and Icier (2010) discussed that the activation energy values for temperature dependency of predicted apparent viscosity were significantly different for ohmic (6.767 ± 0.377 kJ/mol) and conventional heating (8.214 ± 0.370 kJ/mol) ($p < 0.05$). Since they had matched the thermal history of both methods, they concluded that the liquid whole egg was more sensitive to temperature changes during conventional heating.

Studies conducted on fruit juices showed that ohmic heating did not result in different effects on change in rheological behavior of fruit juice, probably due to lower concentrations of degradable constituents by thermal treatments. Ohmic heating was applied to quince nectar by matching the same heating curve of the conventional method by changing the voltage gradient (10–40 V/cm) at 50 Hz (Bozkurt and Icier, 2009). The change in rheological constants of quince nectar was determined in the temperature range of 65–75°C for different holding times (0, 10, 15, 20, and 30 min). Quince nectar showed time-independent, non-Newtonian, pseudoplastic behavior during heating in the range of 20–75°C, independent of heating method. The activation energy values were 9.88±3.24 kJ/mol and 10.08±2.53 kJ/mol for ohmic heating and conventional heating, respectively. Results showed that there was no electrical effect other than the thermal effect of ohmic heating since similar rheological constants were obtained with both methods. Similarly, Yildiz *et al.* (2009) heated pomegranate juice up to 90°C and held it at this temperature for different treatment times (0, 3, 6, 9, 12 min) ohmically (10–40 V/cm) and conventionally. They observed that ohmic heating did not cause different effects on the rheological behavior of pomegranate juice than those of conventional heating.

11.5.8 Sensory Properties

Sensorial tests for liquid food products are conducted to measure the changes in flavor, texture, appearance, and the overall acceptance of foods during processing. It is important to ensure that ohmic and other heating treatments are applied at the same thermal history to create a basis for comparison. However, there are limited studies conducted on sensory evaluation of ohmic treated samples.

Leizerson and Shimoni (2005a) applied sensory evaluation (triangle test for $n = 25$, $\alpha = 0.05$) for fresh, conventionally pasteurized, and ohmically pasteurized orange juices. Their results indicated that a panelist could distinguish between fresh and pasteurized samples, and between pasteurized and ohmic-heated orange juice ($p < 0.05$), however tasters could not differentiate between fresh and ohmic-heated orange juice ($p > 0.05$). They explained this situation by the fact that tasters were only able to determine by flavor since there was no visible change in color. Thus, due to the similar flavor profiles observed for fresh and ohmic-heated orange juice, panelists could not distinguish between them, as the two juices resembled each other.

The shelf-life of fluid foods treated thermally can be determined by the Weibull–Hazard sensory shelf-life method in which the shelf life of a sample is defined as the time when 50% of panelists would reject the sample (Labuza and Schmidl, 1998). Leizerson and Shimoni (2005b) used this method in the shelf-life determination of orange juice samples pasteurized ohmically or conventionally. In their study, sensorial quality was the limiting factor for the shelf-life of conventionally pasteurized juice, at 50 days. The ohmic-heated juice revealed superior sensorial quality, and thus its shelf-life was limited by the vitamin C concentration and was 79 days.

11.6 CURRENT STATUS AND FUTURE TRENDS

Ohmic heating is applied to liquid foods such as fruit and vegetable products (juices, purees, pulps, etc.), milk, ice-cream mix, egg, whey, soups, stews, and fruits in syrup, heat sensitive liquids, soymilk, etc., for the purpose of heating, pasteurization and sterilization. Several other applications of ohmic treatment in the food industry include blanching, thawing, cooking, fermentation, peeling of fruits, and processing of protein-rich liquid foods for formation of protein films and gels. It is also applied as a pre-treatment method for dehydration and extraction (Icier, 2003; Sastry, 2008).

Ohmic heating is an excellent alternative food-processing technique that shows much promise in the food-manufacturing industry. Since the early 1990s, the technology involving ohmic heating has increased dramatically. As time goes on, this technology will become even more streamlined and efficient. In the future, it is possible that even "green" energy, such as water, solar, or wind power could create the electricity needed for ohmic heating systems. (Anderson, 2008)

There are limited commercial ohmic heating applications throughout the world. Such systems have the capacities of 3–6 tons per hour for processing of: Japanese breadcrumbs (UK); tofu (Japan); low-acid meats and vegetables in bags (UK); whole fruits (Japan); liquid egg (USA); baby food, artichokes, carrots, mushrooms, ketchup, fruit nectars, fruit juice, peppers, cauliflower, tomato paste, sausages, pâté, and fruit purée (Italy); fruit slices and large fruit pieces in syrup (Europe, Japan, Mexico, South America, etc.) (Icier, 2003; Anderson, 2008; Sastry, 2009).

Special design of ohmic heaters can be created for different foods, depending on their electrical conductivity range. In addition, a variety of novel products can be developed by controlling their electrical conductivity changes. The potential application areas of ohmic heating could be expanded for a wider variety of foods in future. Ohmic heating would be of benefit to fermented products like beer or wine by decreasing the processing time since it may decrease the lag period of fermentative bacteria. There is still a lack of knowledge on the efficacy of ohmic heating for pathogenic bacteria inactivation.

Research is needed on methods for identification, measurement, and testing for cold-spots and overheated regions during ohmic heating of multiphase foods. Studies are needed for modeling, measurement, and verification of the heating of solid–liquid mixtures (Sastry, 2008)

Commercialization of ohmic heating technology depends, in part, on the development of adequate safety and quality-assurance protocols in order to obtain an approved filing of the process with the FDA for all possible food materials (Ye *et al.*, 2003).

The effects of electrical application to food at the molecular, cellular, and tissue levels need to be studied. The possibility of the occurrence of mutagenic compounds within the food during ohmic heating should also be emphasized.

References

Akyol, C., Bayindirli, A., Alpas, H., 2004. Effect of combined treatment of high hydrostatic pressure and mild heat on peroxidase inactivation in green beans, peas and carrots. Presented at IFT Annual Meeting, July 12–16, Las Vegas, NV.

Allali, H., Marchal, L., Vorobiev, E., 2008. Blanching of strawberries by ohmic heating: effects on the kinetics of mass transfer during osmotic dehydration. Food Bioprocess Technol. (DOI: 10.1007/s11947-008-0115-5, in press)

Amatore, C., Berthou, M., Hebert, S., 1998. Fundamental principles of electrochemical ohmic heating of solutions. J. Electroanal. Chem. 457, 191–203.

An, H.J., King, J.M., 2006. Pasting properties of ohmically heated rice starch and rice flours. J. Food Sci. 71 (7), C437–C441.

An, H.J., King, J.M., 2007. Thermal characteristics of ohmically heated rice starch and rice flours. J. Food Sci. 72 (1), C84–C88.

Anderson, A.K., Finkelsten, R., 1919. A study of the electro-pure process of treating milk. J. Dairy Sci. 2, 374–406.

Anderson, D.R., 2008. Ohmic Heating as an Alternative Food Processing Technology. (Msc. Thesis). Kansas State University, Food Science Institute College of Agriculture, Manhattan, 45 p.

Anonymous, June 2, 2000. Kinetics of Microbial Inactivation for Alternative Food Processing Technologies. U. S. Food and Drug Administration, Center for Food Safety and Applied Nutrition. <http://www.cfsan.fda.gov/~comm/ift-pref.html#review> (01.04.2009).

Assiry, A.M., Sastry, S.K., Samaranayake, C., 2003. Degradation kinetics of ascorbic acid during ohmic heating with stainless steel electrodes. J. Appl. Electrochem. 33, 187–196.

Assiry, A.M., Sastry, S.K., Samaranayake, C., 2006. Influence of temperature, electrical conductivity, power and pH on ascorbic acid degradation kinetics during heating using stainless steel electrodes. Bioelectrochem. 68 (2006), 7–13.

Avila, I.M.L.B., Silva, C.L.M., 1999. Modeling kinetics of thermal degradation of color in peach puree. J. Food Eng. 39, 161–166.

Ayadi, M.A., Bouvier, L., Chopard, F., Berthou, M., Leuliet, J.C., 2003. Heat treatment improvement of dairy products via ohmic heating process: thermal and hydrodynamic effect on fouling. In: Watkinson, P., Muller-Steinhagen, H., Reza Malayeri, M. (Eds.), ECI Conference on Heat Exchanger Fouling and Cleaning—Fundamentals and Applications, vol. RP1. Article 19, 136–144, New Mexico, Santa Fe, USA.

Ayadi, M.A., Leuliet, J.C., Chopard, F., Berthou, M., Lebouche, M., 2004a. Continuous ohmic heating unit under whey protein fouling. Innov. Food Sci. Emerging. Technol. 5, 465–473.

Ayadi, M.A., Leuliet, J.C., Chopard, F., Berthou, M., Lebouche, M., 2004b. Electrical conductivity of whey protein deposit: xanthan gum effect on temperature dependency. Food Bioprod. Process. 82 (4), 320–325.

Ayadi, M.A., Leuliet, J.C., Chopard, F., Berthou, M., Lebouche, M., 2005. Experimental study of hydrodynamics in a flat ohmic cell—impact on fouling by dairy products. J. Food Eng. 70, 489–498.

Bansal, B., Chen, X.D., 2006a. Effect of temperature and power frequency on milk fouling in an ohmic heater. Food Bioprod. Process. 84 (4), 286–291.

Bansal, B., Chen, X.D., 2006b. A critical review of milk fouling in heat exchangers. Crit. Rev. Food Sci. Food Safety 5, 27–33.

Baysal, A.H., Icier, F., 2007. Effects of ohmic heating on *Alicyclobacillus acidoterrestris* in some fruit juices and concentrates. In: Turkish Scientific and Technology Committee Project Report, TOVAG 106 O 240, Ankara, Turkey, *unpublished*.

Baysal, A.H., Icier, F., 2010. Inactivation Kinetics of Alicyclobacillus acidoterrestris Spores in Orange Juice by Ohmic Heating: the effects of voltage gradient and temperature on the inactivation. J. Food Prot. 73 (2), 299–304.

Bhale, S.D., 2004. Effect of Ohmic Heating on Color, Rehydration and Textural Characteristics of Fresh Carrot Cubes. (MSc Thesis). Louisiana State University, Baton Rouge, LA, 56 p.

Biss, C.H., Coombes, S.A., Skudder, P.J., 1989. The development and application of ohmic heating for the continuous heating of particulate foodstuffs. In: Field, R.W., Howell, J.A. (Eds.), Process Engineering in the Food Industry. Developments and Opportunities. Elsevier, New York, pp. 17–25.

Bozkurt, H., Icier, F., 2009. Rheological characteristics of quince nectar during ohmic heating. Int. J. Food Prop. 12 (4), 844–859.

Bozkurt, H., Icier, F., 2010. The change of apparent viscosity of liquid whole egg during ohmic and conventional heating. J. Food Process Eng. (DOI:10.1111/j.1745–4530.2010.00575.x, in press)

Castro, I., Teixeira, J.A., Salengke, S., Sastry, S.K., Vicente, A.A., 2003. The influence of field strength, sugar and solid content on electrical conductivity of strawberry products. J. Food Process Eng. 26 (1), 17–30.

Castro, I., Teixeira, J.A., Salengke, S., Sastry, S.K., Vicente, A.A., 2004a. Ohmic heating of strawberry products: electrical conductivity measurements and ascorbic acid degradation kinetics. Innov. Food Sci. Emerging. Technol. 5, 27–36.

Castro, I., Macedo, B., Teixeira, J.A., Vicente, A.A., 2004b. The effect of electric field on important food-processing enzymes: comparison of inactivation kinetics under conventional and ohmic heating. J. Food Sci. 69 (9), C696–C701.

Chen, C., Abdelrahim, K., Beckerich, I., 2010. Sensitivity analysis of continuous ohmic heating process for multiphase foods. J. Food Eng. 98 (2), 257–265.

Cho, H.-Y., Yousef, A.E., Sastry, S.K., 1996. Growth kinetics of *Lactobacillus acidophilus* under ohmic heating. Biotechnol. Bioeng. 49, 334–340.

Cho, H.Y., Yousef, A.E., Sastry, S.K., 1999. Kinetics of inactivation of *Bacillus subtilis* spores by continuous or intermittent ohmic and conventional heating. Biotechnol. Bioeng. 62 (3), 368–372.

Cogné, C., Andrieu, J., Laurent, P., Besson, A., Nocquet, J., 2003. Experimental data and modelling of thermal properties of ice creams. J. Food Eng. 58, 331–341.

Corredig, M., Kerr, W., Wicker, L., 2001. Particle size distribution of orange juice cloud after addition of sensitized pectin. J. Agric. Food Chem. 49, 2523–2526.

Davies, L.J., Kemp, M.R., Fryer., P.J., 1999. The geometry of Shadows: effects of inhomogeneties in electrical field processing. J. Food Eng. 40, 245–258.

de Alwis, A.A.P., Fryer, P.J., 1990. The use of direct resistance heating in the food industry. J. Food Eng. 11 (1), 3–27.

de Alwis, A.A.P., Fryer, P.J., 1992. Operability of the ohmic heating process: electrical conductivity effects. J. Food Eng. 15, 21–48.

Ditchfield, C., Tadini, C.C., Singh, R., Toledo, R.T., 2004. Rheological properties of banana puree at high temperatures. Int. J. Food Prop. 7 (3), 571–584.

Eliot-Goderéaux, S.C., Fairhurst, P.G., Goullieux, A., Pain., J.P., 2001. Passage time distributions of cubes and spherical particles in an ohmic heating pilot plant. J. Food Eng. 47, 11–22.

Fellows, P.J., 2000. Food Processing Technology—Principles and Practice. second ed. Woodhead Publishing, Cambridge, 610 p.

Fryer, P.J., de Alwis, A.A.P., Koury, E., Stapley, A.G.F., Zhang., L., 1993. Ohmic processing of solid–liquid mixtures: heat generation and convection effects. J. Food Eng. 18, 101–125.

Garza, S., Ibarz, A., Pagan, J., Giner, J., 1999. Non-enzymatic browning in peach puree during heating. Food Res. Int. 32, 335–343.

Geveke, D.J., 2008. UV inactivation of E. coli in liquid egg white. Food Bioprocess. Technol. 1 (2), 201–206.

Goullieux, A., Pain, J.P., 2005. Ohmic heating. In: Sun, D.W. (Ed.), Emerging Technologies for Food Processing. Elsevier Academic Press, United Kingdom, pp. 469–506. (Chapter.18)

Guillou, S., Murr, N.E., 2002. Inactivation of *Saccharomyces cerevisiae* in solution by low-amperage electric treatment. J. Appl. Microbiol. 92, 860–865.

Herrick, J.P., Sastry, S.K., Clyde, G.F., Wedral, E.R., 2000. On-demand direct electrical resistance heating system and method thereof. United States Patent 6,130,990.

Huang, L., Chen, Y., Morrissey, M.T., 1997. Coagulation of fish proteins from fish mince wash water by ohmic heating. J. Food Process Eng. 20, 285–300.

Ibarz, A., Pagan, J., Garza, S., 1999. Kinetic models for color changes in pear puree during heating at relatively high temperatures. J. Food Eng. 39, 415–422.

Ibrahim, S., 1999. Corrosion control in electric power systems. Electr. Pow. Syst. Res. 52 (1), 9–17.

Icier, F., 2003. Experimental Investigation and Mathematical Modelling of Ohmic Heating of Foods. (PhD thesis). Ege University, Izmir, 245 p.

Icier, F., 2004. The use of ohmic heating as a preheating method in milk processing. In: CD Proceedings of International Dairy Symposium, Dairy 2004, May 24–28, Isparta: Turkey.

Icier, F., 2009. Influence of ohmic heating on rheological and electrical properties of reconstituted whey solutions. Food Bioprod. Process. 87, 308–316.

Icier, F., 2010. Ohmic blanching effects on drying of vegetable byproduct. J. Food Process Eng. 33, 661–683.

Icier, F., Bozkurt, H., 2010. Ohmic heating of liquid whole egg: rheological behaviour and fluid dynamics. Food Bioprocess Technol. (DOI 10.1007/s11947-009-0229-4, in press)

Icier, F., Ilicali, C., 2004. Electrical conductivity of apple and sourcherry juice concentrates during ohmic heating. J. Food Process Eng. 27 (3), 159–180.

Icier, F., Ilicali, C., 2005a. Temperature dependent electrical conductivities of fruit purees during ohmic heating. Food Research. Int. 38 (10), 1135–1142.

Icier, F., Ilicali, C., 2005b. The effects of concentration on electrical conductivity of orange juice concentrates during ohmic heating. Eur. Food Res. Technol. 220 (3–4), 406–414.

Icier, F., Ilicali, C., 2005c. The use of tylose as a food analog in ohmic heating studies. J. Food Eng. 69, 67–77.

Icier, F., Tavman, S., 2006. Ohmic heating behaviour and rheological properties of ice cream mixes. Int. J. Food Prop. 9 (4), 679–689.

Icier, F., Yildiz, H., Baysal, T., 2006. Peroxidase inactivation and color changes during ohmic blanching of pea puree. J. Food Eng. 74, 424–429.

Icier, F., Yildiz, H., Baysal, T., 2008. Polyphenoloxidase deactivation kinetics during ohmic heating of grape juice. J. Food Eng. 85, 410–417.

Jun, S., Sastry., S., 2005. Modeling and optimization of ohmic heating of foods inside a flexible package. J. Food Process Eng. 28, 417–436.

Jun, S., Sastry., S., 2007. Reusable pouch development for long term space missions: a 3D ohmic model for verification of sterilization efficacy. J. Food Eng. 80, 1199–1205.

Jun, S., Sastry, S.K., Samaranayake, C., 2007. Migration of electrode components during ohmic heating of foods in retort pouches. Innov. Food Sci. Emerging. Technol. 8, 237–243.

Kanjanapongkul, K., Tia, S., Wongsa-Ngasri, P., Yoovidhya, T., 2009. Coagulation of protein in surimi wastewater using a continuous ohmic heater. J Food Eng 91, 341–346.

Kim, H.J., Choi, Y.M., Yang, T.C.S., Taub, I.A., Tempest, P., Skudder, P., et al., 1996. Validation of ohmic heating for quality enhancement of food products. Food Technol. (May), 253–255, 257–261.

Kulshrestha, S.A., Sastry, S.K., 2006. Low-frequency dielectric changes in cellular food material from ohmic heating: effect of end point temperature. Innov. Food Sci. Emerg. Technol. 7, 257–262.

Labuza, T.P., Schmidl, M.K., 1998. Use of sensory data in the shelf life testing of foods: principles and graphical methods for evaluation. Cereal Food World 33 (2), 193–206.

Larkin, J.W., Spinak, S.H., 1996. Safety considerations for ohmically heated, aseptically processed, multiphase low-acid food products. Food Technol. 50, 242–245.

Lebovka, N.I., Praporscic, I., Ghnimi, S., Vorobiev, E., 2005. Does electroporation occur during the ohmic heating of food. J. Food Sci. 70 (5), E308–E311.

Lee, C.H., Yoon, S.W., 1999. Effect of ohmic heating on the structure and permeability of the cell membrane of *Saccharomyces cerevisae*. In: Proceedings of 1999 IFT Annual Meeting. July 24–28 1999, Chicago.

Legrand, A., Leuliet, J.C., Duquesne, S., Kesteloot, R., Winterton, P., Fillaudeau, L., 2007. Physical, mechanical, thermal and electrical properties of cooked red bean (*Phaseolus vulgaris* L.) for continuous ohmic heating process. J. Food Eng. 81, 447–458.

Lei, L., Zhi, H., Xiujin, Z., Takasuke, I., Zaigui, L., 2007. Effects of different heating methods on the production of protein–lipid film. J. Food Eng. 82, 292–297.

Leizerson, S., Shimoni, E., 2005a. Effect of ultrahigh-temperature continuous ohmic heating treatment on fresh orange juice. J. Agr. Food Chem. 53, 3519–3524.

Leizerson, S., Shimoni, E., 2005b. Stability and sensory shelf life of orange juice pasteurized by continuous ohmic heating. J. Agr. Food Chem. 53, 4012–4018.

Li, F.D., Li, L.T., Li, Z., Tatsumi, E., 2004. Determination of starch gelatinization temperature by ohmic heating. J. Food Eng. 62, 113–120.

Lima, M., Sastry, S.K., 1999. The effects of ohmic heating frequency on hot-air drying rate and juice yield. J. Food Eng. 41, 115–119.

Lima, M., Heskitt, B.F., Sastry, S.K., 1999. The effect of frequency and wave form on the electrical conductivity–temperature profiles of turnip tissue. J. Food Process Eng. 22, 41–54.

Lima, M., Heskett, B.F., Sastry, S.K., 2001. Diffusion of beet dye during electrical and conventional heating at steady-state temperature. J. Food Process Eng. 24, 331–340.

Liu, W.K., Brown, M.R.W., Elliott., T.S.J., 1997. Mechanisms of the bactericidal activity of low amperae electric current. J. Antimicrob. Chemoth. 39, 687–695.

Loghavi, L., Sastry, S.K., Yousef, A.E., 2007. Effect of moderate electric field on the metabolic activity and growth kinetics of *Lactobacillus acidophilus*. Biotechnol. Bioeng. 98, 872–880.

Loghavi, L., Sastry, S.K., Yousef, A.E., 2008. Effect of Moderate Electric Field frequency on growth kinetics and metabolic activity of Lactobacillus acidophilus. Biotechnol. Prog. 24 (1), 148–153.

Lojewska, Z., Farkas, D.L., Ehrenberg, B., Loew., L.M., 1989. Analysis of the effect of medium and membrane conductance on the amplitude and kinetics of membrane potentials induced by externally applied electric fields. Biophys. J. 56, 121–128.

Mainville, I., Montpetit, D., Durand, N., Farnworth, E.R., 2001. Deactivating the bacteria and yeast in kefir using heat treatment, irradiation and high pressure. Int. Dairy J. 11, 45–49.

Marcotte, M., Piette, J.P.G., Ramaswamy, H.S., 1998. Electrical conductivities of hydrocolloid solutions. J. Food Process Eng. 21, 503–520.

Marcotte, M., Trigui, M., Ramaswamy., H.S., 2000a. Effect of salt and citric acid on electrical conductivities and ohmic heating of viscous liquids. J. Food Process. Preserv. 24, 389–406.

Marcotte, M., Trigui, M., Tatibouet, J., Ramaswamy, H.S., 2000b. An ultrasonic method for assessing the residence time distribution of particulate foods during ohmic heating. J. Food Sci. 65 (7), 1180–1186.

Marra, F., Zell, M., Lyng, J.G., Morgan, D.J., Cronin, D.A., 2009. Analysis of heat transfer during ohmic processing of a solid food. J. Food Eng. 91, 56–63.

Mizrahi, S., 1996. Leaching of soluble solids during blanching of vegetables by ohmic heating. J. Food Eng. 29, 153–166.

Mulvihill, D.M., Kinsella, J.E., 1987. Gelation characteristics of whey proteins and β-lactoglobulin. Food Technol. 41, 102–111.

Palaniappan, S., Sastry, S.K., Richter, E.R., 1990. Effects of electricity on microorganisms: a review. J. Food Process. Preserv. 14, 393–414.

Palaniappan, S., Sastry., S.K., 1991. Electrical conductivity of selected juices: influences of temperature, solids content, applied voltage, and particle size. J. Food Process Eng. 14, 247–260.

Palaniappan, S., Sastry, S.K., Richter, E.R., 1992. Effects of electroconductive heat treatment and electrical pretreatment on thermal death kinetics of selected microorganisms. Biotechnol. Bioeng. 39 (2), 225–232.

Park, J.C., Lee, M.S., Lee, D.H., Park, B.J., Han, D.W., Uzawa, M., et al., 2003. Inactivation of bacteria in seawater by low-amperage electric current. Appl. Environ. Microb. (Apr), 2405–2408.

Parrott, D.L., 1992. Use of ohmic heating for aseptic processing of food particulates. Food Technol. (December), 68–72.

Pereira, C., Martins, J., Mateus, C., Teixeira, J.A., Vicente, A.A., 2007. Death kinetics of Escherichia coli in Goat Milk and Bacillus licheniformis in cloudberry jam treated by ohmic heating. Chem. Paper 61 (2), 121–126.

Prescott, S.C., 1927. The treatment of milk by an electrical method. Am. J. Public Health 17, 405–410.

Ramaswamy, R., Balasubramaniam, V.M., Sastry, S.K., 2005. Ohmic Heating of Foods: Fact Sheet for Food Processors. FSE-4-05. Ohio State University, Colombus, <http://ohioline.osu.edu/fse-fact/0004.html> (01.04.2009).

Reznick, D., 1996. Ohmic heating of fluid foods. Food Technol. (May), 250–251.

Roberts, J.S., Balaban, M.O., Zimmerman, R., Luzuriaga, D., 1998. Design and testing of a prototype ohmic thawing unit. Comput. Electro. Agr. 19, 211–222.

Rowan, N., Scott, J., Macgregor, J., Anderson, J.G., Cameron, D., Farish, O., 2000. Pulsed electric field inactivation of diarrhoeagenic *Bacillus cereus* through irreversible electroporation. Lett. Appl. Microbiol. 31, 110–114.

Ruan, R., Chen, P., Chang, K., Kim, H.J., Taub, I.A., 1999. Rapid food particle temperature mapping during ohmic heating using FLASH MRI. J. Food Sci. 64 (6), 1024–1026.

Salengke, S., Sastry, S.K., 2007a. Experimental investigation of ohmic heating of solid–liquid mixtures under worst-case heating scenarios. J. Food Eng. 83, 324–336.

Salengke, S., Sastry, S.K., 2007b. Models for ohmic heating of solid–liquid mixtures under worst-case heating scenarios. J. Food Eng. 83, 337–355.

Samaranayake, P., Sastry, S.K., 2005a. Electrode and pH effects on electrochemical reactions during ohmic heating. J. Electroanal. Chem. 577, 125–135.

Samaranayake, P., Sastry, S.K., 2005b. Pulsed ohmic heating—A novel technique for minimization of electro-chemical reactions during processing. J. Food Sci. 70 (8), E460–E465.

Sastry, S.K., 1992. A model for heating of liquid–particle mixtures in a continuous flow ohmic heater. J. Food Process Eng. 15, 263–278.

Sastry, S.K., 2008. Ohmic heating and moderate electric field processing. Food Sci. Technol. Int. 14, 419–422.

Sastry, S.K., 2009. Ohmic heating and moderate electric field technologies: present and future. In: Vorobiev, E., Lebovka, N., Hecke, E.V., Lanoiselle, J.L. (Eds.), Proceedings of the International Conference on Bio and Food Electrotechnologies, 22–23 October 2009, BFE2009. CD proceedings, Compiegne, France, pp. 234–238.

Sastry, S.K., Li, Q., 1996. Modelling the ohmic heating of foods. Food Technol. 50, 246–248.

Sastry, S.K., Palaniappan, S., 1992. Mathematical modeling and experimental studies on ohmic heating of liquid–particle mixtures in a static heater. J. Food Process Eng. 15, 241–261.

Sastry, S.K., Salengke, S., 1998. Ohmic heating of solid–liquid mixtures: a comparison of mathematical models under worst-case heating conditions. J. Food Process Eng. 21, 441–458.

Sitzmann, W., 1995. High-voltage pulse techniques for food preservation. In: Gould, G.W. (Ed.), New Methods of Food Preservation. Blackie Academic & Professional, London, pp. 236–252.

Somawat, R., Yousef, A., Chung, Y.K., Sastry, S.K., 2009. Effect of electricity on thermophilic bacterial spores: inactivation kinetics of *Geobacillus stearothermophilus* spores under the effect of pulsed ohmic heating. In: Vorobiev, E., Lebovka, N., Hecke, E.V., Lanoiselle, J.L. (Eds.), Proceedings of the International Conference on Bio and Food Electrotechnologies, 22–23 October 2009, BFE2009. CD proceedings, Compiegne, France, pp. 250–254.

Speller, M.N., Litchfield, J.B., Webb, A., 1999. Temperature mapping during Ohmic heating using proton resonance frequency magnetic resonance imaging. ASAE-CSAE-SCGR Annual International Meeting, 18–21 July, 1999, ASAE Paper No. 99-6036, 25 p, Toronto, Ontario, Canada.

Splittstoesser, D.F., 1996. Microbiology of fruit products. In: Somogyi, L.P., Ramaswamy, H.S., Hui, Y.H. (Eds.), Processing Fruits: Science and TechnologysBiology, Principles and Applications. Technomic Publishing, Lancaster, PA, pp. 261−273.

Sugar, I.P., Neumann, E., 1984. Stochastic model for electric field induced membrane pores-electroporation. Biophys. Chem. 19, 211−225.

Sun, H., Kawamura, S., Himoto, Ji, Itoh, K., Wada, T., Kimura, T., 2008. Effects of ohmic heating on microbial counts and denaturation of proteins in milk. Food Sci. Technol. Res. 14 (2), 117−123.

Stanel, J., Zitny, R., 2010. Milk fouling at direct ohmic heating. J Food Eng. (DOI:10.1016/j.foodeng.2009.11.019, in press)

Tempest, P., 1992. Experience with Ohmic Heating and Aseptic Packaging of Particulate Foods. Technical Report. No. PT/FBD1368. Crawley, UK: APV Baker Ltd., Automation Process Division.

Teng, S.S., Chen, B.H., 1999. Formation of pyrochlorophylls and their derivatives in spinach leaves during heating. Food Chem. 65, 367−373.

Tucker, G.S., Lambourne, T., Adams, J.B., Lach, A., 2002. Application of a biochemical time−temperature integrator to estimate pasteurisation values in continuous food processes. Innov. Food Sci. Emerging. Technol. 3, 165−174.

Tulsiyan, P., Sarang, S., Sastry, S.K., 2009. Measurement of residence time distribution of a multicomponent system inside an ohmic heater using radio frequency identification. J. Food Eng. 93 (3), 313−317.

Qihua, T., Jindal, V.K., Winden, J.van, 1993. Design and performance evaluation of an Ohmic heating unit for liquid foods. Comp. Electro. Agr. 9, 243−253.

Qin, B., Pothakamury, U.R., Vega, H., Martin, O., Barbosa-Canovas, G.V., Swanson, B.G., 1995. Food pasteurization using high-intensity pulsed electric fields. Food Technol. 49 (12), 55−60.

Vikram, V.B., Ramesh, M.N., Prapulla, S.G., 2005. Thermal degradation kinetics of nutrients in orange juice heated by electromagnetic and conventional methods. J. Food Eng. 69, 31−40.

Wang, W.C., Sastry, S.K., 1997. Starch gelatinization in ohmic heating. J. Food Eng. 34, 225−242.

Wang, W.C., Sastry, S.K., 2002. Effects of moderate electrothermal treatments on juice yield from cellular tissue. Innov. Food Sci. Emerging. Technol. 3, 371−377.

Wang, L.J., Li, D., Tatsumi, E., Liu, Z.S., Chen, X.D., Li, L.T., 2007. Application of two-stage ohmic heating to tofu processing. Chem. Eng. Process. 46, 486−490.

Weaver, J.C., Chizmadzhev, Y.A., 1996. Theory of electroporation: a review. Bioelectrochem. Bioenergetics 41 (1), 135−160.

Wu, H., Kolbe, E., Flugstad, B., Park, J.W., Yongsawatdigul, J., 1998. Electrical properties of fish mince during multi-frequency ohmic heating. J. Food Sci. 63 (6), 1028−1032.

Ye, X., Ruan, R., Chen, P., Doona, C., Taub, I.A., 2003. MRI temperature mapping and determination of liquid−particulate heat transfer coefficient in an ohmically heated food system. J. Food Sci. 68 (4), 1341−1346.

Ye, X., Ruan, R., Chen, P., Doona, C., 2004. Simulation and verification of ohmic heating in static heater using MRI temperature mapping. Lebensm.-Wiss. u.-Technol. 37, 49−58.

Yildiz, H., Bozkurt, H., Icier, F., 2009. Ohmic and conventional heating of pomegranate juice: effects on rheology, color and total phenolics. Food Sci. Technol. Int. 15 (5), 503−512.

Yildiz, H., Icier, F., Baysal, T., 2010. Changes in β-carotene, chlorophyll and color of spinach puree during ohmic heating. J. Food Process Eng. 33, 763–779.

Yoon, S.W., Lee, C.Y.J., Kim, K.M., Lee, C.H., 2002. Leakage of cellular materials from Saccharomyces cerevisiae by ohmic heating. J. Microbiol. Biotechnol. 12 (2), 183–188.

Zareifard, M.R., Ramaswamy, R.S., Trigui, M., Marcotte, M., 2003. Ohmic heating behaviour and electrical conductivity of two-phase food systems. Innov. Food Sci. Emerging. Technol. 4, 45–55.

Zell, M., Lyng, J.G., Cronin, D.A., Morgan, D.J., 2009a. Ohmic cooking of whole beef muscle-optimisation of meat preparation. Meat Sci. 81 (4), 693–698.

Zell, M., Lyng, J.G., Morgan, D.J., Cronin, D.A., 2009b. Development of rapid response thermocouples probes for use in a batch ohmic heating systems. J. Food Eng. 93 (3), 344–347.

Zhang, L., Fryer, P.J., 1995. A comparison of alternative formulations for the prediction of electrical heating rates of solid–liquid food materials. J. Food Process Eng. 18, 85–97.

Zhang, L., Liu, S., Pain, J.P., Fryer, P.J., 1992. Heat transfer and flow of particle–liquid food mictures. IChemE Symp. Ser. 126, 79–88.

Zhao, Y., Kolbe, E., 1999. A method to characterise electrode corrosion during ohmic heating. J. Food Proc. Eng. 22, 81–89.

Zimmerman, U., Pilwat, G., Riemann, F., 1974. Dielectric breakdown of cell membranes. Biophys. J. 14, 881–899.

Zoltai, P., Swearingen, P., 1996. Product development considerations for ohmic processing. Food Technol. (May), 263–266.

CHAPTER

Microwave Heating of Fluid Foods

12

H. Umesh Hebbar and Navin K. Rastogi

Department of Food Engineering, Central Food Technological Research Institute, Council of Scientific and Industrial Research, Mysore, India

12.1 INTRODUCTION

Food preservation is the primary objective of most food-processing operations and the challenge is to ensure quality and safety of processed products. Thermal processing, which mainly includes blanching, drying, evaporation, pasteurization, and sterilization, is carried out to increase the shelf-life of foods. The thermal processing conditions normally dictate the product quality, process economics, and environmental impact in most of the food-processing operations. Conventional heating methods rely essentially on conduction or convection mechanisms. However, they suffer from drawbacks including lower energy efficiency, longer processing time, and thermal damage due to overheating, especially at the surface of materials. Some of these aspects are addressed through control and monitoring systems and intelligent design of equipment (Pereira and Vicente, 2009). With the increase in consumer demand for high-quality food products at lower prices and growing environmental concerns, efforts are being made to develop processing technologies that are energy efficient, cost effective, and environmentally friendly. Recently, electromagnetic technologies have gained increased industrial interest in food processing and have shown potential to replace, at least partially, the conventional thermal-preservation techniques.

Dielectric heating, which utilizes electromagnetic radiations such as microwave (MW) and radiofrequency (RF), is gaining popularity in food processing. Amongst these two radiations, MW has shown a great potential (Chan and Reader, 2000) to be used as an alternative to conventional heating. These novel processing technologies are regarded as volumetric forms of heating, wherein the heat is generated from inside, as compared to surface heating with conductive or convective modes of heating. The volumetric heating of materials leads to higher rates of heat and mass transfer, resulting in reduced processing times and uniform product quality.

Novel Thermal and Non-Thermal Technologies for Fluid Foods. DOI: 10.1016/B978-0-12-381470-8.00012-8

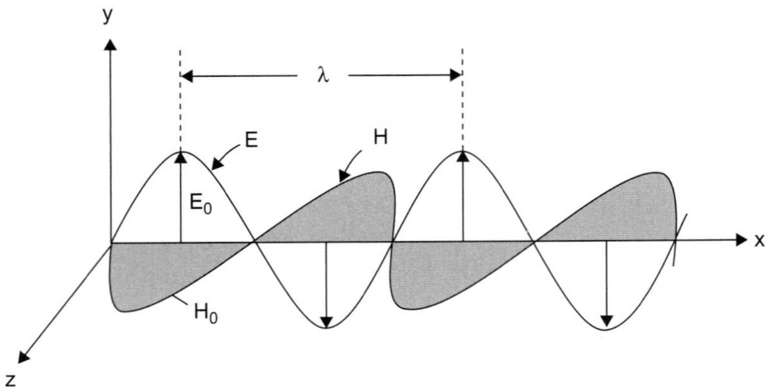

FIGURE 12.1

Propagation of electromagnetic wave.

Radiation may be defined as an energy streaming through space at the speed of light. Electromagnetic radiation consists of alternating electric and magnetic waves, which travel perpendicularly to one another. The electromagnetic waves can be characterized using two alternating vectors, namely, magnetic induction vector (H) and electric field vector (E) that define magnetic and electric fields, respectively (Fig. 12.1). MWs are electromagnetic waves having a frequency band of 300 MHz to 300 GHz (Pereira and Vicente, 2009). These waves propagate with a time interval between peaks during oscillation, ranging from 3×10^{-8} to 3×10^{-11} s (Venkatesh and Raghavan, 2004). This range coincides with the temporal sequence of events at atomic and molecular transitions such as reactions in water, molecular dissociation and, most importantly, dielectric relaxation in water. The dielectric relaxation of water may vary from 100 MHz for bound water to 18 GHz for pure water (Miura *et al.*, 2003) and this is the property that is studied extensively when the heating effects of MWs are investigated. The most effective conversion of MW energy into thermal energy in biological materials or in moist materials will occur in this frequency range.

In the electromagnetic spectrum, MWs occupy the position between infrared and radiofrequency waves. The frequency "f" of electromagnetic wave is linked to the velocity of light "c" (3×10^8 m/s) and corresponding wavelength "λ" by the following equation (Knutson *et al.*, 1987):

$$f = c/\lambda \tag{12.1}$$

According to the wave theory of electromagnetic radiation, waves having shorter wavelength travel with higher frequency, whereas ones with larger wavelength travel at a lower frequency. This results in propagation of all the electromagnetic waves at the same speed, i.e. the speed of light.

MW radiation is considered to be non-ionizing because they have insufficient energy (<10 eV) to ionize atoms (Piyasena *et al.*, 2003). Since, MW is within the radar range, the frequency bands that can be used for applications other than communication are limited by Electromagnetic Compatibility (EMC) regulations. The MW frequency bands that are internationally accepted and known as Industrial and Scientific and Medical (ISM) bands used for heating are 890 (UK) or 915 MHz (for other countries); 2450 MHz; 5800 MHz, and 24,125 MHz (Pereira and Vicente, 2009). While most domestic MW ovens have a frequency of 2450 MHz, many of the industrial units operate at a frequency of 915 MHz, as it has considerable advantages over 2450 MHz.

12.2 GENERAL PRINCIPLES OF MICROWAVE HEATING

When MW impinges on a material, part of the energy is transmitted, part is reflected, and the remaining is absorbed by the material which is dissipated as heat. The heating effect of MW arises from the interaction of the electric field component of the wave with charged particles in the material. Two major effects are responsible for the development of heat in the material exposed to MW due to this interaction. If the charged particles are free to travel through the material, a current will travel in phase with the field, which is termed as ionic polarization or ionic conduction. If, on the other hand, the charged particles are bound within regions of the material, the electric field component will cause them to move until opposing forces balance the electric force. The result is a dipolar polarization in the material. The molecular movement is extremely fast due to the high frequency of the field, e.g., at MW frequency of 2450 MHz, the polarity changes 2.45 billion cycles/second. At this frequency, it creates an intense heat that can escalate as quickly as 10°C per second (Lew *et al.*, 2002). As water is a dipolar molecule and a prominent component of biological materials, its content directly influences heating. The dielectric heating leads to volumetric heating of the product. Schematic representation of ionic polarization and dipolar rotation is shown in Figure 12.2.

Apart from dielectric heating, there are contributions from a host of other factors influencing the heating efficiency of MW (Schiffmann, 1986) such as specific heat, thermal conductivity, and food structure Specific heat is an important property in the thermal behavior of a food subjected to MWs. Hence, produce with low specific heat such as oil may heat very rapidly, and even faster than water of the same weight. The MW-heating rates are a function of the heating system (frequency and equipment design) and the load characteristics (size, shape, dielectric properties, etc.). Any change in these parameters significantly affects the MW-heating process (Peyre *et al.*, 1997). In solids, the molecular dipoles are no

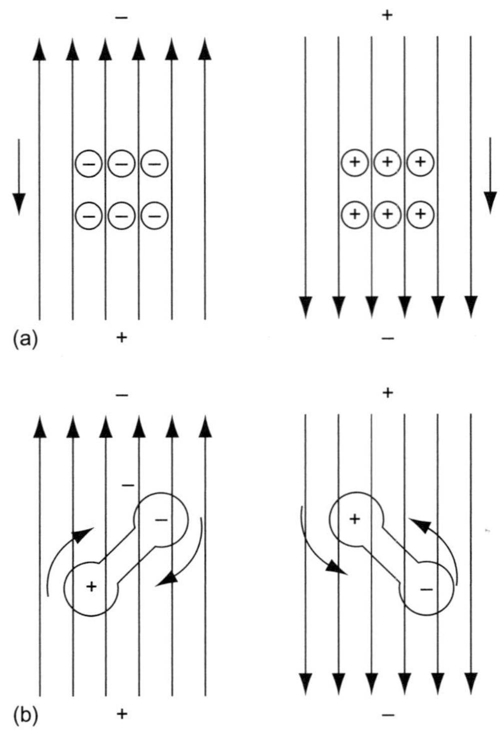

FIGURE 12.2

Schematic representation of (a) ionic polarization, and (b) dipolar rotation in alternating MW field.

longer free to rotate as they are in liquids, but are restricted to a number of equilibrium positions, separated by potential barriers. Theoretical treatments of this behavior have been reported to be similar to those developed for liquids.

12.3 ROLE OF DIELECTRIC PROPERTIES IN MICROWAVE HEATING

The efficiency of MW processing depends on many factors. These can be broadly classified as system parameters and product parameters. MW power output/rate of heating, frequency of MWs, use of auxiliary heat forms, distribution of MW field are regarded as some of the system parameters. However, electrical/dielectric properties, water content, specific heat, thermal conductivity, shape and size, temperature, composition are related to product parameters. The role of dielectric properties, one of the significantly influencing parameters on MW heating is discussed in the following sections.

12.3.1 Electrical/Dielectric Properties

The distribution of electromagnetic energy in MW-heating systems is governed by Maxwell's equations, as shown below (Metaxas and Meredith, 1983).

$$\nabla.\vec{D} = \rho \tag{12.2}$$

$$\nabla \times \vec{E} = -\frac{\delta B}{\delta t} \tag{12.3}$$

$$\nabla.\vec{B} = 0 \tag{12.4}$$

$$\nabla.\vec{H} = j + \frac{\delta D}{\delta t} \tag{12.5}$$

The dielectric properties of materials are the important parameters of Maxwell's equations and significantly influence the efficiency of electromagnetic energy coupling into the material, and its distribution and conversion into thermal energy (Tang, 2005).

The dielectric properties, or complex relative permittivity, are composed of dielectric constant (ε') and dielectric loss factor (ε'') (Mudgett, 1995). The dielectric constant reflects the ability of the material to store electric energy and the dielectric loss factor influences the conversion of electromagnetic energy into thermal energy. The dielectric properties of a material are described by the complex relative permittivity (ε^* relative to that of free space) in the following relationship (Tang, 2005):

$$\varepsilon'' = \varepsilon' - j\varepsilon'' \tag{12.6}$$

where $j = \sqrt{-1}$. The ratio of the above two properties ($\varepsilon'/\varepsilon''$) represents another important parameter, the tangent loss angle (tan δ) which, along with thedielectric constant, determines the attenuation of MW power in foods.

Mechanisms that contribute to the dielectric loss in biological materials include polar, electronic, atomic and Maxwell—Wagner responses of those materials in electromagnetic fields (Metaxas and Meredith, 1983). In foods, the dominant loss mechanism at MW frequencies of practical importance to industrial dielectric heating of foods are ionic conduction and dipole rotation, which are given by the following equation (Rynanen, 1995):

$$\varepsilon'' = \varepsilon''_d + \varepsilon''_\sigma = \varepsilon''_d + \frac{\sigma}{\varepsilon_0 \omega} \tag{12.7}$$

where subscripts d and σ stand for contributions of dipole rotation and ionic conduction, respectively, σ (S/m) is the ionic conductivity, ω (rad/s) is the angular frequency, and ε_0 is the permittivity of free space or vacuum (8.854×10^{-12} F/m).

12.3.2 Factors Influencing Dielectric Properties of Foods

The dielectric properties of foods are influenced by many factors such as MW frequency, material temperature, moisture content, and food compositions, in particular salt and fat contents. A few of these parameters are discussed in the following sections.

12.3.2.1 Frequency Effects

The contribution of various mechanisms of the dielectric loss factor of materials as a function of frequency is shown in Figure 12.3 (Tang *et al.*, 2002). At lower frequencies (< 200 MHz) the ionic conductivity of material plays a major role, whereas both ionic conductivity and the dipole rotation of free water are important at higher MW frequencies. The charge build-up in the interface between components in heterogenous systems is responsible for the Maxwell–Wagner polarization effect, which peaks at 0.1 MHz. For pure liquids with polar molecules, such as alcohol or water, polar dispersion dominates the frequency characteristics of dielectric properties. The polar dispersion characteristics of pure water at three different temperatures (0°C, 25°C, and 50°C) are shown in Figure 12.4. Dielectric constant will be relatively higher at lower frequencies ($<10^3$ MHz) for

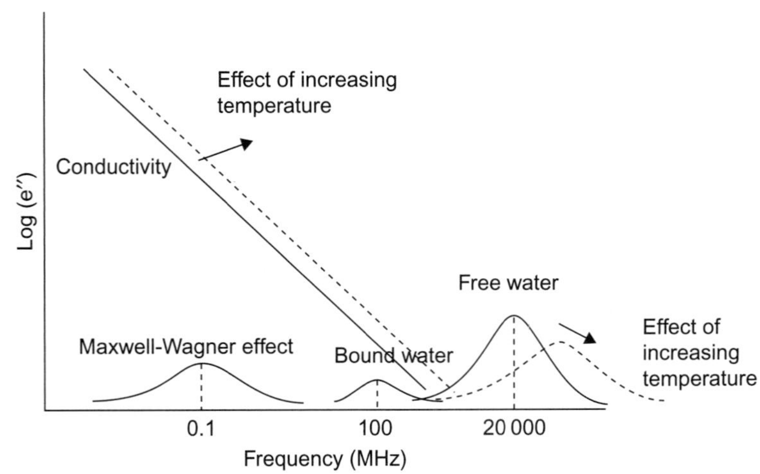

FIGURE 12.3

Variation of dielectric loss factor with MW frequency (Tang *et al.*, 2002).

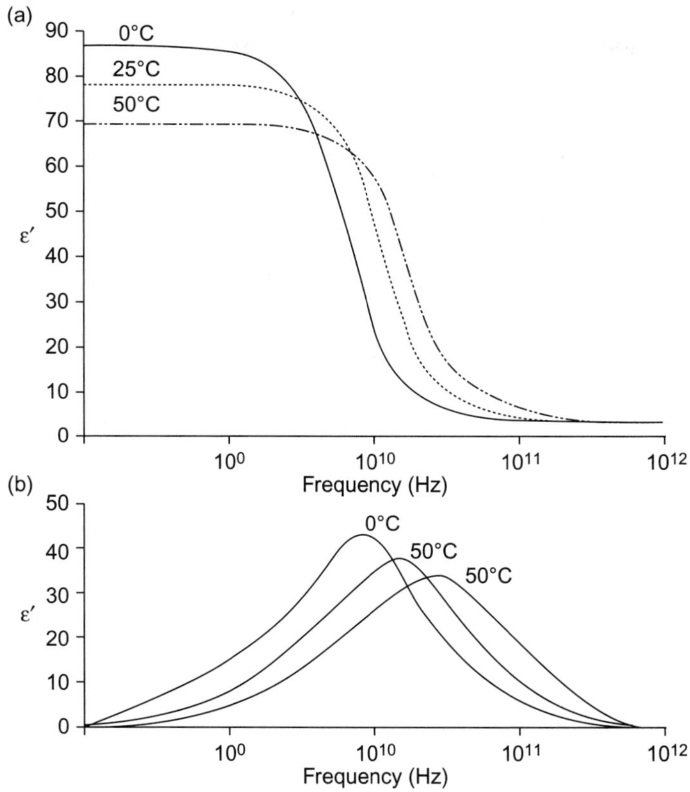

FIGURE 12.4

Variation of dielectric properties with frequency at different temperatures (Mudgett, 1985).

water having lower temperature. Aqueous solutions having charged ions behave differently from pure water with respect to dielectric characteristics, especially at low frequencies (e.g., <20,000 MHz at room temperature) and the deviation depends on the concentration of dissolved ions. The frequency and temperature-dependent dielectric characteristics of pure water and aqueous solutions determine the dielectric characteristics of moist foods. For moist foods with dissolved salt, ionic conduction plays a major role in the lower end of the MW-frequency range. This was illustrated by Guan *et al.* (2004) while studying the mashed potato system with different salt contents.

12.3.2.2 Temperature and Salt Effects

The effect of temperature on the dielectric properties of foods depends on many factors, including food composition, especially moisture and salt content, and

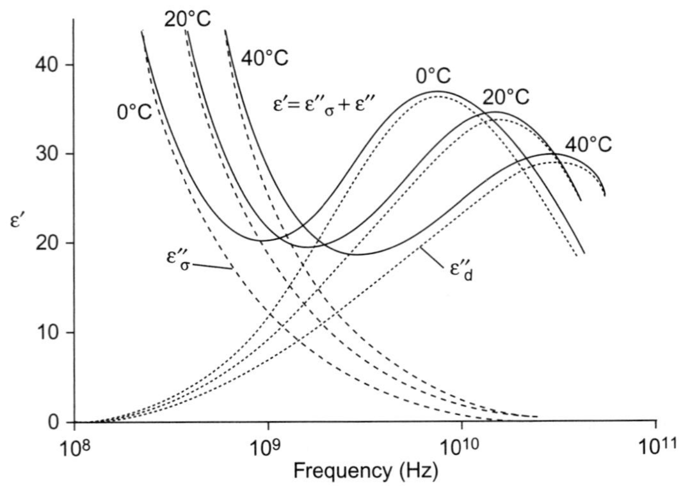

FIGURE 12.5

Effect of temperature on the dielectric loss factor of 0.5 N sodium chloride solution (Roebuck and Goldblith, 1972).

MW frequency. In moist foods with lower salt content, dielectric characteristics are mainly dominated by water. The electric conductivity "σ" in ionic solutions increases with temperature because of reduced viscosity and increased mobility of the ions (Trump, 1954; Stogryn, 1971). The combined contribution of ionic conduction and dipole dispersion of water molecules to the dielectric characteristics of a 0.5 N sodium chloride solution as influenced by temperature is shown in Figure 12.5. Below a frequency band of 2000 MHz, increasing temperature raises the dielectric loss factor of the solution because of the predominant role of ionic conduction at lower frequencies. Between 2000 and 10,000 MHz, increasing temperature reduces the solution's dielectric loss factor as the peak of dielectric loss constant moves towards higher-frequency bands. This band of frequency corresponding to the transition moves to a higher-frequency range in case of solutions having increasing ionic concentrations (Tang, 2005). The temperature and frequency-dependent dielectric characteristics of water are clearly reflected in moist foods. For moist foods with salt, loss factors generally increase with increasing temperatures at lower MW frequencies, which often results in a phenomenon commonly referred to as "thermal runway" (Metaxas and Meredith, 1983). That is, a preferentially heated part of the food in an electromagnetic field accelerates its heating, often causing non-uniform heating.

12.3.2.3 Moisture Effects

Water in moist foods can be divided in to three general categories: (1) free water in intracellular spaces; (2) multilayer water with mobility between free and bound water; and (3) monolayer water tightly bound to the polar sites of solid food components. Free-water molecules in intercellular spaces have dielectric properties similar to those of liquid water, whereas bound water exhibits ice-like dielectric properties. Dielectric properties of food, in general, decrease rapidly with decreasing moisture content to a critical moisture level. Below this moisture level, e.g., about 12% for diced apples at 22°C, the reduction in loss factor is significantly less due to the bound water. High temperature can, however, increase the mobility of bound water, reducing this critical moisture level.

Water molecules in ice are immobilized in well-defined matrices and behave similarly to bound water. Because the dielectric constant and loss factor of frozen moist foods are relatively low, their values depend, to a larger extent, on the amount of water in the unfrozen state and the ionic conductivity of the free water. All frozen foods have a very low loss factor, but after thawing the loss factor increases sharply, close to that of free water. Because of the reduced loss factor with decreasing moisture content, dehydrated foods have less ability to convert electromagnetic energy in to thermal energy. Conversely, during a MW drying process, the wet part of the product is able to convert more MW energy into thermal energy compared to the dry part, which tends to level off the uneven moisture distribution commonly experienced in a hot air drying process where the food particle interior is wetter than the surface. This would also significantly shorten drying time (Feng *et al.*, 2001).

12.4 INTERACTIONS OF MICROWAVE WITH FOOD COMPONENTS

The major food components such as water, carbohydrates, lipids, proteins and salts (minerals) interact differently with MW. Since, the primary mechanisms of MW heating is dipole rotation, water selectively absorbs the energy in intermediate- and high-moisture food products (Mudgett, 1989). Low-moisture products generally heat more evenly due to their low heat capacity (Schiffmann, 1986). Alcohols and the hydroxyl groups on sugars and carbohydrates are capable of forming hydrogen bonds and undergo dipolar rotation in an electrical field. Low levels of alcohol and sugar in solution have little effect on the interaction of MW with water and dissolved ions. However, at higher concentrations (jellies and candies), sugars can alter the response of water with MW (Mudgett, 1989). Proteins have ionizable surface regions that may bind water or salts, giving rise to various effects associated with free surface charge. Lipids, other than the charged carboxyl groups of the fatty acids that are usually unavailable due to their participation in the ester linkages of triglycerides or hydrophobic in nature, interact little with MW, if water is present.

12.4.1 Estimation of Heat Generation

The rate of heat generation per unit volume in a food material during MW heating (Q) is characterized by the following equation (Pereira and Vincente, 2009) and is mainly attributed to dipolar rotation phenomenon:

$$Q = 2\pi f \varepsilon_0 \varepsilon'' |\Delta V|^2 \tag{12.8}$$

where $|\Delta V|$ is the strength of the electric field, "f" is the MW frequency, ε_0 is the permittivity of free space and ε'' is the dielectric loss factor.

The heating rate due to collision of ions with the surroundings in an electric field, which is termed as ionic conduction, is expressed as below (Li and Ramaswamy, 2008).

$$\frac{P_\mu}{V_\mu} = E^2 q n \mu \tag{12.9}$$

where E is electric field, P_μ is power, V_μ is volume of the material, q is the electric charge of the ions, μ is the level of mobility of the ions and n is the number of charges.

12.4.2 The Depth of Penetration of Microwaves

The penetration depth of MWs can be defined as the depth where the dissipated power is reduced to 1/e (Euler's number e $\sim$2.718) of the power entering the surface. The penetration depth d_p in meters of MW energy in food can be estimated using dielectric properties by the following equation:

$$d_p = \frac{c}{2\pi f \sqrt{2\varepsilon' \left[1 + \sqrt{\left(\frac{\varepsilon''}{\varepsilon'}\right)^{2}} - 1\right]}} \tag{12.10}$$

where c is the speed of light in free space (3×10^8 m/s).

The penetration depth of MW into a material is inversely proportional to the frequency. In addition, electromagnetic waves do not penetrate deeply into moist foods (Metaxas and Meredith, 1983), where both the dielectric constants and loss factors are high. For even quasi-volumetric heating, the size of a target material should be of the same order as that of Dp, or preferably, less than Dp.

12.5 DIELECTRIC PROPERTIES OF SELECTED FOODS

The dielectric properties of a few liquid foods are presented in Table 12.1 (Tang *et al.*, 2002). The data demonstrate the effect of two MW frequencies (915 MHz and 2450 MHz) on dielectric constants and loss factors of different

Table 12.1 Dielectric Properties and Penetration Depth of a Few Selected Foods

Foods	Temperature (°C)	915 MHz			2450 MHz		
		ε′	ε″	d_p (mm)	ε′	ε″	d_p (mm)
Water							
Distilled/deionized	20	79.5	3.8	122.4	78.2	10.3	16.8
0.5% salt solution	23	77.2	20.8	22.2	75.8	15.6	10.9
Ice	−12	—	—	—	3.2	0.003	11615
Corn oil	25	2.6	0.18	467	2.5	0.14	220
High-Protein Products							
Yoghurt (pre-mixed)	22	71	21	21.2	68	18	9.0
Whey protein gel	22	51	17	22.2	40	13	9.6

Adapted from Tang et al. *(2002).*

foods. As can be seen in Table 12.1, distilled and deionized water has a much smaller dielectric loss factor at 915 MHz than at 2450 MHz, because 2450 MHz is closer to the frequency (~16,000 MHz) corresponding to the relaxation time of water molecules at room temperature. However, with the addition of 0.5% NaCl, the ionic conduction sharply raised the dielectric loss factor of water, much more at 915 MHz than at 2450 MHz. Ice, having a very small loss factor, is almost transparent to MWs. Oils are esters of long-chain fatty acids which have much less mobility compared to water molecules in response to oscillating electromagnetic fields. The dielectric constant and loss factor of oil are therefore very small compared to free water, as shown by the data for corn oil in Table 12.1. Wang *et al.* (2003) estimated the dielectric properties of a few selected liquid foods at RF and MW frequencies. Table 12.2 provides the dielectric parameters of a cheese sauce at 915 and 1800 MHz.

Penetration depth of MWs at 2450 MHz should be an important consideration when selecting the product size and geometrical requirements of packaging for MW-processed foods. The experimental data provided by Tang *et al.* (2002) demonstrated the variations in dielectric properties and penetration depths of MW in foods and high sensitivity of those properties to food MW frequency, composition etc. The depth of penetration of MWs and RF for whey protein gel and a liquid whey protein mixture reported by Wang *et al.* (2003) is presented in Table 12.3.

García *et al.* (2001) studied the dielectric constant, loss factor, and electrical conductivity of red wine grape juice samples at 0.2 and 3 GHz. The three main varieties (merlot, tempranillo, and cabernet sauvignon) normally used for wine preparation were chosen in the study. It was observed that there was a difference

Table 12.2 Dielectric Properties of Cheese Sauce

Temperature (°C)	Dielectric Properties	915 MHz	1800 MHz
20	ε′	42.90 ± 1.91	39.37 ± 3.37
	ε″	46.17 ± 4.84	27.90 ± 2.79
30	ε′	41.00 ± 2.43	38.97 ± 3.65
	ε′	52.70 ± 4.44	30.47 ± 2.78
40	ε′	40.13 ± 1.68	38.27 ± 3.06
	ε″	60.77 ± 4.98	34.03 ± 2.91
50	ε′	39.10 ± 1.68	37.40 ± 3.06
	ε″	69.03 ± 4.98	37.80 ± 2.91
60	ε′	37.87 ± 1.33	36.37 ± 2.70
	ε″	78.43 ± 6.19	42.27 ± 3.51
70	ε′	36.87 ± 1.06	35.40 ± 2.35
	ε″	87.47 ± 5.61	46.47 ± 3.05
80	ε′	34.87 ± 0.84	33.37 ± 2.14
	ε″	97.00 ± 5.55	51.07 ± 3.22
90	ε′	34.87 ± 0.74	33.37 ± 1.99
	ε″	106.43 ± 4.19	55.60 ± 2.72
100	ε′	34.07 ± 0.76	32.57 ± 1.74
	ε″	114.97 ± 4.40	59.67 ± 2.78
110	ε′	33.20 ± 0.96	31.70 ± 1.51
	ε″	124.17 ± 4.05	64.20 ± 2.70
121.1	ε′	31.70 ± 0.49	31.60 ± 0.21
	ε″	136.80 ± 0.14	70.20 ± 1.70

Adapted from Wang et al. *(2003).*

between varieties in relation to the values of dipolar losses at 3 GHz. The authors also attempted to describe the grape juice based on dielectric parameters, as well as to find out a possible relationship between these properties and the variety.

12.6 MICROWAVE PROCESSING OF FOODS

Applications of MWs for food processing have been researched for the past six decades with various degrees of success. MW processing has become very popular at the level of household applications in many countries. Over 60% in the UK and over 90% of homes in the USA and Canada use MW ovens for processing foods (Orsat *et al.*, 2005), and it is rapidly making inroads in developing countries too. However, industrialization of the technique is not as widespread as domestic

Table 12.3 Depth of Penetration of Microwave and Radiofrequency for Whey Protein Gel and Whey Protein Mixture

Product	*T* (°C)	Power Penetration Depth (mm)				
		27 MHz	40 MHz	915 MHz	1800 MHz	2450 MHz
Whey protein gel	20	45.9	38.2	12.0	8.8	6.5
	30	42.0	34.9	10.6	8.2	6.0
	40	38.5	31.9	9.5	7.6	5.6
	50	35.6	29.5	8.5	6.9	5.1
	60	33.1	27.4	7.8	6.4	4.7
	70	31.0	25.6	7.0	5.8	4.3
	80	29.1	24.1	6.4	5.3	3.9
	90	27.6	22.7	6.0	5.0	3.6
	100	26.2	21.6	5.6	4.5	3.3
	110	25.0	20.6	5.2	4.2	3.1
	121.1	24.0	19.7	4.9	3.9	2.8
Liquid whey protein mixture	20	44.2	37	12.6	9.0	6.6
	25	42.2	35.3	11.9	8.7	6.4
	30	40.0	33.4	11.1	8.4	6.1
	40	36.6	30.5	9.7	7.7	5.6
	50	33.7	28.0	8.6	6.9	5.1
	60	31.3	26.0	7.7	6.3	4.6
	70	29.6	24.6	7.0	5.8	4.2
	80	28.1	23.3	6.3	5.2	3.8
	90	26.8	22.2	5.8	4.8	3.5
	100	25.3	20.9	5.4	4.4	3.2
	110	24.2	20.0	5.0	4.1	3.0
	121.1	23.0	19.0	4.7	3.8	2.8

Reproduced from Wang et al. *(2003).*

applications. Applications at industrial level are popular for drying, cooking, pasteurization, thawing, and tempering of foods. MW vacuum-drying, MW freeze-drying, blanching, and pasteurization are some of the operations in which MWs have been used for processing liquid foods. Some of the operations in which MWs are used for processing foods, including liquid foods, are listed in Table 12.4 (Fito *et al.*, 2005). Liquid foods such as milk, honey, fruit juices, green coconut water, and purees have been processed with MWs for various objectives and are discussed in the following sections.

Table 12.4 Applications of Microwave Heating in Food Processing

Application	Objective	Food Product
Tempering	Raise temperature below freezing	Meat, fish, butter
Vacuum drying	Reduce moisture content	Seeds, grains, citrus juice
Freeze drying	Reduce moisture content	Meat, vegetables, fruits
Drying	Reduce moisture content	Pasta, rice, snacks
Cooking	Modify flavor and texture	Bacon, meat patties, potatoes
Blanching	Inactivate spoilage enzymes	Fruit, corn, potatoes
Baking	Heat and activate leavening agents	Bread, pastry, donuts
Roasting	Heat and promote heat reactions	Nuts, coffee, cocoa
Pasteurization	Inactivate vegetative microorganisms	Dairy products, prepared foods
Sterilization	Inactivate microbial spores	Pastry, prepared dishes, soft cheeses, milk, fruit juices

Reproduced from Fito et al. *(2005).*

12.6.1 Microwave Processing of Liquid Foods

12.6.1.1 Milk

Villamiel *et al.* (1996) studied the effect of MW heating of raw milk in a domestic oven (2450 MHz) on chemical changes such as lactose isomerization, Maillard reaction and protein denaturation. Lactulose, epilactose, furosine, and undenaturated whey proteins were measured as indicators of the heat damage in milk. Comparison with control samples treated by conventional heating showed a rate enhancement of the studied reactions during MW treatment. These differences were attributed, at least to some extent, to uneven heating of the milk in the MW oven.

Valero *et al.* (1999) estimated the effect of MWs on volatile contents of milk. Raw milk was heated using MWs by two different modes, namely, at 120°C in closed vessels, and at 70–90°C using a continuous-flow device. Conventional heating was carried out at the same temperatures, using a glycerol bath or a water bath for comparison. Volatile composition was similar in both the heating systems, although milk heated in closed vessels had some quantitative differences from conventional heating. The latter results were attributed to the unevenness of MW heating inside the vessels, whereas the continuous flow device seemed to allow a very uniform heating.

In another study, chemical and sensorial changes in milk pasteurized by MW and conventional systems during cold storage were studied by Valero *et al.* (2000). Raw milk was subjected to a continuous-flow MW treatment at 80 or 92°C for 15 s and compared with milk pasteurized under similar conditions in a conventional heat exchanger. After pasteurization, samples were bottled in sterile containers and stored at for up to 15 days. Samples stored for 15 days at low temperature (4.5 ± 0.5°C) were analyzed for pH, sensory properties, volatile compounds, and monosaccharides. Volatiles (including aldehydes, ketones, alcohols, esters, and aromatic hydrocarbons) in freshly pasteurized samples were very similar to those found in raw milk and showed that MW heating had no adverse effects on flavor.

Thompson and Thompson (1990) demonstrated that a domestic MW oven can be used effectively to reduce aerobic plate counts in raw goats' milk by up to 6 log-cycles without impairing the organoleptic quality. Good keeping quality of the irradiated product was demonstrated by the 7-day holding standard plate count.

The effect of MWs of 2450-MHz frequency and 600 W power on *Pseudomonas aeruginosa, Escherichia coli, Enterobacter sakazakii, Klebsiella pneumoniae, Staphylococcus aureus, Candida albicans, Mycobacterium terrae* and poliomyelitis vaccine–virus suspended in five infant formula preparations was investigated by Kindle *et al.* (1996). The samples were brought to the boil in 85–100 s depending on milk type. They had reached average temperatures of 82–93°C at this point. Most of the vegetative organisms were killed. In those samples where growth was still detectable after MW treatment, a significant reduction in viable microorganisms (at least 5000-fold) was noticed. It was concluded that MW heating to boiling is a convenient and fast method of reducing microbial contamination of infant feeds.

Sieber *et al.* (1996) reviewed the application of MW for heating of milk using household equipment having a frequency of 2450 MHz. The reports revealed that the effect of MW heat treatment depends on the quantity of the product and the geometry of the vessel used. In some cases, infant formulae heated to high temperature by MWs led to scalding and burning. The literature on the effects of MW heat treatment on microorganisms and nutrients in milk, such as proteins, enzymes, and vitamins, were found to be inconsistent.

Nunes *et al.* (2006) studied the permittivity of milk at room temperature (17–20°C) over the frequency range of 1–20 GHz. Ultra High Temperature-(UHT-) processed whole, low-fat, and skim milk were examined, fresh from the container, as well as over a period of two weeks while they were allowed to spoil at room temperature. The permittivity was found to be function of frequency over the range studied.

Wang and Chen (2005) and Wang *et al.* (2005) reported a heat and mass transfer simulation model for dielectric material (Silicon carbide) assisted microwave freeze-drying (MFD) of skim milk. Simulation results showed that the dielectric material can significantly enhance the MFD process (Fig. 12.6). The drying time was 33.1% shorter than that of ordinary MFD under typical operating conditions. MW heating was less effective when the solid content of the solution to be freeze-dried was very low, or the solid product had a very small loss factor.

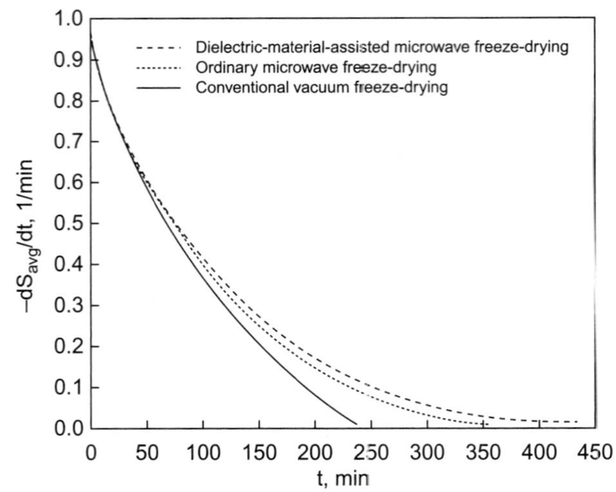

FIGURE 12.6

Drying curves for different freeze-drying methods (Wang and Chen, 2005).

12.6.1.2 Honey

Ghazali *et al.* (1994) studied the effect of MW processing of star-fruit honey for its storage stability. Their study showed that the relatively short time taken to reach the required processing temperature ensured little change in the chemical properties. Honey was heated to 71°C using a MW oven and stored at two different storage conditions (room temperature, 28 ± 2°C and 4°C) for 16 weeks. The physico-chemical properties of unheated and heated honey were measured before and during storage (Table 12.5). Spoilage was noticed in unheated honey (control) irrespective of the storage temperature. Heated samples were more resistant to spoilage. The spoilage of untreated honey was attributed to the yeast count in honey, which was much higher (1.02×10^5 cfu/g) in the unheated honey compared to heated honey samples (5.90×10^2 cfu/g). No appreciable variation in yeast count was noticed in treated honey during storage. No granulation was observed in treated honey samples during storage. Storing of unheated honey at room temperature was also free from granulation, in agreement with the general observation related to dextrose-to-water ratio and storage temperature. Darkening of honey was observed during storage in both heated and unheated honey from light golden color to golden brown. However, the honey stored at 4°C was considerably lighter in color than honey stored at room temperature. Heating did not alter the moisture content of honey (20.8%). Also, no noticeable changes in ash, nitrogen contents, pH, or acidity were observed in MW-heated honey samples. Heat processing of honey led to a 37.5% loss in diastase activity. Storage at 4°C had no effect on subsequent activity. However, room-temperature storage led to

Table 12.5 The Physico-Chemical Properties of Unheated and Microwave-Heated Star-fruit Honey Before And After Storage for 16 Weeks

Property	Unheated	Heated	Storage Temperature	
			4°C	Room Temperature
Moisture (%)	20.80	20.80	21.00 (22.00)	20.80 (21.20)
Diastase Number	4.00	2.51	2.32 (3.28)	1.69 (2.58)
Fructose (%)	32.10	32.12	32.10 (32.07)	30.08 (30.12)
Glucose (%)	30.14	30.10	30.02 (24.75)	26.85 (24.90)
Maltose (%)	2.31	2.30	2.42 (5.33)	5.47 (7.47)
Sucrose (%)	5.96	6.01	5.92 (5.48)	1.67 (0.00)
Yeast (cfu/g)	1.02×10^5	5.90×10^2	4.00×10^2	3.45×10^2

Values indicated in the parentheses are for unheated honey (Reproduced from Ghazali et al., *1994).*

further loss (33%) in activity. Heating did not show any effect on the sugar contents. However, the concentration of glucose, maltose and sucrose changed during storage depending on the storage temperature and whether the sample had been heated or not.

Bath and Singh (2001) studied the effect of MW heating on hydroxymethylfurfural (HMF) formation and browning in two types of honey (*Helianthus annuus* and *Eucalyptus lanceolatus*). The formation of HMF and browning increased with MW power levels as well as with heating duration, the former showing greater effect. Both types of honey differed significantly with respect to HMF formation and browning under similar MW-heating conditions, which was attributed to the difference in chemical composition.

Hebbar *et al.* (2003) conducted studies on MW heating in a micro-convective oven (2450 MHz, maximum power of 800 W). Experiments were carried out at different power levels (PL) ranging from 10 to 100 (175–800 W) and for different heating periods from 15 to 90 s. The extent of change in properties (HMF, diastase activity, moisture content, and yeast count) mainly depended on the power level (power intensity) and duration of heating. The changes in these properties were prominent in samples that were heated at higher power levels and for longer durations. The peak temperature attained by the sample depended on the power level used as well as duration of heating. The reduction in yeast count was observed to the extent of commercially acceptable levels (<500 cfu/ml) at power

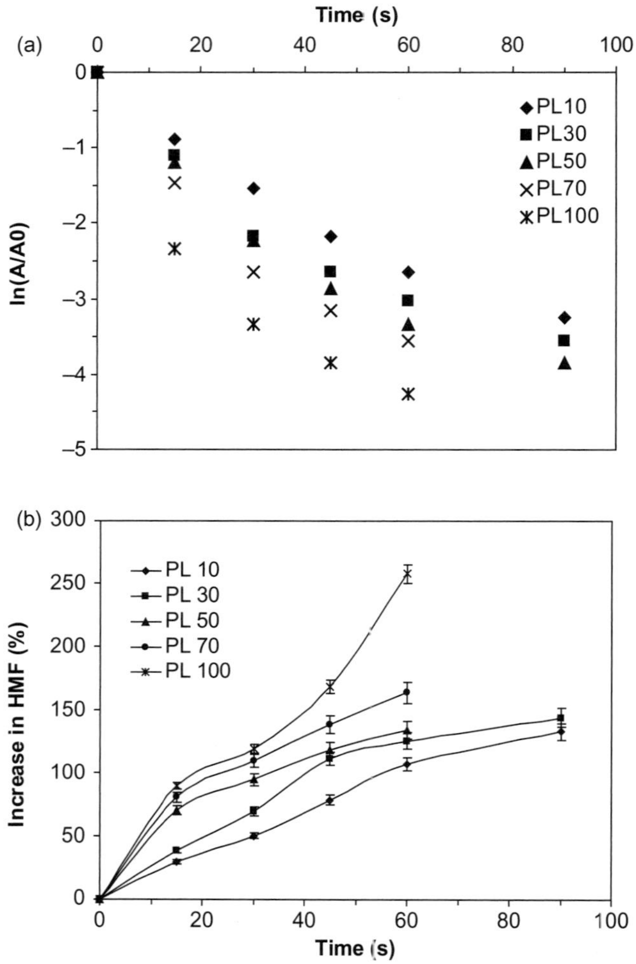

FIGURE 12.7

Reduction in yeast count (a) and increase in HMF (b) with time at different power levels (PL) of MW heating (Hebbar *et al.*, 2003).

levels of 10, 30, and 50 when the samples were heated for more than 45 s. At higher power levels of 70 and 100, heating durations of 30 and 15 s, respectively, were sufficient to achieve the same level of reduction in yeast count. The reductions in yeast count were rapid, generally during the first 20–30 s, and the rate of yeast reduction was directly related to the input power intensity (Fig. 12.7a). The reduction in yeast count is attributed to the rapid increase in sample temperature due to MW exposure, leading to the rupture of yeast cell walls. The increase in HMF value was gradual with heating duration at power levels of 10, 30, 50, and

70 (Fig. 12.7b); but HMF level increased sharply in the samples heated for a longer duration at the maximum power level of 100. However, these values were far below the maximum-permissible statutory level of 40 mg/kg for honey. The trends in the variation of HMF values of the samples clearly depicted the sensitivity of honey to the period of heating and temperature (power level). Heating affects the enzyme activity and the diastase activity showed a decline with heating under all conditions employed. Long heating periods of 60 to 90 s duration at power levels of 30, 50 and 70 reduced the diastase activity of honey by ~50% of its original value (Fig 12.8a). At a power level of 100, heating above 45 s resulted in reduction of the diastase activity to a level lower than the minimum permissible Diastase Number (DN) of 8. A reduction in moisture content above 9% at power levels of 50, 70, and 100 when the samples were heated for 60 s (Fig. 12.8b) was found. Significant reductions in moisture content was not observed at lower power levels. The final moisture content in most of the samples was in the range of 19.8 to 21.2%, which is below the acceptable level (22%) for commercially processed honey. The authors concluded that MW heating at power intensity of 16 W/g (PL 100), for 15 s resulted in substantial reduction in yeast count (450 cfu/ml) besides minimum HMF development (3.8 mg/kg) and higher retention of diastase activity (DN 12).

Cui *et al.* (2008) investigated MW vacuum-drying of liquid honey as a potential method for obtaining high-quality dried honey. Liquid honey was dehydrated in a MW vacuum-dryer to a moisture content less than 2.5% in less than 10 min. A sample thickness of less than 8 mm and a vacuum pressure of 30 mbar were reported as the better parameters for the drying. Darkening of the honey was not observed and there were no significant changes in the contents of fructose, glucose, maltose, or and sucrose of honey after drying. The volatile acids, alcohols, aldehydes, and esters made up the bulk of the identified aroma compounds of liquid honey and the content of alcohols and the esters changed slightly. The acids decreased markedly whereas the aldehydes and the ketones increased remarkably in the dehydrated honey.

12.6.1.3 Fruit Juices

Preservation of apple juice by MW radiation was explored by Canumir *et al.* (2002). Exposure of *E. coli* to MW treatments resulted in a reduction of the microbial population in apple juice. This work determined the effect of pasteurization at different power levels (270–900 W) on the microbial quality of apple juice (Fig. 12.9), using a home 2450-MHz MW. Data obtained were compared with conventional pasteurization (83°C for 30 s). Apple juice pasteurization at 720–900 W for 60–90 s resulted in a 2–4-log population reduction. Using a linear model, the D-values were estimated and they ranged from 0.42 ± 0.3 min at 900 W to 3.88 ± 0.26 min at 270 W. The value for z was 652.5 ± 2.16 W (58.5 ± 0.4°C). These observations indicated that inactivation of *E. coli* is due to heat.

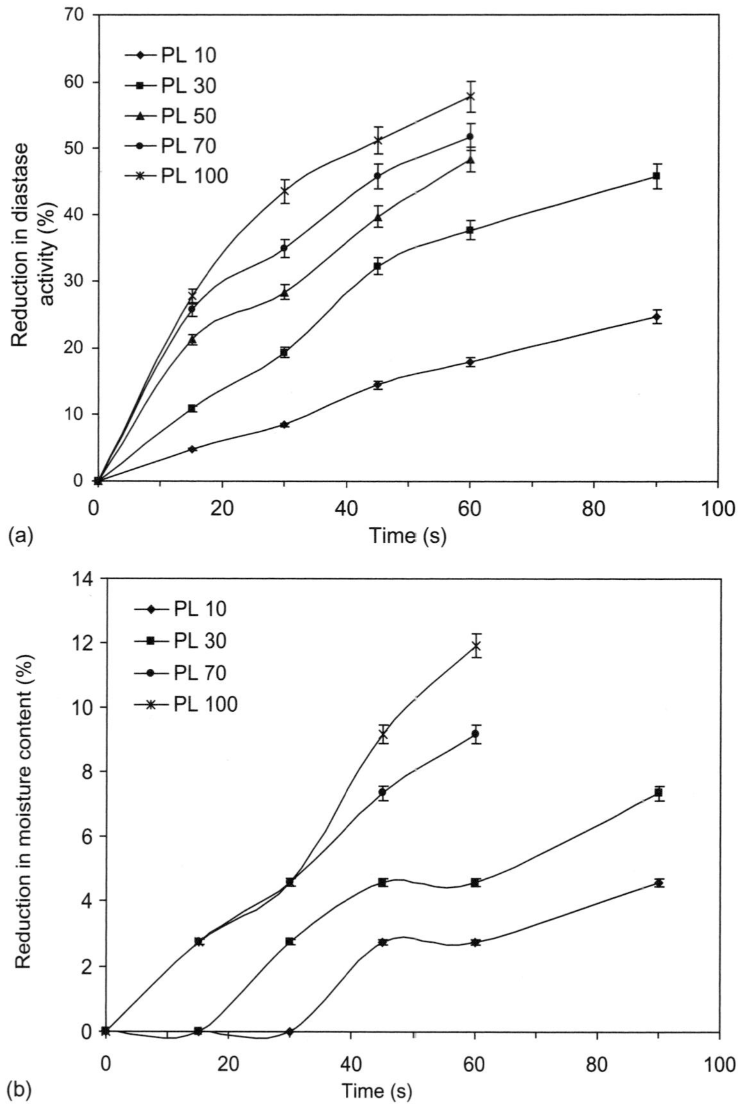

FIGURE 12.8

Reduction in diastase activity (a) and moisture content (b) with time at different power levels (PL) of MW heating (Hebbar *et al.*, 2003).

Gerard and Roberts (2004) studied the effects of four heat treatments of apple mash on juice yield and quality were evaluated and compared to juice produced from unheated apple mash at 21°C. Fuji and McIntosh apple mashes were heated to bulk temperatures of 40°C, 50°C, 60°C and 70°C in a 2450-MHz MW oven at

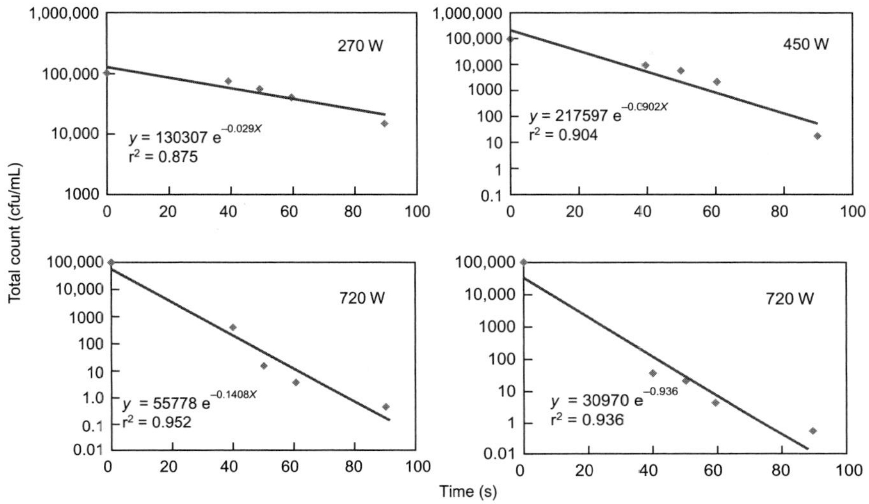

FIGURE 12.9

Inactivation curves of *E. coli* in apple juice heated with MW at different power levels (Canumir *et al.*, 2002).

1500 W. The juice yield increased when mash was heated before pressing. Cider produced from the heated mashes had comparable pH, titratable acidity, and sensory characteristics to cider produced from room-temperature mashes; however, total phenolic and flavonoid content of the juice increased with increasing mash temperature. Soluble solids and turbidity also increased as treatment temperature increased.

Degradation of carotenoids in orange juice was monitored during MW heating at different time/temperature conditions (Fratianni *et al.*, 2010). Degradation rate of carotenoids was influenced by MW-heating temperatures. At 60°C and 70°C for 10 min, violaxanthin and antheraxanthin were the compounds most unstable, while lutein and provitamin A carotenoids were more stable. At 85°C a decrease of about 50% was observed for almost all carotenoids after 1 min of MW heating. Temperature sensitivity (z value) for total carotenoids was 14.2°C, for single compounds z values ranged between 10.9°C for β-carotene and 16.7°C for antheraxanthin. These results indicate that an adequate choice temperature conditions for the quality control of carotenoids and the related nutritional values during MW treatment of orange juices is essential.

The status of the vitamin C during thermal treatment of orange juice heated by different methods (MW, infrared, Ohmic, and water-bath heating) was reported by Vikram *et al.* (2005). A comparative study of kinetics of vitamin degradation and changes in visual color as an index of carotenoids was carried out. The degradation kinetics of vitamin C and color in terms of reaction rate constants, destruction kinetics, enthalpy, and entropy for different methods of heating were studied. The destruction of vitamin C was influenced by the method of heating and temperature of processing (Fig. 12.10). The degradation was highest during MW heating due to uncontrolled temperature generated during processing. Out of the four methods studied, ohmic heating gave the best results facilitating better vitamin retention at all temperatures. The z values were within the literature values of 20–30°C for vitamin destruction, except for MW heating.

Destruction kinetics of two selected spoilage microorganisms, *Saccharomyces cerevisiae* and *Lactobacillus plantarum* in apple juice were evaluated under continuous-flow MW heating conditions and compared with conventional batch heating in a water bath (Tajchakavit *et al.*, 1998). Inoculated apple juice was heated in a MW oven (700 W, 2450 MHz) under continuous-flow conditions to select exit temperatures (52.5–65°C). Samples were also subjected to batch thermal treatments (50–80°C) in a well-stirred water bath. The time-corrected D-values and z-values were estimated for all the microbial species. Results showed that microbial destruction occurred much faster under MW heating than under thermal heating, suggesting some contributory enhanced effects to be associated with MW heating.

Preservation of color of a product is an important quality consideration. MW blanching of strawberry juice and concentrate had been shown to improve color stability and protect anthocyanin pigments, reactive phenolics, and ascorbic acid during 8 weeks of storage (Wrolstad *et al.*, 1980).

Kozempel *et al.*, (1997) reported a MW-based batch-flow process to substantially reduce bacteria in liquid foods such as apple juice, sugar solution, and brine solution. A logarithmic cycle reduction of 3 was achieved for *Pediococcus* sp. in water, 100 g/l sugar solutions, and brine solution, and 2-log reductions in cells in apple juice. The temperature of the process fluid was maintained at a nominal 35°C or less for a total MW exposure time of less than 9 min.

12.6.1.4 Other Liquid Foods

The effect of MW heating, and zinc and copper chloride addition on the color of avocado puree was studied by Guzmán *et al.* (2002). Results showed that color was best preserved when avocado purees either with or without zinc and copper salts were MW treated for 30 s, due to inhibition of enzymatic browning. Samples with copper chloride added showed a better color than samples with zinc chloride added, both showing a better color than the blank. This effect was attributed to the formation of chlorophyll complexes with copper and zinc cations. The study also used response surface analysis to optimize concentrations of zinc and copper salts along with various pH values.

Minimal processing of a Granny Smith apple purée by MW heating was attempted by Picouet *et al.*, (2009). Apple purée, with natural additives, was prepared and subjected to a short MW treatment for 35 s at 652 W. The effects of the treatment were evaluated in terms of stability of vitamin C, total polyphenol content, viscosity, color, and titratable acidity over a period of 2 weeks. Its effects

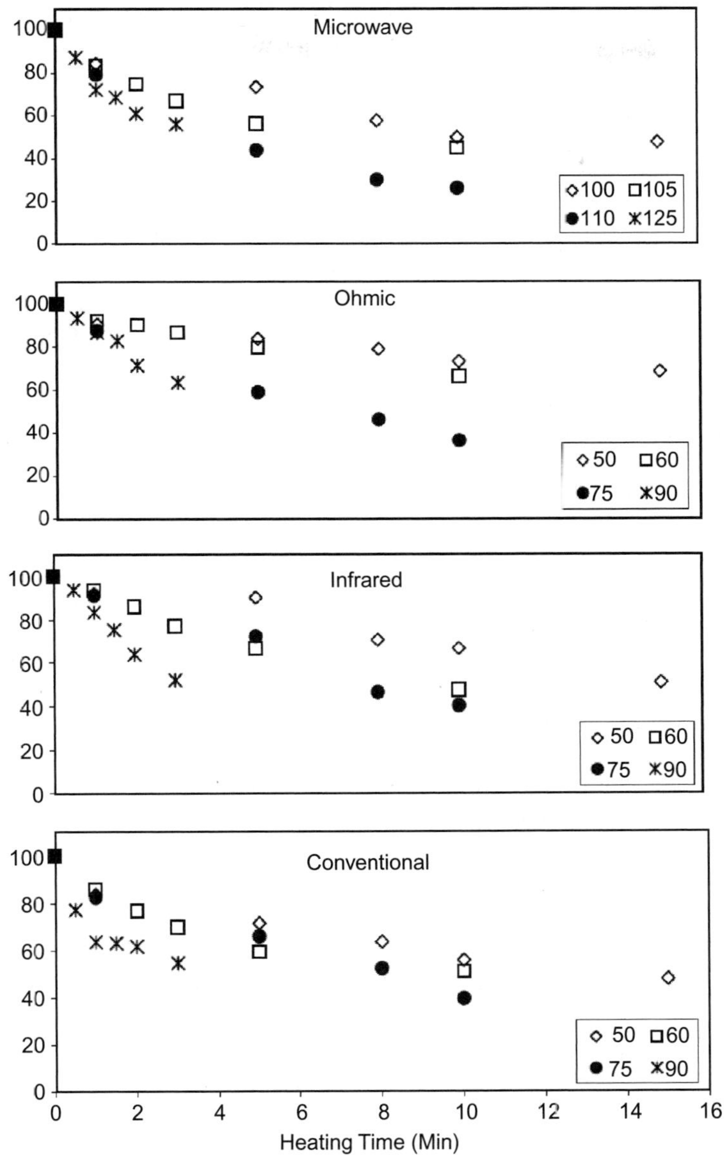

FIGURE 12.10

Vitamin C retention during heating by different methods at different temperatures (Vikram *et al.*, 2005).

on *E. coli, L. innocua* and indigenous microbiota were evaluated. Results showed that while a reduction of microorganisms was achieved, the treatment could not inactivate the enzymes present in the product and prevent vitamin C degradation. However, the treatment did not affect the viscosity and titratable acidity during storage.

MW blanching of kiwi fruit puree was shown to inactivate peroxidase (POD) and polyphenol oxidase (PPO), with minimal loss of bright-green color and only moderate degradation of chlorophylls (de Ancos *et al.*, 1999).

Wang *et al.* (2010) studied the effect of three different food ingredients, namely NaCl content (1–15 g/100 g water), sucrose content (5–25 g/100 g water) and sodium glutamate content (0.5–10 g/100 g water) on MFD of instant vegetable soup. Response surface methodology (RSM) was employed to optimize their addition levels in instant vegetable soup for MFD. Finally, a comparison of MFD for instant vegetable soup with and without added ingredients was carried out. The results showed that NaCl content and sucrose content had significant influences on the drying time and sensory quality, while sodium glutamate content had an insignificant effect. The optimum ingredients of the instant vegetable soup for MFD were: NaCl content of 3.2–5.3 g/100 g water, sucrose content of 2.0–6.8 g/100 g water and sodium glutamate content of below 4.5 g/100 g water. The comparison of MFD vegetable soup with a product without these ingredients indicated that the addition of condiments could increase drying rate and shorten drying time.

Drouzas *et al.* (1999) used a laboratory MW vacuum-drier for studying the drying kinetics of model fruit gels, simulating orange juice concentrate. The system was operated in the vacuum range of 30–50 mbar and MW power of 640–710 W. The distribution of the electromagnetic field in the cavity of the oven was determined from the drying rate of samples, placed at five different locations. The rate constant (K) of the single-layer model of drying was estimated by regression analysis of the experimental data. An empirical model was proposed for estimating the drying constant (K) as a function of the absolute pressure and the MW power of the system.

Premakumar and Khurdiya (2002) studied the effect of MW blanching on the nutritional qualities of banana puree. A domestic MW oven was used for the blanching of bananas. In another study, Premakumar *et al.* (2004) studied the effect of MW blanching on banana pulp. MW-blanched samples were compared with conventionally (water blanching) blanched pulp and activities of POD and PPO and pectin methyl esterase (PME) were estimated for the adequacy of blanching. Results showed that blanching for 3 min in a domestic oven completely inactivated all three enzymes in peeled bananas. Combination of water and MW blanching was used for unpeeled bananas. MW-blanched peeled banana pulp had better organoleptic properties as compared to water blanched.

Thermal inactivation of POD and PPO by MW in green coconut water-simulated solutions was reported by Matsui *et al.* (2007). The time–temperature profiles for different simulated solutions (with PPO/POD) are shown in Figure 12.11.

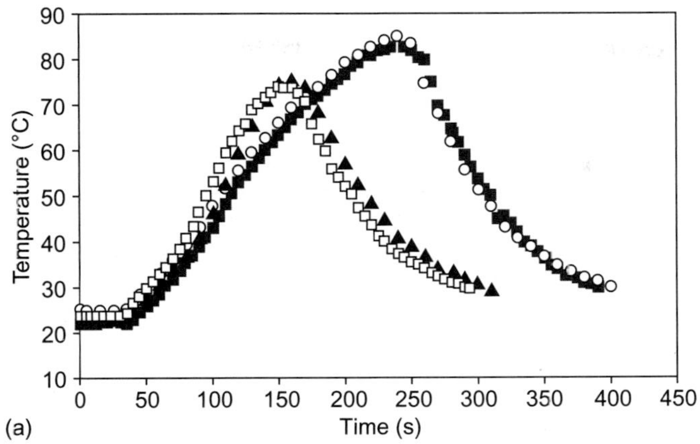

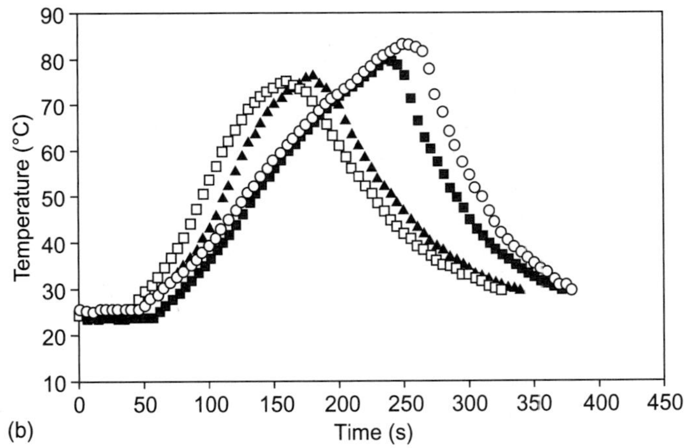

FIGURE 12.11

Time–temperature profiles for PPO-simulated solutions (a) [(▲) PPO/water; (■) PPO/salts; (□) PPO/sugars; (o) PPO/salts/sugars] and for POD-simulated solutions (b) [(▲) POD/water; (■) POD/salts; (□) POD/sugars; (o) POD/salt/sugars] subjected to MW heating. (Matsui *et al.*, 2007).

The contact between salts and enzymes promoted a drastic reduction of the initial activity. After the incidence of MW energy at temperatures above 90°C, enzymes activity was not detected. These results were used to process coconut water (Matsui *et al.*, 2008) in a batch MW processor. Inactivation of commercial PPO and POD added to sterile coconut water was also investigated. The time–temperature profiles of the experimental runs were used for the determination of kinetic parameters D-value and z-value: PPO ($D_{92.20°C} = 52$ s and $z = 17.6$ °C); POD ($D_{92.92°C} = 16$ s and $z = 11.5$°C); PPO/sterile coconut water: ($D_{84.45°C} = 43$ s and $z = 39.5°C$) and POD/sterile coconut water: ($D_{86.54°C} = 20$ s and $z = 19.3°C$). The enzymes naturally present in coconut water showed a higher resistance when compared to those added to the sterilized medium. The thermal inactivation of PPO and POD during MW processing of green coconut water was significantly faster in comparison with conventional processes reported in the literature.

MW-assisted thawing of maple sap was reported by Aider and Halleux (2008). Maple sap concentrates produced by cryoconcentration technology were thawed by two modes (passive and MW assisted). The temperatures to which the solution was frozen and thawing mode did not show any significant effect on total sugar content and other solution parameters. MW-assisted thawing mode was found to be an effective method because it permitted a decrease in the time needed for thawing as compared to passive thawing. In another study, it is reported that MW-assisted thawing mode was an efficient procedure and was faster than gravitational mode for thawing frozen milk whey concentrate (Aider *et al.*, 2008).

Design and evaluation of a continuous-flow MW-pasteurization system for apple cider was reported by Gentry and Roberts (2005). The aim of the work was to design a continuous-flow MW-pasteurization system and to evaluate process parameters such as volume load size, input power, and inlet temperature. Water and two apple ciders, one from a cold press and the other from a hot press extraction, were the fluids used to study the heating characteristics. The temperature profiles are shown in Figure 12.12. Process lethality was verified based on

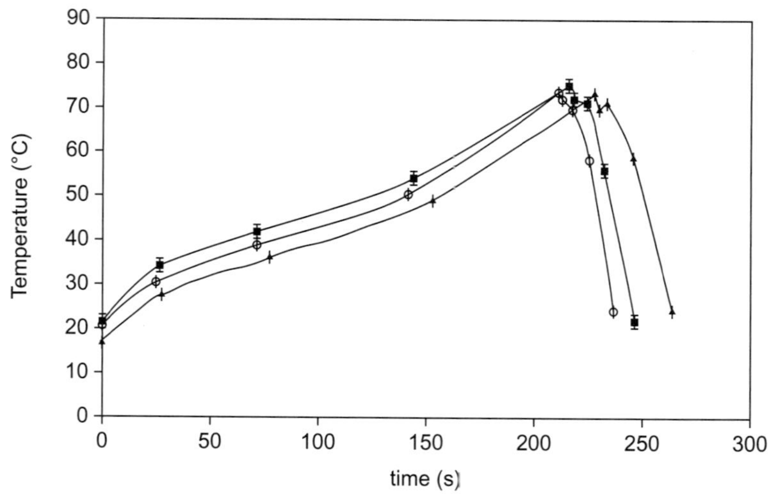

FIGURE 12.12

Time–temperature profiles for MW pasteurization using 2000 W with flow rates optimized for water (■), hot pressed cider (o), and cold-pressed cider (▲), with an inlet temperature of 20°C. (Gentry and Roberts, 2005).

inoculation of *E. coli* 25922 in apple cider, in which the pasteurization process resulted in a 5-$\log_{10}$ reduction.

Studies on sterilization solutions for aseptic processing in a continuous-flow MW system were reported by Coronel *et al.* (2008). Since the conventional method of sterilization of an aseptic-processing system by recirculating hot water can not be applied to a continuous-flow MW heating system, a model solution with dielectric and flow properties similar to those of the food product was designed. Dielectric properties of milk and sweet potato puree were compared to those of salt–sugar–CMC (carboxymethyl cellulose) mixtures and a model solution that closely matched the properties of each product was prepared and experimentally tested in a continuous-flow MW system operating at 915 MHz. The temperature profile at the exit of the heating section during heating of the sterilization solution was compared to the temperature profile during heating of the product. It was observed that the temperatures during transition from sterilization solution to the product did not change appreciably and authors concluded that these model solutions can be used as sterilization solutions for aseptic processing using a continuous flow MW system.

Zhu *et al.* (2007) presented a numerical model to study heat transfer in liquids as they flow continuously in a circular duct that is subjected to MW heating. Three types of food liquids are investigated, namely apple sauce, skim milk, and tomato sauce. The transient Maxwell's equations were solved by the finite difference time domain (FDTD) method to describe the electromagnetic field in the MW cavity and the waveguide. Numerical results showed that the heating pattern strongly depends on the dielectric properties of the fluid in the duct and the geometry of the MW-heating system.

The heating of a liquid layer (water and NaCl) by MW with a rectangular wave guide was investigated numerically and experimentally Ratanadecho *et al.* (2002). The proposed mathematical model used comprehensive two-dimensional heat and momentum equations to describe unsteady temperature and fluid-flow fields. The study also examined the effects of electric conductivity (which corresponds to the percentage by weight of NaCl in the liquid layer) and MW power level on the heating process. Based on a model combining the Maxwell and heat transport and fluid-flow equations, the results showed that the heating kinetic strongly depends on the dielectric properties.

12.6.2 Microwave Pasteurization: Thermal and Non-Thermal Effects

Microbial inactivation of food products, including liquid foods, using MW has been evaluated by many researchers. The rapid development of heat in the product was expected to reduce the sterilization and pasteurization time of food materials. However, MW sterilization has a few problems such as unpredictable and non-uniform heating of food, which made the industrialization of this technique relatively slow as compared to pasteurization. Even though research results point to the thermal effect for microbial inactivation, some reports such as that of Anantheswaran and Ramaswamy (2001), have considered MW effects as non-thermal when they are observed independently of sample temperature. Dipolar rotation of molecules is responsible for the heat development in moist foods, which is normally termed as a thermal effect. However, it is reported that some structures such as hydrogen-bonds, in which protons may be displaced easily in biological materials could be affected by MW. Chromosomes are also susceptible to MW disturbances. Another effect is the orientation of subcellular particles, which line up (pearl chain formation) under the influence of MWs. Many of the early works reported in the review by Rosenberg and Bogl (1987b) mentioned both thermal as well as non-thermal effects on microorganisms. From the practical point of view, the non-thermal effects, found in early work could either not be reproduced or could be explained later by the influence of inhomogeneous temperature distributions.

12.7 MICROWAVE PROCESSING AND EQUIPMENT

The application of MW energy for food processing was widely recognized in 1980s. Subsequently, the application of MWs for domestic application grew up significantly with 90% and 60% of the homes in the US and the UK, respectively, having MW ovens (Orsat *et al.*, 2005). The great increase in the use of domestic MW ovens has resulted in a considerable reduction in the cost of lower-power magnetrons. Nevertheless the cost of MW generators grows exponentially for industrial-scale units. This has led to further studies in the area of MW applications for food processing with a special emphasis on equipment–product interactions (Fito *et al.*, 2005). Schiffmann (2001) reported that nearly 1000 industrial MW ovens were in use in the US.

The major obstacle to the industry in the development of MW-heating applications is the high cost of the process. Electrical power is more expensive than combustion energy and lower degree of conversion of electrical energy to heat energy are the major reasons for the higher cost of processing. Hence, many researchers have recommended the use of MWs in combination with other techniques (MW with vacuum or hot air) and also to use at the final stages of processing, e.g. finish drying. The following sections briefly discuss the MW-heating system components and industrial-processing systems.

12.7.1 Microwave Systems

MW systems consists of three basic parts, namely MW generators, the waveguide, and the applicator. The magnetron tube is by far the most commonly used MW source or generator for domestic and industrial applications. The magnetron consists of a vacuum tube with a central electron-emitting cathode of highly negative

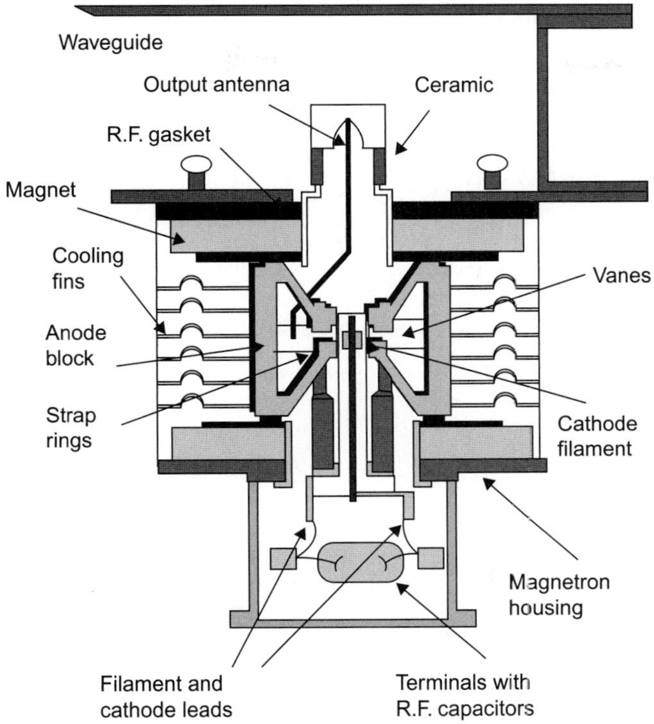

FIGURE 12.13

Schematic view of a magnetron.

potential (Fig. 12.13). This cathode is surrounded by a structured anode that forms cavities, which are coupled by the fringing fields and have the intended MW resonant frequency (Regier and Schubert, 2005). When the high electric (DC) power is applied, the emitted electrons are accelerated radially, and are deflected by the magnetic field, yielding a spiral motion. The electromagnetic energy produced is coupled out by a circular loop antenna in one of the cavities into a waveguide. Waveguides are hollow conductors, normally of rectangular or circular cross-section. MW applicators are classified as near-field, single-mode and multimode applicators (Fig. 12.14). In case of near-field applicators the MW originating from a horn antenna or slot arrays directly hit the product to be heated. Applicators with resonant modes, which enhance the electric field at certain positions, are better suited for low-dielectric properties and are used in single-mode applicators. Multi-mode applicators are used in both domestic as well as industrial MW ovens. Conveyor-belt-tunnel applicators and domestic MW ovens are of multi-mode type due to their typical dimensions.

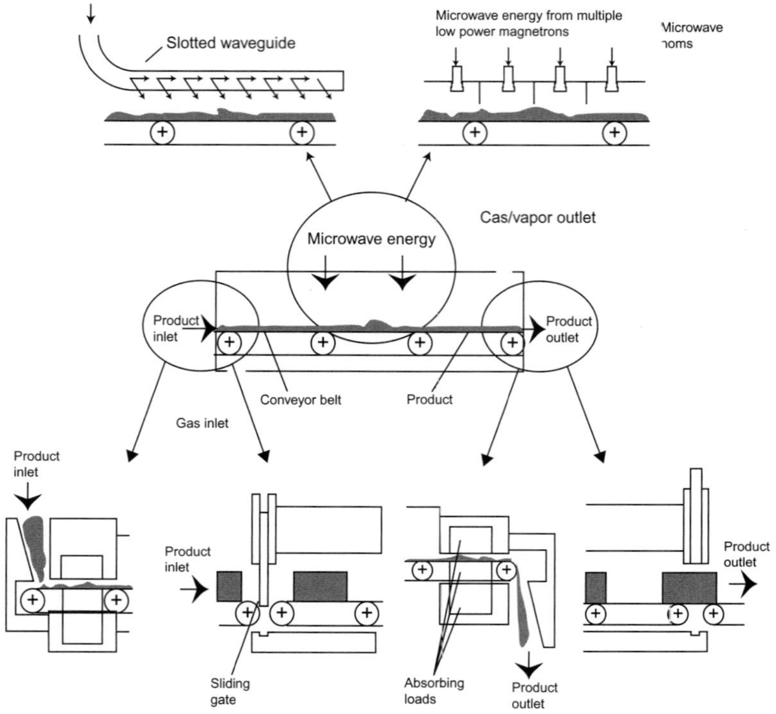

FIGURE 12.14

Different types of magnetrons and parts of conveyorized MW systems (Regier and Schubert, 2001).

12.7.1.1 Industrial Microwave Systems

The typical MW heating system utilizing a conveyorized cavity applicator is shown in Figure 12.15. Schematic representation of typical MW dryers used for food dehydration in combination with other drying techniques is presented in Figures 12.16–12.18. Gentry and Roberts (2007), reported a continuous pasteurization system for handling liquid foods. The spiral heating system was provided for continuous heating with MW, followed by cooling section. Typical MW combined freezing and vacuum systems can also be used for processing liquid foods. Li and Ramaswamy (2008) reported the following industrial MW systems for food processing.

- Linn High Therm GmbH (Germany): in this pass-through MW system the MW generators are arranged in a spiral form around the longitudinal axis of the cylinder chamber to achieve a more uniform distribution of the field. The system is equipped with secondary radiators for concentrating the field.

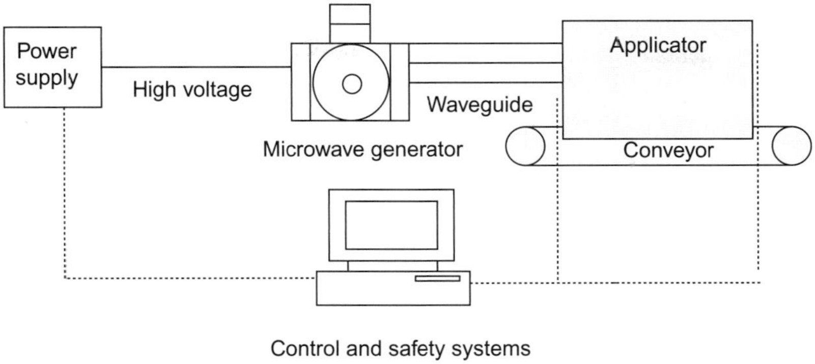

FIGURE 12.15

Typical MW-heating system using a conveyorized cavity applicator (Li and Ramaswamy, 2008).

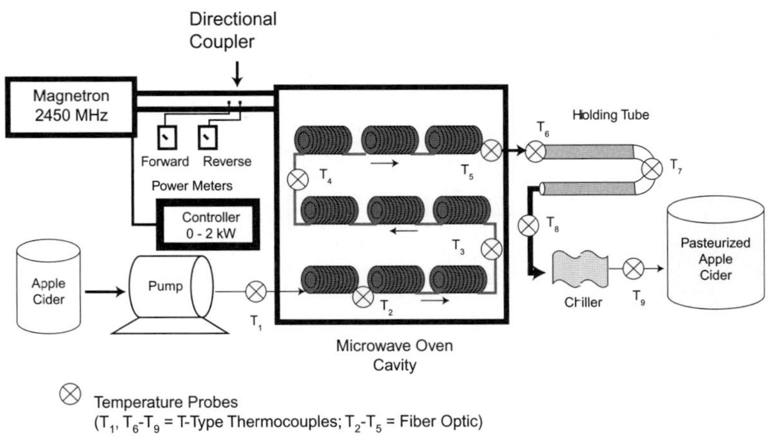

FIGURE 12.16

Continuous-flow MW pasteurization system (Gentry and Roberts, 2005).

Special absorbers are provided at the openings to prevent leakage of radiation. Air used for cooling the magnetron is circulated in the chamber to absorb moisture from the material.

- Gisip's (Sweden) dryer: a modular dryer for continuous processing of bulk materials or for handling small pieces. The drying process is adapted to the to the drying characteristics diagram of the material which is determined by shape and size of the material, water content and temperature.

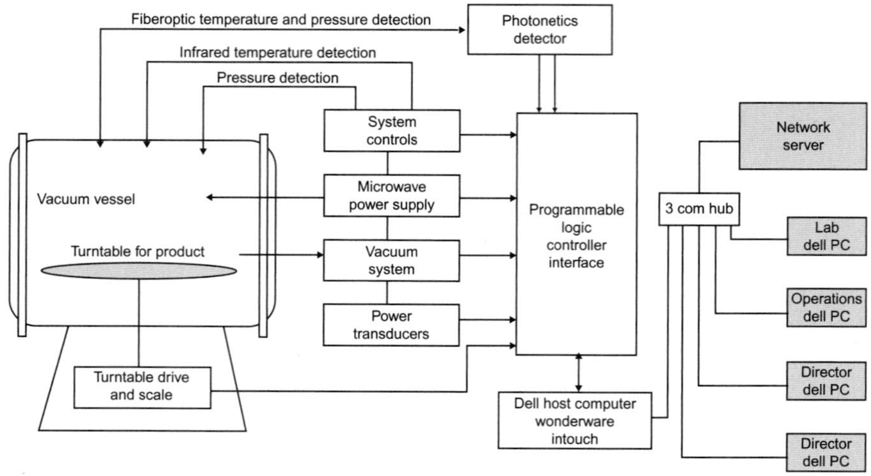

FIGURE 12.17

Schematic diagram of laboratory MW vacuum dryer (Clary *et al.*, 2005).

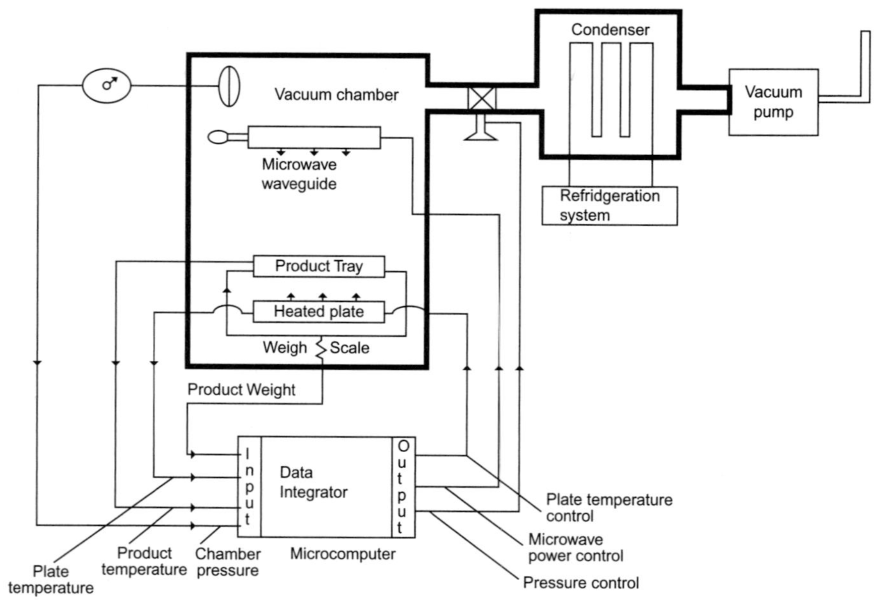

FIGURE 12.18

A typical MW-freeze drying apparatus (Cohen and Yang, 1995).

- Colpitt (Holland): a semi-continuous dryer operating at 2450 MHz frequency for drying of food materials. The Programmable Logic Controller (PLC)-controlled system with facility to vary MW power is also designed to operate in intermittent mode.
- Industrial MW Systems (USA): the patented design for continuous drying of food materials. With higher depth of penetration (2.5–5.0 inches) the system is reported to be highly efficient.
- MIVAP; Hans Binder Maschinenbau (Germany): industrial MW vacuum-drier for drying heat-sensitive products.
- Combined MW–freeze-drying system for various food products such as coffee, beef slices, mushrooms, and various other products. "Booster microwave freeze dryer" combined with infrared is reported to be used at industrial scale for drying coffee.

12.8 CHALLENGES IN MICROWAVE PROCESSING

The rapidly changing scenario in the global food market with complex customer demands, competitive market and environmental regulations has necessitated the application food processing techniques that can yield high quality products, in an efficient and environmental friendly manner. In this context, future prospects of MW radiation in food processing are bright, if some of the key issues of MW processing are adequately answered. With further research in the area of processing and equipment development to overcome the limitations of MW processing, its application in food processing is expected to increase. Some of the aspects that are being addressed/need to be addressed in order to popularize MW heating are discussed in the following sections.

12.8.1 Properties of Material

As materials undergo thermal MW processing, they experience physical and structural transformations that, in turn, change their dielectrical properties and behavior. The ability of MWs to generate thermal energy thus varies during the process. Understanding the generation and interactions of MWs with materials is consequently important for the development of any agricultural/food MW application. The physicochemical composition of the material being processed and the physics of the MW/material interactions govern their success. The data over a temperature range that covers freezing and sterilization temperatures, which reflect physicochemical changes during cooking and reheating processes, and data over a wide range of temperature and moisture content experienced in MW drying are essential. This needs to be generated for food material in order to improve results in computer simulation, product development, and system control.

12.8.2 Addition of Ingredients/Change in Food Compositions

Successful application of MWs for a given food depends on its composition. The food-product composition can be suitably modified to improve its dielectric properties or with the addition of ingredients that provide desirable product quality during MW processing. One of the most interesting areas of development in MW baking is coming from what is called a "MW-emulsion system" (Mast, 2000). The micro-emulsion system is designed to develop a coating for bakery products by creating interfacial tension sufficient to separate water and a surfactant coating to brown and crisp. This coating forms an edible susceptor capable of crisping and browning and has the ability to provide even heating of the product by focusing on the dielectric properties of the micro-emulsion system.

12.8.3 Microwave Processing Equipment

The increasing knowledge on the distribution of electromagnetic field during MW processing and the interactions between the factors will help in improving equipment design and its performance. Furthermore, the ongoing, international standardization of performance and testing methods will lead to improved heating performance. Although, today's MW-heating units have a reliability, which is well comparable to that of alternative processing equipment, improvements in oven components will reduce size, weight, and production costs.

With the changing trend in the food industry towards continuous processing lines and on-line process control, there will be a growing demand for MW heating for operations wherein rapid heating of very thick material is needed. With the development of sophisticated applicators, improved design for uniform distribution of MWs in the cavity and development of low cost magnetrons, the application of MW for food processing is expected to increase. Development of MW heaters with 915 MHz frequency will be more popular for industrial application as it has higher penetration depth, generator power, and efficiency. Research and development is underway to produce new magnetrons for widespread applications at 5.8 GHz. This will offer new horizons for product and applications development. Developments are also underway for scaling up MW-assisted extraction processes, which will have benefits for functional and nutraceutical food-product development.

12.8.3.1 Heat Distribution and Uniformity in Microwave Processing

One of the limitations of MW heating of foods is uneven heating, due to the "hot and cold spots" which may be present in the food product after heating. The temperature distribution in MW-heated foods is determined by the thermophysical properties of the food product as well as of the distribution of the absorbed MW energy. The MW heating distribution in the food is, in turn, determined by the electric and the magnetic fields in the MW cavity or applicator, by the dielectric properties of the food but also by the MW frequency. The electromagnetic field pattern can be controlled by the oven design (e.g., size and shape of the cavity or applicator), including the design of the wave guide system. Other aspects related to techniques for improving heating uniformity include moving conveyor belts in industrial appliances and mode stirrers which may result in improved uniformity. There are a few reported patents and solutions for exciting the MWs in a way which results in a more uniform field pattern. Even though rotating turntables and mode stirrers may to some extent often result in more leveled-out heating patterns in the oven, the need for a proper design of the food and package, as well as the relevant oven parameters is very much needed.

The heating uniformity is also influenced by composition, position and geometry of the food as well as of the package during heating. Heating uniformity in MW processing can be improved in different ways. One such technique could be the modification of relevant food or package parameters in order to optimize the heating uniformity based on modeling and suitable schemes for describing and evaluating the selected optimization parameters (Wappling-Raaholt *et al.*, 1999, 2001; Wappling-Raaholt, 2000). These tools are important today in food product development and the same concept may also be used in MW-oven design (Wappling-Raaholt and Risman, 2000). In order to increase uniformity in processing, phase-control heating, variable-frequency heating systems, and hybrid units may be studied.

12.8.3.2 Hybrid Processes

Research has revealed that hybrid or combination drying processes provide unique drying opportunities in food processing. The combinations wherein MWs are combined with hot air, vacuum, osmotic, freezing, and with other electromagnetic radiations such as infrared and RF, have been attempted for various food-processing operations either in combination or in sequential modes. The results of the study are reported to be encouraging and there is a great scope for further research in this area. Although, a few reports on industrial-scale applications are available, most of the reports are limited to laboratory studies and further work is needed to scale-up to industrial levels. Determination of optimal combination order, type, and conversion point (moisture content and drying time) from one method to the other is essential to make the process economically viable. Development of suitable models for optimizing processing conditions is essential to improve the operational performances for different combination-drying processes. Improvement in process equipment process efficiency, cost, and process automation for continuous operation are key to success in many of the food-processing applications. In the field of MW baking, halogen lamp–MW combination baking is a promising new technology, offering the speed of MW heating and the browning and crisping benefits of halogen lamps (Keskin *et al.*, 2004).

12.8.3.3 Modeling and Simulation

Modeling-based product development in combination with packaging (including different material, shielding, susceptors, geometry, etc.) has increased rapidly in the food industry during the last five years. The modeling tools will be invaluable for modern product design of microwavable foods. By combining today's optimization tools with knowledge on MW engineering and food science, unique possibilities for optimizing and developing MW food products can be generated. These tools will also enable modeling of combination processes, where MWs are combined with other heating techniques. The tools will reduce the time from product development to market for food products within the microwavable segment, with enhanced possibilities to meet the increased requirements from consumers and to respond to market needs. The development of mathematical models and computer simulation will be useful in studying field distribution and heating patterns in foods subjected to MW heating. Oliveira and Franca (2002) studied the simulation of heat transfer due to MWs applied to food products. The results of the study showed that heating is quite dependent on radiation frequency, power absorption and radiation penetration are more effective at lower frequencies than at higher ones. Since the heating migration in MW processing occurs from the initial and hottest locations in the interior of the food, it is difficult to locate and asses the cold point, as in traditional thermal methods. The use of specific software to estimate the parameters of the process will help to achieve a higher efficiency (Rodriguez *et al.*, 2003).

12.9 CONCLUSIONS

Increasing demand for high-quality foods as well as cost competitiveness has necessitated the application of employing electromagnetic radiation, namely, MW-based processes for thermal processing. The efficiency of MW processing depends on many factors, which are broadly classified as system parameters and product parameters. Dielectric properties of foods, one of the important product parameters, are influenced by many factors, such as MW frequency, temperature, moisture content, and other food compositions. MWs have found applications in the areas of pasteurization, blanching, thawing, and concentration for liquid foods such as milk, fruit juice, puree, pulp, etc. Dehydration, roasting, cooking, tempering, baking and extraction are some of the other applications of MW in food processing. Hybrid drying techniques wherein MW is combined with other processes have been reported to be highly efficient as compared to processing with MWs alone. Various reports have indicated that food product quality such as color, texture, and nutraceutical retention is higher with MW or hybrid processes. With the rapidly changing scenario, the future prospects of MWs in food processing are bright. However, some of the key issues such as improvement in equipment design, reduction in equipment cost, modification of food composition and selection of appropriate hybrid processes are to be adequately addressed to make MW processing more attractive.

Acknowledgments

The authors sincerely thank Dr V. Prakash, Director, Central Food Technological Research Institute, Mysore, India for constant encouragement. We wish to thank Dr K.S.M.S. Raghavarao, Head, Food Engineering Department, CFTRI, for support.

References

Aider, M., Halleux, D., 2008. Passive and microwave-assisted thawing in maple sap cryoconcentration technology. J. Food Eng. 85, 65–72.

Aider, M., Halleux, D., Melnikova, I., 2008. Gravitational and microwave-assisted thawing during milk whey cryoconcentration. J. Food Eng. 88, 373–380.

Anantheswaran, R.C., Ramaswamy, H.S., 2001. Bacterial destruction and enzyme inactivation during microwave heating. In: Datta, A.K., Anantheswaran, R.C. (Eds.), Handbook of Microwave Technology for Food Applications. Marcel Dekker, New York, pp. 191–210.

Bath, P.K., Singh, N., 2001. Effect of Microwave Heating on Hydroxymethylfurfural Formation and Browning in Helianthus annuus and Eucalyptus lanceolatus Honey. Food Sci. Technol. 38, 366–368.

Canumir, J.A., Celis, J.E., de Bruijn, J., Vidal, L.V., 2002. Pasteurisation of apple juice by using microwaves. Lebensm.-Wiss. u.-Technol. 35, 389–392.

Chan, T.V.C.T., Reader, H.C., 2000. Understanding Microwave Heating Cavities. Artech House Publishers.

Clary, C.D., Wang, S.J., Petrucci, V.E., 2005. Fixed and incremental levels of microwave power application on drying grapes under vacuum. J. Food Sci. 70 (5), 344–349.

Cohen, J.S., Yang, T.C.S., 1995. Progress in food dehydration. Trends Food Sci. Technol. 6, 20–25.

Coronel, P., Simunovic, J., Sandeep, K.P., Cartwright, G.D., Kumar, P., 2008. Sterilization solutions for aseptic processing using a continuous flow microwave system. J. Food Eng. 85, 528–536.

Cui, Z.W., Sun, L.J., Chen, W., Sun, D.W., 2008. Preparation of dry honey by microwave-vacuum drying. J. Food Eng. 84, 582–590.

de Ancos, B., Cano, M.P., Hernandez, A., Monreal, M., 1999. Effects of microwave heating on pigment composition and color of fruit purees. J. Sci. Food Agric. 79, 663–670.

Drouzas, A.E., Tsami, E., Saravacos, G.D., 1999. Microwave/vacuum drying of model fruit gels. J. Food Eng. 39, 117–122.

Feng, H., Tang, J., Cavalieri, R.P., Plumb, O.A., 2001. Heat and mass transport in microwave drying of hygroscopic porous materials in a spouted bed. AICHE J. 47 (7), 1499–1512.

Fito, P., Chiralt, A., Martin, M.E., 2005. Current state of microwave applications to Food Processing. In: Barbosa-Canovas, G.V., Tapia, S.M., Cano, M.P. (Eds.), Novel Food Processing Technologies. CRC press/Marcel Dekker, NY, USA, pp. 525–537.

Fratianni, A., Cinquanta, L., Panfili, G., 2010. Degradation of carotenoids in orange juice during microwave heating. Food Sci. Technol. 43, 867–871.

García, A., Torres, J.L., Prieto, E., De Blas, M., 2001. Dielectric properties of grape juice at 0.2 and 3 GHz. J. Food Eng. 48, 203–211.

Gerard, K.A., Roberts, J.S., 2004. Microwave heating of apple mash to improve juice yield and quality. LWT-Food Sci. Technol. 37, 551–557.

Gentry, T.S., Roberts, J.S., 2005. Design and evaluation of a continuous flow microwave pasteurization system for apple cider. LWT-Food Sci. Technol. 38, 227–238.

Ghazali, H.M., Ming, T.C., Hashim, D.M., 1994. Effect of microwave heating on the storage and properties of starfruit honey. ASEAN Food J. 9, 30–35.

Guan, D., Cheng, M., Wang, Y., Tang, J., 2004. Dielectric properties of mashed potatoes relevant to microwave and radio-frequency pasteurization and sterilization processes. J. Food Sci. 69, 30–37.

Guzmán, G.R., Dorantes, A.L., Hernández, U.H., Hernández, S.H., Ortiz, A., Mora, E.R., 2002. Effect of zinc and copper chloride on the color of avocado puree heated with microwaves. Innov. Food Sci. Emerging. Technol. 3, 47–53.

Hebbar, U.H., Nandini, K.E., Lakshmi, M.C., Subramanian, R., 2003. Microwave and infrared heat processing of honey and its quality. Food Sci. Technol. Res. 9, 49–53.

Keskin, S.O., Sumnu, G., Sahin, S., 2004. Bread baking in halogen lamp microwave combination baking. Food Res. Int. 37, 489–495.

Kindle, G., Busse, A., Kampa, D., Meyer-König, U., Daschner, F.D., 1996. Killing activity of microwaves in milk. J. Hosp. Infect. 33, 273–278.

Knutson, K.M., Marth, E.H., Wagner, M.K., 1987. Microwave heating of foods. Lebensm.-Wiss. u.-Technol. 20, 101–110.

Kozempel, M., Annous, B.A., Cook, R., Scullen, O.J., Whiting, R., 1997. Inactivation of micro-organisms with microwaves at reduced temperatures. J. Food Prot. 61, 582–585.

Lew, A., Krutzik, P.O., Hart, M.E., Chamberlin, A.R., 2002. Increasing rate of reaction: microwave assisted organic synthesis for combinatorial chemistry. J. Comb. Chem. 4 (2), 95–105.

Li, H., Ramaswamy, H.S., 2008. Microwave drying. In: Hui, Y.H., Clary, C., Farid, M.M., Fasina, O.O., Noomhorm, A., Welti-Chanes, J. (Eds.), Food Drying Science and Technology: Microbiology Chemistry, Applications. DEStech Publications, Inc., USA, pp. 127–156.

Mast, R., 2000. Application of microwave technology to bakery products. British Society of Baking BSB Conference, October 2000, paper #441.

Matsui, K.N., Granado, L.M., de Oliveira, P.V., Tadini, C.C., 2007. Peroxidase and polyphenol oxidase thermal inactivation by microwaves in green coconut water simulated solutions. LWT-Food Sci. Technol. 40, 852–859.

Matsui, K.N., Wilhelms Gut, J.A., de Oliveira, P.V., Tadin, C.C., 2008. Inactivation kinetics of polyphenol oxidase and peroxidase in green coconut water by microwave processing. J. Food Eng. 88, 169–176.

Metaxas, A.C., Meredith, R.J., 1983. Industrial Microwave Heating. Peter Peregrinus, London.

Miura, N., Yagihara, S., Mashimo, S., 2003. Microwave dielectric properties of solid and liquid foods investigated by time-domain reflectormetry. J. Food Sci. 68, 1396–1403.

Mudgett, R.E., 1985. Dielectric properties of foods. In: DeCareau, R.V. (Ed.), Microwaves in the Food Processing Industry. Academic Press, NY.

Mudgett, R.E., 1989. Microwave food processing: scientific status summary. Food Technol. 42 (1), 117–126.

Mudgett, R.E., 1995. Electrical properties of foods. In: Rao, M.A., Rizci, S.S.H. (Eds.), Engineering Properties of Foods. Macel Dekker Inc., New York, pp. 389–455.

Nunes, A.C., Bohigas, X., Tejada, J., 2006. Dielectric study of milk for frequencies between 1 and 20 GHz. J. Food Eng. 76 (2), 250–255.

Oliveira, M.E.C., Franca, A.S., 2002. Microwave heating of food stuffs. J. Food Eng. 53, 347–359.

Orsat, V., Raghavan, V., Meda, V., 2005. Microwave technology for food processing: an overview. In: Schubert, S., Regier, M. (Eds.), The Microwave Processing of Foods. CRC Press/Woodhead Publishing Limited, Cambridge, UK.

Pereira, N.R., Marsaioli Jr., A., Ahrne, L.M., 2007. Effect of microwave power, air velocity and temperature on the final drying of osmotically dehydrated bananas. J.Food Eng. 81, 79–87.

Pereira, N.R., Vicente, A.A., 2009. Environmental impact of novel thermal and non-thermal technologies in food processing. Food Res. Int. 43, 1936–1943.

Peyre, F., Datta, A., Seyler, C., 1997. Influence of the dielectric property on microwave heating patterns: application to food materials. J. Microw. Power Electromagn. Energy 32 (1), 3–14.

Picouet, P.A., Landl, A., Abadias, M., Castellari, M., Viñas, I., 2009. Minimal processing of a Granny Smith apple purée by microwave heating. Innov. Food Sci. Emerg. Technol. 10, 545–550.

Piyasena, P., Dassault, C., Koutchma, T., Ramaswamy, H., Awuah, G., 2003. Radio frequency heating of foods: Principles and applications and related properties—a review. Crit. Rev. Food Sci. Nutr. 43 (6), 587–606.

Premakumar, K., Khurdiya, D.S., 2002. Effect of microwave blanching on the nutritional qualities of banana puree. J. Food Sci. Technol. 39 (3), 258–260.

Premakumar, K., Khurdiya, D.S., Chatterjee, S.R., July–August, 2004. Microwave blanching: effect on polyphenol oxidase, peroxidase and pectin methyl esterase activities of banana pulp. In: Indian Food Packer. pp. 54–57.

Ratanadecho, P., Aoki, K., Akahori, M., 2002. A numerical and experimental investigation of the modeling of microwave heating for liquid layers using a rectangular wave guide (effects of natural convection and dielectric properties). Appl. Math. Model 26, 449–472.

Regier, M., Schubert, H., 2001. Microwave processing. In: Richardson, P. (Ed.), Thermal Technologies in Food Processing. pp. 178–207.

Regier, M., Schubert, H., 2005. Introducing microwave processing of food: principles and technologies. The Microwave Processing of Foods. Woodhead publishing Limited/CRC press, London/USA.

Rodriguez, J.J., Barbosa-Canovas, G.V., Gutierrez-Lopez, G.F., Dorantes-Alvarez, L., Won-Yeom, H., Zhang, H.Q., 2003. An update on some key alternative food processing technologies: microwave, pulsed electric field, high hydrostatic pressure, irradiation and ultrasound. In: Gutierrez-Lopez, G.F., Barbosa-Canovas, G.V. (Eds.), Food Science and Biotechnology. CRC Press, Washington DC, pp. 279–304.

Roebuck, B.D., Goldblith, S.A., 1972. Dielectric properties of carbohydrate eater mixtures at microwave frequencies. J. Food Sci. 37, 199–204.

Rosenberg, U., Bogl, W., 1987. Microwave thawing, drying and baking in the food industry. Food Technol. 41, 85.

Schiffmann, R.F., 1986. Food product development for microwave processing. Food Technol. 94–105.

Schifmann, R.F., 2001. Microwave processes for the food industry. In: Datta, A.K., Anantheswaran, R.C. (Eds.), Handbook of Microwave Technology for Food Applications. Marcel Dekker, New York, pp. 299–337.

Sieber, R., Eberhard, P., Gallmann, P.U., 1996. Heat treatment of milk in domestic microwave ovens. Int. Dairy J. 6, 231–246.

Stogryn, A., 1971. Equations for calculating the dielectric constant of saline water. IEEE Trans. Microw. Theory Tech., MTT 19, 733–736.

Tang, M., 2005. Dielectric properties of foods. In: Regier, M., Schubert, H. (Eds.), The Microwave Processing of Foods. Woodhead publishing Limited/CRC press, London/USA.

Tang, J., Feng, H., Lau, M., 2002. Microwave heating in food processing. In: Yang, X.H., Tang, J. (Eds.), Advances in Bioprocess Engineering. World Scientific, Hackensack, NJ.

Tajchakavit, S., Ramaswamy, H.S., Fustier, P., 1998. Enhanced destruction of spoilage microorganisms in apple juice during continuous flow microwave heating. Food Res. Int. 31, 713–722.

Thompson, J.S., Annemarie Thompson, A., 1990. In-home pasteurization of raw goat's milk by microwave treatment. Int. J. Food Microbiol. 10, 59–64.

Trump, J.G., 1954. Dielectric materials and their applications. In: von Hippel, A.R. (Ed.), Dielectric and Waves. Wiley, New York.

Valero, E., Sanz, J., Martinez-Castro, I., 1999. Volatile components in microwave- and conventionally-heated milk. Food Chem. 66, 333–338.

Valero, E., Villamiel, M., Sanz, J., Martinez-Castro, I., 2000. Chemical and sensorial changes in milk pasteurized by microwave and conventional systems during cold storage. Food Chem. 70, 77–81.

Venkatesh, M.S., Raghavan, G.S.V., 2004. An overview of microwave processing and dielectric properties of agri-food materials. Biosyst. Eng. 88 (1), 1–18.

Villamiel, M., Lopez-Fandino, R., Corzo, N., Martinez-Castro, I., Olano, A., 1996. Effects of continuous flow microwave treatment on chemical and microbiological characteristics of milk. Zeitsrift fur Lebensmittel Untersuchung und Forschung 202, 15–18.

Vikram, V.B., Ramesh, M.N., Prapulla, S.G., 2005. Thermal degradation kinetics of nutrients in orange juice heated by electromagnetic and conventional methods. J. Food Eng. 69, 31–40.

Wang, R., Zhang, M., Mujumdar, A.S., 2010. Effect of food ingredient on microwave freeze drying of instant vegetable soup. LWT-Food Sci. Technol. 43, 1144–1150.

Wang, S., Wig, T., Tang, J., Hallberg, L.M., 2003. Dielectric properties of food relevant to RF and microwave pasteurization and sterilization. J. Food Eng. 57 (3), 257–268.

Wang, W., Chen, G., 2005. Heat and mass transfer model in dielectric-material-assisted microwave freeze drying of skim milk, a porous hygroscopic material. Chem. Eng. Sci. 60, 6542–6550.

Wang, W., Chen, G.H., Gao, F.R., 2005. Effect of dielectric material on microwave freeze drying of skim milk. Drying Technol. 23 (1–2), 317–340.

Wappling-Raaholt, B., 2000. Finite difference time domain analysis of heating of foods in household microwave ovens. In: Improving the Heating Uniformity of Microwave Heated Foods. Thesis for the degree of Licenitate of Engineering, Chalmers University of Technology.

Wappling-Raaholt, B., Galt, S., Ohlsson, T., 1999. FDTD simulation of a microwave heating process: effects of food parameters., In: Proceedings of the Seventh International Conference on Microwave and High Frequency Heating, Valencia, Spain.

Wappling-Raaholt, B., Risman, P.O., 2000. FDTD simulation of a microwave heating process: effects of oven parameters on heating uniformity. Third International Conference on Predictive Modeling, Leuven.

Wappling-Raaholt, B., Ohlson, T., Risman, P.O., 2001. Microwave heating of ready meals FDTD simulation tools for improving the heating uniformity. In: Proceedings of the Eight International Conference on Microwave and High Frequency Heating, Bayreuth, Germany.

Wrolstad, R.E., Dee, D., Poei, M.S., 1980. Effect of microwave blanching on the colour and composition of strawberry concentrate. J. Food Sci. 45 (6), 1573–1577.

Zhu, J., Kuznetsov, A.V., Sandeep, K.P., 2007. Mathematical modeling of continuous flow microwave heating of liquids (effects of dielectric properties and design parameters). Int. J. Therm. Sci. 46, 328–341.

CHAPTER

Infrared Heating of Fluid Foods

13

Navin K. Rastogi

Department of Food Engineering, Central Food Technological Research Institute, Council of Scientific and Industrial Research, Mysore, India

13.1 INTRODUCTION

Processing of food products is a necessary requirement for extending their shelf-life. However, such processing generally involves heat treatment that can enhance the safety of the food, but reduce organoleptic quality. Over the years, researchers have looked for many technologies to optimize time and temperature profiles in order to minimize the exposure of food to heat. The newer food-processing technologies may have potential to supplement or even eliminate the use of heat treatment. A number of potential opportunities exist for exploiting the benefits of electromagnetic radiations in food processing, which include technologies like ohmic, infrared (IR), and microwave heating. Depending on the requirement of the process and demand of the consumer, these technologies can be applied in a diversity of ways. Conventional pasteurization requires longer heating time leading to deterioration of product quality. Electromagnetic heating, however, has been successfully used for the efficient pasteurization of food products in the recent years.

IR radiation is part of the electromagnetic spectrum in the wavelength range between 0.5 and 1000 mm (Rosenthal *et al.*, 1996), which is mainly utilized for food processing because of the several advantages such as higher heat-transfer capacity, instant heating because of direct heat penetration, high energy efficiency, faster heat treatment, fast regulation response, better process control, no heating of surrounding air, equipment compactness, uniform heating, preservation of vitamins, and less chance of flavor losses from burning of foods (Dagerskog and Osterstrom 1979; Afzal and Abe, 1998; Skjoldebrand, 2001). IR radiation falls between the region of visible light (0.38–0.78 μm) and microwaves (1–1000 mm). Based on the wavelength, it can be divided into three regions

Novel Thermal and Non-Thermal Technologies for Fluid Foods. DOI: 10.1016/B978-0-12-381470-8.00013-X

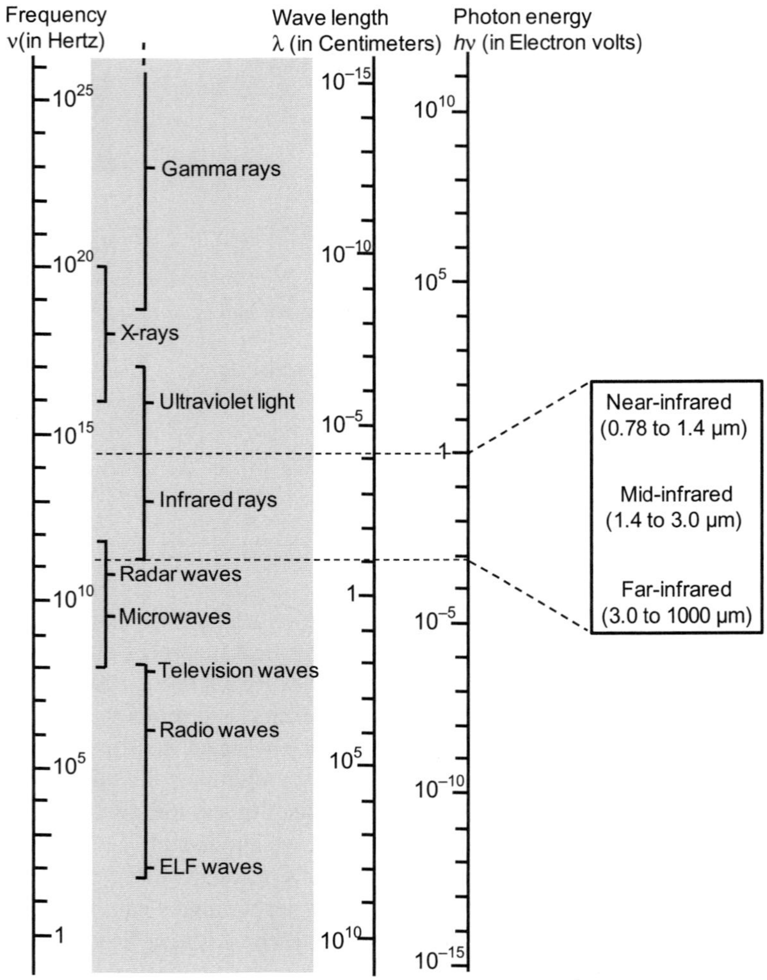

FIGURE 13.1

The electromagnetic spectrum.

namely near- (0.78–1.4 μm), mid- (1.4–3.0 μm) and far-IR (3.0–1000 μm) (Fig. 13.1). In general, far-IR radiation is more advantageous for food processing because most food components absorb radiative energy in this region (Sandu, 1986).

There is always an increasing interest in the applicability of IR radiation in food processing for inactivation of pathogens. The food components absorb energy effectively in the far-IR region (3–1000 mm), resulting in heating of food systems leading to the inactivation of microorganisms. IR heating inactivates microorganisms by damaging intracellular components such as DNA, RNA, ribosomes, cell envelopes, and/or proteins in the cell (Sawai *et al.*, 1995).

IR radiation has numerous applications in the case of solid-food-processing operations such as drying, freeze-drying, thawing, roasting, blanching, baking, and cooking. These applications have been summarized in Table 13.1. In spite of the several advantages of IR heating, its use in the case of liquid foods is not widespread. It has been indicated in the literature that it can be used to full potential in the case of several liquid foods such as orange juice, honey, milk, beer, as well as for microbial suspensions. Vasilenko (2010) developed a method of preserving the quality of the perishable products consisting of near-IR (0.8–1.2 μm) with or without involving a magnetic field (10–500 mTesla).

In the case of fluid foods, the main effect of IR heating is due to heating of a thin layer of food material on the surface. The radiation can not penetrate deep and heats up only a few millimeters below the surface of the sample. The absorbed energy is then transferred by conduction to other areas within the food material. As the sample volume increases, this conduction is limited, and thus the total energy absorbed is limited. As IR heating mainly heats a thin layer from the surface, the food product can be rapidly cooled after IR treatment, and thus provides less change in the quality of food material because of negligible heat conduction (Hamanaka *et al.*, 2000). The penetrative radiation energy does not make significant contribution to internal heating. Additionally, the less convective currents, because of the small depths of the samples, result in low flow velocities. Therefore, the thermal behaviors of the liquid samples seem to fully correspond to internal conduction, following external radiative and convective heat transfer on the top surface.

IR heating has lower energy requirements to achieve the same temperature, leading to lower operational costs. Therefore, IR heating has a potential to be utilized as an effective alternative to the conventional heating methods. Absorption of IR energy by water molecules in microorganisms is one of the important factors for microbial inactivation, because water absorbs readily in the IR region and results in rapid temperature increases (Hamanaka *et al.*, 2006).

There are many literature reviews dealing with the basic principles, capability and limitations of IR processing (Sakai and Hanzawa, 1994; Skjoldebrand, 2001, 2002; Sakai and Mao, 2006; Krishnamurthy *et al.*, 2008a, 2009; Rastogi, 2010). However, most of them have paid attention to the processing of solid foods. This chapter comprehensively reviews the application of IR processing of fluid foods.

13.2 BASIC PRINCIPLES GOVERNING INFRARED RADIATION

IR radiation is a form of energy, specifically known as electromagnetic waves arising from the movement of electrons in atoms and molecules. The average or bulk properties of electromagnetic radiation interacting with matter are

Table 13.1 Key Findings in the Area of Infrared Processing of Solid Foods

Product	Salient Results	References
IR and Drying		
Onion	Combination of IR with vacuum showed high shrinkage and rehydration potential. The effect of drying temperature, slice thickness, inlet air temperature and air velocity on drying kinetics was evaluated. Various models were fitted to drying data and thin-layer model best described the drying behavior. Quality of the product was superior in terms of rehydration ratio, color, and pyruvic acid content.	Mongpraneet *et al.* (2002a, b), Sharma *et al.* (2005a, b), Praveen Kumar *et al.* (2005, 2006), Pathare and Sharma (2006), Hebbar *et al.* (2004), Gabel *et al.* (2006)
Banana	IR radiation modified the structure of the dried bananas by increasing their final porosity. Developed mathematical model to predict moisture content and temperature of banana during combined IR and vacuum-drying. Novel drying technology consisting of low-pressure superheated steam drying with far-IR radiation was proposed.	Sun *et al.* (2007), Leonard *et al.* (2008), Thanit *et al.* (2009), Chatchai *et al.* (2007)
Apple	Comparison of the quality of apple slices dried by IR heating and convectional drying indicated that color parameter, rehydration capacity, and mechanical properties were dependent on final material temperature and not on method by which heat was supplied.	Nowak and Lewicki (2004, 2005), Wesolowski and Glowacki (2003), Togrul (2005, 2006)
Pineapple and potato	Combination of IR with hot air and osmotic pre-treatment resulted in reduction in overall colour change whilst maintaining high drying rates. Drying rates were not influenced by RH but were dependent on radiation intensity levels and were negatively correlated with air velocity.	Afzal and Abe (1998), Tan *et al.* (2001)
Blueberries	Evaluation of the quality and IR drying characteristics of fresh and sugar-infused dried blueberries. IR drying resulted in firmer-texture products in a much shorter drying time as compared to hot-air drying. NaOH pretreatment increased drying rate and reduced the number of broken berries at higher drying temperature.	Junling *et al.* (2008), Shi *et al.* (2008)
IR and Freeze-Drying		
Sweet potato and yam	Combination of IR heating with freeze-drying reduced the processing time to less than half. Drying temperature, distance between sample and IR heater, and thickness of the sample was optimized in view of drying time, rehydration ratio, and total color difference.	Yeu *et al.* (2005, 2007)

(*Continued*)

Table 13.1 (Continued)

Product	Salient Results	References
Banana	Partially IR-dried banana slices had a higher drying rate as compared to the hot air pre-dried sample during freeze-drying. IR pretreatment showed collapse of cellular tissue in the surfaces of the banana slices, forming a crust on the surfaces of the banana slices.	Zhongli *et al.* (2008)
Strawberries	Combination of IR and freeze-drying resulted in high-quality crispy strawberry pieces at less cost. The technology was energy efficient. The product had more desirable color, higher crispness, and more shrinkage, but a lower rehydration ratio than freeze-dried product.	Shih *et al.* (2008)
IR and Thawing		
	Heat transfer in frozen foods heated by IR radiation resulted in less damage during thawing. Heat transfer in frozen tuna heated by far-IR radiation was modeled to prevent overheating so as to control surface temperature by intermittent irradiation treatment. IR combined with air-blast thawing was also shown to have potential for improving thawed meat quality aspects. Temperature distribution inside the sample subjected to IR assisted microwave tempering of frozen potato puree was simulated.	Sakai *et al.* (1995), Geun *et al.* (2009), Seyhun *et al.* (2009)
IR and Roasting		
	The rapid surface heating by IR can be used to seal-in moisture and flavor or aroma compounds without burning the surface black, which results in highly acceptable sensory quality of the product. Cracked wheat, chestnuts, hazelnuts, green tea, coffee beans, and sesame seeds were roasted using IR radiation for their specific advantages.	Ozdemir and Devres (2000), Brown *et al.* (2001), Hee (2006), Kim *et al.* (2006), Poss (2007), Chung (2008), Park *et al.* (2009), Uysal *et al.* (2009), Kumar *et al.* (2009)
IR and Blanching		
	Several studies have shown that IR blanching resulted in products with better quality. IR-blanched peas had comparable ascorbic acid retention and better taste and flavor than hot-water-blanched samples. IR-blanched endive and spinach had firmer texture than steam or hot-water-blanched product. Texture of carrot slices was firmer as compared to hot water blanched due to less extent of tissue damage. Simultaneous IR dry-blanching and dehydration with continuous heating was used to produce high-quality partially dehydrated products for a	Van Zuilichem *et al.* (1985), (Ponne *et al.*, 1994), Gomez *et al.* (2005), Zhongli and McHugh (2006), Cenkowski *et al.* (2006), Yi *et al.* (2007), Yi and Zhongli (2009), Boudhrioua *et al.* (2009)

(Continued)

Table 13.1 (Continued)

Product	Salient Results	References
	variety of fruits and vegetables, such as pears, carrots, sweetcorn kernels, French fries, and apples. IR heating inactivated the peroxidase without discoloring and increased water absorption of oat groat. IR radiation for blanching and drying showed a significant increase in total phenol content and preserving the green color of fresh olive leaves.	
IR and Baking		
	Crust formation and the baking of the crumb took place simultaneously during bread baking using near-IR oven. Shorter wavelength resulted in greater heat penetration into the food samples and the formation of wet crust layers. Longer wavelength resulted in dry crust and faster coloring. Two-stage procedure for baking and crust formation of bakery products such as cakes and breads involving microwave and IR resulted in better product characteristics. Microwave-IR combination baking resulted in surface color development, reduced weight loss as well as firmness, and increased volume. Application of this technology with the addition of xanthan–guar gum blend in bread showed delay in staling. IR-baked tortillas showed good characteristics of rollability, puffing, layering, color, and texture. Sponge-cake baked in IR oven was softer as compared to the cake baked in an electric oven after 7 days storage.	Skjoeldebrand and Andersson (1989), Sato *et al.* (1992), Levinson (1992), Martinez *et al.* (1999), Sumnu *et al.* (2005, 2007), Yung *et al.* (2008), Turabi *et al.* (2008), Ozkoc *et al.* (2009)
IR and Cooking		
	IR results in energy saving and conservation of cooked-out material, which will not be lost to evaporation by high oven temperatures. The ease with which heat can be applied evenly over a broad surface area is an added advantage. Far-IR radiation was more efficient heating source for beef patties than mid-IR radiation. The presence of fat improved heat transfer in the case of far IR radiation giving shortened cooking times. An unsteady-state one-dimensional model for prediction of heat and mass transfers during cooking of beef patties by far-IR radiation was described. A method for determination of thermal diffusivity of beef burger patties during far-IR radiation was developed. Combination of IR-grilling and hot-air cooking was explored for hamburgers. Higher air velocity and air temperature during thermal processing led to shorter cooking times resulting in less-significant differences in moisture and total weight loss of the processed meat patties.	Sheridan and Shilton (1999), Shilton *et al.* (2002), Sheridan and Shilton (2002), Braeckman *et al.* (2009)

systematized in a simple set of rules called radiation laws. These laws apply when the radiating body is a black body radiator. A hypothetical body that completely absorbs all radiant energy falling upon it, reaches some equilibrium temperature, and then re-emits that energy as quickly as it absorbs it. The sum of the IR radiation that impinges on any surface has a spectral dependence because energy coming out of an emitter consists of different wavelengths and the fraction of the radiation in each band is dependent upon the temperature and emissivity of the emitter. The temperature of the IR heating elements governs the wavelength at which the maximum radiation occurs. The basic laws for black body radiation are described in the following section (Skjoldebrand, 2001; Krishnamurthy *et al.*, 2008a).

13.2.1 Planck's Law

Planck's law gives the spectral black body emissive power distribution, E(λ, T), of the radiation emitted by unit surface area into a fixed direction (solid angle) from the black body as a function of wavelength for a fixed temperature. The spectral characteristics of black body radiation from objects at different temperatures indicate that the curves give the maximum possible radiation that can be emitted at a selected temperature (Fig. 13.2). A black body produces maximum

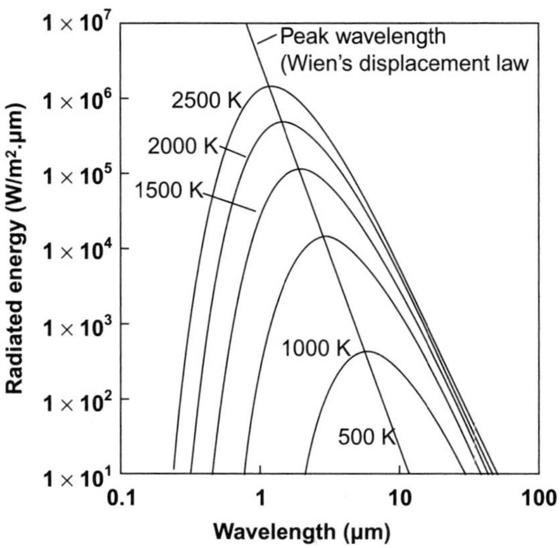

FIGURE 13.2

Spectral characteristics of black body radiation at different temperatures. (Sakai and Hanzawa, 1994).

intensity according to Planck's law, which can be expressed through the following equation:

$$E(\lambda, T) = \frac{2hc^2}{\lambda^5} \frac{1}{\exp\left(\frac{hc}{\lambda kT}\right) - 1} \tag{13.1}$$

where c is the speed of light (3×1010 cm/s), h and k are the Planck ($6.625 \times 10 - 27$ erg-s) and Boltzmann ($1.38 \times 10 - 16$ erg K^{-1}) constants, respectively, λ (cm) and T (K) are the wavelength and temperature, respectively.

13.2.2 Wien's Displacement Law

Wien's displacement law states that the wavelength of the most intense radiation (λ_{max}) emitted by a black body only depends on its temperature (T) according to the following formula:

$$\lambda_{max} = \frac{a}{T} \tag{13.2}$$

where a is 2897 μm K. λ_{max} and T are in μm and Kelvin, respectively. The Wien Law explains the shift of the peak to shorter wavelengths as the temperature increases.

13.2.3 Stefan–Boltzmann's Law

Stefan–Boltzman's law provides the total energy being emitted at all wavelengths (E_T) by the black body (which is the area under the Planck law curve, Fig.13.2) at a specific temperature from an IR source. Radiant heaters are not perfect radiators and foods are not perfect absorbers, although they do emit and absorb a constant fraction of the theoretical maximum. To take account of this, the concept of gray bodies is used, and the Stefan–Boltzmann equation is modified to (Skjoldebrand, 2001):

$$E_T = \varepsilon\sigma AT^4 \tag{13.3}$$

where ε, σ and A are the emissivity (varies from 0 to 1), Stefan–Boltzmann constant (5.670×10^{-8} Wm^{-2} K^{-4}) and surface area (m^2), respectively. Emissivity varies with the temperature of the body and the wavelength of the radiation emitted.

13.3 OPPORTUNITIES FOR THE INFRARED PROCESSING OF LIQUID FOODS

13.3.1 Microbes in Suspension

Far-IR radiation is easily absorbed by water and organic materials, which are the main components of food, and thus offers considerable potential for efficient pasteurization (Van Zuilichem *et al.*, 1986). Far-IR irradiation was shown to be more effective in pasteurizing vegetative bacterial cells in comparison to thermal conductive heating (Hashimoto *et al.*, 1991, 1993; Sawai *et al.*, 1997a). Moreover, Far-IR irradiation resulted in inactivation of *Bacillus subtilis* spores over a temperature range in which thermal conductive heating had no effect on spore viability (Sawai *et al.*, 1997b). The pasteurization effects of far-IR were reportedly due to the absorption of radiative energy by the bacterial suspension in a very thin volume near the surface and due to an increase in the bulk temperature of the suspension (Hashimoto *et al.*, 1991, 1992a). Temperature distribution within the far-IR-treated microbial suspension suggested that the temperature of the surface region was significantly higher than the bulk temperature (Hashimoto *et al.*, 1992b).

Sawai *et al.* (2000) demonstrated that *Escherichia coli* cells in phosphate-buffered saline irradiated with far-IR energy were injured and even killed under the condition where the bulk temperature of the suspension was maintained below the lethal temperature. Far-IR increased the temperature of the bacterial suspension, so it was cooled to maintain the bulk temperature at 40°C, considerably below the lethal temperature for *E. coli* (Fig. 13.3a). The survival ratio of *E. coli*

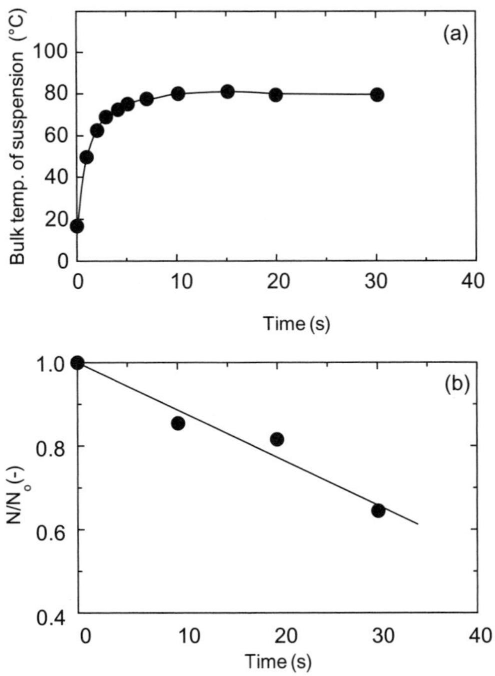

FIGURE 13.3

(a) Transient behavior of bulk temperature of a bacterial suspension irradiated by far-IR. (b) Pasteurization of *E. coli* by far-IR irradiation below the lethal temperature (40°C). (From Sawai *et al.*, 2000).

was found to decrease with an increase in irradiation. As most of the suspension was maintained at 40°C, the pasteurization effect by far-IR irradiation can not be attributed to an elevation in the bulk temperature (Fig. 13.3b).

Heat treatment causes a number of other reactions that may degrade the quality of the processed food. Therefore, it is important to attain the optimum level of pasteurization at as low a temperature as possible. Sawai *et al.* (2003) indicated that far-IR heating can decrease bacterial contamination levels more quickly and maintain greater enzyme activity levels than conductive heating (Fig. 13.4). The use of far-IR heating as a thermal treatment is expected to be highly effective in food processing of liquid foods. It has a potential for maintaining the required level of pasteurization at lower temperatures than conductive heating while maintaining the α-amylase activity and less change in lipase activity.

13.3.2 Orange Juice

Vikram *et al.* (2005) studied the kinetics of degradation of vitamin C in orange juice during IR heating and compared the results with conventional heating. The temperature profile for conventional and IR methods of heating are indicated in Figure 13.5a, b. Under both heating methods, temperature had a significant influence with the degradation of vitamin C being rapid at higher temperatures (Fig. 13.5c, d). The higher k-value or lower D-values in the case of the IR heating indicated that the degradation of vitamin C was higher relative to conventional heating (Table 13.2). In order to optimize the sterilization effect of IR heating in terms of nutrient (vitamin C, carotenoids) and color retention, as well as to arrive at a suitable time–temperature profile, further investigations are needed.

13.3.3 Milk

Krishnamurthy *et al.* (2008b) investigated the potential of IR heating for the processing of milk and demonstrated the efficacy of IR heating for inactivation of *Staphylococcus aureus*, a pathogenic microorganism found in milk. The effects of IR lamp temperatures (536 and 619°C), volumes of treated milk samples (3, 5, and 7 ml) and treatment times (1, 2, and 4 min) on the lethality of targeted microbes were found to be significant. Complete inactivation of *S. aureus* was obtained for 3- and 5-ml samples within 4 min at a 619°C lamp temperature, resulting in 8.41 $\log_{10}$ cfu/ml reduction. Larger sample volumes and lower temperatures resulted in lower reductions (Fig. 13.6). The results demonstrated that IR heating has a potential for effective inactivation of *S. aureus* in milk. To ensure the sterility efficacy, the heating patterns of milk samples under IR radiation were simulated using computational fluid dynamics codes in three dimensions with user-defined functions for radiative internal energy source terms. To validate the thermal behavior of IR-treated milk, the authors simulated in 3-D and compared the model-predicted temperature values with the

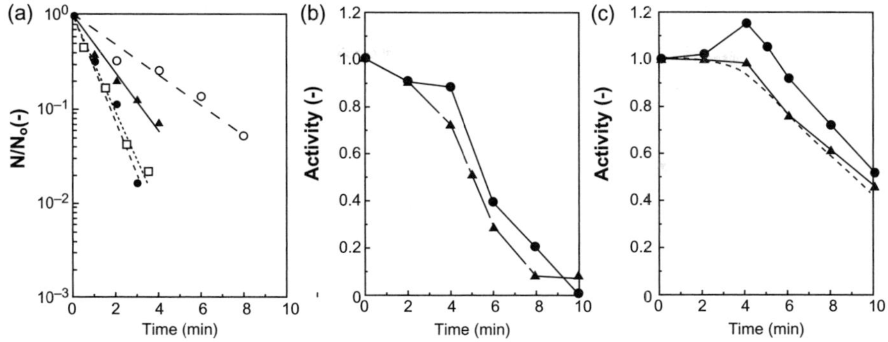

FIGURE 13.4

(a) Inactivation of E. coli by far-infrared (FIR) heating at different temperatures. ○, 56°C; ▲, 58°C; △, 60°C; •, 61°C. (b) Inactivation of lipase by FIR heating and thermal conductive heating. •, FIR; ▲, thermal conduction. (c) Inactivation of α-amylase by FIR heating and thermal conductive heating. •, FIR; ▲, thermal conduction; - - -, calculated from k for thermal conductive heating using a constant temperature water bath. From Sawai *et al.* (2003).

experimental data measured at the center of each sample. The predicted temperature values were in good agreement with the experimental data. The average percentage deviation between simulated and measured temperature values was 5.09% (Fig. 13.7).

Opaqueness of milk decreases the IR penetration depth, and very low milk-flow velocities are developed because of the small depths; thus, the overall heat transfer mechanism inside the food system seems to rely only on the conduction mode. Further, optimization of the process parameters, i.e. temperature and heating time, using the developed model is required to ensure that even under-processed food components will receive desirable doses of lethality, which can result in a commercially successful milk pasteurization method.

13.3.4 Honey

Honey, a natural biological product produced by bees from nectar and of benefit to humans both as medicine and food, is consumed in every country of the world in some form. The unprocessed honey tends to ferment within a few days of storage at ambient temperature because of its high moisture content and yeast count. To prevent fermentation, honey is heat processed before storage. Hebbar *et al.* (2003) explored the application of IR radiation for thermal processing of honey and also studied its effect on the physicochemical characteristics as well as the microbiological quality. IR heating achieved the desired results in a relatively shorter period offering advantages over the conventional method. IR heating caused substantial reductions in yeast count. For instance, heating for 5 min resulted in a

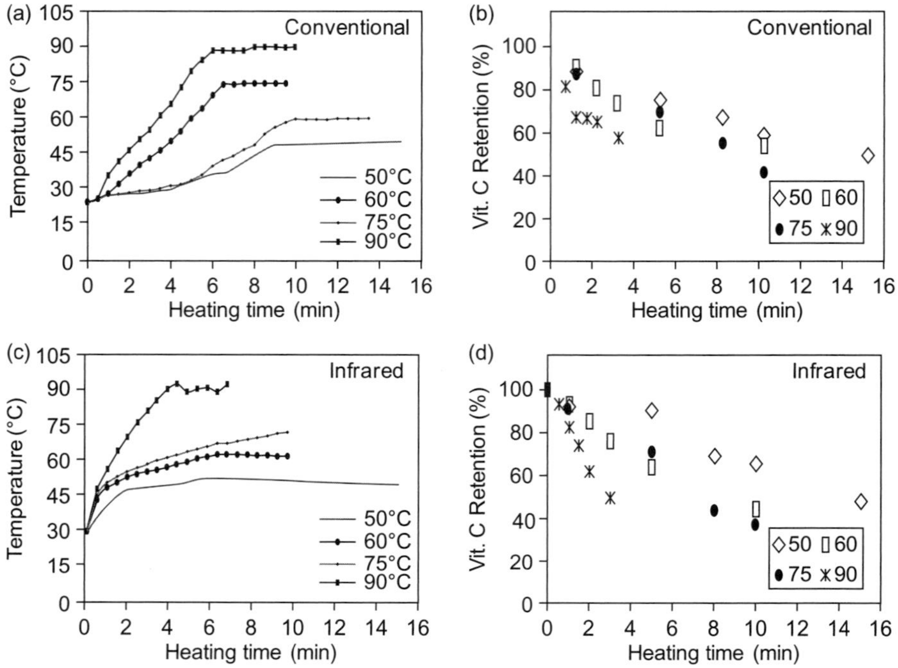

FIGURE 13.5

(a, b) Temperature profiles and (c, d) Vitamin C retention during heating conventional and IR heating at different temperatures. (From Vikram *et al.*, 2005).

Table 13.2 Rate Constant and Thermal Resistance Parameters for Vitamin C Degradation

Heating Method	Temperature(°C)	k-Value (min^{-1})	D-value (min)
Conventional	50	0.0351	65.67
	60	0.0462	49.81
	75	0.0852	27.02
	90	0.1784	12.91
IR	50	0.0444	51.91
	60	0.076	30.32
	75	0.0969	23.76
	90	0.2284	10.08

From Vikram et al. *(2005).*

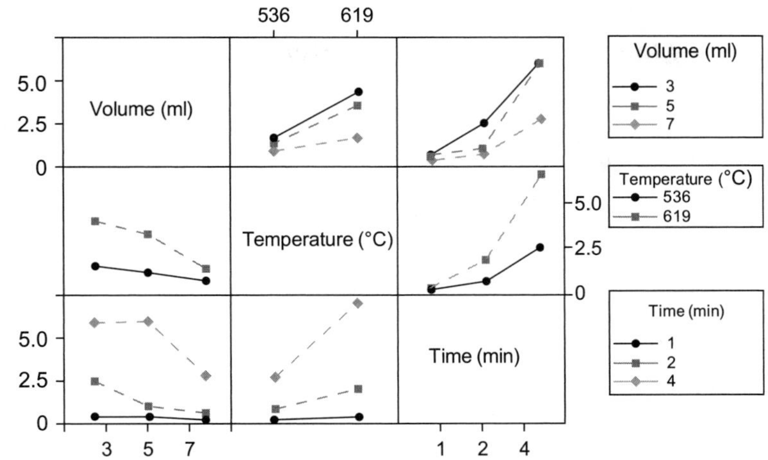

FIGURE 13.6

Interaction plot (fitted means) for $\log_{10}$ reduction. (From Krishnamurthy *et al.*, 2008b).

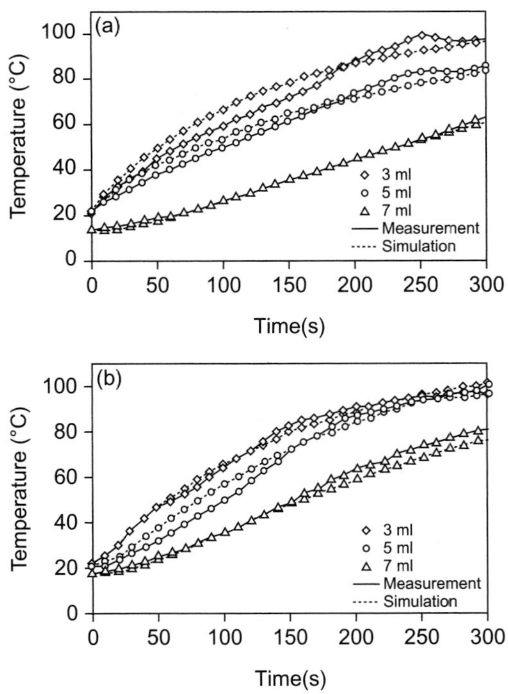

FIGURE 13.7

Comparison of simulation data with experimental measurements under different IR source temperatures: (a) 536, and (b) 619°C. (From Krishnamurthy *et al.*, 2008b).

Table 13.3 Continuous Heating of Honey with Infrared Radiation

S. No.	Time (min)	Temperature (°C)	Moisture (%)	Yeast count (cfu/ml)	HMF (mg/kg)	Diastase number
Control			21.8	7000	2	16.6
1	2	47	20.2	500	3.2	13.8
2	3	61	19.8	300	3.6	12.4
3	4	74	19.8	200	4.6	11.6
4	5	85	19.2	150	6.5	10.5
5	8	110	18.2	Nil	7.9	Traces

From Hebbar et al. *(2003).*

product temperature of 85°C, a hydroxymethylfurfural increase of 220% and a 37% drop in enzyme activity. Moreover, an increase in IR heating time to 8 min led to the complete inactivation of yeast. Consequently, the activity of diastase enzyme was found to decrease along with an increase in product temperature (110°C), which clearly indicated the excessive heating of honey (Table 13.3). A heating period of 3–4 min was reported to be adequate to obtain a commercially acceptable product, which met all the quality requirements in terms of hydroxymethylfurfural (≤ 40 mg/kg), diastase activity (DN ≥ 8), moisture content (19.8%), and yeast count (200–300 CFU/ml). Further studies will strengthen the relationship between processing conditions and honey quality in continuous-flow systems, which may be necessary for industrial adoption.

13.3.5 Beer

In order to extend the shelf-life of beer, it can be either thermally pasteurized or subjected to a sterile microporous filtration. The existing methods may have a negative impact on the quality of the beer. Thermal pasteurization may affect the flavor of the beer. Microporous filtration can trap all microbes that are present in the beer, but may also remove a lot of the aroma, body, and even flavor. Vasilenko (2001) demonstrated that near-IR radiation can be used in non-destructive methods of pasteurization of the beer because it does not cause molecular ionization and is not detrimental to beer quality. The results indicated that short-term exposure of beer to near-IR treatment strongly suppressed the propagation of yeast and inactivated bacteria. The average numbers of yeast and bacterial cells in treated beer were three- and 17-times less than in the control, respectively. The other advantages were in-pack processing, low energy consumption, low prices (cheaper than the commercial methods currently used by breweries worldwide), besides offering a high-quality product. The destructive effect of IR radiation on primitive bacterial membranes was indicated as a possible mechanism of action. Prokaryotes (bacteria) are probably more sensitive to near-IR than eukaryotes (yeast) because they do not possess a well-defined nuclear membrane and, therefore, are more vulnerable to IR radiation. The IR radiation is absorbed strongly by certain protein molecules of the membrane, resulting in local overheating and deterioration of cell structures.

13.4 EQUIPMENT FOR INFRARED PROCESSING OF LIQUID FOODS

IR processing of liquid foods can be an attractive alternative way of treating liquid foods due to the associated simplicity of construction and operation of the equipment. The other advantages include quick transient response, energy savings over other thermal processes, and easy accommodation with other modes of heat transfer such as convective or conductive (Sandu 1986). The radiator is the core element of IR equipment, which may be divided into two categories namely gas and electrically heated radiators. The electrical heaters are further classified as tubular/flat metallic heaters (long waves), ceramic heaters (long waves), quartz tube heaters (medium, short waves) and halogen tube heaters (ultra short waves).

Vikram *et al.* (2005) used an IR heater for the processing of orange juice, which consisted of a heating chamber with IR modules (250 W), which was equipped with reflectors to direct IR waves on to a platform. A temperature controller maintained the set temperature during processing. The provision was made to enable adjustment of the distance between the platform and the IR source in order to vary the intensity (Fig. 13.8). Hebbar *et al.* (2003) employed a near-IR batch oven for honey processing, which was fitted with IR lamps (1.0 kW, peak wavelength 1.1−1.2 μm). The distance between the sample and the source was fixed in such a way that uniform power intensity of 0.2 W/cm^2 was ensured.

Krishnamurthy *et al.* (2008b) employed a laboratory-scale, custom-made IR heating system with a cone-shaped waveguide for IR processing of milk. The IR heating system had six ceramic IR lamps (500 W) with a cast-in K-type thermocouple. All the lamps were fixed inside the closing on top of the waveguide and arranged symmetrically to the central axis of the waveguide (Fig. 13.9).

Hashimoto *et al.* (1991) and Sawai *et al.* (2003) used a far-IR heater consisting of a mulite cylinder and reflector. The reflector was placed at the top of an irradiation chamber made up of an aluminum plate. During irradiation, the enzyme solution was agitated with the help of a rotary shaker (Fig. 13.10).

13.5 CONCLUSIONS AND SUGGESTIONS FOR FUTURE WORK

Due to limited penetration power, IR heating is regarded primarily as a technology for surface-heating applications. To realize the optimum energy usage and

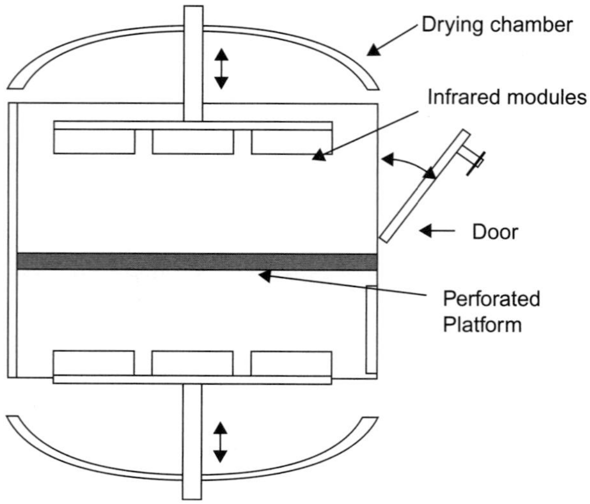

FIGURE 13.8

Batch type IR heater. (From Vikram *et al.*, 2005).

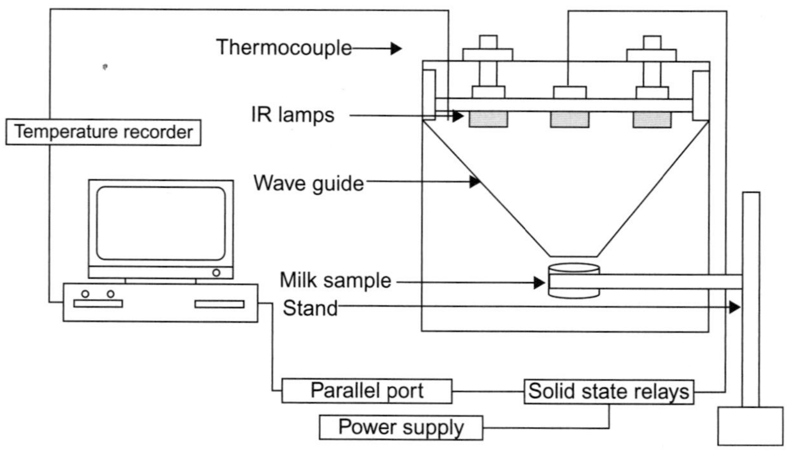

FIGURE 13.9

Schematic view of the laboratory-scale IR heating system. (From Krishnamurthy *et al.*, 2008b).

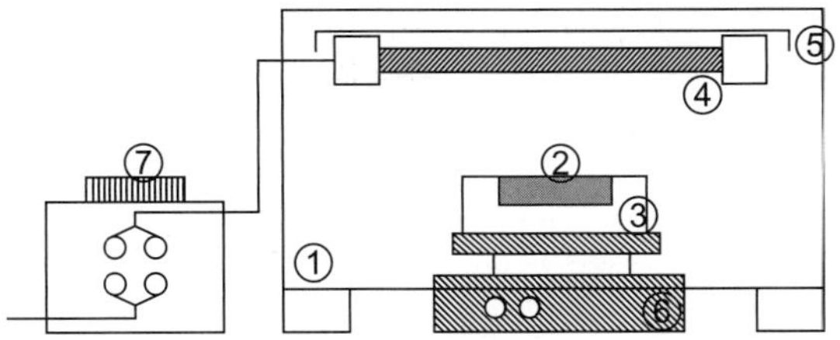

FIGURE 13.10

Schematic apparatus for FIR heating. From Sawai *et al.* (2003).

efficient and potential practical applicability of this technology, a combination of IR heating with microwave and other common conductive and convective modes of heating will be of great potential in the near future. The energy efficiency of IR heating is a beneficial feature, and is a driving force to support its application in the food-processing sector. The potential applications of IR heating in fluid foods (such as milk, honey fruit juice, and honey) discussed in this chapter will encourage researchers to look at the diversified uses of IR radiation in food processing. At the same time, it will also push the food-processing industries with more vigor to look at this alternative way of processing as a highly potential technique.

The various applications in food processing such as yeast reduction in honey without increasing the product temperature above 90°C and pasteurization of liquid food while maintaining the enzyme activity indicate the novel and diverse application of IR in food processing.

Spectral manipulation of IR radiation is expected to result in selective heating of food components. Proper spectral manipulation of IR radiation might result in selective heating of microorganisms in liquid food without actually heating the heat-sensitive components. Consequently, it will lead to fewer quality changes in comparison to conventional heating. Such a controlled radiation can stimulate the maximum optical response of the target object when the emission band of IR and the peak absorbance band of the target object coincide. Specific applications of IR radiation for selective heating of foods could be very useful and open up avenues for future research in this area.

Studies on the kinetics of nutrient (vitamin C or carotenoids) and color degradation during IR processing will facilitate the optimization of the heating process and will help researchers to arrive at a suitable time–temperature profile. The effects of irradiation on nutritional or sensorial characteristics and physicochemical properties, as well as interaction of food components under IR radiation, may further justify the use of IR radiation as a future food-processing option.

The applicability of this technology is expected to grow as food equipment manufacturers begin to realize its full potential.

Acknowledgments

The author is very grateful to Dr V. Prakash, Director, Central Food Technological Research Institute, Mysore, India for constant encouragement. Thanks are also due to Dr K.S.M.S. Raghavarao, Head, Food Engineering Department for support. I would like to express my sincere gratitude to Dr H. Umesh Hebbar for his help in providing useful information.

References

Afzal, T.M., Abe, T., 1998. Diffusion in potato during far infrared radiation drying. J. Food Eng. 3 (7), 353–365.

Boudhrioua, N., Bahloul, N., Slimen, I.B., Kechaou, N., 2009. Comparison on the total phenol contents and the color of fresh and infrared dried olive leaves. Indust. Crops Prod. 29, 412–419.

Braeckman, L., Ronsse, F., Cueva, H.P., Pieters, J., 2009. Influence of combined IR grilling and hot air cooking conditions on moisture and fat content, texture and colour attributes of meat patties. J. Food Eng. 93, 437–443.

Brown, R.B., Rothwell, T.M., Davidson, V.J., 2001. A fuzzy controller for infrared roasting of cereal grain. Can. Biosys. Eng. 43, 3.9–3.15.

Cenkowski, S., Ames, N., Muir, W.E., 2006. Infrared processing of oat groats in a laboratory scale electric micronizer. Can. Biosys. Eng. 48, 3.17–3.25.

Chatchai, N., Sakamon, D., Thanit, S., Somchart, S., 2007. Drying of banana slices using combined low pressure superheated steam and far infrared radiation. J. Food Eng. 81, 624–633.

Chung, D., 2008. Coffee bean roaster and method for roasting coffee beans using the same. US Patent 2008/0268119 A1.

Dagerskog, M., Osterstrom, L., 1979. Infra-red radiation for food processing I. A study of the fundamental properties of infra-red radiation. Lebens. Wissen. Technol 12, 237–242.

Gabel, M.M., Zhongli, P., Amaratunga, K.S.P., Harris, L.J., Thompson, J.F., 2006. Catalytic infrared dehydration of onions. J. Food Sci. 71, E351–E357.

Geun, P.H., Kook, B.S., Mi, J.C., Sang, G.M., 2009. Effects of air blast thawing combined with infrared radiation on physical properties of pork. Korean J. Food Sci. Ani. Resour. 29, 302–309.

Gomez, G.F., Toledo, R.T., Sjoholm, I., 2005. Tissue damage in heated carrot slices. Comparing mild hot water blanching and infrared heating. J. Food Eng. 67, 381–385.

Hamanaka, D., Dokan, S., Yasunaga, E., Kuroki, S., Uchino, T., Akimoto, K., 2000. The sterilization effects on infrared ray of the agricultural products spoilage microorganisms (part 1). An ASAE Meeting Presentation, Milwaukee, WI, July 9–12, No. 00 6090.

Hamanaka, D., Uchino, T., Furuse, N., Han, W., Tanaka, S.I., 2006. Effect of the wavelength of infrared heaters on the inactivation of bacterial spores at various water activities. Int. J. Food Microbiol. 108, 281–285.

Hashimoto, A., Shimizu, M., Igarashi, H., 1991. Effect of far infrared radiation on pasteurization of bacteria suspended in phosphate-buffered saline. Kagaku Kogaku Ronbunshu (in Japanese) 17, 627–633.

Hashimoto, A., Sawai, J., Igarashi, H., Shimizu, M., 1993. Irradiation power effect of IR pasteurization below lethal temperature of bacteria. J. Chem. Eng. Japan 26, 331–333.

Hashimoto, A., Igarashi, H., Shimizu, M., 1992a. Far-infrared irradiation on pasteurization of bacteria on or within wet-solid medium. J. Chem. Eng. Japan 25, 666–671.

Hashimoto, A., Sawai, J., Igarashi, H., Shimizu, M., 1992b. Effect of far-infrared irradiation of pasteurization of bacteria suspended in liquid medium below lethal temperature. J. Chem. Eng. Japan 25, 275–281.

Hebbar, U.H., Nandini, K.E., Lakshmi, M.C., Subramanian, R., 2003. Microwave and infrared heat processing of honey and its quality. Food Sci. Technol. Res. 9, 49–53.

Hebbar, U.H., Vishwanathan, K.H., Ramesh, M.N., 2004. Development of combined infrared and hot air dryer for vegetables. J. Food Eng. 65, 557–563.

Hee, S.C., 2006. Preparation methods of retort roast chestnuts using far infrared ray thawing. WO 2006/065018 A1.

Junling, S., Zhongli, P., McHugh, T.H., Wood, D., Hirschberg, E., Olson, D., 2008. Drying and quality characteristics of fresh and sugar infused blueberries dried with infrared radiation heating. LWT- Food Sci. Technol. 41, 1962–1972.

Kim, S.Y., Jeong, S.M., Jo, S.C., Lee, S.C., 2006. Application of far infrared irradiation in the manufacturing process of green tea. J. Agric. Food Chem. 54, 9943–9947.

Krishnamurthy, K., Khurana, K., Soojin, J., Irudayaraj, J., Demirci, A., 2008a. Infrared heating in food processing, an overview. Comp. Rev. Food Sci. Food Safety 7, 2–13.

Krishnamurthy, K., Soojin, J., Irudayaraj, J., Demirci, A., 2008b. Efficacy of infrared heat treatment for inactivation of *Staphylococcus aureus* in milk. J. Food Proc. Eng. 31, 798–816.

Krishnamurthy, K., Khurana, H.K., Jun, S., Irudayaraj, J.M., Demirci, A., 2009. Infrared radiation for food processing. In: Jun, S., Irudayaraj, J.M. (Eds.), Food Processing Operations Modeling: Design and Analysis. CRC Press, Boca Raton, FL, pp. 115–142.

Kumar, C.M., Rao, A.G.A., Singh, A.S., 2009. Effect of infrared heating on the formation of sesamol and quality of defatted flours from *Sesamum indicum* L. J. Food Sci. 74, H105–H111.

Leonard, A., Blacher, S., Nimmol, C., Devahastin, S., 2008. Effect of far infrared radiation assisted drying on microstructure of banana slices, an illustrative use of X ray microtomography in microstructural evaluation of a food product. J. Food Eng. 85, 154–162.

Levinson, M.L., 1992. Two stage process for cooking/browning/crusting food by microwave energy and infrared energy. WO 92/14369 A1.

Martinez, B.F., Morales, S.E., Chang, Y.K., Herrera, G.A, Martinez, M.J.L., Banos, L., et al., 1999. Effect of infrared baking on wheat flour tortilla characteristics. Cereal Chem. 76, 491–495.

Mongpraneet, S., Abe, T., Tsurusaki, T., 2002a. Accelerated drying of Welsh onion by far infrared radiation under vacuum conditions. J. Food Eng. 55, 147–156.

Mongpraneet, S., Abe, T., Tsurusaki, T., 2002b. Far infrared vacuum and convection drying of Welsh onion. Trans. ASAE 45, 1529–1535.

Nowak, D., Lewicki, P.P., 2004. Infrared drying of apple slices. Innov. Food Sci. Emerg. Technol. 5, 353–360.

Nowak, D., Lewicki, P.P., 2005. Quality of infrared dried apple slices. Drying Technol. 23, 831–846.

Ozdemir, M., Devres, O., 2000. Analysis of color development during roasting of hazelnuts using response surface methodology. J. Food Eng. 45, 17–24.

Ozkoc, S.O., Sumnu, G., Sahin, S., Turabi, E., 2009. Investigation of physicochemical properties of breads baked in microwave and infrared microwave combination ovens during storage. Eur. Food Res. Technol. 228, 883–893.

Park, J.H., Lee, J.M., Cho, Y.J., Kim, C.T., Kim, C.J., Nam, K.C., et al., 2009. Effect of far infrared heater on the physicochemical characteristics of green tea during processing. J. Food Biochem. 33, 149–162.

Pathare, P.B., Sharma, G.P., 2006. Effective moisture diffusivity of onion slices undergoing infrared convective drying. Biosys. Eng. 93, 285–291.

Ponne, C.T., Baysal, T., Yuksel, D., 1994. Blanching leafy vegetables with electromagnetic energy. J. Food Sci. 59, 1037–1041.

Poss, G.T., 2007. Roasting coffee beans. US Patent 7 235 764 B2.

Praveen Kumar, D.G., Hebbar, U.H., Sukumar, D., Ramesh, M.N., 2005. Infrared and hot air drying of onions. J. Food Proc. Preserv. 29, 132–150.

Praveen Kumar, D.G., Hebbar, U.H., Ramesh, M.N., 2006. Suitability of thin layer models for infrared hot air drying of onion slices. LWT-Food Sci. and Technol. 39, 700–705.

Rastogi, N.K., 2011. Application of infrared heating for food processing. Crit. Rev. Food. Sci. Nutr. (in press).

Rosenthal, I., Rosen, B., Bernstein, S., 1996. Surface pasteurization of Cottage cheese. Milchwissenschaft 51, 198–201.

Sakai, N., Hanzawa, T., 1994. Applications and advances in far-infrared heating in Japan. Trends Food Sci. Technol. 5, 357–362.

Sakai, N., Morita, N., Ping, Q., Hanzawa, T., 1995. Two dimensional heat transfer analysis of the thawing process of tuna by far infrared radiation. J. Japanese Soci. Food Sci. Technol. 42, 524–530.

Sakai, N., Mao, W., 2006. Infrared heating. In: Sun, D.W. (Ed.), Thermal Food Processing: New Technologies and Quality Issues. CRC Press, Boca Raton, FL, pp. 493–544.

Sandu, C., 1986. Infrared radiative drying in food engineering: a process analysis. Biotechnol. Prog. 2, 109–119.

Sato, H., Hatae, K., Shimada, A., 1992. Studies on radiative heating condition of foods. I. Effect of radiant characteristics of heaters on crust formation and coloring processes of food surfaces. J. Japanese Soci. Food Sci. Technol. 39, 784–789.

Sawai, J., Sagara, K., Igarashi, H., Hashimoto, A., Kokugan, T., Shimizu, M., 1995. Injury of *Escherichia coli* in physiological phosphate-buffered saline induced by far-infrared irradiation. J. Chem. Eng. Japan 28, 294–299.

Sawai, J., Fujisawa, M., Igarashi, H., Hashimoto, A., Kokugan, T., Shimizu, M., et al., 1997a. Pasteurization of bacterial spores in liquid medium by far-infrared irradiation. J. Chem. Eng. Japan 30, 170–172.

Sawai, J., Sagara, K., Miyagawa, M., Igarashi, H., Hashimoto, A., Kokugan, T., et al., 1997b. Injuries of *Escherichia coli* at exponential phase suspended in physiological phosphate buffered saline by far-infrared irradiation. J. Chem. Eng. Japan 30, 669–676.

Sawai, J., Sagara, K., Kasai, S., Igarashi, H., Hashimoto, A., Kokugan, T., et al., 2000. Far infrared irradiation induced injuries to *Escherichia coli* at below the lethal temperature. J. Indust. Microbiol. Biotechnol. 24, 19–24.

Sawai, J., Sagara, K., Hashimoto, A., Igarashi, H., Shimizu, M., 2003. Inactivation characteristics shown by enzymes and bacteria treated with far infrared radiative heating. Int. J. Food Sci. Technol. 38, 661–667.

Seyhun, N., Ramaswamy, H., Sumnu, G., Sahin, S., Ahmed, J., 2009. Comparison and modeling of microwave tempering and infrared assisted microwave tempering of frozen potato puree. J. Food Eng. 92, 339–344.

Sharma, G.P., Verma, R.C., Pankaj, P., 2005a. Mathematical modeling of infrared radiation thin layer drying of onion slices. J. Food Eng. 71, 282–286.

Sharma, G.P., Verma, R.C., Pankaj, P.B., 2005b. Thin layer infrared radiation drying of onion slices. J. Food Eng. 67, 361–366.

Sheridan, P.S., Shilton, N.C., 1999. Application of far infra red radiation to cooking of meat products. J. Food Eng. 41, 203–208.

Sheridan, P.S., Shilton, N.C., 2002. Determination of the thermal diffusivity of ground beef patties under infrared radiation oven shelf cooking. J. Food Eng. 52, 39–45.

Shi, J., Pan, Z., McHugh, T.H., Wood, D., Zhu, Y., Avena Bustillos, R.J., et al., 2008. Effect of berry size and sodium hydroxide pretreatment on the drying characteristics of blueberries under infrared radiation heating. J. Food Sci. 73, E259–E265.

Shih, C., Pan, Z., McHugh, T., Wood, D., Hirschberg, E., 2008. Sequential infrared radiation and freeze drying method for producing crispy strawberries. Trans ASABE. 51, 205–216.

Shilton, N., Mallikarjunan, P., Sheridan, P., 2002. Modeling of heat transfer and evaporative mass losses during the cooking of beef patties using far infrared radiation. J. Food Eng. 55, 217–222.

Skjoeldebrand, C., Andersson, C., 1989. A comparison of infrared bread baking and conventional baking. J. Microwave Pow. Electromag. Energy 24, 91–101.

Skjoldebrand, C., 2001. Infrared heating. In: Richardson, P. (Ed.), Thermal Technologies in Food Processing. Woodhead Publishing Limited, Abington Hall, Abington, England, pp. 208–228.

Skjoldebrand, C., 2002. Infrared processing. In: Henry, C.J.K., Chapman, C (Eds.), The Nutrition Handbook for Food Processors. Woodhead Publishing Limited, Abington Hall, Abington, England, pp. 423–433.

Sumnu, G., Sahin, S., Sevimli, M., 2005. Microwave, infrared and infrared microwave combination baking of cakes. J. Food Eng. 71, 150–155.

Sumnu, G., Datta, A.K., Sahin, S., Keskin, S.O., Rakesh, V., 2007. Transport and related properties of breads baked using various heating modes. J. Food Eng. 78, 1382–1387.

Sun, J., Hu, X., Zhao, G., Wu, J., Wang, Z., Chen, F., et al., 2007. Characteristics of thin layer infrared drying of apple pomace with and without hot air pre drying. Food Sci. Technol. Int. 13, 91–97.

Tan, M., Chua, K.J., Mujumdar, A.S., Chou, S.K., 2001. Effect of osmotic pre-treatment and infrared radiation on drying rate and color changes during drying of potato and pineapple. Drying Technol. 19, 2193–2207.

Thanit, S., Sakamon, D., Poomjai, S.A., Somchart, S., 2009. Mathematical modeling of combined far infrared and vacuum drying banana slice. J. Food Eng. 92, 100–106.

Togrul, H., 2005. Simple modeling of infrared drying of fresh apple slices. J. Food Eng. 71, 311–323.

Togrul, H., 2006. Suitable drying model for infrared drying of carrot. J. Food Eng. 77, 610–619.

Turabi, E., Sumnu, G., Sahin, S., 2008. Optimization of baking of rice cakes in infrared microwave combination oven by response surface methodology. Food Bioproc. Technol. 1, 64–73.

Uysal, N., Sumnu, G., Sahin, S., 2009. Optimization of microwave infrared roasting of hazelnut. J. Food Eng. 90, 255–261.

Van Zuilichem, D.J., Riet, V.T.K., Stolp, W., 1985. Food engineering and process applications. In: Maguer, L.M., Jelen, P. (Eds.), Transport Phenomena, vol. 1. Elsevier Applied Science. pp. 595–610.

Van Zuilichem, D.J., Vant Reit, K., Stolp, W., 1986. An overview of new infrared radiation processes for various agricultural products. Food Eng. Proc. Appl 1, 595–610.

Vasilenko, V., 2001. The pasteurization effect of laser infrared irradiation on beer. Tech. Q. MBAA 38, 211–214.

Vasilenko, V., 2010. Methods is apparatus for preserving perishable products. US Patent 7682641 B1. Date of filing June 11, 2003, Date of issuing March 23, 2010.

Vikram, V.B., Ramesh, M.N., Prapulla, S.G., 2005. Thermal degradation kinetics of nutrients in orange juice heated by electromagnetic and conventional methods. J. Food Eng. 69, 31–40.

Wesolowski, A., Glowacki, S., 2003. Shrinkage of apples during infrared drying. Pol. J. Food Nutr. Sci. 12/53, 9–12.

Yeu, P.L., Jen, H.T., An Erl King, V., 2005. Effects of far infrared radiation on the freeze drying of sweet potato. J. Food Eng. 68, 249–255.

Yeu, P.L., Tsai, Y.L., Jen, H.T., An Erl King, V., 2007. Dehydration of yam slices using far-IR assisted freeze drying. J. Food Eng. 79, 1295–1301.

Yi, Z., Zhongli, P., 2009. Processing and quality characteristics of apple slices under simultaneous infrared dry blanching and dehydration with continuous heating. J. Food Eng. 90, 441–452.

Yi, Z., Zhongli, P., McHugh, T.H., 2007. Effect of dipping treatments on color stabilization and texture of apple cubes for infrared dry blanching process. J. Food Proc. Preserv. 31, 632–648.

Yung, S.S., Wen, C.S., Ming, H.C., Jean, Y.H., 2008. Effect of far infrared oven on the qualities of bakery products. J. Culinary Sci. Technol. 6, 105–118.

Zhongli, P., and McHugh, T.H. (2006). Novel infrared dry blanching (IDB), infrared blanching, and infrared drying technologies for food processing. US Patent 2006/ 0034981 A1. Filed 8/13/ 2004, published 2/16/2006.

Zhongli, P., Shih, C., McHugh, T.H., Hirschberg, E., 2008. Study of banana dehydration using sequential infrared radiation heating and freeze drying. LWT Food Sci. Technol. 41, 1944–1951.

CHAPTER

Modeling the Kinetics of Microbial and Quality Attributes of Fluid Food During Novel Thermal and Non-Thermal Processes

14

V.P. Valdramidis[1], P.S. Taoukis[2], N.G. Stoforos[3] and J.F.M. Van Impe[4]

[1]*University College Dublin, Biosystems Engineering, School of Agriculture, Food Science & Veterinary Medicine, Belfield, Dublin, Ireland*

[2]*National Technical University of Athens, School of Chemical Engineering, Laboratory of Food Chemistry and Technology, Athens, Greece*

[3]*Agricultural University of Athens, Department of Food Science and Technology, Athens, Greece*

[4]*Katholieke Universiteit Leuven – Department of Chemical Engineering, BioTeC – Chemical and Biochemical Process Technology and Control, Leuven (Heverlee), Belgium*

14.1 INTRODUCTION

During the lifetime of a food product (from harvesting to consumption) it is exposed to many adverse factors. Processing is the step of the food chain that mainly affects its physical, chemical, and microbiological characteristics and determines the safety and shelf-life of the product. Optimal design of food processing is crucial in order to achieve the production of nutritive, safe products of high sensorial quality. Any optimization procedure implies the change of certain conditions, called decision variables, which are defined by the process engineer. These decision variables are process dependent and it is imperative that they are characterized by accurate and robust process diagnostics. The final aim of optimizing a process is the maximization or minimization of pre-defined criteria, the so-called objective functions (Banga *et al.*, 2008). In the case of food production,

Novel Thermal and Non-Thermal Technologies for Fluid Foods. DOI: 10.1016/B978-0-12-381470-8.00014-1

these criteria may include the minimization of product quality deterioration and the maximization of the financial profits. Quality characteristics related to food products include those directly affecting consumer choices (related to appearance, color, aroma, taste, freshness, and in general organoleptically assessed attributes) as well as internal product characteristics not directly detected by the consumers (e.g., vitamin content). The result of these optimization procedures should meet a number of predefined constraints usually imposed by national/international regulations. These can be based on a number of product characteristics, e.g., vitamin content, but they are mainly derived from safety and hygienic requirements. In the case of food production, they are formulated in terms of the level of pathogenic or spoilage microorganisms in the final product for consumption. The essential tools for performing (and achieving) the optimization and design of a process are the mathematical models.

Mathematical models are the main tools to quantitatively describe the impact of a process as well as storage conditions on a food. For that purpose *monitoring and control* of critical parameters are essential factors under consideration within the food chain. The knowledge of the processing parameters combined with the kinetics of quality attributes (e.g., color, enzymes) and target microorganisms is of utmost importance. Thereafter, *flexible* and *versatile* (mechanistic) mathematical models need to be developed. Factors that are considered for the development of models representing a physical system are accuracy of fit, ability to predict untested combinations of factors, incorporation of all relevant variables, possession of the minimum number of parameters, and/or use of *a priori* (micro-) biological, physical, (bio-) chemical knowledge, reparameterization for improving statistical properties (Franceschini and Macchietto, 2008).

A model-based process design in which (total) food quality is optimized, while safety constraints are satisfied simultaneously, is a driving force for the further evaluation of the performance of novel processes. If these studies are combined with quality and safety assessment during the marketing route of the products, then total food chain process engineering and preservation management can be implemented. In this way, novel thermal and non-thermal food processes favoring the current consumer attitudes, and development of high-quality processed products while obeying the National/International Regulations could be more rapidly commercialized.

As reported by the National Advisory Committee on microbiological criteria for foods (NACMCF) all pasteurization processes, including thermal and non-thermal, need to be validated through the use of process authorities, challenge studies, predictive modeling and/or safe harbors (Anonymous, 2006). The objective of this chapter is to discuss the predictive, kinetic modeling approaches dedicated to novel thermal and non-thermal processes when applied to liquid foods. Experimental issues, data-processing approaches, kinetic modeling practises, model structure characterization and validation are reviewed. All of these are tackled focusing on the development of valid modeling structures to be exploited for the design and optimization of thermal and non-thermal technologies.

14.2 *AD HOC* HYPOTHESES FOR KINETIC STUDIES

14.2.1 Are the Applied Processes Really Static?

The majority of thermal and non-thermal technologies are dynamic in their nature. For example, in the case of High Hydrostatic Pressure (HHP) processing, there is a time-period during which pressure increases until the desired level is reached, while the temperature rises due to adiabatic heating (e.g., Balasubramaniam *et al.* (2004), Stoforos *et al.* (2002), Valdramidis *et al.* (2009)). Similar temperature rises are observed during the application of high-pressure carbon dioxide processing (Garcia-Gonzalez *et al.*, 2009). Previous studies on HHP have illustrated that the effect of the pressure build-up time can induce a microbial stress response and hence make the process less effective (Balasubramaniam *et al.*, 2004; Smelt, 1998). This could be minimized experimentally, e.g., by employing high pressurization rates for reaching desired pressure conditions very fast (e.g., Erkmen and Dogan (2004)). Another phenomenon reported for HHP is that products may experience a temperature drop as a result of heat loss to the vessel (Shao *et al.*, 2010). These issues could also be tackled experimentally (e.g., through the development of insulated chambers to control temperature (Shao *et al.*, 2010)).

Another example of a dynamic processing environment is that of Pulsed Electric Field (PEF) processing where temperature inside PEF chambers increases during treatment (Jaeger *et al.*, 2010). These changes depend on the density and the electrical conductivity of the product. It has been illustrated that the chamber set-up influences the uniformity of the electric field strength. Considering a uniform distribution of electric-field strength simplifies mathematical models to predict microbial inactivation by PEF (Min *et al.*, 2007). Similar heat-transfer phenomena during processing of liquid foods are also evident in other thermal and non-thermal technologies. For more details, see the previous chapters of this book.

In order to develop kinetic models that describe the quality and safety attributes under static conditions, the processing parameters should be controlled, if possible, with suitable operating systems. Otherwise *ad hoc* hypotheses should be formulated (e.g., homogenous liquid systems, insignificant variations in the processing conditions), so that constant processing parameters can be considered in each of these studies. Alternatively, the application of fluid dynamics to describe the complete processing variables in time and space during each one of these processes (see also Chapter 2 of this book) will be required for the accurate modeling of the microbial inactivation and the quality deterioration.

14.2.2 The Case of Microbial Non-Linearities: Experimental Artefact or A Challenging Research Objective?

Especially for the case of microbial survival, some authors argued that non-linear kinetics under controlled processing conditions could be induced by heterogenous process conditions in the treatment chamber. These conditions are likely to mask the true inactivation kinetics (Donsi *et al.*, 2007) and can be attributed to: (i) non uniformity of treatment (e.g., heterogeneous distribution of the electric field in PEF treatments); (ii) formation of cell clusters exhibiting increasing resistance to the applied treatment; (iii) existence of distribution of sensitivities to the treatment within the microbial population. The last two phenomena, when described by appropriate kinetic models, can then be assigned to specific model parameters (e.g., shoulder length, specific inactivation rates of subpopulations, tailing). The interpretation of tailing effects in microbial kinetics has been debated extensively in recent years. In several articles, heat transfer inside the clump is mentioned as a cause of tailing. However, it has been demonstrated (Hasting *et al.*, 2001) that the size of a clump is so small that lag in heat transfer can be ignored. Overall, in case the tailing effect described by the applied models is not an artefact, it must be considered as a potential safety issue. Therefore, it could be of high interest to quantify it and account for it in industrial processes (Geeraerd *et al.*, 2004).

14.3 PARAMETERS OF INTEREST FOR NOVEL THERMAL AND NON-THERMAL PROCESSES

The developed models require the assessment of specific parameters on the studied safety and quality attributes. These parameters can be categorized in extrinsic and intrinsic parameters. An overview of these parameters is given on Tables 14.1 and 14.2.

14.4 EXPERIMENTAL DESIGN

Informative experimental data and *a priori* (micro-) biological, enzymatic and (bio-) chemical knowledge are the major ingredients for the development of useful predictive models during thermal and non-thermal processing. In general, the larger the quantity of data collected, the better the accuracy and reliability of the derived models. An example of a protocol for data generation for predictive microbiology recommends that, for a particular combination of conditions a minimum of 10 data-points should be collected, and that the majority of them should be positioned at areas of infection where the rate of change of the microbial kinetics is maximal (Walker and Jones, 1993). In cases of studies where more environmental factors are involved, concepts of full or fractional factorial experiment design are implemented (Juneja and Eblen, 1999; Rasch, 2004; Whiting *et al.*, 1996). Other studies for both microbial and enzyme kinetics have also indicated that not only the quantity of data but also the positioning of these data in time has a substantial influence on the model parameter uncertainty (Poschet *et al.*, 2004, 2005).

Table 14.1 Main Extrinsic Parameters of Novel Thermal and Non-Thermal Processes

PEF	HHP	Ultrasound	Irradiation	UV light	Ozone	Supercritical CO_2	Ohmic Heating	Microwave	Infrared Heating
Electric-field strength (kV/cm)	Temperature (°C)	Amplitude (μm)[1]	Dose (kGy)	Wavelength (nm)	Flow rate (l/min)	Pressure (MPa)	Time (min)	Time (min)	Temperature (°C)
Treatment time (μs)	Pressure (MPa)	Exposure time (min)	Temperature (°C)	Fluence (mJ/cm^2)	Concentration (μg/ml)	Time (min)	Temperature (°C)	Temperature (°C)	Time (min)
Pulse width (μs)	Holding/ decompression/ compression time (min)	Temperature (°C)	Time (min)	Time (ms)	Time (min)	Temperature (°C)	Voltage gradient (V/cm)	Frequency (MHz)	
Pulse shape	Pulse shape frequency (only on pulsed pressures)	Frequency (kHz)				Depressurization rate (MPa/min)			
Treatment temperature (°C)									
Energy applied per ml (J/ml)									

Units mentioned are related to the application scale.
[1]Reported as power too. Amplitude can be calculated from power.

Table 14.2 Main Intrinsic Parameters of Novel Thermal and Non-Thermal Processes

PEF	HHP	Ultrasound	Irradiation	UV Light	Ozone	Supercritical CO_2	Ohmic Heating	Microwave	Infrared Heating
Electric conductivity	pH	Volume	Mass	Transmissivity	pH	Water activity	pH	Ionic density	Composition
Density	Water activity	Composition	Bulk density	Composition (e.g., solid content)	Composition (solid content)	Physicochemical properties	Water activity	Moisture	Moisture
Viscosity	Composition					pH	Fat content	Density	
pH									
Water activity									

A novel methodology for the collection of more informative data via dynamic input profiles (for instance, temperature change(s) during an experiment) is based on concepts of Optimal Experiment Design for Parameter Estimation (OED/PE). This methodology has been applied so far only on traditional thermal processes (sterilization or mild heating) for microbial and chemical inactivation kinetics (Balsa-Canto *et al.*, 2007; Cunha *et al.*, 1997; Versyck *et al.*, 1999). Its application to novel thermal and non-thermal technologies for more accurate parameter estimation of the related kinetic models is of interest because most of these technologies exhibit transient phenomena by nature (e.g., build-up of pressure in HHP).

A very important aspect during data collection is the sampling procedure. This should be based on collecting primary samples in a random manner. Consequently, the series of samples will properly reflect the diversity of types and numbers of organisms associated with the product and the applied process, while the between-sample-variation will be minimized. In the case of microbial data, biologically independent replications (i.e. preparation of fresh inocula) are important to be considered.

14.4.1 Microbial Data for Developing Sound Modeling Approaches

Any novel thermal and non-thermal technology aiming at a pasteurized product should be assessed for the most resistant pathogen of public health concern that is likely to survive the process. Additionally, the initial cell numbers and real-life variations in concentration that occur before the application of the treatment need to be determined (Anonymous, 2006).

Microbiological methods of data collection vary among researchers, but the standard method used for describing the microbial kinetics is the total viable count (viability is defined as the ability to reproduce and form visible colonies) which is very laborious. The use of selective and non-selective media to enumerate injured cells for assessing the efficacy of novel thermal and non-thermal technologies is also of interest (Ponce *et al.*, 1999; Saldana *et al.*, 2010). Another important issue is the ability to assess low levels of microbial counts. A technique that has been widely applied in this respect is the Most Probable Number or MPN technique (also known as Multiple Tube Method). An example of its application for studies performed during pulsed-light treatments is described by Sauer and Moraru (2009). It is a sensitive method for analyzing samples that contain very low numbers of microorganisms ($<$10 CFU per g or ml). The most probable quantity of organisms per test sample is then found by the use of statistical tables (ISO 2007). These data can be incorporated within the classical plate count data and treated together for complete kinetic studies where the detection limit is thus significantly reduced. Consequently, tailing effects can be better identified and kinetic models can be developed for a wide range of microbial responses.

Fourier Transform Infrared has been considered for quantifying viable spores (Subramanian *et al.*, 2006). This very fast technique requires only a few minutes for sample preparation and infrared analysis.

Another method recently applied for evaluating the efficiency of non-thermal processes (HHP, PEF, Supercritical Carbon Dioxide) is flow cytometry (e.g., Ananta *et al.* (2004), Ananta and Knorr (2009), Jaeger *et al.* (2009), Kim *et al.* (2009)). This analytical technique, apart from providing quantitative viability information, yields better insight into the mechanism(s) of microbial inactivation during novel thermal and non-thermal technologies. As Ananta and Knorr (2009) argued, this technique could also contribute to improving industrial decontamination processes as the treatment regimes for minimizing residual metabolic activities can be identified.

Some of the above-mentioned techniques do require calibration with standard plate count measurements, in order to allow for a better interpretation of the results and proper application of kinetic models.

14.4.2 Enzymatic and Chemical Data for Kinetic Studies

Depending on the particular attribute of interest, distinct determination methods are used. For example, a titration method (refer to Boulekou *et al.* (2010), Krapfenbauer *et al.* (2006), Polydera *et al.* (2004), Riahi and Ramaswamy (2003), Rouse and Atkins (1955)) is used for pectin methylesterase activity measurements, a spectrophotometric method is used for polyphenoloxidase activity measurements (e.g., Guerrero-Beltran and Barbosa-Canovas (2006), Krapfenbauer *et al.* (2006), Weemaes *et al.* (1998)), while high-performance liquid chromatography (HPLC) analysis is applied to vitamin C and anthocyanins content determination (Hong and Wrolstad, 1990; Munyaka *et al.*, 2010; Oruña-Concha *et al.*, 1998; Polydera *et al.*, 2005). Depending on the particular food under examination, other chemical data, such as total phenolic content can be of interest (e.g., Folin–Ciocalteu assay (Singleton *et al.*, 1999)), polyphenolic profile based on HPLC analysis or lycopene content (measured spectrophotometrically (Oms-Oliu *et al.*, 2009)) can be of interest. Appearance, color, aroma, taste, freshness, and in general organoleptically determined characteristics rely on colorimetric measurements, taste panels, and appropriate statistical data interpretation. Obviously, the discussion made earlier, concerning the quality and quantity of experimental microbial data needed for proper modeling of the effects of novel thermal and non-thermal processes, also holds true when enzymatic or chemical or general quality changes are under investigation. For example, for proper kinetic modeling at least 50–75% reduction of the initial concentration of the studied index is advisable (Labuza, 1984). Apart from the classical enzymatic and chemical techniques, novel, rapid, specific methods of detection can provide the means for screening the quality dynamics of food and its components. Examples include headspace volatile measurements (Boff *et al.*, 2003), monitoring of off-flavor formation (Siegmund and Pollinger-Zierler, 2007) and determination of sugars and organic acids by the use of spectroscopic techniques coupled with chemometrics (Bureau *et al.*, 2009).

14.4.3 Collection of Data Under Dynamic Conditions

As discussed in Section 14.2.2, most of the thermal and non-thermal processes are dynamic by nature. The implementation of approaches in which parameter estimates are obtained under dynamic processing conditions has been shown to result in more accurate and precise parameter estimation (Valdramidis *et al.*, 2008). This kind of approach is based on the same principles as OED/PE, i.e. application of experimental approaches that aim at system excitation in such a way that the information content with respect to parameter estimation is maximal, while at the same time the model structure validity is not violated. These approaches have been applied so far to classical thermal treatments (e.g., thiamine loss by using 10 dynamic (unsteady) experiments (Banga *et al.*, 1993), or lipoxygenase inactivation (Garrote *et al.*, 2001)). Similar work needs to be performed for the thermal and non-thermal processes described in this book.

14.5 DATA (PRE)PROCESSING

For all the data collected during any kind of process, appropriate transformation of the data to stabilize the measurement error variance is a common practise. As it was reported by Van Impe *et al.* (2001), repeated sampling will not yield identical measurements of the dependent variable. This is due to experimental errors and/or (biological) process variability. Therefore, the variance transformations are required; some examples for microbiological data have been critically reviewed by Van Impe *et al.* (2001). Similar approaches, e.g., logarithmic transformations (Van Eylen *et al.*, 2008), can be applied to quality attribute data (vitamin, enzyme concentrations, etc.).

14.6 MODEL STRUCTURE CHARACTERIZATION OF MICROBIAL DYNAMICS

During recent years, models to describe the microbial dynamics during thermal and non-thermal processes have been developed based on different approaches. A lot of attention has been directed at relating the survival fraction directly to the process parameters. Characteristic examples have been reported by Hulsheger *et al.* (1981) and by Peleg (1995) to describe the microbial kinetics during PEF processing. In the first case, a Fermi's kinetic model representing the survivors fraction as a function of both the electric-field strength and the number of pulses was employed. Another example is the Hulsheger's kinetic model to describe the kinetics of survival curves, assuming a linear relationship between the natural logarithm of the survivors fraction and the electric field strength as well as a linear relationship between the natural logarithm of the survivors fraction and the natural logarithm of the PEF treatment time.

The focus in recent years has been on the collection of so-called complete kinetics studies, an approach typically applied in predictive microbiology. The advantage of this approach is that the developed models (after their validation) can be applied to predict the microbial population at any time of the process within the experimental range of conditions tested.

14.6.1 Primary Models

Microbial inactivation processes are traditionally described as loglinear. In the case of these survival curves, a linear relationship between the logarithm of the microbial population and time is considered (Anonymous, 2000). The implicit formulation of this paradigm is as follows:

$$\frac{dN}{dt} = -kN \quad with \quad k = \frac{ln(10)}{D} \tag{14.1}$$

which yields under static conditions:

$$\log_{10}(N) = \log_{10}(N(0)) - \frac{k_{\max} \cdot t}{ln(10)} \tag{14.2}$$

$$\log_{10}\frac{N(t)}{N_0} = -k' t \tag{14.3}$$

where N represents the microbial cell density (cfu/ml), N_0 is the initial microbial cell density (cfu/ml), k (where $k' = k \log(e)$) (1/min) is the first-order inactivation constant, and D (min) is the decimal reduction time. In this model it is assumed that all cells in a population have equal heat sensitivity and that the death of an individual is dependent upon the random chance that a key molecule or "target" within it receives sufficient heat (Cole *et al.*, 1993). An additional underlying assumption is that the probability of occurrence of an event, i.e. death of a microorganism does not change with time.

When considering the thermal inactivation of microorganisms and spores (Geeraerd *et al.*, 2005; Mafart *et al.*, 2010), there are eight commonly observed types of inactivation or survivor curves: linear curves, curves with a shoulder, linear curves with tailing, sigmoidal-like curves, biphasic curves, biphasic curves with a shoulder, concave curves, and convex curves (see Fig. 14.1). Some of these types have been applied to novel thermal and non-thermal inactivation kinetic modeling (e.g., Patil *et al.*, 2010a). In addition, some researchers assessed the microbial inactivation kinetics of thermal or non-thermal processes inspired by previously developed *growth* models (refer, for example, to Erkmen (2000) for application of High-Pressure Carbon Dioxide, and to Lee *et al.* (2009) who applied the modified Gompertz model and the log-logistic model during microbial inactivation studies at different ultrasound treatments).

These models can be categorized based on some of their features: expressed in a dynamic (implicit) form (i.e. differential equation), dependency on the initial

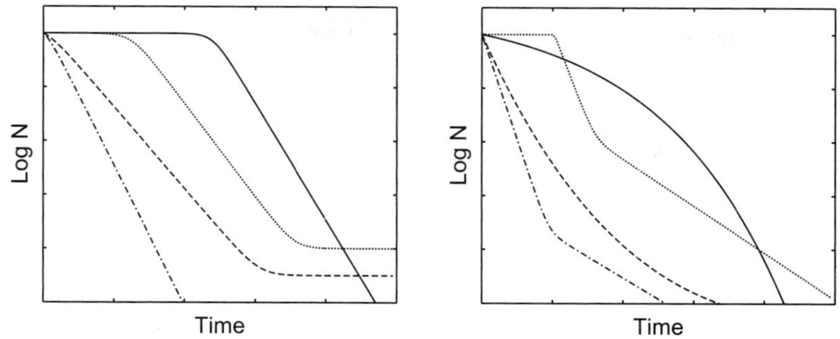

FIGURE 14.1

Commonly observed types of inactivation curves. Left plot: linear curves (— · —), linear curves with tailing (— —), sigmoidal-like curves (...), curves with a shoulder (-). Right plot: biphasic curves (— · —), concave curves (——), biphasic curves with a shoulder (...), and convex curve (-).

value of the dependent variable $N(0) = N_0$, or incorporation of available *a priori* knowledge. For more details on this categorization and the impact of some of these features, refer to Geeraerd *et al.* (2000) and Janssen *et al.* (2008). It should be highlighted that in most of the kinetic studies of thermal and non-thermal processes appearing in the literature, an explicit (or static) version of the kinetic models is considered. So, issues related to their dynamic character are not usually assessed. Nevertheless, the use of differential equations is required whenever the processing parameters are changing during a process, e.g., oscillating High Hydrostatic Pressure.

A useful tool for evaluating the described non-linearities during microbial inactivation is the (freely available) GInaFiT software (Geeraerd *et al.*, 2005) (www.cit.kuleuven.be/biotec). Estimation of the inactivation parameters by the use of this tool permits calculation of the time required to obtain a specific log reduction (refer to the next section) as defined by regulatory authorities for some liquid foods (e.g., FDA requires a 5-log reduction, see also Chapter 16) (Anonymous, 2006).

The types of models included in this software are described here by categorizing them as follows.

- *Class I: curves built up as an extension of the log-linear type model.* These are the biphasic curves, curves displaying a so-called shoulder before a log-linear decrease, curves displaying a so-called tail after a log-linear decrease, curves displaying both shoulder and tailing behavior, and the biphasic inactivation kinetics preceded by a shoulder. These curves will be described hereafter starting with the most extended version, i.e. the biphasic inactivation kinetics preceded by a shoulder

- *Class II: curves built up as an extension of the Weibull-type model.* These include concave curves, convex curves, convex/concave curves followed by tailing. This class will also be described starting with the most complex, i.e. convex/concave curves followed by tailing

14.6.1.1 Class I: Curves Built Up As An Extension of the Log-Linear Type Model

Geeraerd *et al.* (2005) presented a model that describes biphasic inactivation kinetics preceded by a shoulder. This model can be further reduced to different types of non-linear curves and the classical log-linear curve. The model has been inspired by the mathematical expressions of Cerf (1977) and Geeraerd *et al.* (2000). The dynamic version of the model reads as follows:

$$\frac{dN_1}{dt} = -k_{\max 1} \cdot N_1 \cdot \left(\frac{1}{1 + C_c}\right) \tag{14.4}$$

$$\frac{dN_2}{dt} = -k_{\max 2} \cdot N_2 \cdot \left(\frac{1}{1 + C_c}\right) \tag{14.5}$$

$$\frac{dC_c}{dt} = -k_{\max 1} \cdot C_c \tag{14.6}$$

As in the Cerf model (Cerf, 1977), the total population N (CFU/ml) equals the sum of the two subpopulations N_1 and N_2, and f is the fraction of the initial population in a major subpopulation ($f = N_1(0)/N(0)$). Here, the major population N_1 is more sensitive (k_{max1} is larger than k_{max2}) and a shoulder is incorporated. C_c is related to the physiological state of the cells (–) and more specifically ($1/1 + C_c$) describes the so-called shoulder effect (Geeraerd *et al.*, 2000, 2005).

The explicit version of this set of equations, considering that $C_c(0)$ equals $k_{max1} \cdot S_l - 1$ (Valdramidis *et al.*, 2005) with S_l a parameter representing the shoulder length (min) reads as follows (Geeraerd *et al.*, 2006):

$$\log_{10}(N) = \log_{10}(N(0)) + \log_{10} \cdot \left(f \cdot e^{-k_{\max 1} \cdot t} \cdot \left(\frac{e^{k_{\max 1} \cdot S_l}}{1 + (e^{k_{\max 1} \cdot S_l} - 1) \cdot e^{-k_{\max 1} \cdot t}}\right) + \ldots\right.$$
$$\left.\ldots + (1 - f) \cdot e^{-k_{\max 2} \cdot t} \cdot \left(\frac{e^{k_{\max 1} \cdot S_l}}{1 + (e^{k_{\max 1} \cdot S_l} - 1) \cdot e^{-k_{\max 1} \cdot t}}\right)^{\frac{k_{\max 2}}{k_{\max 1}}}\right) \tag{14.7}$$

Note that this model reduces (i) to the log-linear model when $k_{max2} = 0$ and $S_l = 0$, (ii) to the biphasic model when $S_l = 0$, and (iii) to the log-linear model with a shoulder model when $k_{max2} = 0$.

14.6.1.2 Class II: Curves Built Up As An Extension of the Weibull Type Model

The Weibull type curves are described based on the model structure proposed by Albert and Mafart (2005), thus including concave, convex, and linear curves followed by a tailing effect:

$$\log_{10}(N) = \log_{10}\left[(N(0) - N_{res}) \cdot 10^{\left(-\left(\frac{t}{\delta}\right)^{p}\right)} + N_{res}\right] \tag{14.8}$$

where δ (time units) is a scaling parameter denoting the time for the first decimal reduction if $p = 1$ where p (−) is a shape parameter. For $N_{res} = 0$ and $p > 1$, convex curves are obtained, while for $p < 1$, concave curves are described. If $N_{res} = 0$ and $p = 1$ then the model reduces to the log-linear model.

Representative examples of specific applications of these primary models are listed in Table 14.3.

It should be emphasized that these type of models are rather heuristic as they do not reflect an identified microbial inactivation mechanism during thermal or non-thermal treatments. Criteria to select the most appropriate model include: (*a posteriori*) interpretability of the model parameters, statistical performance (with respect to parameter estimation accuracy and accuracy of fit), and model parsimony (reflected in the smallest possible number of parameters).

14.6.2 Assessing the Achieved Log-Reduction

In order to asses the efficacy of a specific processing technology, the estimation of the time to achieve a certain number of log reductions can be applied. This has been previously discussed by some other researchers (Buchanan *et al.*, 1993; Valdramidis *et al.*, 2005; Xiong *et al.*, 1999). An example is included here in the case that microbial inactivation kinetics are described by a Weibull or a shoulder log-linear model. In the case of the Weibull model, the estimated parameters δ and p are used to calculate a desired log reduction. The time required to obtain an x-log reduction (t_{xd}) is calculated using Equation 14.9:

$$t_{xd} = \delta \cdot (x)^{\frac{1}{p}} \tag{14.9}$$

In the case of a shoulder log-linear model, the numerical values of S_l, $\log_{10}(N_O)$ and k_{max} are used to calculate the desired log reduction. The time required to obtain an x-log reduction (t_{xd}) is calculated using Equation 14.10 (Valdramidis *et al.*, 2005).

$$t_{xd} = S_l + (x) \cdot \frac{\ln(10)}{k_{\max}} \tag{14.10}$$

For a specific case study aiming at meeting the FDA requirement of 5-log reduction, x is equal to 5. Observe that for $S_l = 0$, Equation 14.10 reduces to log-linear inactivation. An example of the application for calculating the achieved log-reduction during ozonation of apple juice is given by Patil *et al.* (2010b).

Table 14.3 Representative Examples of Primary Models to Describe Microbial Inactivation Kinetics During Thermal and Non-Thermal Processing of Liquid (Model) Food Systems

Process	Model	Liquid System	References
PEF	Weibull	Media of different pH/ liquid whole egg	Saldana *et al.* (2010), Monfort *et al.* (2010)
	Log-linear	Grape juice	Marselles-Fontanet *et al.* (2009)
HHP	Biphasic	Model liquid system (HEPES/Aces buffer)	Reyns *et al.* (2000), Van Boeijen *et al.* (2008)
	Weibull	Carrot juice	Pilavtepe-Çelik *et al.* (2009)
	Log-linear with tailing	Model system (peptone water)	Koseki and Yamamoto (2007b)
	Quasi-chemical	Model system (whey protein in buffer)	Doona *et al.* (2005)
	Neural Networks	UHT whole milk	Amina *et al.* (2010)
	Other models		Guan *et al.* (2006), Klotz *et al.* (2007)
Ultrasound	Sigmoid-like shape (log-linear and shoulder)	Orange juice	Gomez-Lopez *et al.* (2010)
	Weibull, biphasic	Apple juice	Wang *et al.* (2010)
	Weibull type	Milk	Bermïdez-Aguirre *et al.* (2009)
	Sigmoidal-like model	Reconstituted infant formula	Adekunte *et al.* (2010)
Ultraviolet (Pulsed) light	Log-linear	Orange juice	Oteiza *et al.* (2010)
	Log-linear	Liquid egg	Unluturk *et al.* (2008)
Ozone	Sigmoid-like shape (shoulder log-linear)	Apple juice	Patil *et al.* (2010b)
	Weibull	Orange juice	Patil *et al.* (2009)
Supercritical CO_2	Sigmoid-like (shoulder/ loglinear/tail)	Models system (phosphate buffer)	Garcia-Gonzalez *et al.* (2010a)
	Weibull	Apple juice/model system	Ferrentino *et al.* (2008), Ferrentino *et al.* (2010)
	Other type of models	Model system (BHI)	Erkmen (2000)
Ohmic heating	Log-linear	Model system (phosphate buffers)	Guillou and El Murr (2002)
Microwave	Log-linear	Apple juice	Canumir *et al.* (2002), Tajchakavit *et al.* (1998)

14.6.3 Secondary Models

Secondary models describe the response of primary model parameters to changes in one or more intrinsic, extrinsic, and/or processing conditions (Whiting and Buchanan, 1993). There are different types of secondary models for application to microbial kinetics. All these models incorporate the critical intrinsic, extrinsic, and processing conditions to quantify their influence on the parameters of the primary models. Nevertheless, there are not many studies focusing on the development of secondary models. Some applications include polynomial models (e.g., for HHP (Valdramidis *et al.*, 2007a) and PEF treatments (Marselles-Fontanet *et al.*, 2009)), Arrhenius-type models (e.g., for HHP (Phua and Davey, 2007)), the Bigelow-type model (e.g., for ultrasound application (Adekunte *et al.*, 2010)), a log-logistic equation (for high pressure CO_2 process (Ferrentino *et al.*, 2010, 2008)), and inspired from microbial growth kinetics, the Belehradek/Ratkowsky-type models (e.g., for HHP (Koseki and Yamamoto, 2007b; Phua and Davey, 2007)).

The general format of a polynomial model for a primary model parameter k that describes the interaction between two intrinsic, extrinsic or processing conditions P_1 and P_2 (e.g., temperature and amplitude during ultrasound processing, or temperature and pressure during HHP) reads as follows:

$$k = \beta_0 + \beta_1 \cdot P_1 + \beta_2 \cdot P_1^2 + \beta_3 \cdot P_2 + \beta_4 \cdot P_2^2 + \beta_5 \cdot P_1 \cdot P_2 \tag{14.11}$$

where β_i are the polynomial coefficients. Recent developments suggest the application of a constrained-optimization approach combining the flexibility of polynomial models with the incorporation of *a priori* knowledge on the kinetic process (Valdramidis *et al.*, 2007a).

The Bigelow model was originally built to describe the effect of the temperature on the microbial inactivation rates. This reads as:

$$k_{\max} = \frac{\ln 10}{AsymD_{ref}} \cdot \exp\left(\frac{\ln 10}{z}(T - T_{ref})\right) \tag{14.12}$$

This model has physically interpretable parameters z, AsymD_{ref}. The parameter z indicates the change in temperature (°C) required to achieve a 10-fold change in Asym$D$-value. AsymD_{ref} (min) denotes the asymptotic decimal reduction time at a reference temperature T_{ref} (°C). (The D_{ref} parameter of the original Bigelow equation is replaced by the AsymD_{ref} as it describes the negative inverse of the slope of the log linear part of the inactivation data.)

In the case where the treatments involve more effects than the temperature, the model needs to be extended most often in a multiplicative way (e.g., Adekunte *et al.*, 2010):

$$k_{\max} = k_{\max}(T) \cdot k_{\max}(P_2) \tag{14.13}$$

$$k_{\max}(T, P_2) = \frac{\ln 10}{AsymD_{ref}} \cdot \exp\left(\frac{\ln 10}{z}(T - T_{ref})\right) \cdot \exp\left(\frac{\ln 10}{z_{P_2}}(P_2 - P_{2ref})\right) \tag{14.14}$$

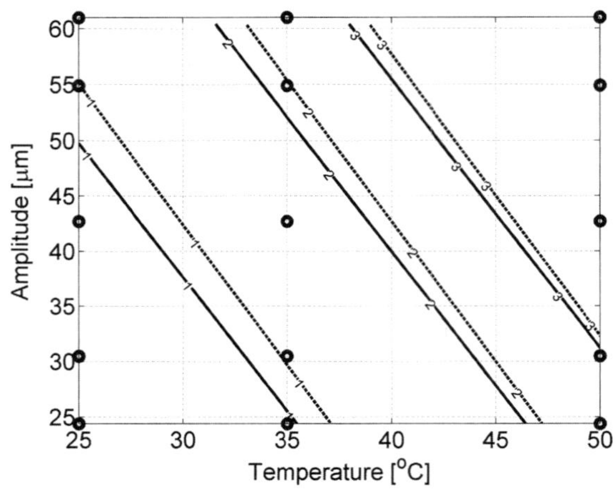

FIGURE 14.2

Specific inactivation rate k_{max} at values equal to 1, 2, 3, 1/min of *C. sakazakii* strain ATCC 11467 (-), strain NCTC 08155 (–) (based on the format of Equation 14.14) and combination of temperature and amplitude at which experimental data were collected (o). (From Adekunte *et al.* (2010)).

P_2 stands for any other processing parameter of a novel process (e.g., high pressure, amplitude during ultrasound).

Plotting iso-rate contour lines for the developed secondary models is very informative to assess the applied (thermal or non-thermal) technology efficiency (for example, refer to Fig. 14.2).

14.6.4 Probabilistic Type of Models

Some probabilistic models have also been developed to assess the efficiency of novel food processes. For that purpose, different logistic models have been constructed to describe microbial survival/death, recovery/no recovery (following a processing step) and spoilage/no spoilage interfaces. These models aim at establishing a correlation between the studied binary response variable (1 or 0, indicating survival or death, spoilage or no spoilage, recovery or no recovery, respectively) and the set of the studied explanatory (i.e. processing, post-processing) variables (see Fig. 14.3 for an example). Currently, the following applications of probabilistic models to liquid foods (e.g., milk, apple juice) after the application of HHP have been developed:

- survival/death of *C. sakazakii* in reconstituted infant formula (Koseki *et al.*, 2009) and liquid model systems (Koseki and Yamamoto, 2007a);
- recovery/no recovery of *Listeria monocytogenes* in milk (Koseki *et al.*, 2008), and of spoilage microorganisms in apple juice (Valdramidis *et al.*, 2009);
- spoilage/no spoilage of apple juice (Valdramidis *et al.*, 2009).

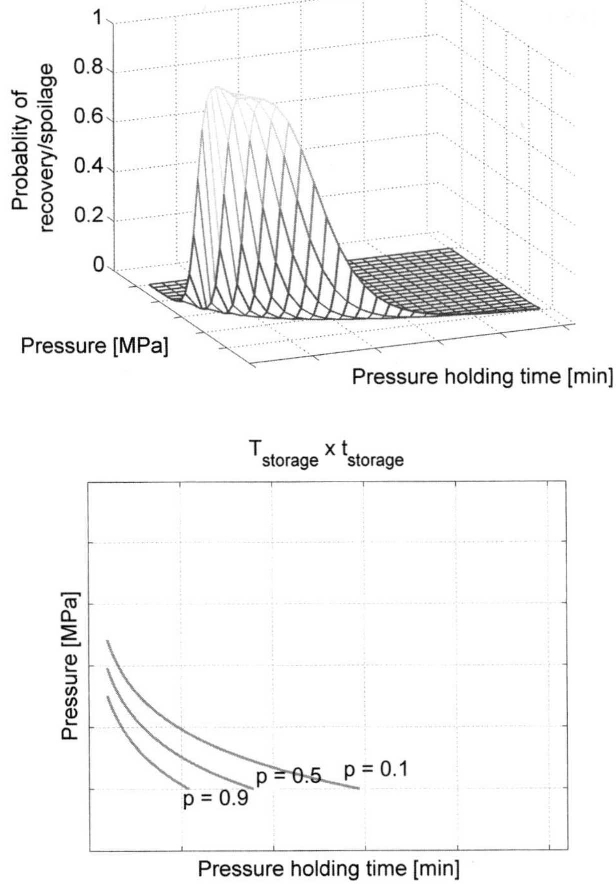

FIGURE 14.3

An illustration of a probabilistic modeling approach in which the probability of recovery/spoilage post-high-pressure treatment at different pressure level and pressure holding time and specific storage time and temperature. Top plot: three-dimensional illustration, bottom plot: two-dimensional projection identifying the interfaces of recovery/spoilage at different probability values, *p*.

The most common forms developed are (i) a simple logistic model without interaction terms and (ii) a polynomial logistic equation. An example of these equations is presented hereunder for a set of four parameters, i.e. high-pressure amplitude (HHP), pressure holding time (t_{HHP}), storage time after pressure ($t_{storage}$), storage temperature after pressure ($T_{storage}$).

$$\text{logit}(p) = \mathrm{b_0 + b_1 \cdot (HHP) + b_2 \cdot log10(t_{HHP}) + b_3 \cdot (t_{storage}) + b_4 \cdot (T_{storage})} \quad (14.15)$$

$$\begin{aligned}\text{logit}(p) = {} & \text{b}_0 + \text{b}_1 \cdot (\text{HHP}) + \text{b}_2 \cdot \log10(\text{t}_{\text{HHP}}) + \text{b}_3 \cdot (\text{t}_{\text{storage}}) \\ & + \text{b}_4 \cdot (\text{T}_{\text{storage}}) + \text{b}_5 \cdot (\text{HHP})^2 + \text{b}_6 \cdot (\log10(\text{t}_{\text{HHP}}))^2 + \text{b}_7 \cdot (\text{t}_{\text{storage}})^2 \\ & + \text{b}_8 \cdot (\text{T}_{\text{storage}})^2 + \text{b}_9 \cdot (\text{HHP}) \cdot \log10(\text{t}_{\text{HHP}}) + \text{b}_{10} \cdot (\text{HHP}) \cdot (\text{t}_{\text{storage}}) \\ & + \text{b}_{11} \cdot (\text{HHP}) \cdot (\text{T}_{\text{storage}}) + \text{b}_{12} \cdot \log10(\text{t}_{\text{HHP}}) \cdot (\text{t}_{\text{storage}}) \\ & + \text{b}_{13} \cdot \log10(\text{t}_{\text{HHP}}) \cdot (\text{T}_{\text{storage}}) + \text{b}_{14} \cdot (\text{t}_{\text{storage}}) \cdot (\text{T}_{\text{storage}})\end{aligned} \tag{14.16}$$

where p is the probability of the studied event (survival, spoilage or recovery) with logit(p) defined as ln $[p/(1-p)]$, and coefficients b_i the parameters of the models.

14.7 MODEL STRUCTURE CHARACTERIZATION OF CHEMICAL DYNAMICS

Food is a physicochemical system of high complexity involving, next to the microbiological dynamics previously discussed, numerous physical and chemical changes. Food quality (Q) changes during and post-processing in general are expressed as a function of composition and environmental factors as:

$$\frac{dQ}{dt} = F(C_i, E_j) \tag{14.17}$$

where C_i are composition factors (such as concentration of reactive compounds, inorganic catalysts, enzymes, reaction inhibitors, pH, water activity) and E_j are processing parameters or environmental factors (such as temperature, relative humidity, total pressure and partial pressure of different gases, light or other frequency radiation and mechanical stresses depending on the applied thermal or non-thermal technology). Even if the complete system could be explicitly expressed in terms of measurable parameters, in general no analytical solution exists and numerical solutions are required for any practical purpose. The established practical methodology consists of first identifying the chemical reactions that influence the quality of the food. Then, through a careful study of the food components and the process, the reactions judged to have the most critical impact on the quality and safety are determined (Taoukis *et al.*, 1997).

14.7.1 Primary Models

In an effort to generalize the description of non-microbial changes, the *Quality Function* concept is followed. Based on the previous analysis and without underestimating the underlying complexity of food systems, food-quality loss is in practice represented by the loss of desirable quality factors A (such as nutrients, characteristic flavors) or the formation of undesirable factors B (such as off-flavors, microbial load, discoloration). In most cases, the rate of loss of A

(correspondingly of formation of *B*) is expressed with an *n*th order reaction kinetics approach:

$$r_A = \frac{-d[A]}{dt} = k \cdot [A]^n \tag{14.18}$$

The quality factors [*A*] are usually quantifiable chemical, physical, or sensory parameters characteristic of the particular food system, *k* the apparent reaction rate constants, and *n* the apparent reaction order. The apparent reaction order and rate constant are determined by fitting the change with time of the experimentally measured values of [*A*] to Equation 14.18. In the integral method, variables are separated and integration is carried out. Thus for Equation 14.18, we have:

$$-\int_{A_o}^{A} \frac{d[A]}{[A]^n} = k \cdot t \tag{14.19}$$

Regardless of the value of *n* (the order of the reaction) Equation 14.19 can be expressed in the form:

$$f(A) = k \cdot t \tag{14.20}$$

where the expression *f*(*A*) is defined as the *quality function* of the food. The determination of the quality function, i.e. the relationship between quality loss and time, refers to the development of the primary model mentioned earlier and is the core of any quality-modeling scheme.

The form of the quality function of the food for an apparent zero-, 1st-, 2nd-, and nth-order reaction can be derived from Equation 14.19, and is shown in Table 14.4 along with the half-life time of the reaction, i.e. the time for the concentration of the quality index *A* to reduce to half its initial value (A_t being the concentration of the quality index *A* at time *t* and A_o the corresponding initial concentration). A more detailed discussion on the definition and methodology of determination of the quality function and the involved caveats can be found in Taoukis *et al.* (1997).

For first-order reactions, similarly to the classical thermobacteriological approach (Ball and Olson, 1957; Stumbo, 1973), the *D*-value can be used instead

Table 14.4 Quality Function and Half-Life Time for Deterioration of Quality Index A

Apparent Reaction Order	Quality Function $f(A)_t$	Half-Life $t_{1/2}$
0	$A_o - A_t$	$A_o/(2 \cdot k_o)$
1	$\ln(A_o/A_t)$	$\ln 2/k_1$
2	$1/A_t - 1/A_o$	$1/(k_2 \cdot A_o)$
n (n≠1)	$\frac{1}{n-1} \cdot (A_t^{1-n} - A_o^{1-n})$	$\frac{2^{n-1}-1}{k_n(n-1)} \cdot A_o^{1-n}$

of the apparent reaction rate constant k, with D and k being related by Equation 14.1. For cases where a remaining stable fraction of the parameter under study exists, a modification of the first-order model has been proposed (Van den Broeck *et al.*, 2000):

$$\ln\left(\frac{A_o - A_\infty}{A_t - A_\infty}\right) = k \cdot t \tag{14.21}$$

where A_∞ is the stable fraction of the parameter under consideration remaining after processing or storage for infinite time.

Deviations from the above-described kinetic approach do exist. For example, for vitamin loss prediction during non-isothermal heat processes and storage, Corradini and Peleg (2006) proposed the following equation:

$$\frac{A_t}{A_o} = \exp(-b(T) \cdot t^{n(T)}) \tag{14.22}$$

where $b(T)$ and $n(T)$ are temperature-dependent parameters.

Furthermore, any of the primary models presented earlier for microbial inactivation are equally applicable to describe quality changes as a function of time, if such a model appears to be appropriate. One of these is the use of a biphasic kinetic model which, for example, can describe two enzyme fractions with different processing stability both inactivating according to a first-order model (Jolie *et al.*, 2009). In cases where the above-described equations are not adequate for modeling the measured quality loss, case dependent, alternative (primary) models have been used. The steps for developing such models have been described (Saguy and Karel, 1980).

14.7.2 Secondary Models

In order to include in the quality function the effect of the processing parameters or the environmental factors, E_j, the commonly used approach is to model this effect into the apparent reaction rate constant, i.e. expressing k of Equation 14.20 as a function of E_j: $k = k(E_j)$. The factor most often considered and studied is temperature, but additional parameters, such as pH, a_w, pressure, or any other parameter depending on the type of thermal and non-thermal process under consideration, may appear equally important. Defining the $k(E_j)$ function for the target safety indices of the food system will allow the determination of the range of required process conditions to achieve a safe food. However, process design optimization requires the knowledge of the relevant quality functions and respective $k = k(E_j)$ values so that the finally selected conditions yield the lowest-possible quality degradation.

For temperature, the Arrhenius relation (Equation 14.23) developed theoretically for chemical and biochemical reactions has been experimentally shown to hold empirically for a number of more complex chemical and physical phenomena:

$$k = k_A \cdot \exp\left(\frac{-E_A}{RT}\right) \tag{14.23}$$

where k_A represents the Arrhenius equation constant, E_A is defined as the activation energy of the reaction that controls quality loss, R is the universal gas constant (8.3144 J/mole · K) and T is the absolute temperature. The activation energy of food-related chemical reactions usually falls within 30–150 kJ/mol.

Food quality-loss reactions described by the aforementioned kinetic models have also been reported widely in the literature to follow an Arrhenius-type behavior for temperatures within the particular temperature range of practical interest for each food product category. Another form, frequently used in the food literature, uses a constant reference temperature T_{ref}, characteristic of the process, and then Equation 14.23 assumes the following form:

$$k_{\mathrm{T}} = k_{\mathrm{T}_{ref}} e^{-\frac{E_A}{R}\left(\frac{1}{T} - \frac{1}{T_{ref}}\right)} \tag{14.24}$$

where k_{Tref} is the reaction rate-constant at T_{ref}.

Again, inspired by microbial destruction experiments, for first-order reactions the z-value is used instead of the E_A parameter to express the temperature effect on the destruction rates. The z-value is the temperature difference required to achieve a decimal change of the D_T-value and it is defined through:

$$D_T = D_{T_{ref}} \cdot 10^{\frac{T_{ref} - T}{z}} \tag{14.25}$$

The above definition applies for any heat-labile substance. Eliminating the k and D parameters between Equations 14.24 and 14.25 (through Equation 14.1), we obtain the following relationship between z and E_A:

$$z = \frac{\ln(10) \cdot R \cdot T_{ref} \cdot T}{E_A} \tag{14.26}$$

Note that in Equation 14.26, the temperatures T ant T_{ref} represen absolute temperatures (i.e. expressed in K) while the z-value is either in °C or K. Furthermore note that although both E_A and z are assumed independent of temperature, Equation 14.26 suggests the opposite. However, as long as there is not extrapolation beyond the experimental temperature range used for collecting thermal destruction data, both models are considered satisfactory for thermal destruction calculations (Johnson *et al.*, 1977; Ramaswamy *et al.*, 1989).

An expression fully analogous to Equation 14.25 has been used (through the z_P-value) to describe the pressure difference required to achieve a decimal change of the D_T-value:

$$D_P = D_{P_{ref}} \cdot 10^{\frac{P_{ref}-P}{z_P}} \tag{14.27}$$

Following the Arrhenius approach, and by combining Equations 14.20 and 14.24, the quality of a food product during and post-processing can be expressed by the measurement of a characteristic quality index, A, as:

$$f(A) = k \cdot t = k_{ref} \cdot \exp\left[-\frac{E_A}{R} \cdot \left(\frac{1}{T} - \frac{1}{T_{ref}}\right)\right] \cdot t \tag{14.28}$$

The value of the quality function $f(A)_t$ at time t after exposure of the food to a known variable temperature $T(t)$, during processing or storage, can be found based on Equation 14.29, by calculating the integral of $k[T(t)]$ over time, from time 0 to t, that is,

$$f(A)_t = k_{ref} \cdot \int_0^t \exp\left[-\frac{E_A}{R} \cdot \left(\frac{1}{T(t)} - \frac{1}{T_{ref}}\right)\right] dt \tag{14.29}$$

The integral can be calculated analytically for simple $T(t)$ functions or numerically for more complex ones.

Similarly to novel thermal processes, kinetic models for non-thermal processes can be built based on the same principles. For example in the case of the combined HHP and temperature processing, assuming that the effect of temperature (under constant pressure) as well as the effect of pressure (under constant temperature) on the reaction rate-constant can be expressed through the activation energy (see Equation 14.24) and the activation volume (V_A) concept, respectively (Johnson and Eyring, 1970), that is,

$$k_P = k_{P_{ref}} \cdot \exp\left(-\frac{V_A}{R} \cdot \frac{(P - P_{ref})}{T}\right) \tag{14.30}$$

and allowing for time varying pressure and/or temperature during processing, and further assuming that a pressure dependence of E_A and a temperature dependence of V_A can exist, then the impact of a particular process on the factor of interest, expressed here with its concentration, A, is given by:

For $n = 1$

$$\log\left(\frac{A}{A_o}\right) = \int_0^t -k_{ref} \cdot \exp\left[-\frac{E_A(P)}{R} \cdot \left(\frac{1}{T} - \frac{1}{T_{ref}}\right) - \frac{V_A(T)}{R} \cdot \frac{(P - P_{ref})}{T}\right] dt \tag{14.31}$$

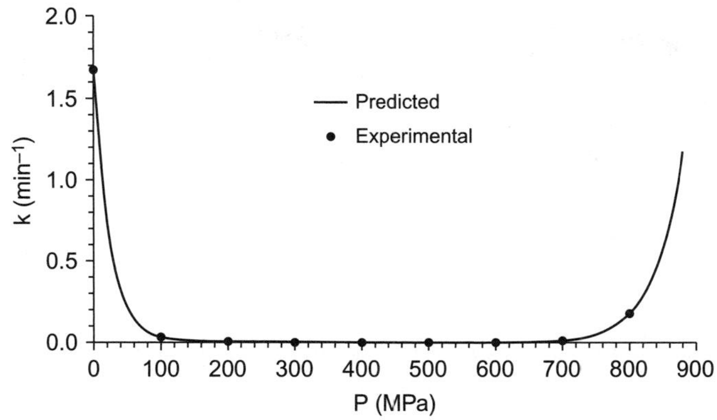

FIGURE 14.4

Effect of pressure on the apparent reaction rate constant for tomato PME inactivation at 75°C. (From Stoforos *et al.* (2002), with permission.)

and for $n \neq 1$

$$A^{1-n} - A_o^{1-n} = \int_0^t (n-1) \cdot k_{ref} \cdot \exp\left[-\frac{E_A(P)}{R} \cdot \left(\frac{1}{T} - \frac{1}{T_{ref}} \right) - \frac{V_A(T)}{R} \cdot \frac{(P - P_{ref})}{T} \right] dt \tag{14.32}$$

Models of inactivation of different enzymes as function of pressure and temperature, based on Equation 14.31, have been reported in several publications (Boulekou *et al.*, 2010; Gogou *et al.*, 2010; Katsaros *et al.*, 2009a, b, c, 2010).

Stoforos *et al.* (2002) suggested an extended version of Equation 14.31 in order to describe the antagonistic effect of the processing parameters on the inactivation rate of pectin methylesterase by temperature and high pressure. Its featuring characteristics are given in Figure 14.4.

Deviations from the Arrhenius equation and alternative approaches for modeling the processing-parameter dependence on the reaction rates have been reported and dealt with in the literature. Empirical models have been developed to relate parameters of the primary models to the processing conditions (e.g., exponential models for PEF treatments (Odriozola-Serrano *et al.*, 2008)). The use of Ratkowsky-type models (Ratkowsky *et al.*, 1983) has been recently proposed for describing the effects of amplitude and temperature with respect to parameters derived from the kinetics of the non-enzymatic browning and ascorbic acid concentrations of orange juice during ultrasound processing (Valdramidis *et al.*, 2010a). Other researchers proposed the development of thermodynamic-based kinetic models for studies focusing on enzyme kinetics during HHP treatments (Ly-Nguyen *et al.*, 2003). These models encompass numerous parameters (up to nine) and although they have a thermodynamic background, they are empirical in nature. Response surface methodology appears to be another secondary-modeling approach applied in thermal and non-thermal processes for a number of different quality indices (e.g., Rawson *et al.* (2011), Tiwari *et al.* (2010), Torres *et al.* (2011)).

14.8 MODEL PARAMETER ESTIMATION

Next to the mathematical model structures, suitable kinetic parameters should be estimated in order to compare any type of inactivation–deterioration data used to design novel thermal and non-thermal processes. The final aim is to ensure microbial safety and quality preservation of a food product. The parameter estimation step is the result of the minimization of some identification cost as was discussed in the introduction. This cost quantifies the deviation between the model predictions and the measured output. The identification cost that is commonly used is the sum of squared errors, *SSE*. Minimization of this criterion yields parameter values p that minimize the sum of the squares of the differences between the model predictions $y(ti,p)$ and the experimental observations $y_{exp}(t,j)$:

$$\mathrm{SSE} = \sum_{i=1}^{n_{ts}} \sum_{j=1}^{n_{ri}} (y_{\mathrm{exp}}(t_{i,j}) - y_{\mathrm{exp}}(t_i, p))^2 \tag{14.33}$$

where n_{ts} denotes the number of sampling times, n_{ri} is the number of replicate measurements (if available) at the sampling time t_i, and p is the vector of model parameters. It should be mentioned that this criterion implies that the noise is additive, independently and normally distributed $N(0,\sigma^2)$ with a constant variance σ^2 (Walter and Pronzato, 1997).

The variance of the residual errors can be approximated by the mean sum of squared errors, *MSE*. Ratkowsky (2003) argued about the appropriateness of using classical R^2 and R^2_{adj} indices to quantify the goodness of fit in non-linear regressions. He supported that the most simple and the most informative measure of the goodness of fit for regression models, both linear and non-linear, is the root mean squared error (RMSE),

$$RMSE = \sqrt{\frac{\sum_{i=1}^{n_t} (y_{\mathrm{exp}}(t_i) - y(t_i, p_{ls}))^2}{n_t - n_p}} \tag{14.34}$$

where $y_{exp}(t_i)$ denotes the experimental observations, $y(t_i,p_{ls})$ the predicted values, n_t the total number of data-points, n_p the number of estimated model parameters.

Concerning parameter optimization, different methods can be applied. For example, Valdramidis *et al.* (2008) assessed three different optimization techniques, i.e. (i) the Gauss–Newton method, (ii) the Levenberg–Marquardt method, and (iii) a subspace trust-region method based on the interior-reflective Newton method (Coleman and Li, 1994, 1996). In the Gauss–Newton method, a first and second derivative of the change in the sum of squared errors (SSE) with respect to the parameter value is used to estimate the direction and distance of the next iteration step within the optimization algorithm. The Levenberg–Marquardt algorithm interpolates between the Gauss–Newton algorithm and the method of gradient descent (in which steps proportional to the approximate gradient of the function at the current point are taken). In the third optimization technique, each iteration involves the approximate solution of a large linear system using the method of preconditioned conjugate gradients (Coleman and Li, 1994; Coleman and Li, 1996). While it is generally known that the Levenberg–Marquardt method can be more time consuming, it is more robust than the Gauss–Newton algorithm as it can deal with the possibly ill-conditioned or rank-deficient linear least-squares subproblem. Stochastic optimization techniques (e.g., swarm, evolutionary algorithms) are other methods that have been developed and applied in other fields of research and may have a potential in the kinetic studies applied during optimization in food processing.

Another important issue is to assess the uncertainty of the estimated model parameters. Parameter estimation quality can be evaluated by calculation of the standard errors, *SE*. The calculations for this statistical value can be derived based on the parameter variance–covariance matrix:

$$\mathrm{SE} = \sqrt{(J^T \cdot J)^{-1} \cdot MSE}$$

where J is the Jacobian matrix containing the partial derivatives of the model output with respect to the model parameters evaluated at each measurement point and MSE is the mean squared error defined with respect to the sum of squared errors (SSE) as $\mathrm{SSE}/(\mathrm{n_t} - \mathrm{n_p}) \cdot \mathrm{n_t}$ represents the number of sampling times (based on each sample result) and $\mathrm{n_q}$ the number of parameters.

14.9 MODEL VALIDATION

The application of a mathematical model in more complex systems (different from the laboratory conditions used to develop the model structure and identify its parameters) entails an increase in the error of the predictions (Pin *et al.*, 1999). These errors, which occur for microbial, chemical, and enzymatic kinetics, can be attributed to differences between predicted responses and responses generated (i) in (more realistic) model systems and/or (ii) in artificially contaminated foods, instead of the real liquid food.

For these reasons, an appropriate number of inactivation experiments to determine the model parameters is essential, and has to be followed by testing the accuracy of the model with new data using combinations of the examined environmental factors. The step of validation is an auxiliary methodology to evaluate the goodness of fit, to decide on re-identifying the proposed model structure/parameters to improve model accuracy, and to consider the necessity of generating additional data. Models can not be used with confidence to make predictions in foods until this validation step is successfully passed (Whiting, 1995).

Various statistical indices are suggested in the literature in order to compare competing models. Among these indices, Jeyamkondan *et al.* (2001) referred to graphical plots, mean relative percentage residual (*MRPR*), mean absolute relative residual (*MARR*), and root-mean-squared residual (*RMSR*). Similarly, Ross (1996) suggested the use of the bias (B_f) and accuracy factors (A_f) for assessing the predictive performance of secondary models (e.g., kinetic parameters of primary models in function of process parameters). Additionally, Campos *et al.* (2005) discussed the use of the robustness index (*RI*) and the mean relative error (*RE*) for comparing experimental and predicted microbial counts, computed after integration of the secondary models into the primary ones. Among these statistics the *MRPR*, *MARR* and *RE* should be considered with caution as a normalization takes place (residuals are divided by the observed or predicted data). This normalization is built on the assumption that the variability is larger for larger values of the observed variable. If this would be the case (which can be proven experimentally, see for example Alber and Schaffner (1992), Zwietering *et al.* (1994)) then the use of these statistics entails the necessity of using also a relevant cost function (e.g., Weighted Sum of Squared Errors, *WSSE*) during parameter identification. In other words, if such variability increases would be reality, then this phenomenon should also be taken into account during parameter estimation, and not only during model prediction.

The accuracy and bias factors (which are consistent with the least-squares algorithm of fitting models) presented by Baranyi *et al.* (1999) can be considered in order to evaluate the performance capability of inactivation models. The following example is proposed by Valdramidis *et al.* (2007b):

$$A_f = 10^{\sqrt{\frac{\sum_{i=1}^{n} (\log_{10}\hat{N}_i - \log_{10}N_i)^2}{n}}} \tag{14.35}$$

$$B_f = 10^{\frac{\sum_{i=1}^{n} (\log_{10}\hat{N}_i - \log_{10}N_i)}{n}} \tag{14.36}$$

where $\log_{10}\hat{N}_i$ denotes the predicted microbial load and n is the number of the experimental measurements.

On the one hand, the accuracy factor indicates the spread of results around the simulation or, in other words, how close on average the simulations are to the observations (values close to 1 are indicative of small deviations). On the other hand, the bias factor evaluates whether the observed values lie above the simulation line (i.e. $B_f < 1$, under-prediction of microbial load or fail-dangerous) or below the simulation line (i.e. $B_f > 1$, over-prediction of microbial load or fail-safe) (Ross, 1996).

14.10 APPLICATIONS ON THE DESIGN AND OPTIMIZATION OF THERMAL AND NON-THERMAL TECHNOLOGIES

Design of any thermal and non-thermal process ensuring safety and quality of a food (model) system simultaneously is imperative in food technology. Microbial, quality data and their associated models, if integrated, can describe the iso-rate contour plots of the studied kinetics and define the standards for an effective process design in the context of process optimization. Valdramidis *et al.* (2007a) applied a so-called model based design of the combined high pressure and mild heat ensuring safety and quality of a liquid food system (Fig. 14.5). Katsaros *et al.* (2010) tested and kinetically modeled target enzyme and microbial inactivation and defined the optimal pressure/temperature process conditions for Valencia orange juice cold pasteurization. Processing at 360 MPa and 35°C for 2 min was

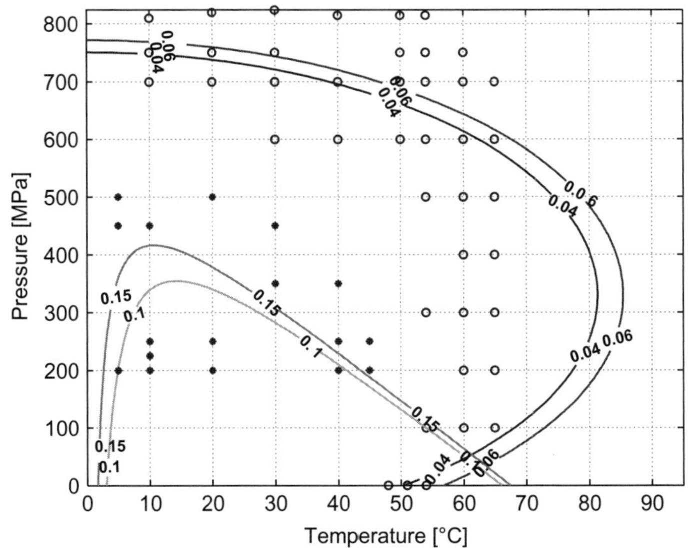

FIGURE 14.5

Microbial and enzymatic iso-rate contour plots for $k_m = 0.1, 0.15$ (min^{-1}) and $k = 0.04, 0.06$ (min^{-1}), respectively. Open circles: treatment combinations for enzymatic data, closed circles: treatment combinations for microbial data. (Reprinted from Valdramidis *et al.* (2007) with permission from Elsevier.)

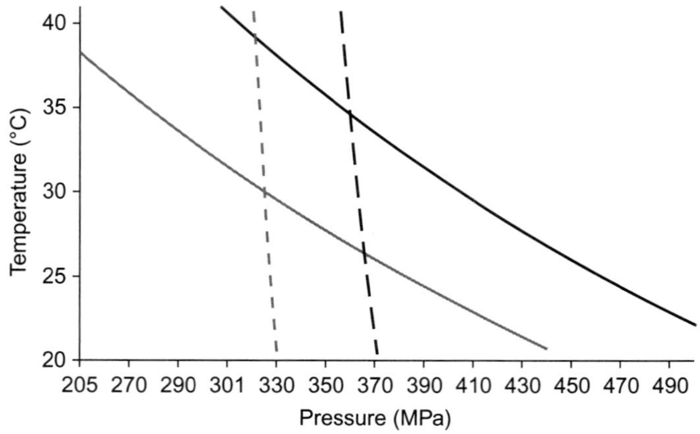

FIGURE 14.6

Microbial (LAB) and enzymatic (PME) iso-reduction plots for achieving a 7-microbial-log reduction and 90% PME inactivation in Valencia orange juice after high-pressure processing for 2 and 5 min. Dashed lines represent a 7D LAB destruction and solid lines represent 90% PME inactivation. Black lines show processing for 2 min, while gray lines show processing for 5 min. (From Katsaros *et al.* (2010), with permission.)

recommended (Fig. 14.6). Another study was undertaken to assess processing effects of high pressure during controlled storage conditions of apple juice (Valdramidis *et al.*, 2009). On the same principle, Krebbers *et al.* (2003) included qualitative studies of different indices for liquid products (tomato puree) at sterilization conditions. In this study, among the different quality characteristics were polygalacturonase and pectin methylesterase activity, water binding capacity, color appreciation, and viscosity of the product. Similar treatments for studying more quality characteristics in carrot juice products may require evaluation of the effects on carotenes, lipoxygenase, polyphenol oxidase (e.g., Bull *et al.* (2004), Kim *et al.* (2001)). All these different quality characteristics should be considered as new add-ins for developing model-based (novel thermal and non-thermal) process designs that can approach more realistically a desired final liquid product in which (total) quality and safety is ensured.

On the same basis the process lethality as given by the *F*-value (Ball, 1923) and defined as the time required in order to achieve a specific (microbial, chemical, enzymatic) reduction (following linear kinetics) at a given process condition can be calculated. This value then represents the total time-process variable (e.g., temperature, pressure) combination received by a food. Some applications have already been discussed by Stolt *et al.* (1999) in the case of quality kinetics. More recently, Corradini *et al.* (2006), Mafart *et al.* (2002), Sant'Ana *et al.* (2009), and Valdramidis *et al.* (2010b) have suggested the estimation of alternative lethality values based on considering microbial non-linearities.

14.11 CONCLUSION AND FUTURE

Several researchers have identified the need for creating databases containing kinetic model parameters for various target pathogenic and spoilage microbes (Balasubramaniam and Farkas, 2008). Although the authors were referring to HHP, this need appears valid for all the thermal and non-thermal processes described in this book. Experience gained from the field of predictive microbiology in the creation of web-based software (e.g., www.combase.cc) (Baranyi and Tamplin, 2004) to make data and predictive tools on microbial responses to food environments freely available, also should be considered in the kinetic studies discussed in this book.

Food-quality modeling is a valuable tool that contributes to process optimization and shelf-life management. The complexity of the food can be handled with carefully proposed assumptions. Modeling approaches applied with representative kinetic data are useful for establishing methodologies and for demonstrating their applicability, but system specificity and variability can not be overlooked for food product design and optimization. There is still a lack of reliable data concerning the kinetics of quality deterioration during and after processing, as well as relating a measurable response with particular quality aspects. Contrary to safety-microbial-modeling, there are no user-friendly models referring to food-quality changes in the public domain. Integration of safety and quality quantitative approaches will permit the application of multi-objective optimization techniques as described and applied in the field of (bio-) chemical process control (Logist *et al.*, 2010).

As discussed in Section 14.4.1 different alternative viability techniques as well as chemical and other quality analytic techniques need to be tested for some thermal and non-thermal processes. Novel microbial techniques such as fluorescence techniques (Breeuwer and Abee, 2000), real-time PCR, etc. can also be considered. New experimental techniques in which cell viability is quantified taking into account membrane damage, release of proteins (Klotz *et al.*, 2010) should equally be considered. Other issues of safety concern are to quantitatively assess the ability of microorganisms to adapt to non-thermal processes and how this effects kinetic studies. Examples of adaptations during HHP, HHP and light pulses of inoculated liquid systems have already been discussed in the literature (Garcia-Gonzalez *et al.*, 2010b; Hauben *et al.*, 1997; Karatzas and Bennik, 2002; Rajkovic *et al.*, 2009) .

Finally, as discussed in Section 14.2 the *ad hoc* hypotheses that are considered should be clearly specified for every study (e.g., homogenous liquid systems, insignificant variations of the processing conditions); alternatively, advanced modeling approaches that are developed with respect to time and space are required. Overall, the application of kinetic models should aim at ascertaining the safety risk of novel technologies via optimization aided techniques. Such an essential scientific and technical knowledge can be used in the multidisciplinary area of risk assessment. Therefore, these applications will provide the basis for safer and more competitive products and services from the food industry.

Acknowledgments

This research is supported in part by project PFV/10/002 (OPTEC–Center of Excellence Optimization in Engineering) of the KULeuven Research Council, Knowledge Platform KP/09/005 (SCORES4CHEM) of the KULeuven Industrial Research Fund, and the Belgian Program on Interuniversity Poles of Attraction initiated by the Belgian Federal Science Policy Office. Author J.F.M. Van Impe holds the chair Safety Engineering sponsored by the Belgian Chemistry and Life Sciences Federation essenscia.

References

Adekunte, A., Valdramidis, V.P., Tiwari, B.K., Slone, N., Cullen, P.J., Donnell, C.P.O., et al., 2010. Resistance of *Cronobacter sakazakii* in reconstituted powdered infant formula during ultrasound at controlled temperatures: a quantitative approach on microbial responses. Int. J. Food Microbiol. 142, 53–59.

Alber, S.A., Schaffner, D.W., 1992. Evaluation of data transformations used with the square root and schoolfield models for predicting bacterial growth rate. Appl. Environ. Microbiol. 58, 3337–3342.

Albert, I., Mafart, P., 2005. A modified Weibull model for bacterial inactivation. Int. J. Food Microbiol. 100, 197–211.

Amina, M., Panagou, E.Z., Kodogiannis, V.S., Nychas, G.J.E., 2010. Wavelet neural networks for modelling high pressure inactivation kinetics of Listeria monocytogenes in UHT whole milk 103, 170–183Chemom. Intell. Lab. Syst. 103, 170–183.

Ananta, E., Heinz, V., Knorr, D., 2004. Assessment of high pressure induced damage on *Lactobacillus rhamnosus* GG by flow cytometry. Food Microbiol. 21, 567–577.

Ananta, E., Knorr, D., 2009. Comparison of inactivation pathways of thermal or high pressure inactivated *Lactobacillus rhamnosus* ATCC 53103 by flow cytometry analysis. Food Microbiol. 26, 542–546.

Anonymous, 2000. Special supplement: kinetics of microbial inactivation for alternative food processing technologies—IFT' s response to task order #1, US Food and Drug Administration: how to quantify the destruction kinetics of alternative processing technologies. J. Food Sci. 4–108.

Anonymous, 2006. Requisite scientific parameters for establishing the equivalence of alternative methods of pasteurization—national advisory committee on microbiological criteria for foods. J. Food Prot. 69, 1190–1216.

Balasubramaniam, V.M., Farkas, D., 2008. High-pressure food processing. Food Sci. Technol. Int. 14, 413–418.

Balasubramaniam, V.M., Ting, E.Y., Stewart, C.M., Robbins, J.A., 2004. Recommended laboratory practices for conducting high-pressure microbial inactivation experiments. Innov. Food Sci. Emerg. 5, 299–306.

Ball, C.O., 1923. Thermal process time for canned food. Natl. Res. Counc. Bull. 7.

Ball, C.O., Olson, F.C.W., 1957. Sterilization in Food Technology. Theory, Practice and Calculations. McGraw-Hill Book Co., New York.

Balsa-Canto, E., Rodriguez-Fernandez, M., Banga, J.R., 2007. Optimal design of dynamic experiments for improved estimation of kinetic parameters of thermal degradation. J. Food Eng. 82, 178–188.

Banga, J.R., Alonso, A.A., Gallardo, J.M., Perezmartin, R.I., 1993. Mathematical modeling and simulation of the thermal-processing of anisotropic and non-homogeneous conduction heated canned foods-application to canned tuna. J. Food Eng. 18, 369–387.

Banga, J.R., Balsa-Canto, E., Alonso, A.A., 2008. Quality and safety models and optimization as part of computer-integrated manufacturing. Compr. Rev. Food Sci. F. 7, 168–174.

Baranyi, J., Pin, C., Ross, T., 1999. Validating and comparing predictive models. Int. J. Food Microbiol. 48, 159–166.

Baranyi, J., Tamplin, M.L., 2004. ComBase: a common database on microbial responses to food environments. J. Food Prot. 67, 1967–1971.

Bermïdez-Aguirre, D., Corradini, M.G., Mawson, R., Barbosa-Cαnovas, G.V., 2009. Modeling the inactivation of *Listeria innocua* in raw whole milk treated under thermosonication. Innov. Food Sci. Emerg. 10, 172–178.

Boff, J.M., Truong, T.T., Min, D.B., Shellhammer, T.H., 2003. Effect of thermal processing and carbon dioxide-assisted high-pressure processing on pectin methylesterase and chemical changes in orange juice. J. Food Sci. 68, 1179–1184.

Boulekou, S.S., Katsaros, G.J., Taoukis, P.S., 2010. Inactivation kinetics of peach pulp pectin methylesterase as a function of high hydrostatic pressure and temperature process conditions. Food Bioprocess Tech. 3, 699–706.

Breeuwer, P., Abee, T., 2000. Assessment of viability of microorganisms employing fluorescence techniques. Int. J. Food Microbiol. 55, 193–200.

Buchanan, R.L., Golden, M.H., Whiting, R.C., 1993. Differentiation of the effects of pH and lactici acid or acetic–acid concentration on the kinetics of *Listeria monocytogenes* inactivation. J. Food Prot. 56, 474–478.

Bull, M.K., Zerdin, K., Howe, E., Goicoechea, D., Paramanandhan, P., Stockman, R., et al., 2004. The effect of high pressure processing on the microbial, physical and chemical properties of Valencia and Navel orange juice. Innov. Food Sci. Emerg. 5, 135–149.

Bureau, S., Ruiz, D., Reich, M., Gouble, B., Bertrand, D., Audergon, J.M., et al., 2009. Application of ATR-FTIR for a rapid and simultaneous determination of sugars and organic acids in apricot fruit. Food Chem. 115, 1133–1140.

Campos, D.T., Marks, B.P., Powell, M.R., Tamplin, M.L., 2005. Quantifying the robustness of a broth-based *Escherichia coli* O157 : H7 growth model in ground beef. J. Food Prot. 68, 2301–2309.

Canumir, J.A., Celis, J.E., de Bruijn, J., Vidal, L.V., 2002. Pasteurisation of apple juice by using microwaves. Food Sci. Technol.-Leb. 35, 389–392.

Cerf, O., 1977. Tailing of survival curves of bacterial-spores. J. Appl. Bacteriol. 42, 1–19.

Cole, M.B., Davies, K.W., Munro, G., Holyoak, C.D., Kilsby, D.C., 1993. A vitalistic model to descrie the thermal inactiation of *Listeria monocytogenes*. J. Ind. Microbiol. 12, 232–239.

Coleman, T., Li, Y., 1994. On the convergence of reflective Newton methods for largescale nonlinear minimization subject to bounds. Math. Program. 67, 189–224.

Coleman, T., Li, Y., 1996. An interior, trust region approach for nonlinear minimization subject to bounds. SIAM J. Optimiz. 6, 418–445.

Corradini, M.G., Normand, M.D., Peleg, M., 2006. Expressing the equivalence of non-isothermal and isothermal heat sterilization processes. J. Sci. Food Agric. 86, 785–792.

Corradini, M.G., Peleg, M., 2006. Prediction of vitamins loss during non-isothermal heat processes and storage with non-linear kinetic models. Trends Food Sci. Technol. 17, 24–34.

Cunha, L.M., Oliveira, F.A.R., Brandao, T.R.S., Oliveira, J.C., 1997. Optimal experimental design for estimating the kinetic parameters of the Bigelow model. J. Food Eng. 33, 111–128.

Donsi, G., Ferrari, G., Pataro, G., 2007. Inactivation kinetics of *Saccharomyces cerevisiae* by pulsed electric fields in a batch treatment chamber: the effect of electric field unevenness and initial cell concentration. J. Food Eng. 78, 784–792.

Doona, C.J., Feeherry, F.E., Ross, E.W., 2005. A quasi-chemical model for the growth and death of microorganisms in foods by non-thermal and high-pressure processing. Int. J. Food Microbiol. 100, 21–32.

Erkmen, O., 2000. Predictive modelling of *Listeria monocytogenes* inactivation under high pressure carbon dioxide. LWT-Food Sci. Technol. 33, 514–519.

Erkmen, O., Dogan, C., 2004. Kinetic analysis of *Escherichia coli* inactivation by high hydrostatic pressure in broth and foods. Food Microbiol. 21, 181–185.

Ferrentino, G., Balaban, M.O., Ferrari, G., Poletto, M., 2010. Food treatment with high pressure carbon dioxide: *Saccharomyces cerevisiae* inactivation kinetics expressed as a function of CO2 solubility. J. Supercrit. Fluids. 52, 151–160.

Ferrentino, G., Ferrari, G., Poletto, M., Balaban, M.O., 2008. Microbial inactivation kinetics during high-pressure carbon dioxide treatment: nonlinear model for the combined effect of temperature and pressure in apple juice. J. Food Sci. 73, E389–E395.

Franceschini, G., Macchietto, S., 2008. Model-based design of experiments for parameter precision: state of the art. Chem. Eng. Sci. 63, 4846–4872.

Garcia-Gonzalez, L., Geeraerd, A.H., Elst, K., Van Ginneken, L., Van Impe, J.F., Devlieghere, F., 2009. Inactivation of naturally occurring microorganisms in liquid whole egg using high pressure carbon dioxide processing as an alternative to heat pasteurization. J. Supercrit. Fluids. 51, 74–82.

Garcia-Gonzalez, L., Geeraerd, A.H., Mast, J., Briers, Y., Elst, K., Van Ginneken, L., et al., 2010a. Membrane permeabilization and cellular death of *Escherichia coli, Listeria monocytogenes* and *Saccharomyces cerevisiae* as induced by high pressure carbon dioxide treatment. Food Microbiol. 27, 541–549.

Garcia-Gonzalez, L., Rajkovic, A., Geeraerd, A.H., Elst, K., Ginneken, L.V., Impe, J.F.V., et al., 2010b. The development of *Escherichia coli* and *Listeria monocytogenes* variants resistant to high-pressure carbon dioxide inactivation. Lett. Appl. Microbiol. 50, 653–656.

Garrote, R.L., Silva, E.R., Bertone, R.A., 2001. Kinetic parameters for thermal inactivation of cut green beans lipoxygenase calculated using unsteady-state methods. Int. J. Food Sci. Technol. 36, 377–385.

Geeraerd, A., Juneja, V.K., Baranyi, J., Cerf, O., Datta, N., Joosten, H., et al., 2004. Working group 3: predictive microbiology. Bull. Int. Dairy Fed. 392, 124–125.

Geeraerd, A.H., Herremans, C.H., Van Impe, J.F., 2000. Structural model requirements to describe microbial inactivation during a mild heat treatment. Int. J. Food Microbiol. 59, 185–209.

Geeraerd, A.H., Valdramidis, V., Van Impe, J.F., 2005. GInaFiT, a freeware tool to assess non-log-linear microbial survivor curves. Int. J. Food Microbiol. 102, 95–105.

Geeraerd, A.H., Valdramidis, V.P., Van Impe, J.F., 2006. GInaFiT, a freeware tool to assess non-log-linear microbial survivor curves (vol 102, pg 95, 2005). Int. J. Food Microbiol. 110, 297.

Gogou, E., Katapodis, P., Taoukis, P.S., 2010. High pressure inactivation kinetics of a thermomyces lanuginosus xylanase evaluated as a process indicator. J. Food Sci. 75, E379–E386.

Gomez-Lopez, V.M., Orsolani, L., Martinez-Yepez, A., Tapia, M.S., 2010. Microbiological and sensory quality of sonicated calcium-added orange juice. LWT-Food Sci. Technol. 43, 808–813.

Guan, D.S., Chen, H.Q., Ting, E.Y., Hoover, D.G., 2006. Inactivation of *Staphylococcus aureus* and *Escherichia coli* O157 : H7 under isothermal-endpoint pressure conditions. J. Food Eng. 77, 620–627.

Guerrero-Beltran, J.A., Barbosa-Canovas, G.V., 2006. Inactivation of *Saccharomyces cerevisiae* and polyphenoloxidase in mango nectar treated with UV light. J. Food Prot. 69, 362–368.

Guillou, S., El Murr, N., 2002. Inactivation of *Saccharomyces cerevisiae* in solution by low-amperage electric treatment. J. Appl. Microbiol. 92, 860–865.

Hasting, A.P.M., Blackburn, C.D.W., Crowther, J.S., 2001. *Mycobacterium paratuberculosis* and the commercial pasteurization of milk. Food Bioprod. Process. 79, 83–88.

Hauben, K.J.A., Bartlett, D.H., Soontjens, C.C.F., Cornelis, K., Wuytack, E.Y., Michiels, C.W., 1997. *Escherichia coli* mutants resistant to inactivation by high hydrostatic pressure. Appl. Environ. Microbiol. 63, 945–950.

Hong, V., Wrolstad, R.E., 1990. Use of HPLC separation photodiode array detection for characterization of anthocyanins. J. Agric. Food Chem. 38, 708–715.

Hulsheger, H., Potel, J., Niemann, E.G., 1981. Killing of bacteria with electric pulses of high-field strength. Radiat. Environ. Bioph. 20, 53–65.

ISO, 2007. I. S., Microbiology of food and animal feeding stuffs — General requirements and guidance for microbiological examinations.

Jaeger, H., Meneses, N., Moritz, J., Knorr, D., 2010. Model for the differentiation of temperature and electric field effects during thermal assisted PEF processing. J. Food Eng. 100, 109–118.

Jaeger, H., Schulz, A., Karapetkov, N., Knorr, D., 2009. Protective effect of milk constituents and sublethal injuries limiting process effectiveness during PEF inactivation of *Lb. rhamnosus*. Int. J. Food Microbiol. 134, 154–161.

Janssen, M., Verhulst, A., Valdramidis, V., Devlieghere, F., Van Impe, J.F., Geeraerd, A. H., 2008. Inactivation model equations and their associated parameter values obtained under static acid stress conditions cannot be used directly for predicting inactivation under dynamic conditions. Int. J. Food Microbiol. 128, 136–145.

Jeyamkondan, S., Jayas, D.S., Holley, R.A., 2001. Microbial growth modelling with artificial neural networks. Int. J. Food Microbiol. 64, 343–354.

Johnson, F.H., Eyring, H., 1970. The kinetic basis of pressure effects. In: Zimmerman, A. M. (Ed.), Biology and Chemistry. Academic Press, New York, pp. 1–44.

Jolie, R.P., Duvetter, T., Houben, K., Clynen, E., Sila, D.N., Van Loey, A.M., et al., 2009. Carrot pectin methylesterase and its inhibitor from kiwi fruit: Study of activity, stability and inhibition. Innov. Food Sci. Emerg. 10, 601–609.

Jonsson, U., Snygg, B.G., Härnulv, B.G., Zachrisson, T., 1977. Testing two models for the temperature dependence of the heat inactivation rate of *Bacillus stearothermophilus* spores. J. Food Sci. 42, 1251–1252, 1263.

Juneja, V.K., Eblen, B.S., 1999. Predictive thermal inactivation model for *Listeria monocytogenes* with temperature, pH, NaCl, and sodium pyrophosphate as controlling factors. J. Food Prot. 62, 986–993.

Karatzas, K.A.G., Bennik, M.H., 2002. Characterization of a *Listeria monocytogenes* Scott A isolate with high tolerance towards high hydrostatic pressure. Appl. Environ. Microbiol. 68, 3183–3189.

Katsaros, G.I., Giannoglou, M.N., Taoukis, P.S., 2009a. Kinetic study of the combined effect of high hydrostatic pressure and temperature on the activity of Lactobacillus delbrueckii ssp bulgaricus Aminopeptidases. J. Food Sci. 74, E219–E225.

Katsaros, G.I., Katapodis, P., Taoukis, P.S., 2009b. High hydrostatic pressure inactivation kinetics of the plant proteases ficin and papain. J. Food Eng. 91, 42–48.

Katsaros, G.I., Katapodis, P., Taoukis, P.S., 2009c. Modeling the effect of temperature and high hydrostatic pressure on the proteolytic activity of kiwi fruit juice. J. Food Eng. 94, 40–45.

Katsaros, G.I., Tsevdou, M., Panagiotou, T., Taoukis, P.S., 2010. Kinetic study of high pressure microbial and enzyme inactivation and selection of pasteurisation conditions for valencia orange juice. Int. J. Food Sci. Technol. 45, 1119–1129.

Kim, H.T., Choi, H.J., Kim, K.H., 2009. Flow cytometric analysis of *Salmonella enterica* serotype Typhimurium inactivated with supercritical carbon dioxide. J. Microbiol. Meth. 78, 155–160.

Kim, Y.S., Park, S.J., Cho, Y.H., Park, J., 2001. Effects of combined treatment of high hydrostatic pressure and mild heat on the quality of carrot juice. J. Food Sci. 66, 1355–1360.

Klotz, B., Pyle, D.L., Mackey, B.M., 2007. New mathematical modeling approach for predicting microbial inactivation by high hydrostatic pressure. Appl. Environ. Microbiol. 73, 2468–2478.

Klotz, B., Manas, P., Mackey, B.M., 2010. The relationship between membrane damage, release of protein and loss of viability in *Escherichia coli* exposed to high hydrostatic pressure. Int. J. Food Microbiol. 137, 214–220.

Koseki, S., Matsubara, M., Yamamoto, K., 2009. Prediction of a required log reduction with probability for *Enterobacter sakazakii* during high-pressure processing, using a survival/death interface model. Appl. Environ. Microbiol. 75, 1885–1891.

Koseki, S., Mizuno, Y., Yamamoto, K., 2008. Use of mild-heat treatment following high-pressure processing to prevent recovery of pressure-injured *Listeria monocytogenes* in milk. Food Microbiol. 25, 288–293.

Koseki, S., Yamamoto, K., 2007a. Modelling the bacterial survival/death interface induced by high pressure processing. Int. J. Food Microbiol. 116, 136–143.

Koseki, S., Yamamoto, K., 2007b. A novel approach to predicting microbial inactivation kinetics during high pressure processing. Int. J. Food Microbiol. 116, 275–282.

Krapfenbauer, G., Kinner, M., Gossinger, M., Schonlechner, R., Berghofer, E., 2006. Effect of thermal treatment on the quality of cloudy apple juice. J. Agric. Food Chem. 54, 5453–5460.

Krebbers, B., Matser, A.M., Hoogerwerf, S.W., Moezelaar, R., Tomassen, M.M.M., van den Berg, R.W., 2003. Combined high-pressure and thermal treatments for processing of tomato puree: evaluation of microbial inactivation and quality parameters. Innov. Food Sci. Emerg. 4, 377–385.

Labuza, T.P., 1984. Application of chemical kinetics to deterioration of foods. J. Chem. Educ. 61, 348–358.

Lee, H., Zhou, B., Liang, W., Feng, H., Martin, S.E., 2009. Inactivation of *Escherichia coli* cells with sonication, manosonication, thermosonication, and manothermosonication: microbial responses and kinetics modeling. J. Food Eng. 93, 354–364.

Logist, F., Sager, S., Kirches, C., Van Impe, J.F., 2010. Efficient multiple objective optimal control of dynamic systems with integer controls. J. Process Control. 20, 810–822.

Ly-Nguyen, B., Van Loey, A.M., Smout, C., Ozcan, S.E., Fachin, D., Verlent, I., et al., 2003. Mild-heat and high-pressure inactivation of carrot pectin methylesterase: a kinetic study. J. Food Sci. 68, 1377–1383.

Mafart, P., Couvert, O., Gaillard, S., Leguerinel, I., 2002. On calculating sterility in thermal preservation methods: application of the Weibull frequency distribution model. Int. J. Food Microbiol. 72, 107–113.

Mafart, P., Leguerinel, I., Couvert, O., Coroller, L., 2010. Quantification of spore resistance for assessment and optimization of heating processes: a never-ending story. Food Microbiol. 27, 568–572.

Marselles-Fontanet, A.R., Puig, A., Olmos, P., Minguez-Sanz, S., Martin-Belloso, O., 2009. Optimising the inactivation of grape juice spoilage organisms by pulse electric fields. Int. J. Food Microbiol. 130, 159–165.

Min, S., Evrendilek, G.A., Zhang, H.Q., 2007. Pulsed electric fields: Processing system, microbial and enzyme inhibition, and shelf life extension of foods. IEEE T. Plasma Sci. 35, 59–73.

Monfort, S., Gayan, E., Saldana, G., Puertolas, E., Condon, S., Raso, J., et al., 2010. Inactivation of *Salmonella* Typhimurium and *Staphylococcus aureus* by pulsed electric fields in liquid whole egg. Innov. Food Sci. Emerg. 11, 306–313.

Munyaka, A.W., Oey, I., Van Loey, A., Hendrickx, M., 2010. Application of thermal inactivation of enzymes during vitamin C analysis to study the influence of acidification, crushing and blanching on vitamin C stability in Broccoli (Brassica oleracea L var. italica). Food Chem. 120, 591–598.

Odriozola-Serrano, I., Soliva-Fortuny, R., Gimeno-Ano, V., Martin-Belloso, O., 2008. Kinetic study of anthocyanins, vitamin C, and antioxidant capacity in strawberry juices treated by high-intensity pulsed electric fields. J. Agric. Food Chem. 56, 8387–8393.

Oms-Oliu, G., Odriozola-Serrano, I., Soliva-Fortuny, R., Martin-Belloso, O., 2009. Effects of high-intensity pulsed electric field processing conditions on lycopene, vitamin C and antioxidant capacity of watermelon juice. Food Chem. 115, 1312–1319.

Oruña-Concha, M.J., Gonzalez-Castro, M.J., Lopez-Hernandez, J., Simal-Lozano, J., 1998. Monitoring of the vitamin C content of frozen green beans and Padrón peppers by HPLC. J. Sci. Food Agric. 76, 477–480.

Oteiza, J.M., Giannuzzi, L., Zaritzky, N., 2010. Ultraviolet treatment of orange juice to inactivate *E. coli* O157:H7 as affected by native microflora. Food Bioprocess Tech. 3, 603–614.

Patil, S., Torres, B., Tiwari, B.K., Wijngaard, H.H., Bourke, P., Cullen, P.J., et al., 2010a. Safety and quality assessment during the ozonation of cloudy apple juice. J. Food Sci. 75, M437–M443.

Patil, S., Valdramidis, V.P., Cullen, P.J., Frias, J., Bourke, P., 2010b. Inactivation of *Escherichia coli* by ozone treatment of apple juice at different pH levels. Food Microbiol. 27, 835–840.

Patil, S., Bourke, P., Frias, J.M., Tiwari, B.K., Cullen, P.J., 2009. Inactivation of *Escherichia coli* in orange juice using ozone. Innov. Food Sci. Emerg. 10, 551–557.

Peleg, M., 1995. A model of microbial survival after exposure to pulsed electric-fields. J. Sci. Food Agric. 67, 93–99.

Phua, S.T.G., Davey, K.R., 2007. Predictive modelling of high pressure (<= 700 MPa)-cold pasteurisation (<= 25°C) of *Escherichia coli, Yersinia enterocolitica* and *Listeria monocytogenes* in three liquid foods. Chem. Eng. Process. 46, 458–464.

Pilavtepe-Çelik, M., Buzrul, S., Alpas, H., Bozoglu, F., 2009. Development of a new mathematical model for inactivation of *Escherichia coli* O157:H7 and *Staphylococcus aureus* by high hydrostatic pressure in carrot juice and peptone water. J. Food Eng. 90, 388–394.

Pin, C., Sutherland, J.P., Baranyi, J., 1999. Validating predictive models of food spoilage organisms. J. Appl. Microbiol. 87, 491–499.

Polydera, A.C., Galanou, E., Stoforos, N.G., Taoukis, P.S., 2004. Inactivation kinetics of pectin methylesterase of greek navel orange juice as a function of high hydrostatic pressure and temperature process conditions. J. Food Eng. 62, 291–298.

Polydera, A.C., Stoforos, N.G., Taoukis, P.S., 2005. Quality degradation kinetics of pasteurised and high pressure processed fresh navel orange juice: nutritional parameters and shelf life. Innov. Food Sci. Emerg. 6, 1–9.

Ponce, E., Pla, R., Sendra, E., Guamis, B., Mor-Mur, M., 1999. Destruction of *Salmonella enteritidis* inoculated in liquid whole egg by high hydrostatic pressure: comparative study in selective and non-selective media. Food Microbiol. 16, 357–365.

Poschet, F., Bernaerts, K., Geeraerd, A.H., Scheerlinck, N., Nicolai, B.M., Van Impe, J.F., 2004. Sensitivity analysis of microbial growth parameter distributions with respect to data quality and quantity by using Monte Carlo analysis. Math. Comput. Simulat. 65, 231–243.

Poschet, F., Geeraerd, A.H., Van Loey, A.M., Hendrickx, M.E., Van Impe, J.F., 2005. Assessing the optimal experiment setup for first order kinetic studies by Monte Carlo analysis. Food Control. 16, 873–882.

Rajkovic, A., Smigic, N., Uyttendaele, M., Medic, H., de Zutter, L., Devlieghere, F., 2009. Resistance of *Listeria monocytogenes, Escherichia coli* O157:H7 and *Campylobacter jejuni* after exposure to repetitive cycles of mild bactericidal treatments. Food Microbiol. 26, 889–895.

Ramaswamy, H.S., van de Voort, F.R., Ghazala, S., 1989. An analysis of TDT and Arrhenius methods for handling process and kinetic data. J. Food Sci. 54, 1322–1326.

Rasch, M., 2004. Experimental design and data collection. In: McKellar, R.C., Xuewen, L. (Eds.), Modeling Microbial Responses in Food. CRC Press.

Ratkowsky, D.A., 2003. Model fitting and uncertainty. In: McKellar, R., Lu, X. (Eds.), Modelling Microbiology Food Responses. CRC Press.

Ratkowsky, D.A., Lowry, R.K., McMeekin, T.A., Stokes, A.N., Chandler, R.E., 1983. Model for bacterial culture-growth rate throughout the entire biokinetic temperature range. J. Bacteriol. 154, 1222–1226.

Rawson, A., Tiwari, B.K., Patras, A., Brunton, N., Brennan, C., Cullen, P.J., et al., 2011. Effect of thermosonication on bioactive compounds in watermelon juice. Food Res. Int. In Press.

Reyns, K., Soontjens, C.C.F., Cornelis, K., Weemaes, C.A., Hendrickx, M.E., Michiels, C.W., 2000. Kinetic analysis and modelling of combined high-pressure–temperature inactivation of the yeast *Zygosaccharomyces bailii*. Int. J. Food Microbiol. 56, 199–210.

Riahi, E., Ramaswamy, H.S., 2003. High-pressure processing of apple juice: kinetics of pectin methyl esterase inactivation. Biotechnol. Prog. 19, 908–914.

Ross, T., 1996. Indices for performance evaluation of predictive models in food microbiology. J. Appl. Bacteriol. 81, 501–508.

Rouse, A.H., Atkins, C.D., 1955. Pectinesterase and pectin in commercial orange juice as determined by methods used at Citrus Experiment Station. Lake Alfred, Fla. Agr. Exp. Sta. Tech. Bul. 570.

Saguy, I., Karel, M., 1980. Modeling of quality deterioration during food processing and storage. Food Technol.-Chicago 34, 78–85.

Saldana, G., Puertolas, E., Condon, S., Alvarez, I., Raso, J., 2010. Modeling inactivation kinetics and occurrence of sublethal injury of a pulsed electric field-resistant strain of *Escherichia coli* and *Salmonella* Typhimurium in media of different pH. Innov. Food Sci. Emerg. 11, 290–298.

Sant'Ana, A.S., Rosenthal, A., Massaguer, P.R., 2009. Heat resistance and the effects of continuous pasteurization on the inactivation of *Byssochlamys fulva* ascospores in clarified apple juice. J. Appl. Microbiol. 107, 197–209.

Sauer, A., Moraru, C.I., 2009. Inactivation of *Escherichia coli* ATCC 25922 and *Escherichia coli* O157:H7 in apple juice and apple cider, using pulsed light treatment. J. Food Prot. 72, 937–944.

Shao, Y.W., Zhu, S.M., Ramaswamy, H., Marcotte, M., 2010. Compression heating and temperature control for high-pressure destruction of bacterial spores: an experimental method for kinetics evaluation. Food Bioprocess Tech. 3, 71–78.

Siegmund, B., Pollinger-Zierler, B., 2007. Growth behavior of off-flavor-forming microorganisms in apple juice. J. Agric. Food Chem. 55, 6692–6699.

Singleton, V.L., Orthofer, R., Lamuela-Raventos, R.M., 1999. Analysis of total phenols and other oxidation substrates and antioxidants by means of Folin–Ciocalteu reagent. Oxidants and Antioxidants, Pt A, vol. 299. Academic Press Inc, San Diego, pp. 152–178.

Smelt, J., 1998. Recent advances in the microbiology of high pressure processing. Trends Food Sci. Technol. 9, 152–158.

Stoforos, N.G., Grelier, S., Robert, M.-C., Taoukis, P.S., 2002. Kinetics of tomato pectin methylesterase inactivation by temperature and high pressure. J. Food Sci. 67, 1026–1031.

Stumbo, C.R., 1973. Thermobacteriology in Food Processing. second ed. Academic Press, Inc., New York.

Stolt, M., Stoforos, M.G., Taoukis, P.S., Autio, K., 1999. Evaluation and Modelling of Rheological Properties of High Pressure Treated Waxy Maize Starch Dispersions. J. Food Eng. 40, 293–298.

Subramanian, A., Ahn, J., Balasubramaniam, V.M., Rodriguez-Saona, L., 2006. Determination of spore inactivation during thermal and pressure-assisted thermal processing using FT-IR spectroscopy. J. Agric. Food Chem. 54, 10300–10306.

Tajchakavit, S., Ramaswamy, H.S., Fustier, P., 1998. Enhanced destruction of spoilage microorganisms in apple juice during continuous flow microwave heating. Food Res. Int. 31, 713–722.

Taoukis, P., Labuza, T.P., Saguy, I., 1997. Kinetics of food deterioration and Shelf-Life prediction. In: Valentas, K.J., Rotstein, E., Singh, R.P. (Eds.), The Handbook of Food Engineering Practice. CRC Press, Boca Raton, FL.

Tiwari, B.K., Patras, A., Brunton, N., Cullen, P.J., O' Donnell, C.P., 2010. Effect of ultrasound processing on anthocyanins and color of red grape juice. Ultrason. Sonochem. 17, 598–604.

Torres, B., Tiwari, B.K., Patras, A., Wijngaard, H.H., Brunton, N., Cullen, P.J., et al., 2011. Effect of ozone processing on the colour, rheological properties and phenolic content of apple juice. Food Chem. 124, 721–726.

Unluturk, S., Atilgan, M.R., Baysal, A.H., Tari, C., 2008. Use of UV-C radiation as a non-thermal process for liquid egg products (LEP). J. Food Eng. 85, 561–568.

Valdramidis, V.P., Bernaerts, K., Van Impe, J.F., Geeraerd, A.H., 2005. An alternative approach to non-log-linear thermal microbial inactivation: modelling the number of log cycles reduction with respect to temperature. Food Technol. Biotechnol. 43, 321–327.

Valdramidis, V.P., Cullen, P.J., Tiwari, B.K., O' Donnell, C.P., 2010a. Quantitative modelling approaches for ascorbic acid degradation and non-enzymatic browning of orange juice during ultrasound processing. J. Food Eng. 96, 449–454.

Valdramidis, V.P., Geeraerd, A.H., Bernaerts, K., Van Impe, J.F.M., 2008. Identification of non-linear microbial inactivation kinetics under dynamic conditions. Int. J. Food Microbiol. 128, 146–152.

Valdramidis, V.P., Geeraerd, A.H., Poschet, F., Ly-Nguyen, B., Van Opstal, I., Van Loey, A.M., et al., 2007a. Model based process design of the combined high pressure and mild heat treatment ensuring safety and quality of a carrot simulant system. J. Food Eng. 78, 1010–1021.

Valdramidis, V.P., Geeraerd, A.H., Tiwari, B.K., Cullen, P.J., Kondjoyan, A., Van Impe, J.F., 2010b. Estimating the efficacy of mild heating processes taking into account microbial non-linearities: a case study on the thermisation of a food simulant. Food Control. 22, 137–142.

Valdramidis, V.P., Geeraerd, A.H., Van Impe, J.F., 2007b. Stress-adaptive responses by heat under the microscope of predictive microbiology. J. Appl. Microbiol. 103, 1922–1930.

Valdramidis, V.P., Graham, W.D., Beattie, A., Linton, M., McKay, A., Fearon, A.M., et al., 2009. Defining the stability interfaces of apple juice: Implications on the optimisation and design of high hydrostatic pressure treatment. Innov. Food Sci. Emerg. 10, 396–404.

Van Boeijen, I.K.H., Moezelaar, R., Abee, T., Zwietering, M.H., 2008. Inactivation kinetics of three *Listeria monocytogenes* strains under high hydrostatic pressure. J. Food Prot. 71, 2007–2013.

Van den Broeck, I., Ludikhuyze, L.R., Van Loey, A.M., Hendrickx, M.E., 2000. Inactivation of orange pectinesterase by combined high-pressure and -temperature treatments: a kinetic study. J. Agric. Food Chem. 48, 1960–1970.

Van Eylen, D., Oey, I., Hendrickx, M., Van Loey, A., 2008. Effects of pressure/temperature treatments on stability and activity of endogenous broccoli (*Brassica oleracea* L. *cv. Italica*) myrosinase and on cell permeability. J. Food Eng. 89, 178–186.

Van Impe, J.F., Bernaerts, K., Geeraerd, A.H., Poschet, F., Versyck, K.J., 2001. Modelling and Prediction in an Uncertain Environment. Food Process Modelling. Woodhead Publishing Limited, Cambridge, pp. 156–179.

Versyck, K.J., Bernaerts, K., Geeraerd, A.H., Van Impe, J.F., 1999. Introducing optimal experimental design in predictive modeling: a motivating example. Int. J. Food Microbiol. 51, 39–51.

Walker, S.J., Jones, J.E., 1993. Protocols for data generation for predictive modeling. J. Ind. Microbiol. 12, 273–276.

Walter, E., Pronzato, L., 1997. Identification of Parametric Models from Experimental Data. Springer-Verlag, Masson.

Wang, J., Hu, X.S., Wang, Z.F., 2010. Kinetics models for the inactivation of Alicyclobacillus acidiphilus DSM14558(T) and Alicyclobacillus acidoterrestris DSM 3922(T) in apple juice by ultrasound. Int. J. Food Microbiol. 139, 177–181.

Weemaes, C., Ludikhuyze, L., Van den Broeck, I., Van Loey, A., Hendrickx, M., 1998. High pressure inactivation of polyphenoloxidases. J. Food Sci. 63, 873–877.

Whiting, R.C., 1995. Microbial modeling in foods. Crit. Rev. Food Sci. 35, 467–494.

Whiting, R.C., Buchanan, R.L., 1993. A classification of models in predictive microbiology—reply. Food Microbiol. 10, 175–177.

Whiting, R.C., Sackitey, S., Calderone, S., Morely, K., Phillips, J.G., 1996. Model for the survival of *Staphylococcus aureus* in nongrowth environments. Int. J. Food Microbiol. 31, 231–243.

Xiong, R., Xie, G., Edmondson, A.E., Sheard, M.A., 1999. A mathematical model for bacterial inactivation. Int. J. Food Microbiol. 46, 45–55.

Zwietering, M.H., Cuppers, H., Dewit, J.C., Vantriet, K., 1994. Evaluation of data transformations and validation of a model for the effect of temperature on baterial-growth. Appl. Environ. Microbiol. 60, 195–203.

CHAPTER

Regulatory and Legislative Issues for Non-Thermal Technologies: An EU Perspective 15

Dominic Watkins

Bond Pearce LLP, New London House

6 London Street, London, EC3R7LP, UK

15.1 INTRODUCTION

With over 495 million Europeans becoming increasingly skeptical about food production and a preference for "natural" and organic products developing, the need to boost public confidence in novel foods and emerging technologies is essential. The roots of the European Union (EU) are in free trade and making intra-community trade simpler and without restriction. For manufacturers trading within the EU, the attempts at harmonization of standards, together with the concept of free movement of goods should lead to more consistent standards of safety and of the composition of products across all Member States. In reality, the legislative framework often requires local governments to create domestic legislation to implement EU law, which often gives rise to variations and uncertainty between the approaches taken in member states.

The challenge is one that goes to the very heart of the EU and its purpose. It has its own executive and legislative bodies creating its own laws, which must then interact with domestic legislation. When one then adds to that equation bodies such as the European Food Safety Agency which has a remit covering the entirety of the EU, and considers that there are similar agencies in all other member states, for example the Food Standards Agency in the UK, an exceptionally complicated regulatory structure, with a number of inter-related players starts to emerge.

All of this is predicated on the laudable objective of consumer safety. The European Commissioner and her domestic counterparts in the member states consider the needs of the consumer as being paramount. Some might argue that there

Novel Thermal and Non-Thermal Technologies for Fluid Foods. DOI: 10.1016/B978-0-12-381470-8.00015-3

is regulation for regulation's sake but where consumer safety is at risk, it is often hard to make a compelling argument against it.

For non-thermal technologies for fluid foods, much regulation exists. There are the product-specific requirements for many types of fluid foods such as fruit juice or milk, together with requirements applying to the food contact materials containing these foods, novel foods, and the treatments applied to those foods. This chapter considers how the EU makes law and then explores the key elements of the regulation impacting this area.

15.2 AN OVERVIEW OF HOW LAW IS CREATED IN THE EU

The European Union (the EU) is a political and legislative body bringing together the affairs of 30 member states of "Europe". Since its inception, gradually more and more issues that were previously governed domestically, in each of the member states, are now centrally legislated upon by the EU. The primary rationale for this is difficult to dispute: to ensure that there is a consistently high level of consumer protection across the EU and to enable the free movement of goods by ensuring that the same broad legal standards apply to each member state.

It is estimated that 95% of all food legislation in the UK now originates in the EU. Accordingly, while there remains important domestic legislation in each member state, it is EU legislation that is of paramount importance in understanding the regulation of food technology.

15.2.1 The Institutions of the European Union

There are four main European institutions involved in the development of EU law. These are the European Parliament, the European Commission, the European Council, and the European Court of Justice. Each of these institutions plays a key role in the creation or enforcement of the EU food law regime, and it is important to understand the role played by each institution in order to properly understand the regulatory framework in which legislation is created.

While not an institution of the EU, the European Food Safety Authority (EFSA) is also very important to the development of EU food law and its regulation. Following a number of food crises, EFSA was created to provide an independent source of scientific advice and communication on the risks in the food chain. To do this, EFSA works closely with national competent authorities such as the UK's Food Standards Agency and will consult on its opinions.

There is an important delineation between EFSA, the body with the scientific experience and knowledge to be able to carry out a the risk assessment of the process, and the European Commission and the European Parliament which that make the policy decisions on how to manage that risk, based on EFSA's assessment. Accordingly, EFSA may identify that a novel food or a particular process poses a risk and it is then for the legislature, either on an EU or a domestic level, to consider this assessment and if appropriate exercise its power to set policy or law to manage that risk and so protect the public.

EFSA is split into four directorates, one dealing with administration and three others dealing with risk assessment, scientific cooperation and assistance and communication. EFSA aims to provide an objective science-based advice from latest information and data. The scientific committee and panels of EFSA are tasked with carrying out this work and are comprised of experts from across Europe who are appointed on the basis of scientific excellence. The panels carry out risk assessment across 12 topic areas including nutrition, food contact materials, enzymes, flavorings, biological hazards, and food additives and nutrient sources. These risk assessments are published as part of the food-safety regime and inform the decision making of the European Commission.

EFSA has currently over 1,200 scientific experts in place on its Scientific Committees and Panels and has adopted almost 1,000 scientific opinions on foods or practices. A recent example of this is the Scientific Opinion on Nanotechnology,[1] published in early 2009 following a request by the European Commission. The Opinion considers the use of nanotechnologies and assesses whether current approaches to risk assessment can be applied to nanotechnologies. The Scientific Committee recommended developing and validating routine methods to detect, characterize and quantify engineered nanomaterials in food.

On the basis of this assessment, in November 2009 the European Commission asked EFSA to prepare a guidance document on how to assess risks related to certain food-related uses of nanotechnology. A final draft of the proposed guidance document was published for public consultation by EFSA on 14 January 2011. In addition to the risk-assessment work, another core function of EFSA's is to communicate the risks it has assessed in relation to particular foodstuffs. EFSA's stated aim is to "provide appropriate, consistent and accurate and timely communications on food and food safety to all interested parties based on scientific findings".

In order to ensure that it achieves its objectives, it looks not only at the risks but also at the public perception of those risks. In this regard, there is much interesting work being undertaken; for example, a study commissioned by the FSA in the UK in relation to attitudes to nanotechnology and other immerging food technology.[2] The research tends to show a general lack of knowledge from the general public and ambivalence in relation to the issues which makes the work of EFSA in communicating the risks more important. EFSA faces challenges in this area, particularly with regard to utilizing the media to communicate risks. It is difficult

[1]Scientific Opinion of the Scientific Committee on a request from a European Commission on the potential risks arising from nano-science and nano-technologies on food and feed safety. European Food Standards Agency, The EFSA Journal (2009) 958, 1–39.

[2]An Evidence Review of Public Attitudes to Emerging Food Technologies" Brook Lyndhurst, March 2009.

to think of an example where the scientific opinions of EFSA have been communicated via the national and European media, as food-scare stories appear to be much more attractive to the media (particularly in the UK) than a balanced scientific opinion. Accordingly, while EFSA may be achieving its objectives by ensuring that the European Parliament and European Commission have the best-possible scientific advice upon which to make informed policy decisions, it struggles to achieve the other of its objectives.

15.2.2 Types of European Legislation

The EU is created and governed by the rule of law. Accordingly, the legislative scope and development of the EU is agreed by all Member States signing up to Treaties. The EU is now the primary legislator on most food-related issues. The decision to make EU law is not taken lightly and a number of checks and balances are applied to ensure that it is absolutely necessary to pass that legislation.

One example of those checks and balances is what is known as the subsidiarity principle.[3] This principle means that each time that the EU legislates, it must positively consider whether taking action is necessary, and unless the area to be legislated upon falls within its exclusive competence or it will be more effective than any action in the member state, it will not legislate. An example of this can be seen in the proposal for a new Regulation on Novel Foods,[4] where it was concluded that while the proposal does not fall within the exclusive competency of the EU, community action will better achieve the objectives with a centralized procedure to improve the efficiency of novel food authorization. Individual action by member states could lead to differing levels of food safety and protection of human health and potentially confuse consumers.

Whether the proposed action is proportionate is also considered. The principle of proportionality[5] is closely linked to that of subsidiarity and they will be considered together to determine whether the proposed legislation is both necessary and proportionate. In order to be consistent with the principle of proportionality, the proposed action by EU Institutions must be necessary in order to meet the objectives of the Treaties. It also means that where there are two options available to the EU, it must take the option that provides for the greatest level of freedom for its citizens.

15.2.3 The Supremacy of the European Union

The obligations set out in EU legislation will always take precedence over any obligations in domestic legislation, and where there is any conflict between the two, the obligation set out in the EU legislation will be preferred.

[3]See Article 5 of the Treaty establishing the European Community OJ C 325, 24.12.2002, p. 33.

[4]Proposal for a Regulation of the European Parliament and of the Council on novel foods and amending Regulation (EC) XXXXX COM 2008/02 (COD) COM(2007)872 Final.

[5]Article 5 of the Treaty establishing the European Community.

The supremacy of EU enables the EU to protect one of the primary freedoms created by the EU, the free movement of goods. This principle means that the EU operates without national frontiers as one territory for the purposes of trade. The consequence of this is that import or export restrictions which disadvantage those entering the market from elsewhere in the EU are expressly prohibited.[6] There is, however, derogation from that principle where there is a public health issue at stake. Accordingly, any domestic provision that puts in place a measure that is equivalent to a quantitative restriction will be unlawful as it harms intra-community trade.

15.3 OVERVIEW OF THE REGULATION OF FOOD IN THE EUROPEAN UNION

There is a wide range of regulation governing the placement of food on the EU market. This regulation covers a food's composition, its labeling, the claims that might be made about it as well, as the materials that might come into contact with it.

All of this legislation has consumer protection at its core. By setting clear standards for manufacturers, the EU has tried to create a uniformly safe market in EU food, whilst simultaneously breaking down barriers to trade. This objective has largely been met, though in areas where domestic legislation is required to give effect to the intention of the EU, some variance remains between the requirements in different member states. Given the concern which arose in the EU about genetically modified foods, the application of new and emerging novel technologies is an area which requires careful consideration from a regulatory perspective. Consumers' attitude towards novel food technologies in the EU "is one of wariness, unease, uncertainty and sometimes outright negativity" (Brook Lyndhurst, 2009). Accordingly, the regulation of these technologies has an important role to play to ensure that the public's confidence in the technology grows and its benefits are recognized and adopted.

The safety of European consumers is paramount and therefore careful consideration must be given in order to ensure they are protected from the risks posed by emerging technology. This does not mean that it is right, necessary, or proportionate to legislate for that reason alone, without considering whether the existing legislation is fit for purpose.

The EU regulatory regime contains the overarching protection that only safe products and foods are placed on the market. To a great extent, it does not matter whether the technology in question is nanotechnology, pulsed-UV light, ozone, irradiation, or a more established technology, the overarching duty of safety will

[6] Article 34 Treaty establishing the European Community and "Cassis de Dijon" Rewe-Zentralfinanze GmbH v Landwirtschaftskammer [1975] ECR 843, Dassonville and Keck and Mithouard.

pervade and should protect the public. The effect of the technology on the product must be considered; if it changes the product at a molecular or in the case of nanotechnology, the nanometer level, then regulation is likely to be considered to control the use of the technology to the extent that it could cause a safety issue. The novel food regime is, however, in the short term at least, likely to be considered the best solution for regulating this area by the EU regulators, and in Section 15.4 we discuss how it might apply. There are, however, other technologies such as pulsed-light or pulsed-electric field, which only affect the surface of the product and therefore are outside of the scope of the novel food regime, although an appropriate regulatory framework must be found for those products. These are discussed in Section 15.5.

Even though the current overarching regulatory framework was not designed to deal with all of the new technologies it now regulates, it is however capable of dealing with most emerging technologies. Very few emerging technologies actually require new regulations as the general overarching controls are sufficient; however, where the risk or controls are unusual further regulation may be necessary. Nevertheless there may always be arguments that certain technologies are not regulated as they should be. In other cases, it is simply not possible to regulate until more is known about the risks posed by a technology—nanotechnology being a case in point.

In the sections that follow, we shall first consider in Section 15.4 the framework for emerging technologies created by the novel food regime, and then go on to consider, in Section 15.5, how two emerging technologies are regulated, and then in Section 15.6, the other regulatory requirements that are placed on fluid foods.

15.4 THE NOVEL FOOD REGIME

It is human nature that whenever there is change, people become apprehensive. This is particularly true of food where the idea of changing food, particularly at a molecular level, is not fully understood and therefore often met with mistrust. For this reason the area is very closely regulated (akin to the regulation of medicines) and risk assessed to ensure the safety of consumers; this regulation is achieved by the novel food regime.

The Novel Foods Regulation[7] applies to any food or food ingredient that has not been sold in the EU to a significant degree before May 1997 and relates to foods that are:

(a) new or intentionally modified in their primary molecular structure;
(b) isolated from microorganisms, fungi or algae; or

[7]Regulation (EC) No 258/97 of the European Parliament and of the Council of 27 January 1997 concerning novel foods and novel food ingredients, OJ L 43, 14.2.1997, pp. 1–6.

(c) isolated from plants or food ingredients or to which a new production process that gives rise to significant changes in the composition or structure of the food or food ingredients has been applied.

The Novel Foods Regulation does not apply to food additives regulated by Council Directive 89/107/EEC[8] or the flavorings falling within Council Directive 88/388.[9]

The novelty of the food means that it must undergo a risk assessment before it can be placed on the market in order to ensure that it is safe.

It is likely that non-thermal technologies, depending upon the technology involved, will have the effect of changing the molecular structure of a product or giving rise to a significant change in its composition and, therefore, will cause the food or ingredient to fall within the novel food regime. Accordingly, an understanding of the novel food regime is important.

Article 3(1) sets out the standards that novel products must meet and requires that products falling within this Regulation do not:

- "present a danger for the consumer
- mislead the consumer,
- differ from foods or food ingredients which they are intended to replace to such an extent that their normal consumption would be nutritionally disadvantageous for the consumer."

These criteria form the test that is applied by the EU Commission in determining whether or not to grant permission for the product or ingredient to be marketed in the EU. The Novel Foods Regulation puts in place a clear approval system for novel food, requiring it to be assessed and approved by both the member states and the Commission (though the risk analysis is done by EFSA). It is a slow and considered process.

In order for a product to be considered to be authorized for marketing as a novel food, a full technical file must first be presented. That technical file should include a proposal for presentation and labeling, a copy of the study (or studies) carried out and any other material which is available to demonstrate the food or food ingredient meets Article 3(1) criteria. The technical file will initially be assessed by the member states' Competent Authority and if it approves the application, then it will send it together with its own recommendation to the Commission. The Commission will then assess the application and publish a decision setting out whether or not the product is authorized for sale in the EU.

[8]Council Directive 89/107/EEC of 21 December 1988 on the approximation of the laws of the Member States concerning food additives authorized for use in foodstuffs intended for human consumption. OJ L 040 , 11.02.1989, pp. 0027–0033.

[9]Council Directive 88/388/EEC of 22 June 1988 on the approximation of the laws of the Member States relating to flavourings for use in foodstuffs and to source materials for their production. OJ L 184, 15.7.1988, pp. 61–66.

The EU reports that a total of 53 applications were made in respect of novel foods between 1997 and 2004.[10] In the five years that followed, to the end of September 2009, that number rose to 111. This indicates that there is an increase in the number of products being submitted and a recognition of the opportunities offered by innovations in novel foods. As would be expected, the scope of these applications varies significantly and covers the full range of innovation across the industry. This is demonstrated by comparing the 110th and 111th applications, which relate to guar gum[11] and Pectine-derived Acidic Oligosaccharide,[12] respectively.

While clearly there are commercial opportunities in novel foods, the application process is slow and complicated, with the typical application process taking 2–3 years. Unfortunately, it is not uncommon for the process to last significantly longer: as at January 2011, the oldest outstanding application originates from March 2000 and 14 of the 38 live applications are from November 2007 or earlier. In some cases, this may be as a result of the outcome of further studies being awaited.

Also, as at January 2011, only three applications had been refused;[13] there have however been a further 17 applications withdrawn and another six that were either held to be regulated by different legislation or were found not to be novel. Nevertheless, taking the worst case scenario —the assumption that all of these 23 applications did not continue because they would ultimately fail—it means that 63% of all applications have resulted in a marketing authorization being granted to allow sale in the EU.

Overall, the EU has published 45 Commission Decisions[14] authorizing the placement of a particular novel food on the market. Where authorization is granted, it will state the conditions of usage for the food or food ingredient. The

---

[10]http://ec.europa.eu/food/food/biotechnology/novelfood/index_en.htm.

[11]An application made on 11 September 2009 by Agence Française de Securite Sanitaire des Aliments (AFSSA), the French competent agency, on behalf of Danone Vitapole.

[12]An application made on 4 September 2009 by the Novel Foods Unit of the Netherlands on behalf of Environ Netherlands B.V.

[13]The three refusals were:

- Commission Decision 2005/580/EC: Commission Decision of 25 July 2005 refusing the placing on the market of betaine as a novel food or novel food ingredient under Regulation (EC) No 258/97 of the European Parliament and of the Council (notified under document number C (2005) 2770). OJ L 199, 29.7.2005, pp. 89–89.
- Commission Decision 2001/17/EC: Commission Decision of 19 December 2000 on refusing the placing on the market of "Nangai nuts" (as a novel food or novel food ingredient under Regulation (EC) No 258/97 of the European Parliament and of the Council (notified under document number C(2000) 3888). OJ L 004, 09.01.2001, p. 35.
- Commission Decision 2000/196/EC: Commission Decision of 22 February 2000 refusing the placing on the market of Stevia rebaudiana Bertoni: plants and dried leaves as a novel food or novel food ingredient under Regulation (EC) No 258/97 of the European Parliament and of the Council (notified under document number C(2000) 77). OJ L 61, 8.3.2000, p. 14.

[14]http://ec.europa.eu/food/food/biotechnology/novelfood/authorisations_en.htm.

authorization will also include any of the other specific labeling requirements provided for by Article 8 as are necessary to ensure that the customer is fully informed. In particular, the packaging and labeling must inform the consumer of:

(a) any characteristic of the food, to include its composition, nutritional value, all nutritional effects and intended use which makes this product novel and not equivalent to an existing food ingredient;
(b) that a novel ingredient is present which may have health implications, or health implications for a particular section of society;
(c) that the product may give rise to ethical concerns;
(d) that the product was genetically modified.

The information required will vary depending upon the type of novel food covered by the application. Any nutritional or health claims will nevertheless need to be permitted (or in the case of health claims authorized) under Regulation No. 1924/2006/EC on Nutrition and Health Claims.

15.4.1 The Expedited Process: Concept of Substantial Equivalence

Article 5 of the Novel Food Regulation creates an expedited procedure for obtaining approval to market foods that are "substantially equivalent" to existing foods, or food ingredients. The reason for this is that the potential risk posed by the novel product has already been assessed in relation to the existing product on the market, and therefore the substantially equivalent product can be treated in the same way for the purposes of safety.

Certain foods are outside of this derogation; this includes food containing genetically modified organisms, foods with intentionally modified primary molecular structure, and foods that have a significant change in their composition or structure. There have been significantly more of these types of notifications than novel food applications. There were, as at January 2010, 140 notifications listed by the EU[15] covering a wide range of food types. This demonstrates the clear attraction for business as the process is usually significantly faster than a full novel application. That said, it is still a fairly lengthy procedure, especially if issues are raised by other member states.

The assessment of substantial equivalence is made by the Competent Authority of a member state, and if it is concluded that the product is substantially equivalent then, rather than going through the full process described above, it is only necessary to notify the Commission of its conclusion. All notifications are then circulated to other member states within 60 days of receipt and if no objections are received or requests for additional information, its approval will feature in the Official Journal of the European Union.

In order to determine whether a product is substantially equivalent, the Competent Authority will consider the equivalence of the composition, nutritional

[15]http://ec.europa.eu/food/food/biotechnology/novelfood/notif_list_en.pdf.

value, metabolism, intended use and the level of undesirable substances in both products. A dossier providing evidence substantiating each point must be submitted by the applicant to the Competent Authority. Guidance from the UK's Advisory Committee on Novel Foods (ACNFP),[16] which advises the UK Food Standards Agency in relation to novel foods, indicates that comparisons should be drawn against one other existing product. It explains that the product should be described in the same level of detail, in terms of compositional information and the other criteria set out above, as the ingredient that is the subject matter of the application.

The meaning of substantial equivalence will vary depending upon the product. The novel and existing products should be derived from the same or very similar species, grown and harvested in similar conditions, and the novel product should not contain significant levels of substances that are not present in the existing product. The most important criterion, however, is the intended use. It would generally be acceptable for a novel ingredient that is to be used in one type of product to be compared against one that is used in a food supplement, or *vice versa*. While there is a cogent reason for this in terms of the risk assessment undertaken, it means that different uses of what is otherwise a "substantially equivalent" product will need to go through the full process.

A recent example of a food that has gone through this process includes an application of Zenith-Group in relation to Argan oil. Here the French Competent Authority, the DGCCRF, confirmed that the previously issued opinion by AFSSA was also applicable to the Argan oil placed on the market by the company Zenith-Group.

Another area of particular concern for the application of non-thermal technologies is the level of undesirable substances. If the application of those technologies to an existing novel product will change the level of "undesirable" substances in that product then the substantial equivalence route is no longer available.

Finally, in terms of labeling, products that are able to use this route are still required to comply with the labeling requirements set out in Article 8 of the Novel Foods Regulations and the requirements of Regulation 1924/2006 on nutrition and health claims.

15.4.2 Proposal for New Novel Food Regulation

As at January 2011, there is a draft new Novel Food Regulation[17] in circulation that is intended to update the existing regime in due course. Recital 6 to the draft Regulation makes clear that the definition of novel foods includes "... foods modified by new production processes, such as nano-technology and nanoscience

[16]ACNFP guidelines for the presentation of data to demonstrate substantial equivalence between a novel food or food ingredient and an existing counterpart, The Advisory Committee on Novel Foods and Processes, 2005.

[17]Currently under review with a proposal for a new updated Regulation [Proposal for a Regulation of the European Parliament and of the Council on novel foods and amending Regulation (EC) XXXXX COM 2008/02 (COD) COM(2007)872 Final].

which might have an impact on food". While the Recitals are silent on other forms or types of non-thermal technology, this is a clear indication that nanotechnology is intended to be covered by the novel food regime.

The draft Regulation will create a new centralized procedure for dealing with the safety, evaluation and approval of novel foods. The new centralized procedure means that all applications for the approval of novel foods will be sent to the European Commission, with EFSA carrying out the risk assessment on its behalf, rather than any evaluation at member state level. A consequence of this is the assessment of substantial equivalence will no longer take place at member state level and therefore the expedited process will be lost. EFSA is however required to compare the product with a comparable product already on the market in order to determine its safety and therefore, where appropriate, this may have the effect of expediting the approval process. However, given EFSA's current workload, it is likely that the approval process will speed up in the short term under the new regime.

As before, the draft Regulation will include the ability to prescribe additional labeling required for consumer information, together with conditions of use and other requirements such as maximum usage levels for the novel food. The language of the previous Regulation has also been updated and streamlined. This can be seen in the definition of novel food. In the current regulation, the definition can be determined by one working one's way through the regulation and reaching a conclusion as to whether or not the product falls within one of the headings cited.

The draft Regulation contains a much clearer definition:

(i) "food that has not been used for human consumption to a significant degree within the Community before 15 May 1997;

The use of a food exclusively as or in a food supplement shall not be sufficient to show whether it has been used for human consumption to a significant degree within the Community before 15 May 1997. However, if a food has been used exclusively as or in a food supplement prior that date, it can be placed on the Community market after that date for the same use without being considered as novel food. Further criteria for assessing if a food has been used for human consumption to a significant degree within the Community before 15 May 1997, which are designed to amend non-essential elements of this Regulation, inter alia by supplementing it, may be adopted in accordance with the regulatory procedure with scrutiny referred to in Article 14(3).

(ii) food of plant or animal origin when to the plant and animal is applied a non-traditional breeding technique not used before 15 May 1997; and

(iii) food to which is applied a new production process, not used before 15 May 1997, where that production process gives rise to significant changes in the composition or structure of the food which affect its nutritional value, metabolism or level of undesirable substances."

All novel foods that have been authorized under the current regime would continue to be authorized when the new Regulation comes into force. However, new uses of the novel food would still require a new authorization.

15.5 HOW DOES THIS APPLY TO SPECIFIC EMERGING NON-THERMAL TECHNOLOGIES?

At present, most emerging technologies are regulated through the framework of general food law or by the novel food regime to ensure consumer safety. In the next two sections we shall consider irradiation and high pressure, where there has been further work by EFSA and the EU authorities in order to consider if further regulation was required.

15.5.1 Irradiation

Food irradiation is often used instead of pasteurization or heat treatment to process food and has the advantage of being a "cold" technology which means it will have less effect on the food itself than heat treatment, resulting in the foodstuff retaining more of its desired look, feel, and taste.

Council Directive 1999/2/EEC[18] governs the circumstances in which irradiated foods may be used. As with all food law, this Directive has public protection at its core. This is necessary as the word "radiation" alone is enough to cause public disquiet, despite there having been many years' worth of research demonstrating that the technology is safe and effective. In the opinion of the World Health Organization,[19] irradiation should be more widely used, and the UK Food Standards Agency describes food irradiation as a consumer labeling and not a safety issue. Nevertheless, Directive 1999/2/EEC's recognizes that the irradiation needs to be "within the limits required for the protection of health and of economic and technical needs".

Directive 1999/2/EEC broadly mirrors the requirements of the Codex Alimentarius in respect of irradiation and prescribes that there should be an authorization system for allowing its use. This is a two-stage process which requires the use of irradiation to be authorized for specific foodstuffs and also for premises undertaking irradiation to be registered to use irradiation. Irradiation may only be authorized for use in the limited circumstances set out in Annex 1 to Directive 1999/2/EEC, i.e. only when:

- there is a reasonable technological need;
- it presents no health hazard and is carried out under the conditions proposed;

[18]Directive 1999/2/EC of the European Parliament and of the Council of 22 February 1999 on the approximation of the laws of the Member States concerning foods and food ingredients treated with ionising radiation. OJ L 66, 13.3.1999, pp. 16–23.

[19]High-Dose Irradiation: Wholesomeness of Food Irradiated with Doses above 10 kGy. Joint FAO/IAEA/WHO Study Group, WHO technical report series: 890 (1997: Geneva, Switzerland).

- it is of benefit to the consumer; or
- it is not used as a substitute for hygiene or health practice or for good manufacturing or agricultural practice.

Irradiation may only be used for reducing the incidence of food-borne disease by destroying pathogenic bacteria, reducing spoilage and premature ripening, and to rid foodstuffs of organisms harmful to plant products. The facilities at which radiation will be administered must also be approved and registered with the national Competent Authority. That Competent Authority then has a duty to pass that information to the Commission, so that it can maintain a central record. In addition to registration, there are also detailed requirements regarding the records that must be kept by those administering irradiation. While not overly onerous, or significantly beyond what is expected of most food production sites in terms of due diligence documentation, the requirements are prescribed in law and reflect the public nervousness around this technology. The records that must be maintained and kept, for a period of five years, for each batch of foodstuffs irradiated are:

- the nature and quantity of foodstuffs irradiated;
- the batch number;
- the number of the person ordering the irradiation treatment;
- the recipient of the treated foodstuffs;
- the date of irradiation;
- the packaging materials used during treatment.

The dose of radiation that is acceptable is also defined in the Regulation. It can be assumed that an overall dose of 10 kGy or less would be proportionate and appropriate as a dose over 10 kGy will start to adversely affect the nutrient content of the food.[20] In addition to the informational requirements, the materials used for packaging foodstuffs that will be irradiated must also be demonstrably suitable for that purpose.

Irradiated products are required to be labeled with the word: "irradiated" or "treated with ionizing radiation", together with both the address and name of the city in which the irradiation was carried out, or the facility's reference number. This is in addition to the core labeling information required under the labeling Directive, 2000/13/EC.

If it is an ingredient rather than the product as a whole that has been irradiated, the same "irradiated" descriptor should appear next to the ingredient on the ingredients list. The same principle applies to compound ingredients, even if they constitute less than 25% of the finished product.

The ultimate intention is to create an authorized list of foodstuffs that can be treated with radiation, however at present the only product category authorized

[20]Revision of the opinion of the Scientific Committee on Food on the irradiation of food, European Commission, HEALTH and Consumer Protection Directorate-General, Directorate C—Scientific Opinions, SCF/CS/NF/IRR/24 Final 24 April 2003.

for trade within the EU are "dried, aromatic herbs, spices and vegetable seasonings".[21]

Prior to being approved for inclusion on the EU list of products authorized for irradiation, a favorable opinion must be provided. A number of favorable opinions[22] have been given, including for fruit, vegetables, cereals, starchy fish, shellfish, fresh meat, poultry, frogs' legs, gum arabic, casein and caseinates, amongst other products. For a typical example, in relation to casein and caseinates, the EU concluded:

> "Microbiological analysis of acid casein, rennet casein and caseinates, manufactured in dehydrated form under European conditions has shown, that despite good hygienic control the final products may still contain *Enterobacter sakazakii, Escherichia coli, Salmonella spp* and *Klebsiella pneumoniae*. To avoid cross-contamination of other food products not pasteurised as a final step before being sold to the general public, it is necessary to decontaminate these casein products by irradiation with a dose up to 6 kGy to avoid any incidence of toxic infection of the consumer. At that dosage, it has been shown that no significant chemical or physico-chemical changes occur in casein products which would constitute a health risk through ingestion of food products containing these products as ingredients".

Nevertheless, there still remains only one class of product category on the EU list.

Until such time as an EU-wide list of products authorized for irradiation comes into force, Article 4(6) allows national authorizations for irradiation to be granted and compiled into a Commission list. This list[23] demonstrates that, for example, fruit in Belgium and the Czech Republic may be irradiated at a maximum dose of 2 kGy; vegetables including pulses may be irradiated in the UK, Belgium, and the Czech Republic at a dose of 1 kGy; and potatoes at a dose of 0.2 kGy in the UK and Czech Republic, a dose of 0.15 kGy in Belgium and Italy and a dose of 0.1 kGy in Poland. Clearly it is undesirable for there to be inconsistencies across the EU, as this hinders cross-border trade and is an example of where specific regulation for a particular technology has not been effective in ensuring harmonization.

A similar regulatory system exists in the US. The 1958 Food Additives Amendment prohibits irradiated food from being marketed unless that irradiation

---

[21]Directive 1999/3/EC of the European Parliament and of the Council of 22 February 1999 on the establishment of a Community list of foods and food ingredients treated with ionizing radiation. OJ L 66, 13.3.1999, p. 24.

[22]Most recently: Opinion of the Scientific Committee on Food on the irradiation of eight foodstuffs (expressed on 17 September 1998) http://ec.europa.eu/food/fs/sc/scf/out15_en.html.

[23]List of Member States' authorizations of food and food ingredients which may be treated with ionizing radiation (According to Article 4(6) of Directive 1999/2/EC of the European Parliament and of the Council on the approximation of the laws of the Member States concerning foods and food ingredients treated with ionizing radiation) OJ C 112, 12.5.2006, pp. 6–7.

has been carried out in accordance with the specific rules in relation to safe conditions of use[24] and it has successfully completed the food additive petition process and has consequently been published in the Federal Register. Products such as wheat, potatoes, fruit and vegetables, herbs and spices, and some meats appear on this register.

Food irradiated outside of the EU is permitted for sale in the EU provided certain criteria are met. In the UK those criteria are set out in the Food Irradiation Regulations 2009. The criteria are quite restrictive requiring that:

- It is "properly" irradiated food, though what "properly" means in this context is not defined.
- The food is in one of the seven categories of irradiated food to be imported (as discussed above in the UK those are: fruit, vegetables, cereals, bulbs and tubers, dried aromatic herbs, spices and vegetable seasonings, fish and shellfish, and poultry).
- The irradiation took place at one of the facilities approved by the European Community and, with the exception of dried aromatic herbs, spices, and vegetable seasonings, the facility has been approved by its local competent authority in relation to the food it is importing into the UK. The website of the UK Competent Authority indicates that there are seven approved facilities in non-EU countries.
- It complies with the labeling requirements, and is labeled "irradiated" or "treated with ionizing radiation".
- The product is accompanied by full due diligence documentation showing the location where the irradiation took place and also the range of information set out in Part 3 of Schedule 2 to the Regulations.

15.5.2 High Pressure

High-pressure technology is not particularly new. It has its roots in food preservation in discoveries made in the late 19th century; however, it is really in the last 10 years that it has started to develop swiftly as it has become more commercially viable. High-pressure technology has a number of uses, particularly in relation to food preservation and has been shown to be effective at extending the shelf-life of milk and fruit without altering its state or taste in preference to the other more traditional methods that can significantly alter the food. The challenge has been finding a route to apply the technology in a commercially viable way. This now appears to have been resolved as there are now a number of products that are reported as using this technology; they include most commonly fruit juices, jams, and fruit preparations.

[24]Section 403 Federal Food, Drug, and Cosmetic Act.

In May 2001 the European Commission authorized the use of fruit preparations pasteurized by high-pressure treatment as a novel ingredient,[25] following an application made on behalf of Groupe Danone to the competent authority in France. The initial assessment report from the French competent authority concluded that a high-pressure treatment (8 kbar for 6 min at 20°C) may be safely used instead of the generally used pasteurization process which involved significantly higher heat for a longer period of time. However, as a result of some of the objections raised to the application, experts from Groupe Danone were called to the Commission to provide information on the process. In particular, they were asked to demonstrate that the high-pressure treatment offered the same level of safety as the traditionally used pasteurization process. They clearly were able to do so as the Commission concluded that high-pressure treatment was not likely to have a negative effect on public health and accordingly could be placed on the market.

The intended benefit of high-pressure pasteurization over regular heat treatment is that it offers a better-quality product as a result. Now that the technology has been authorized for this use, it paves the way for similar applications and demonstrates that the novel foods regime is flexible enough to regulate this area.

15.6 OTHER EXISTING REGULATORY REQUIREMENTS

There are a wide range of more general regulatory requirements that will also apply to any fluid food marketed in the EU. These include:

- the general labeling requirements for providing consumers with the information that they need to make decisions about a food stuff;
- the regulation of nutrition and health claims in relation to food;
- specific regulations that control the composition of fluid foods such as fruit juices and foods for energy-restricted diets; and
- the regulation of materials that will come into contact with fluid foods.

15.6.1 Specific Requirements for Particular Fluid Foods

There are a number of food groups for which specific requirements have been prescribed under EU legislation including, for example, foods for particular nutritional uses, long-life milk, recast natural mineral water, "aromatized" drinks and fruit juices. There are also a series of Council Decisions[26] laying down methods for analysis and testing of heat-treated milk for direct human consumption which

[25]Commission Decision 2001/424/EC of 23 May 2001 authorizing the placing on the market of pasteurized fruit-based preparations produced using high-pressure pasteurization under Regulation (EC) No. 258/97 of the European Parliament and of the Council. OJ L 152, 7.6.2001, p. 42.

[26]Council Decision 92/608/EEC of 14 November 1992 laying down methods for the analysis and testing of heat-treated milk for direct human consumption. OJ L 407, 31.12.1992, pp. 29–46.

specifically govern the use of thermal technologies for milk. This chapter will not consider all of this legislation in detail but will comment on just some of the additional product-specific legislative requirements that apply to certain fluid foods.

15.6.1.1 Fruit Juices

Council Directive 2001/112/EC[27] regulates fruit juices, concentrated fruit juice, dehydrated fruit juice, and fruit nectar. The appropriate nomenclature for the product is determined according to its composition. For example, concentrated fruit juice must be used to describe a product that is "obtained from fruit juice of one or more kinds by the physical removal of a specific proportion of the water content. Where the product is intended for direct consumption that removal will be of at least 50%".

The term "fruit juice" may only be used when the product is "the fermentable but unfermented product obtained from the fruit which is sound and ripe, fresh or preserved by chilling, of one or more kinds mixed together having the characteristic colour, flavour, and taste typical of the fruit juice of the fruit from which it comes". It is quite common for flavors, pulp, and cells from the fruit juice which are separated during the processing to be restored to the same juice, and this is expressly allowed.

In the case of citrus fruits, the fruit juice must come from the endocarp. Lime juice, however, may be obtained from the whole fruit by similar production processes whereby the proportion of the constituents of the outer part of the fruit is reduced to a minimum.

The Regulations also permit certain specified ingredients to be added. This includes vitamins and minerals and restoring flavor, pulp, and cells to the fruit juice. For juices other than pear or grape juice, the addition of sugars is authorized in order to regulate acidic taste and also for sweetening purposes, provided that the amount of added sugar does not exceed 150 g/l. In addition, for certain products, addition of lemon juice and for all products, addition of carbon dioxide is permitted. In relation to fruit nectar, the minimum fruit content for all types of fruit is prescribed.

The Annex to Council Directive 2001/112/EC also details the authorized treatments for fruit juices. It permits mechanical extraction processes together with the "usual physical processes". Chemical infiltration adjuvant and precipitation agents which comply with Community Directives are allowed, as are chemically inert absorption adjuvant. There is no mention of any new or emerging technologies in this section, though provided the technology in question does not have an effect on the composition of the fruit juice or affect its safety, this is permitted.

[27] Council Directive 2001/112/EC of 20 September 2001 relating to fruit juices and certain similar products intended for human consumption. OJ L 10, 12.1.2002, pp. 58–66.

15.7 IT'S A SMALL WORLD, BUT IS ONE SET OF GLOBAL RULES REALISTIC?

In an ideal world there would be one global set of rules governing food manufacture and food safety. However, this is clearly not the case and it seems there can not be a "one size fits all" approach. There are a number of obvious and practical reasons for this, not least that the rules need to meet the demands of each jurisdiction. What may be acceptable or usual practice in one jurisdiction may not be in another, the cultural difference between the Far East and Western Europe being a case in point.

More practically, there is no true concept of international law for these purposes. Similarly, there is no international court in which these regulations might be enforced or international parliament in which the laws may be debated before they came into force. Accordingly, we are left with a system of local rules and governments driving legislation in each individual jurisdiction, or in the case of the EU a collective of countries.

While international standards do exist in a number of areas and the Codex Alimentarius and reports from bodies such as the World Health Organization provide a framework for much of the legislation in Europe and around the world, these "international" requirements are not enforceable. Concepts such as HACCP provide an excellent example of this and demonstrate how a concept, having its inception in the Codex Alimentarius, is now globally recognized as a key part of all food-safety management systems. The need for a global regulation of all food processing activities has been ongoing for some time (Rogers, 1999). With the creation of World Trade Organization (WTO) in 1995, the search for global harmonization in food safety regulations became an important issue needing to be addressed by many countries (Barbosa-Canovas and Bermúdez-Aguirre, 2010). The Global Harmonization Initiative (GHI) is a network of scientific organizations and individual scientists working together to promote harmonization of global food safety regulations and legislation. It was founded in 2004 as a joint activity of the US-based Institute of Food Technologists (IFT) International Division and the European Federation of Food Science and Technology (EFFoST).

The GHI is an attempt to eliminate hurdles; a network of scientific organizations has launched a global initiative to facilitate harmonization of food safety regulations and legislation. The objective is to discuss, globally, the scientific issues that buttress the decisions made by individual governments and international regulatory bodies in order to achieve global scientific consensus on such issues. The GHI Working Group anticipates that elimination of the regulatory differences will make it more attractive for the private sector to invest in food safety R&D, consequently strengthening the competitiveness of each nation's food industry and of the industries supplying the food sector. Harmonizing global regulations will facilitate the application of new technologies, encouraging the food industry to invest in such technologies to ensure better safety and quality for consumers. The GHI's objective is to achieve consensus on the science of food regulations and legislation to ensure the global availability of safe and wholesome food products for all consumers. The GHI facilitates global discussion about the scientific issues that support decisions made by national governments and international regulatory bodies by:

1. Providing the foundation for sound, sensible, science-based regulations.
2. Creating a forum for scientists and technologists to interact with regulatory authorities, globally.
3. Providing industry, regulators and consumers an independent, authoritative information resource.

In the EU, the freedom of movement of goods (alongside the other fundamental freedoms), has gone some way to ensuring that businesses are not prejudiced by higher standards of quantitative restrictions imposed on imported goods into another EU jurisdiction. Accordingly, the gradual harmonization of food law across the EU has brought a level of consistency across the 27 member states of the European Union that would be difficult to achieve in other circumstances.

On most issues there is some nexus between jurisdictions. On many occasions, because the position adopted by the EU broadly represents best practice or the leading scientific thinking on a topic, that position is also found in other jurisdictions such as the USA or Australia. Similarly regulatory positions differ in response to food-borne outbreaks. In the US, the Food and Drug Administration (FDA) recently passed a requirement for fruit juice processors to achieve a 5-log reduction in the numbers of the most resistant pathogens in their finished products (USFDA, 2001). Comparably, the lack of such a ruling in the EU has continued a trend toward minimal processing. This is not, however, always the case and research on consumer perceptions of emerging technologies has demonstrated[28] that consumers interpret the issues in relation to those technologies very differently, depending on whether they live in Manchester, Minnesota, or Melbourne, and elected representatives of the EU must ensure that regulation reflects this.

15.8 CONCLUSIONS

As technology advances, regulation fights to keep up and to ensure that EU citizens are protected and kept safe. These dual objectives form a central thread flowing through all EU food law and are balanced against the EU's intention to make trade easier, and without borders, within the EU.

The primary concern regarding new technology is the effect that it will have on the food in question and in turn whether that renders the food unsafe. For most technologies, the existing regime will suffice, if the technology has an effect on

[28]Brook Lyndhurst, March 2009.

the composition of the product and changes it, then it will fall within the novel food regime. Only where there is either a social or safety need for there to be more legislation (see irradiation, for example), do we see additional layers of regulation added.

The supremacy of the EU means that it is able to, at least in theory, control the standards that must be met to ensure that the balance between safety and free trade is maintained. It allows the EU to legislate on a particular issue knowing that, where there is a conflict, its laws will be the ones adopted and applied. Moreover, where higher domestic standards are allowable, then the concept of the free movement of goods prohibits quantitative restrictions that disproportionately affect goods entering the market, again with the important derogation of public health.

Sometimes the EU will not immediately know the answer or the risk that a technology poses. In those cases, further studies will be necessary in order to determine risk and the likely issues that might arise. Nanotechnology provides a perfect example, at the time of writing, despite having carried out a general risk assessment, EFSA has not been able to fully evaluate the technology and as a result it is dealt with through the existing novel food regime, rather than by a specific regulatory regime.

Overall, the current regime generally works and presents the opportunities for innovation and development of new products.

Citations

Legislation and Case law

Council Directive 2000/13/EC of the European Parliament and of the Council of 20 March 2000 on the approximation of the laws of the Member States relating to the labelling, presentation and advertising of foodstuffs. Off. J. L 109, 6.5.2000, 29–42.

Council Directive 89/107/EEC on the approximation of the laws of the Member States concerning food additives authorized for use in foodstuffs intended for human consumption. Off. J. L 40, 11.2.1989, 27.

Federal Food, Drug, and Cosmetic Act.

Food Hygiene (England) Regulation 2006.

List of Member States' authorisations of food and food ingredients which may be treated with ionising radiation (According to Article 4(6) of Directive 1999/2/EC of the European Parliament and of the Council on the approximation of the laws of the Member States concerning foods and food ingredients treated with ionising radiation). Off. J. C 112, 12.5.2006, 6–7.

Regulation (EC) No 258/97 of the European Parliament and of the Council of 27 January 1997 concerning novel foods and novel food ingredients. Off. J. L 43, 14.2.1997, 1–6.

Revision of the opinion of the Scientific Committee on Food on the irradiation of food, European Commission, HEALTH and Consumer Protection Directorate-General, Directorate C—Scientific Opinions, SCF/CS/NF/IRR/24 Final 24 April 2003.

Treaty establishing the European Community. Off. J. C 325, 24.12.2002, 33.

Treaty on the European Union. Off. J. C 191, 29.7.1992.

Council Decision 92/608/EEC of 14 November 1992 laying down methods for the analysis and testing of heat- treated milk for direct human consumption. Off. J. L 407, 31.12.1992, 29–46.

Council Directive 1999/2/EC of the European Parliament and of the Council of 22 February 1999 on the approximation of the laws of the Member States concerning foods and food ingredients treated with ionising radiation. Off. J. L 66, 13.3.1999, 16–23.

Council Directive 1999/3/EC of the European Parliament and of the Council of 22 February 1999 on the establishment of a Community list of foods and food ingredients treated with ionising radiation. Off. J. L 66, 13.3.1999, 24–25.

Commission Decision 2000/196/EC of 22 February 2000 refusing the placing on the market of *Stevia rebaudiana Bertoni*. Off. J. L 61 8.3.2000 14.

Commission Decision 2001/17/EC of 19 December 2000 refusing the placing on the market of "Nangai nuts" *Canarium indicum.* Off. J. L 4, 9.1.2001, 35.

Commission Decision 2001/424/EC of 23 May 2001 authorising the placing on the market of pasteurised fruit based preparations produced using high pressure pasteurisation under Regulation (EC) No. 258/97 of the European Parliament and of the Council. Off. J. L 152, 7.6.2001, 42.

Council Directive 2001/112/EC of 20 September 2001 relating to fruit juices and certain similar products intended for human consumption. Off. J. L 10, 12.1.2002, 58–66.

Commission Decision 2005/580/EC of 25 July 2005 refusing the placing on the market of betaine as a novel food or novel food ingredient under Regulation (EC) No 258/97 of the European Parliament and of the Council. Off. J. L 199, 29.7.2005, 89.

Published Works

Advisory Committee on Novel Foods and Proceses guidelines for the presentation of data to demonstrate substantial equivalence between a novel food or food ingredient and an existing counterpart, The Advisory Committee on Novel Foods and Processes, 2005.

An Evidence Review of Public Attitudes to Emerging Food Technologies" Brook Lyndhurst, March 2009, 6.

Proposal for a Regulation of the European Parliament and of the Council on novel foods and amending Regulation (EC) XXXXX COM 2008/02 (COD) COM(2007)872 Final.

European Commission, Health and Consumers Directorate-General: Summary report of the Working Groups of the Advisory Group on the Food Chain and Animal and Plant Health on nutrient profiles. Held in Brussels on 8 July and 28 November 2008.

European Food Standards Agency, 2009. Scientific opinion of the scientific committee on a request from a european commission on the potential risks arising from nano-science and nano-technologies on food and food safety. The EFSA J. 958, 1–39.

Guidance on the Implementation of Regulation No 1924/2006 on Nutrition and Health Claims made on Foods. Conclusions of the Standing Committee on Foods Chain and Animal Health.

High-Dose Irradiation: Wholesomeness of Food Irradiated with Doses above 10 kGy, Joint FAO/IAEA/WHO Study Group, WHO technical report series: 890 (1997: Geneva, Switzerland).

Opinion of the Scientific Committee on Food on the irradiation of eight foodstuffs (expressed on 17 September 1998).

The Setting of Nutrient Profiles for Foods Bearing Nutrition and Health Claims pursuant to Article 4 of the Regulation (EC) No 1924/2006, 2008. Scientific opinion of the panel on dietetic products, nutrition and allergies. The EFSA J. 644, 1–44.

United States Food and Drug Administration (USFDA), 2001. Hazard analysis and critical point (HACCP); procedures for the safe and sanitary processing and importing of juice; final rule. Fed. Reg. 66, 6137–6202.

Food Science and Technology International Series

Amerine, M.A., Pangborn, R.M., and Roessler, E.B., 1965. Principles of Sensory Evaluation of Food.

Glicksman, M., 1970. Gum Technology in the Food Industry.

Joslyn, M.A., 1970. Methods in Food Analysis, second ed.

Stumbo, C. R., 1973. Thermobacteriology in Food Processing, second ed.

Altschul, A.M. (Ed.), New Protein Foods: Volume 1, Technology, Part A—1974. Volume 2, Technology, Part B—1976. Volume 3, Animal Protein Supplies, Part A—1978. Volume 4, Animal Protein Sup-plies, Part B—1981. Volume 5, Seed Storage Proteins—1985.

Goldblith, S.A., Rey, L., and Rothmayr, W.W., 1975. Freeze Drying and Advanced Food Technology.

Troller, J.A., and Christian, J.H.B., 1978. Water Activity and Food.

Bender, A.E., 1975. Food Processing and Nutrition.

Osborne, D.R., and Voogt, P., 1978. The Analysis of Nutrients in Foods.

Loncin, M., and Merson, R.L., 1979. Food Engineering: Principles and Selected Applications.

Vaughan, J. G. (Ed.), 1979. Food Microscopy.

Pollock, J. R. A. (Ed.), Brewing Science, Volume 1—1979. Volume 2—1980. Volume 3—1987.

Christopher Bauernfeind, J. (Ed.), 1981. Carotenoids as Colorants and Vitamin A Precursors: Technological and Nutritional Applications.

Markakis, P. (Ed.), 1982. Anthocyanins as Food Colors.

Stewart, G.G., and Amerine, M.A. (Eds.), 1982. Introduction to Food Science and Technology, second ed.

Iglesias, H.A., and Chirife, J., 1982. Handbook of Food Isotherms: Water Sorption Parameters for Food and Food Components.

Dennis, C. (Ed.), 1983. Post-Harvest Pathology of Fruits and Vegetables.

Barnes, P.J. (Ed.), 1983. Lipids in Cereal Technology.

Pimentel, D., and Hall, C.W. (Eds.), 1984. Food and Energy Resources.

Regenstein, J.M., and Regenstein, C.E., 1984. Food Protein Chemistry: An Introduction for Food Scientists.

Gacula Jr. M.C., and Singh, J., 1984. Statistical Methods in Food and Consumer Research.

Clydesdale, F.M., and Wiemer, K.L. (Eds.), 1985. Iron Fortification of Foods.

Decareau, R.V., 1985. Microwaves in the Food Processing Industry.

Herschdoerfer, S.M. (Ed.), Quality Control in the Food Industry, second edition. Volume 1—1985. Volume 2—1985. Volume 3—1986. Volume 4—1987.

Cunningham, F.E., and Cox, N.A. (Eds.), 1987. Microbiology of Poultry Meat Products.

Urbain, W.M., 1986. Food Irradiation.

Bechtel, P.J., 1986. Muscle as Food.

Chan, H.W.-S., 1986. Autoxidation of Unsaturated Lipids.

McCorkle Jr. C.O., 1987. Economics of Food Processing in the United States.

Japtiani, J., Chan Jr., H.T., and Sakai, W.S., 1987. Tropical Fruit Processing.

Solms, J., Booth, D.A., Dangborn, R.M., and Raunhardt, O., 1987. Food Acceptance and Nutrition.

Macrae, R., 1988. HPLC in Food Analysis, second ed.

Pearson, A.M., and Young, R.B., 1989. Muscle and Meat Biochemistry.

Penfield, M.P., and Campbell, A.M., 1990. Experimental Food Science, third ed.

Blankenship, L.C., 1991. Colonization Control of Human Bacterial Enteropathogens in Poultry.

Pomeranz, Y., 1991. Functional Properties of Food Components, second ed.

Walter, R.H., 1991. The Chemistry and Technology of Pectin.

Stone, H., and Sidel, J.L., 1993. Sensory Evaluation Practices, second ed.

Shewfelt, R.L., and Prussia, S.E., 1993. Postharvest Handling: A Systems Approach.

Nagodawithana, T., and Reed, G., 1993. Enzymes in Food Processing, third ed.

Hoover, D.G., and Steenson, L.R., 1993. Bacteriocins.

Shibamoto, T., and Bjeldanes, L., 1993. Introduction to Food Toxicology.

Troller, J.A., 1993. Sanitation in Food Processing, second ed.

Hafs, D., and Zimbelman, R.G., 1994. Low-fat Meats.

Phillips, L.G., Whitehead, D.M., and Kinsella, J., 1994. Structure-Function Properties of Food Proteins.

Jensen, R.G., 1995. Handbook of Milk Composition.

Roos, Y.H., 1995. Phase Transitions in Foods.

Walter, R.H., 1997. Polysaccharide Dispersions.

Barbosa-Cánovas, G.V., Marcela Góngora-Nieto, M., Pothakamury, U.R., and Swanson, B.G., 1999. Preservation of Foods with Pulsed Electric Fields.

Jackson, R.S., 2002. Wine Tasting: A Professional Handbook.

Bourne, M.C., 2002. Food Texture and Viscosity: Concept and Measurement, second ed.

Caballero, B., and Popkin, B.M. (Eds.), 2002. The Nutrition Transition: Diet and Disease in the Developing World.

Cliver, D.O., and Hans Riemann, P. (Eds.), 2002. Foodborne Diseases, second ed.

Kohlmeier, M., 2003. Nutrient Metabolism.

Stone, H., and Sidel, J.L., 2004. Sensory Evaluation Practices, third ed.

Han, J.H., 2005. Innovations in Food Packaging.

Sun, D.-W. (Ed.), 2005. Emerging Technologies for Food Processing.

Riemann, H., and Cliver, D. (Eds.), 2006. Foodborne Infections and Intoxications, third ed.

Arvanitoyannis, I.S., 2008. Waste Management for the Food Industries.

Jackson, R.S., 2008. Wine Science: Principles and Applications, third edition. 2008.

Sun, D.-W. (Ed.), 2008. Computer Vision Technology for Food Quality Evaluation.

David, K., and Thompson, P., (Eds.), 2008. What Can Nanotechnology Learn From Biotechnology?.

Arendt, E.K., and Bello, F.D. (Eds.), 2008. Gluten-Free Cereal Products and Beverages.

Bagchi, D. (Ed.), 2008. Nutraceutical and Functional Food Regulations in the United States and Around the World.

Paul Singh, R., and Heldman, D.R., 2008. Introduction to Food Engineering, fourth ed.

Berk, Z., 2009. Food Process Engineering and Technology.

Thompson, A., Boland, M., and Singh, H. (Eds.), 2009. Milk Proteins: From Expression to Food.

Florkowski, W.J., Prussia, S.E., Shewfelt, R.L., and Brueckner, B. (Eds.), 2009. Postharvest Handling, second ed.

Gacula Jr., M., Singh, J., Bi, J., and Altan, S., 2009. Statistical Methods in Food and Consumer Research, second ed.

Shibamoto, T., and Bjeldanes, L., 2009. Introduction to Food Toxicology, second ed.

BeMiller, J., and Whistler, R. (Eds.), 2009. Starch: Chemistry and Technology, third ed.

Jackson, R.S., 2009. Wine Tasting: A Professional Handbook, second ed.

Sapers, G.M., Solomon, E.B., and Matthews, K.R. (Eds.), 2009. The Produce Contamination Problem: Causes and Solutions.

Heldman, D.R., 2011. Food Preservation Process Design.

Tiwai, B.K., Gowen, A., and Mckenna, B. (Eds.), 2011. Pulse Foods: Processing, Quality and Nitraceutical Applications.

Index

Note: Page numbers with "f" and "t" refer to figures and tables, respectively.

A

Absidia coerulea, 280–281
absorption, light, 189–193, 192f
acceleration electron beams (e-beams)
  characteristics, 169, 170t
  packaging requirements, 174–177
acoustic cavitation, 137, 138f, 142t
acrylic acid, 179t
acrylonitrile, 179t
adenosine triphosphate (ATP), 244–246
adenovirus type-2 (HAdV-2), 244–246
adenovirus type 40 (AD40), 242–243
Advisory Committee on Novel Foods (ACNFP), 481–482
Aeromonas hydrophila
  carbonation-based preservation, 263–264
  pulsed light technologies, 210–211
aflatoxin, 246–247
aglycon, 250–251
albumin, 237–243
Alcaligenes faecalis, 237–243
alcohol dehydrogenase, 244–246
alcohols, 377–378, 382–383
aldehydes, 94f, 248–249, 294, 382–383
Alicyclobacillus acidocaldarius, 237–243
Alicyclobacillus acidoterrestris, 330
alkaline phosphatase (ALP)
  dense-phase carbon dioxide (DPCD), 290t
  ohmic heating, 348–352, 349t
  ozone processing, 244–246
  pulsed electric field (PEF) processing, 81t, 84
  ultrasound processing, 152, 153t
alkene formation, 151
α-amylase, 15, 290t
aluminum stearate, 180t
amino acids
  Maillard reaction, 250–251
  ozone processing, 243–244
  pulsed electric field (PEF) processing, 69
antheraxanthin, 389
anthocyanin
  analytical measurement techniques, 440
  dense-phase carbon dioxide (DPCD), 293–294
  high-pressure processing (HPP), 118–126
  irradiation processes, 173–174
  microwave heating, 390
  ozone processing, 249–250
  pulsed electric field (PEF) processing, 88, 89t
  ultrasound processing, 152–156
antimicrobial treatments, 78
antioxidant capacity, 126–128, 212, 293–294
apple juice/apple cider
  antimicrobial treatments, 78
  carbonation-based preservation, 263–264
  cloud stability, 250–251
  color changes, 152–156, 213
  dense-phase carbon dioxide (DPCD)
    physicochemical and sensory effects, 294
    research background, 286t
    solubility measures, 279
  furan formation, 211–213
  high-pressure processing (HPP), 111t, 117, 122t
  irradiation processes
    quality and nutritional effects, 173–174
    shelf-life extension, 172–173
  light absorptivity, 190, 192f
  microbial inactivation, 145
  microwave heating, 41, 387–390, 389f, 394–395, 394f
  ohmic heating, 313–317, 314t, 317f
  ozone processing
    color changes, 248–249
    enzyme inactivation, 250–251
    liquid food processing, 227–228
    microbial inactivation, 237–243, 238t
    physical properties, 250–251
    quality parameter effects, 247–251, 248t
    temperature effects, 232–233
  physical properties, 193–195, 194t, 195t
  primary kinetic models, 446t
  pulsed electric field (PEF) processing
    effectiveness, 72t
    food-quality-related enzyme inactivation, 80, 81t
    nutritional parameter effects, 87–91
    quality parameter effects, 86
    shelf-life extension, 91–95
  ultrasound processing, 141, 147–150, 148t
  ultraviolet (UV) light processing, 195–196
    background and functional role, 185–186
    microbial inactivation, 206–209, 206t, 207t
    physical properties, 193–196, 194t, 195t
    quality and nutritional effects, 211–213
    reactor designs, 214–215

apple puree, 390–395
apples, 414t
apple sauce, 314t, 318, 395
apricot juice, 173–174
apricot puree, 314t, 319
aromatic hydrocarbons, 382–383
Arrhenius equation, 453
artichoke by-product puree, 341t, 343–344, 352
artichokes, 359
ascorbic acid
  dense-phase carbon dioxide (DPCD), 293–294
  high-pressure processing (HPP), 127
  infrared heating, 414t
  microwave heating, 390
  ohmic heating
    color changes, 345–348
    degradation, 340–344, 341t, 343f
  ozone processing, 236, 249–250
  pulsed electric field (PEF) processing, 87–91
  ultrasound processing, 152–156
  ultraviolet (UV) light processing, 212
aseptic packaging, 174–177, 310, 327–333, 395
Aspergillus niger, 203–204, 227–228, 275t, 290t
Aspergillus niger conidia, 275t
autoclaves, 11, 12f
avocado
  heat of compression value, 115t
  microwave heating, 390–395

B

baby food, 359
Bacillus cereus
  carbonation-based preservation, 263–264
  dense-phase carbon dioxide (DPCD)
    depressurization rate, 281–282
    pressure and temperature parameters, 272t, 275t
  ozone processing, 227–228, 238t, 244–246
Bacillus coagulans
  dense-phase carbon dioxide (DPCD), 275t
  pulsed electric field (PEF) processing, 76
Bacillus fulva, 275t
Bacillus licheniformis
  dense-phase carbon dioxide (DPCD), 275t
  ohmic heating, 332
  ozone processing, 237–243
Bacillus megaterium, 275t
Bacillus polymixa, 275t
Bacillus pumilis, 275t
Bacillus spp., 237–243
Bacillus subtilis
  α-amylase, 15, 420
  dense-phase carbon dioxide (DPCD)
    depressurization rate, 281–282
    enzyme inactivation, 290t
    pressure and temperature parameters, 271, 272t, 275t
  far-infrared (FIR) radiation, 418–420
  ohmic heating, 329–330, 330f
  ozone processing, 232–233, 234f
  photochemical effect, 199–203, 201f
  pulsed electric field (PEF) processing, 67, 76, 84
  thermal shock treatments, 77–79
  ultrasound processing, 145, 148–149
Bacillus thermosphacta, 272t
bacteria
  carbonation-based preservation, 263–264
  dense-phase carbon dioxide (DPCD), 288–289
  far-infrared (FIR) radiation, 419–420, 419f, 421f
  high-pressure processing (HPP), 117–118
  irradiation processes, 172–173
  microwave heating, 383
  ohmic heating, 327–333
  ozone processing
    fluid foods, 227–228
    inactivation mechanisms, 243–246
    microbial inactivation, 225–226, 237–243, 238t
    organic matter, 237f
    structural changes, 244f
  pulsed electric field (PEF) processing, 68–69, 71
  ultrasound processing, 141–146, 142t
  ultraviolet (UV) light processing, 185–186, 193–196, 206–209, 206t
bacteriophage f2, 242–246
baking processes, 380–396, 382t, 414t
bananas/banana puree
  infrared heating, 414t
  microwave heating, 392
batch systems, 43–44
beans, 177–178
beef, ground
  heat of compression value, 115t
  infrared heating, 414t
beer
  dense-phase carbon dioxide (DPCD) treatments, 56–57, 285, 286t, 293–294
  infrared heating, 424–425
Beer*Lambert Law, 139–140, 189–193, 196–197
beetroot juice, 111t
Belgium, 486

β-carotene
dense-phase carbon dioxide (DPCD), 294
high-pressure processing (HPP), 127
irradiation processes, 173–174
microwave heating, 389
ohmic heating, 343–344
pulsed electric field (PEF) processing, 87–88, 89t
ultraviolet (UV) light processing, 212
β-cryptoxanthin
high-pressure processing (HPP), 127
pulsed electric field (PEF) processing, 87–88
β-galactosidase, 244–246, 348–352, 349t
β-glucosidase
ozone processing, 250–251
pulsed electric field (PEF) processing, 81t, 85
β-lactoglobulin (β-LG), 49, 122t, 319, 335–336
Better Than Fresh™ (BTF), 295–296
Bifidobacteria, 140–152
Bigelow model, 447–448
Bingham fluids, 36–38, 37f
biphasic models, 442, 443f, 446t
blackberries, 249–250
blackberry juice
ozone processing, 248–250, 248t
ultrasound processing, 147–150, 152–156
black body radiation
characteristics, 413–418
Planck's law, 417–418
spectral characteristics, 417f
Stefan*Boltzman's law, 417f, 418
Wien's displacement law, 418
black currant juice, 173–174
blanching processes, 380–396, 382t, 414t
blueberries, 414t
bovine serum albumin (BSA), 122t
brain*heart infusion broth, 272t, 275t
bread, 414t
Brix levels
dense-phase carbon dioxide (DPCD), 293–294
fluid foods, 194t
ozone processing, 250–251
ultraviolet (UV) light processing, 213
broccoli, 111t, 126
Brochothrix thermosphacta, 263–264
browning
dense-phase carbon dioxide (DPCD), 289–293
high-pressure processing (HPP), 118–126
irradiation processes, 173–174
microwave heating, 384–387, 390–395
ohmic heating, 345–348
ozone processing, 250–251
pulsed electric field (PEF) processing, 85–87
ultrasound processing, 149–150, 152–154, 156
ultraviolet (UV) light processing, 211–213
brown shrimp, 290t, 292
bubbles *See* cavitation phenomena
bulk viscosity, 8–10
burgers, 414t
butylated hydroxyanisole (BHA), 180t
butylated hydroxytoluene (BHT), 180t

C

cabbage, 122t
cakes, 414t
calcium propionate, 180t
calcium stearate, 180t
calorimetry, 139
Candida albicans
microwave heating, 383
ozone processing, 227–228, 237–243
Candida parapsilosis, 236, 238t
Candida rugosa, 290t
Candida tropicalis, 238t
Candida utilis, 272t
carbohydrates
heat of compression value, 114, 115t
microwave heating, 377–378
ultrasound technologies, 51–52
carbonation *See* dense-phase carbon dioxide (DPCD)
carbon dioxide (CO_2), pressurized
carbon dioxide (CO_2) properties, 264–266, 265f, 276–277
carbon dioxide (CO_2) solubility
microbial inactivation, 277–278
solubility measures, 278–279
thermodynamic models, 279–280
food preservation, 263–264
microbial inactivation, 282–285, 283f
carboxylic acids, 55–56, 249–250
carboxyl methyl cellulose (CMC), 55–56
carotenoids
high-pressure processing (HPP), 118–126
infrared heating, 420
kinetic modeling, 459–460
microwave heating, 389
pigments, 248–249
pulsed electric field (PEF) processing, 67–68, 69, 87–91, 89t
ultraviolet (UV) light processing, 213
carrageenan, 314t

carrot juice
  dense-phase carbon dioxide (DPCD), 286t, 290t
  high-pressure processing (HPP), 111t, 119t
  irradiation processes, 172, 174
  kinetic modeling, 446t, 459–460
  physical properties, 193–195, 194t
  pulsed electric field (PEF) processing
    effectiveness, 72t
    food-quality-related enzyme inactivation, 81t
    nutritional parameter effects, 87–91, 89t
  ultraviolet (UV) light processing, 206–209, 206t
carrot puree
  microwave heating, 42–43
  ohmic heating, 314t
carrots
  high-pressure processing (HPP) effects, 126
  infrared heating, 414t
  ohmic heating, 359
casein
  dense-phase carbon dioxide (DPCD), 56–57
  high-pressure processing (HPP), 48–49, 50f
  irradiation processes, 486
  ozone processing, 237–243
  ultrasound technologies, 51–52
castor oil, 112
catalase
  high-pressure processing (HPP), 118–126
  ozone processing, 244–246
cauliflower, 334, 359
cavitation phenomena, 44, 51–52, 137, 138f, 142t, 229
cell lysis, 243–246, 244f
cellophane, 179t
cereals, 486
Cerf model, 444
chalcone, 152–156
cheese *See* milk/milk products
cheese sauce, 380t
chemical dosimetry, 139–140
chestnuts, 414t
chicken breast, 115t
chicken fat, 115t
Chinese irradiation processes
  irradiated wine, 180–181
  regulatory and consumer aspects, 177–179
chitosan, 55–56
chlorogenic acid, 89t
chlorophyll
  high-pressure processing (HPP), 118–126
  microwave heating, 390–395
  ohmic heating, 346–348
cholecalciferol, 87–91
chymosin, 151
cinnamon oil, 78, 79f, 95f
cis-3-hexenol, 94f
citric acid, 79f, 91–92, 95f, 119t, 279
citrus juices
  high-pressure processing (HPP), 111t
  irradiation processes, 172–173
  ohmic heating, 344–345
  pulsed electric field (PEF) processing, 85–87
  regulatory requirements, 489
Citrus paradisi, 288
Clostridium botilinum, 193–195
Clostridium perfringens, 233–235
Clostridium sporogenes, 26–27, 211
cloudberry jam, 332
cloud stability, 154–155, 250–251, 293–294, 352
coconut oil, 56
coconut water, 286t, 392–394, 393f
Codex Alimentarius, 484–485, 490
coffee beans, 414t
cold plasma method, 227
coliform bacteria, 237–243
color changes
  dense-phase carbon dioxide (DPCD), 293–294
  microwave heating, 390–395
  ohmic heating, 345–348, 347t
  ozone processing, 248–249
  ultrasound processing, 152–156
  ultraviolet (UV) light processing, 213
colostrum, 111t
computational fluid dynamics (CFD) simulations, 45
conidia, 185–186
conservation of energy, 8–10
conservation of mass, 8–10
conservation of momentum, 8–10
consumer acceptibility, 4
containers *See* packaging requirements
continuous ohmic heating systems, 308–310, 309f
continuous-wave UV (CW UV), 186
cooking processes, 380–396, 382t, 414t
copper salts, 390–395
corn, 414t
corn oil, 378–379
corn starch suspension, 314t
corona discharge method, 226–227
Council Directive 1999/2/EEC, 484–487
Council Directive 2001/112/EC, 489
Coxiella burnetii, 193–195

Coxsackie viruses, 238t
cracked wheat, 414t
cranberry juice, 206t
cream cheese, 115t
Criegee mechanism, 55–56, 249–250
Cronobacter sakazakii, 448, 448f
crude beef fat, 115t
Cryptosporidium parvum, 185, 206t, 232–233, 234f, 238t
cucumbers, 293–294
cyanidin-3-glucoside, 152–156
cyclopiazonic acid, 246–247
Czech Republic
  high-pressure processing (HPP), 111t
  irradiation processes authorization, 486

D

dairy products *See* milk/milk products
dark repair, 199
Dean flow, 217
Dean numbers, 41–43
decadienal, 94f
decanal, 344
Dekkera bruxellensis, 78
deMan, Rogosa and Sharpe broth, 272t, 275t
dense-phase carbon dioxide (DPCD), 263–304
  basic concepts
    carbon dioxide (CO_2) properties, 264–266, 265f
    engineering aspects
      batch systems, 266–270, 266f
      continuous flow-through systems, 268, 269f
      continuous high-pressure systems, 270, 270f
      multi-batch systems, 267, 267f
      semi-continuous systems, 267–268, 268f, 269f
    pressurized carbon dioxide (CO_2), 263–264
  commercialization efforts
    background information, 295–296
    potential advantages, 295
    potential difficulties, 296
  extrinsic parameters, 437t
  food treatment effects, 56–57
  intrinsic parameters, 438t
  primary kinetic models, 446t
  process control parameters
    carbon dioxide (CO_2) physicochemical properties, 276–277
    carbon dioxide (CO_2) solubility
      microbial inactivation, 277–278
      solubility measures, 278–279
      thermodynamic models, 279–280
    depressurization rate, 280–282
    microbial inactivation, 282–285, 283f
    mixing speeds, 280
    pressure and temperature, 270–276, 272t, 275t
    spores, 271, 275t
    vegetative forms, 272t
  quality and nutritional parameter effects
    enzyme inactivation, 289–293, 290t
    inactivation and stability effects, 288–289
    physicochemical and sensory effects, 293–294
    research background, 285–294, 286t
density (pulsed electric field (PEF) processing), 69–70
deoxyribonucleic acid (DNA), 244–246
Diastase Number (DN), 385–387, 385t, 388f, 421–424
dielectric heating, 369 *See also* microwave heating
dilatant fluids, 36–38, 37f
dimethyl dicarbonate (DMDC), 228
direct electrical resistance heating *See* ohmic heating
disinfectants, 225–226
dissolved organic matter (DOM), 236
D parameters, 451–452
dried herbs and spices, 177–178, 178t, 485–486
drying processes, 380–396, 382t, 414t

E

e-beams
  characteristics, 169, 170t
  packaging requirements, 174–177
eggs/egg products
  dense-phase carbon dioxide (DPCD), 288
  fluid rheology
    high-pressure processing (HPP), 51
    irradiation effects, 52–53
    pulsed electric fields (PEFs), 45–47
  heat of compression value, 115t
  irradiation processes, 178t
  light absorptivity, 192f
  ohmic heating, 312, 314t, 329, 355–356, 357f, 359
  physical properties, 193–195, 194t, 195t
  primary kinetic models, 446t
  pulsed electric field (PEF) processing
    effectiveness, 71–77, 72t
    fluid rheology, 45–47

eggs/egg products (*Continued*)
research background, 63
ultrasound processing
microbial inactivation, 147–148
research background, 140–152
ultraviolet (UV) light processing, 185–186
electrical conductivity (pulsed electric field (PEF) processing), 69–70, 85–87
electrical discharge method, 226–227
electrical resistance heating *See* ohmic heating
electrochemical (cold plasma) method, 227
electroconductive heating *See* ohmic heating
electroheating *See* ohmic heating
Electrolytic Non-Random Two Liquids (ELECNRTL) model, 279–280
electromagnetic radiation, 370, 370f, 412f
electroporation, 71, 327–333
ellagic acid, 88, 89t
endive, 414t
enteric viruses, 238t
Enterobacter aerogenes, 237–243
Enterobacter cloacae, 237–243
Enterobacter sakazakii
irradiation processes, 486
microwave heating, 383
Enterococci sp., 145
Enterococcus faecalis
dense-phase carbon dioxide (DPCD), 272t
ozone processing, 227–228
enterohemorrhagic *E. coli* infections, 167–168
environmental factors, 452–456
environmental impacts, 5
environmental scanning electron microscopy (ESEM), 143, 145f
enzymatic browning *See* browning
enzymes
dense-phase carbon dioxide (DPCD), 282–285, 289–293, 290t
high-pressure processing (HPP)
combined pressure*heat effects, 118–126, 119t
enzyme activity outcomes, 14, 14f
microbiological safety, 117–118
irradiation processes, 178t
ohmic heating, 348–352, 349t, 350f, 351t, 353t
ozone processing, 243–244, 250–251
pulsed electric field (PEF) processing, 69, 80–85, 81t
ultrasound processing
enzyme inactivation, 136, 142t, 147
fruit juices, 147–150
ultraviolet (UV) light processing, 213
epilactose, 382–383
erucic acid, 180t
Escherichia coli
carbonation-based preservation, 263–264
dense-phase carbon dioxide (DPCD), 272t, 286t
enterohemorrhagic infections, 167–168
far-infrared (FIR) radiation, 419–420, 419f, 421f
high-pressure processing (HPP), 15, 16f
irradiation processes, 172–173, 486
light pulse technologies, 210–211
microwave heating
apple puree, 390–392
continuous flow-through systems, 41–43
fruit juices, 387–390, 389f
milk/milk products, 383
ohmic heating, 329
ozone processing
inactivation mechanisms, 243–246
microbial inactivation, 227–228, 237–243, 238t
organic matter, 236
ozone concentration, 229–230, 231f, 232t
pH levels, 233–235
structural changes, 244f
temperature effects, 230–233, 232t
voltage effects, 232f
photochemical effect, 199–203, 201f, 203f
pulsed electric field (PEF) processing, 72t, 74–75
ultrasound processing
fruit juices, 147–150, 148t
microbial inactivation, 141, 145f, 146f
ultraviolet (UV) light processing, 193–195, 206–209, 206t
esters, 382–383
ethanol
heat of compression value, 115t
high-pressure processing (HPP), 112
ethylene*vinyl acetate copolymers, 179t
Eucalyptus lanceolatus, 385
European Commission, 474–476
European Council, 474–476
European Court of Justice, 474–476
European Federation of Food Science and Technology (EFFoST), 490
European Food Safety Agency (EFSA), 474–476
European Parliament, 474–476
European Union (EU)
background information, 474–477

regulatory and legislative issues
consumer safety, 473–474, 477–478, 491–492
European Food Safety Agency (EFSA), 474–476
international standards, 490–491
law creation and enforcement
institutional role, 474–476
legislative characteristics, 476
supremacy factor, 476–477
novel food regime
application refusals, 480
Novel Foods Regulation, 476, 478–484
substantial equivalence concept, 481–482
update proposal draft, 482–484
regulatory framework, 477–478
regulatory requirements
fruit juices, 489
high-pressure processing (HPP), 487–488
irradiation processes, 484–487
labeling and health claims, 488–489
milk/milk products, 488–489
extracted beef fat, 115t

F

far-infrared (FIR) radiation, 418–420, 419f, 421f
farm-to-fork food chain, 4
fatty acids
ozone processing, 56
pulsed electric field (PEF) processing, 69
feline calicivirus (FCV), 242–243
fermented foods, 178
Feta cheese brines, 210
fish
fish gelatin, 54–55, 54f
fish wastewater, 339
irradiation processes, 486
5-methlytetrahydrofolic acid, 127
flavonols, 88
flavor changes, 167–168, 344–345
flow cytometry, 440
fluence, 193, 196–197
fluence rate, 191–193, 196–197
fluid dynamics
basic concepts, 8–10
food treatment processes, 7–8
high-pressure processing (HPP), 109–134
basic concepts
enzyme activity outcomes, 13–14, 14f
heat loss effects, 15
influencing factors, 10–17
inhomogeneous inactivation, 16
packaging effects, 15, 16f
process time-scale, 16–17
time*pressure*velocity relationship, 12f
combined pressure*heat effects
nutrient content, 126–128
quality attributes, 118–126, 119t, 122t
extrinsic parameters, 437t
government regulations, 128
intrinsic parameters, 438t
kinetic modeling, 435, 446t, 448–450, 460f, 461
microbiological safety, 117–118
operational characteristics
food material response, 112–117, 113f
governing principles, 110–117
heat of compression values, 114, 115t, 116f
packaging requirements, 111–112
pH shifts, 117
pressure*temperature response, 112–117, 113f
pressure-transmitting fluids, 112
process equipment, 110–111
regulatory and legislative issues, 487–488
research background, 109
research summary, 128
solid*liquid systems, 45
stability effects, 4
treatment effects, 47–51, 50f
light pulse technologies, 185–224
background and functional role, 185–186
basic concepts, 26–27
industrial scale-up
challenges, 218–219
future outlook, 218–219
reactor designs, 214–218
microbial inactivation
antimicrobial effects, 210–211
background information, 185–186, 199–206
honey, 211
juices, 206–209, 206t
kinetic mechanisms, 204–206
liquid sweeteners, 211
milk/milk products, 210–211
photochemical effect, 199–203, 201f, 203f
photophysical effect, 204, 205f
photothermal effect, 203–204
operation principles
design parameters, 189–193, 192f

fluid dynamics (*Continued*)
gas discharges, 188–189
generation mechanisms, 187–188
lamp emission spectra, 192f
light absorption, 189–193, 192f
pulsed power energization, 189
spectral characteristics, 187–193
process control parameters
dose delivery, 196–197
flow dynamics, 197
fluid food parameters, 193–196, 194t
lamp emission spectra, 198f, 200t
light sources, 197–199, 200t
quality and nutritional effects, 211–213
reactor designs
annular reactors, 214–215, 214f
coiled-tube reactors, 216–217, 216f
dynamic mixers, 217–218, 217f
static reactors, 216–217, 216f
thin-film reactors, 214–215, 215f
research summary, 219
ohmic heating, 305–368
advantages, 310
aseptic food processing, 310, 327–333
basic concepts, 27–28, 305–306, 306f
continuous flow-through systems, 40–41
disadvantages, 310–311
extrinsic parameters, 437t
future outlook, 359
intrinsic parameters, 438t
microbial inactivation, 327–333, 330f, 331f
operation principles, 306–311, 308f, 309f
primary kinetic models, 446t
process control parameters, 311–327
quality and nutritional parameter effects, 327–333
research background, 306–307
research summary, 359
rheological properties, 57, 353–358, 354f, 355t, 356f, 357f
solid*liquid systems, 44–45, 46f
pulsed electric fields (PEFs), 63–108
basic concepts, 17–24
combination preservation strategies, 77–79, 79f
continuous flow-through systems, 43, 44f
electrohydrodynamic phenomena simulations, 18–19, 23f, 24f
extrinsic parameters, 437t
food-quality-related enzyme inactivation, 69, 80–85, 81t
functional foods, 4
health-related compounds, 69
intrinsic parameters, 438t
kinetic modeling, 435, 446t
microbial inactivation
effectiveness, 68–69, 71–77, 72t
inactivation mechanisms, 71
nutritional-related properties, 69, 87–91, 89t
operation principles, 64–65, 65f
process control parameters, 66
product parameters, 69–70
quality-related properties, 69, 85–87, 93f
research background, 63
research summary, 95–96
shelf-life extension, 91–95, 92f, 93f, 94f
treatment chamber geometries, 22–23, 22f
treatment effects, 45–47
treatment parameters, 66–68
research summary, 28
ultrasound technologies, 135–166
basic concepts, 25–26
batch systems, 44
extrinsic parameters, 437t
food processing, 135
functional foods, 4
generation mechanisms, 136–137, 138f
intrinsic parameters, 438t
liquid food sterilization
enzyme inactivation, 147
fruit juices, 147–150, 148t
microbial inactivation, 141–146, 142t
milk/milk products, 151–152, 153t
research background, 140–152
measurement techniques
calorimetry, 139
characteristics, 137–140
chemical dosimetry, 139–140
primary kinetic models, 446t
quality parameter effects, 152–156
research applications, 135–136
research challenges, 156–157
research summary, 157
treatment effects, 51–52
ultraviolet (UV) light processing, 185–224
background and functional role, 185–186
basic concepts, 26–27
extrinsic parameters, 437t
food treatment effects, 54–55, 54f
industrial scale-up
challenges, 218–219
future outlook, 218–219
reactor designs, 214–218

intrinsic parameters, 438t
microbial inactivation
background information, 185–186, 199–206
juices, 206–209, 206t
kinetic mechanisms, 204–206
milk/milk products, 210
photochemical effect, 199–203, 201f, 203f
operation principles
design parameters, 189–193, 192f
gas discharges, 188–189
generation mechanisms, 187–188
lamp emission spectra, 192f
light absorption, 189–193, 192f
pulsed power energization, 189
spectral characteristics, 187–193
primary kinetic models, 446t
process control parameters
dose delivery, 196–197
flow dynamics, 197
fluid food parameters, 193–196, 194t
lamp emission spectra, 198f, 200t
light sources, 197–199, 200t
quality and nutritional effects, 211–213
reactor designs
annular reactors, 214–215, 214f
coiled-tube reactors, 216–217, 216f
dynamic mixers, 217–218, 217f
static reactors, 216–217, 216f
thin-film reactors, 214–215, 215f
research summary, 219
fluid rheology
basic concepts, 36–38
fluid foods, 324–325
food treatment challenges, 35–36
ohmic heating, 353–358, 354f, 355t, 356f, 357f
process efficacy
batch systems, 43–44
continuous flow-through systems
governing equations, 40–43
microwave heating, 41–43, 42f
ohmic heating, 40–41, 44–45, 46f
pulsed electric fields (PEFs), 43, 44f
solid*liquid systems, 44–45, 46f
viscous fluids, 39–45
research summary, 57–58
shear flow, 36–38, 37f
treatment effects
dense-phase carbon dioxide (DPCD), 56–57
general discussion, 45–57
high-pressure processing (HPP), 47–51, 50f
irradiation processes, 52–53, 53f
ohmic heating, 57
ozone processing, 55–56
pulsed electric fields (PEFs), 45–47
ultrasound technologies, 51–52
ultraviolet (UV) light, 54–55, 54f
folic acid, 212
Food Additives Amendment Act (1958), 486–487
Food and Drug Administration (FDA), 2–3, 491
Food Drug and Cosmetic Act (FD&C Act), 128
Food Irradiation Regulations (UK, 2009), 487
food-safety management systems, 490
Food Standards Agency (UK), 473
Fourier transform infrared (FTIR) spectroscopy, 439
France, 111t
free radical formation, 141, 142t, 171–172, 226–227, 243–246
freeze drying, 380–396, 382t, 383, 384f, 400f, 414t
fresh fruits and vegetables
dense-phase carbon dioxide (DPCD), 289–293
high-pressure processing (HPP), 487–488
infrared heating, 414t
irradiation processes, 177–178, 178t, 486
ohmic heating, 359
frogs' legs, 486
frozen foods, 414t
fructose
dense-phase carbon dioxide (DPCD), 279
furan formation, 211–213
light pulse technologies, 211
microwave heating, 385t
physical properties, 194t, 195–196, 195t
fruit gels, 392
fruit juices *See* juices
fruit nectars, 359, 489
fumonisin, 246–247
functional foods, 4, 126–128
fungi
irradiation processes, 172–173
ozone processing, 225–226, 243–246
furans, 168–171, 211–213
furosine, 382–383

G

gallic acid, 212
gamma-glutamyl transpeptidase (GGTP), 152, 153t
gamma rays
characteristics, 170t, 171

gamma rays (*Continued*)
- mode of action, 171–172
- packaging requirements, 174–177
- wavelengths and frequencies, 370f

gas discharges, 188–189
Gauss*Newton method, 456–457
gazpacho, 81t, 87–91, 127
gelatin/gelatin mixtures
- high-pressure processing (HPP), 48
- ohmic heating, 314t
- ozone processing, 237–243

Geobacillus stearothermophilus, 275t, 332
Giardia lamblia, 238t
Giardia muris, 238t
Giardia sp., 185
Gibbs free-energy model, 279–280
GInaFiT software, 443
ginger juice, 111t
glassine paper, 179t
Global Harmonization Initiative (GHI), 490
glucose
- dense-phase carbon dioxide (DPCD), 279
- microwave heating, 384–387, 385t
- physical properties, 194t

glutathione, 244–246
glutathione disulfide reductase, 244–246
glyceraldehyde-3-phosphate dehydrogenase, 244–246
goat's milk
- microwave heating, 383
- ohmic heating, 332
- ultraviolet (UV) light processing, 206t, 212–213

goat's milk cheese, 122t
government regulations, 2–3
grains, 177–178
Gram-positive/gram-negative bacteria
- high-pressure processing (HPP), 117–118
- ozone processing, 227–228, 237–243
- pulsed electric field (PEF) processing, 68–69, 71, 76
- ultrasound processing, 145

grapefruit juice
- dense-phase carbon dioxide (DPCD)
  - inactivation and stability effects, 288
  - research background, 286t
  - solubility measures, 279
- irradiation processes, 173–174
- ultraviolet (UV) light processing, 206t

grape juice
- dense-phase carbon dioxide (DPCD), 286t, 288–289, 290t, 293–294
- dielectric properties, 379–380
- irradiation processes, 173–174
- ohmic heating
  - electrical conductivity, 314t
  - enzyme inactivation, 349, 350f
- physical properties, 194t, 195
- primary kinetic models, 446t
- pulsed electric field (PEF) processing, 80, 81t
- regulatory requirements, 489
- ultrasound processing, 147–150
- ultraviolet (UV) light processing, 206t, 212

grape must, 119t
graphite electrodes, 322, 323f
gravy, beef, 115t
green coconut water, 392–394, 393f
greenhouse gases, 296
green pea puree
- microwave heating, 42–43, 42f
- ohmic heating, 345–346, 347t, 352, 353t
- pulsed electric field (PEF) processing
  - effectiveness, 72t
  - food-quality-related enzyme inactivation, 81t, 84

green peas, 414t
green table olive solutions (GTOS), 247–248
green tea
- infrared heating, 414t
- pulsed electric field (PEF) processing, 72t, 74–75

growth media (dense-phase carbon dioxide (DPCD)), 272t
guacamole, 118
guar gum
- infrared heating, 414t
- irradiation effects, 53, 53f
- ozone processing, 55–56

guava juice
- color changes, 152–156
- high-pressure processing (HPP), 122t
- irradiation processes, 173–174
- physical properties, 193–195, 194t
- ultrasound processing, 147–150, 148t
- ultraviolet (UV) light processing, 206–209, 206t, 213

guava puree
- dense-phase carbon dioxide (DPCD), 279
- high-pressure processing (HPP), 122t

gum arabic, 486

H

H1N1 virus, 238t, 244–246
H5N1 virus, 238t

hamburgers, 414t
Hass avocado, 115t
hazelnuts, 414t
health-related compounds (pulsed electric field (PEF) processing), 69
heat transfer processes *See* ohmic heating
Helianthus annuus, 385
hemolytic uremic syndrome, 167–168
Henry's law, 265–266
hepatitis A virus, 242–246
herbs, 177–178, 178t, 485–486
herpes simplex virus type-1 (HHV-1), 244–246
Herschel*Bulkley model, 36–38, 37f
hesperetin, 88
hexanal, 94f
Hibiscus sabdariffa, 279
high-performance liquid chromatography (HPLC), 440
high-pressure processing (HPP), 109–134
  basic concepts
    enzyme activity outcomes, 13–14, 14f
    heat loss effects, 15
    influencing factors, 10–17
    inhomogeneous inactivation, 16
    packaging effects, 15, 16f
    process time-scale, 16–17
    time*pressure*velocity relationship, 12f
  combined pressure*heat effects
    nutrient content, 126–128
    quality attributes, 118–126, 119t, 122t
  extrinsic parameters, 437t
  government regulations, 128
  intrinsic parameters, 438t
  kinetic modeling, 435, 446t, 448–450, 460f, 461
  microbiological safety, 117–118
  operational characteristics
    food material response, 112–117, 113f
    governing principles, 110–117
    heat of compression values, 114, 115t, 116f
    packaging requirements, 111–112
    pH shifts, 117
    pressure*temperature response, 112–117, 113f
    pressure-transmitting fluids, 112
    process equipment, 110–111
  regulatory and legislative issues, 487–488
  research background, 109
  research summary, 128
  solid*liquid systems, 45
  stability effects, 4
  treatment effects, 47–51, 50f
high-temperature short-time (HTST) sterilization processes, 327–333
honey
  heat of compression value, 115t
  infrared heating, 421–424, 424t
  irradiation effects, 52–53
  light pulse technologies, 211
  microwave heating, 384–387, 385t, 386f, 388f
  ultraviolet (UV) light processing, 185–186
honeydew melons, 293–294
horchata, 81t, 83–84
horseradish, 290t
hue angle ratio, 346, 347t
human immunodeficiency virus (HIV), 244–246
humic acid, 236
hurdle technology, 5, 7–8
hydrocolloids, 314t
hydrodynamic boundary layer, 11–12
hydrogen peroxide (H_2O_2), 139–140, 142t, 180–181, 228, 242–243
hydroperoxide liase (HPL), 85
hydroxyl radicals, 171–172, 248–249, 377–378
hydroxymethylfurfural (HMF), 385, 386f, 421–424

I

ice, 378–380, 379t
iceberg lettuce, 178t
ice cream mix, 312, 314t, 319, 325
infant formula, 383, 446t
influenza A virus, 244–246
infrared heating, 411–432
  advantages, 411–413
  black body radiation
    characteristics, 413–418
    Planck's law, 417–418
    spectral characteristics, 417f
    Stefan*Boltzman's law, 417f, 418
    Wien's displacement law, 418
  extrinsic parameters, 437t
  future outlook, 425–428
  intrinsic parameters, 438t
  microbial inactivation
    bacterial suspensions, 418–420, 419f, 421f
    beer, 424–425
    honey, 421–424, 424t
    milk/milk products, 420–421, 423f
    orange juice, 420, 422f
  processing equipment, 425, 426f, 427f
  research summary, 425–428
  solid-food processing applications, 413, 414t
  vitamin degradation, 391f

infrared heating (*Continued*)
wavelengths and frequencies, 370f, 411–412, 412f
insect pests, 178, 178t
Institute of Food Technologists (IFT), 490
insulated gate bipolar transistor (IGBT), 322
intracellular cavitation, 141, 142t
iodine dosimetry, 139–140
ionizing radiation, 167–168
irradiation processes, 167–184
benefits, 167–168
case studies
Chinese irradiated wine, 180–181
packaging requirements, 174–177, 175f, 176f
China
irradiated wine, 180–181
regulatory and consumer aspects, 177–179
extrinsic parameters, 437t
food safety concerns, 172–173
food treatment effects, 52–53, 53f
government regulations
China, 177–179
United States
applications and dose limits, 178t
background information, 177–179
packaging requirements, 179t, 180t
intrinsic parameters, 438t
irradiation technologies and characteristics, 168–171, 170t
mode of action, 171–172
packaging requirements, 174–177, 175f, 176f, 179t, 180t
primary kinetic models, 446t
quality and nutritional effects, 173–174
radura symbol, 169, 175f
regulatory and legislative issues, 484–487
research summary, 181
shelf-life extension, 172–173
isostatic principle, 110–117
itaconic acid, 179t
Italy, 486

J

jams, 487–488
Japan, 111t
Japanese breadcrumbs, 359
jellies, 118
Joule heating *See* ohmic heating
juices
antimicrobial treatments, 78
cloud stability, 250–251, 293–294
color changes, 152–156, 213, 293–294
dense-phase carbon dioxide (DPCD)
enzyme inactivation, 290t
inactivation and stability effects, 288–289
physicochemical and sensory effects, 293–294
research background, 286t
solubility measures, 279
flavor changes, 167–168, 344–345
fluid rheology
dense-phase carbon dioxide (DPCD), 56–57
high-pressure processing (HPP), 47–51
pulsed electric fields (PEFs), 45–47
ultrasound technologies, 51–52
heat of compression value, 115t
high-pressure processing (HPP), 47–51, 111, 111t, 118, 487–488
infrared heating, 420, 422f, 422t
irradiation processes
quality and nutritional effects, 173–174
shelf-life extension, 172–173
kinetic modeling, 446t, 459–460, 460f
light absorptivity, 190
microbial inactivation, 145
microwave heating, 387–390, 389f
ohmic heating
color changes, 345–348, 347t
commercial applications, 359
electrical conductivity, 313, 314t, 317f, 318f
flavor changes, 344–345
microbial inactivation, 329
rheological properties, 357–358
sensory properties, 358
vitamin degradation, 340–344, 341t
voltage gradient, 312
ozone processing
color changes, 248–249
liquid food processing, 227–228
microbial inactivation, 237–243
nutritional parameter effects, 249–250
organic matter, 236
quality parameter effects, 247–251, 248t
pasteurization effects, 167–168
physical properties, 193–195, 194t, 195t
pulsed electric field (PEF) processing
effectiveness, 71–77, 72t
fluid rheology, 45–47
food-quality-related enzyme inactivation, 80–85, 81t
nutritional parameter effects, 87–91, 89t

quality parameter effects, 85–87, 93f
research background, 63
shelf-life extension, 91–95, 92f, 93f, 94f
treatment parameters, 67
regulatory requirements, 489
ultrasound processing
microbial inactivation, 141–150, 148t
research background, 140–152
ultraviolet (UV) light processing
background and functional role, 185–186
flow dynamics, 197
microbial inactivation, 206–209, 206t, 207t
physical properties, 193–196, 194t, 195t
quality and nutritional effects, 211–213
reactor designs, 214–215, 214f, 215f

K

kaempferol, 88
kale juice, 172, 174
kefir, 331–332, 331f
ketchup, 359
ketones, 55–56, 248–249, 382–383
kiwi fruit puree, 392
Klebsiella pneumoniae
irradiation processes, 486
microwave heating, 383
Kluyveromyces fragilis, 272t
Kraft paper, 179t

L

lactate dehydrogenase, 244–246
Lactobacillus acidophilus
ohmic heating, 332
ultrasound processing, 141, 146f
Lactobacillus brevis
dense-phase carbon dioxide (DPCD), 272t
pulsed electric field (PEF) processing, 72t, 74–75
ultraviolet (UV) light processing, 206–209, 206t
Lactobacillus plantarum
dense-phase carbon dioxide (DPCD), 271–276, 272t
microwave heating, 390
pulsed electric field (PEF) processing, 72t, 75
Lactobacillus sp
dense-phase carbon dioxide (DPCD), 275t, 281–282
ultrasound processing, 140–152
lactoperoxidase
high-pressure processing (HPP), 118–126, 119t
ultrasound processing, 152, 153t
lactulose, 382–383
Lebanon, 111t
Le Chatelier's principle, 110–117
Legionella dunnifii, 272t
Legionella pneumophila, 238t
lemon juice
high-pressure processing (HPP), 122t
regulatory requirements, 489
ultraviolet (UV) light processing, 213
lettuce
irradiation processes, 178t
ozone processing, 250–251
Leuconostoc mesenteroides, 237–243
Levenberg*Marquardt method, 456–457
light absorption, 189–193, 192f
light irradiance distribution (LID), 196–197
light pulse technologies, 185–224
background and functional role, 185–186
basic concepts, 26–27
industrial scale-up
challenges, 218–219
future outlook, 218–219
reactor designs
annular reactors, 214–215, 214f
coiled-tube reactors, 216–217, 216f
dynamic mixers, 217–218, 217f
static reactors, 216f, 216–217
thin-film reactors, 214–215, 215f
microbial inactivation
antimicrobial effects, 210–211
background information, 185–186, 199–206
honey, 211
juices, 206–209, 206t
kinetic mechanisms, 204–206
liquid sweeteners, 211
milk/milk products, 210, 210–211
photochemical effect, 199–203, 201f, 203f
photophysical effect, 204, 205f
photothermal effect, 203–204
operation principles
design parameters, 189–193, 192f
gas discharges, 188–189
generation mechanisms, 187–188
lamp emission spectra, 192f
light absorption, 189–193, 192f
pulsed power energization, 189
spectral characteristics, 187–193
process control parameters

light pulse technologies (*Continued*)
- dose delivery, 196–197
- flow dynamics, 197
- fluid food parameters, 193–196, 194t
- lamp emission spectra, 198f, 200t
- light sources, 197–199, 200t

quality and nutritional effects, 211–213
research summary, 219

limonene, 344
linalool, 94f
linoleic acid, 180t
linolenic acid, 115t
lipase
- dense-phase carbon dioxide (DPCD), 290t
- high-pressure processing (HPP), 118–126
- pulsed electric field (PEF) processing, 81t
- ultrasound processing, 151–152

lipids, 339–340, 377–378
lipoxygenase (LOX)
- dense-phase carbon dioxide (DPCD), 289–293, 290t
- high-pressure processing (HPP), 14, 14f, 118–126
- high-pressure processing (HPP) effects, 119t
- kinetic modeling, 459–460
- ohmic heating, 348–352, 349t
- pulsed electric field (PEF) processing, 70, 81t
- ultrasound processing, 147, 149–150

liquid sweeteners
- light pulse technologies, 211
- physical properties, 194t, 195–196, 195t

Listeria innocua
- dense-phase carbon dioxide (DPCD), 272t
- microwave heating, 390–392
- ozone processing, 227–228
- pulsed electric field (PEF) processing, 72t, 75
- thermal shock treatments, 78–79
- ultrasound processing, 151–152

Listeria monocytogenes
- dense-phase carbon dioxide (DPCD), 272t, 286t
- irradiation processes, 172–173
- kinetic modeling, 448
- light pulse technologies, 210–211
- ozone processing, 227–228, 237–243, 238t
- pulsed electric field (PEF) processing, 72t
- ultrasound processing, 141, 147–150, 148t
- ultraviolet (UV) light processing, 193–195, 202–203, 206t

lobsters, spiny, 290t, 292
log-linear models, 442, 443f, 444, 446t
lutein
- high-pressure processing (HPP), 127
- microwave heating, 389
- pulsed electric field (PEF) processing, 87–88

lychee, 119t
lycopene, 87–88, 89t, 122t, 126

M

magnesium stearate, 180t
Maillard reaction
- high-pressure processing (HPP), 118–126
- microwave heating, 382–383
- ohmic heating, 346
- ozone processing, 250–251
- ultrasound processing, 156
- ultraviolet (UV) light processing, 211–213

malate dehydrogenase, 244–246
malic acid, 279
maltose, 384–387, 385t
mandarin juice, 286t, 293–294
mango nectar, 206–209, 206t, 213
mango pulp, 172–173
manothermosonication, 147, 148–149
Mano-Thermo-Sonication (MTS) treatment, 51–52
maple sap, 394
Maras-type ice-cream mix, 312, 314t, 319, 325
mathematical models, 433–472
- chemical dynamics
  - model structure characterization, 450–456
  - primary models, 450–452
  - quality functions, 450–452, 451t
  - secondary models, 452–456
- data pre-processing guidelines, 441
- experimental design
  - data collection guidelines, 436–441
  - dynamic conditions, 441
  - enzymatic and chemical data, 440
  - microbial data, 439–440
- extrinsic parameters, 436, 437t
- future outlook, 461
- intrinsic parameters, 436, 438t
- kinetic modeling
  - *ad hoc* hypotheses, 435–436
  - nonlinear kinetics, 435–436
  - static versus dynamic processes, 435
- microbial and enzymatic iso-contour plots, 459f
- microbial and enzymatic iso-reduction plots, 460f
- microbial dynamics
  - achieved log-reduction assessments, 445–446

model structure characterization, 441–450
primary models
basic concepts, 442–445
Class I curves, 443, 444, 446t
Class II curves, 444, 445, 446t
inactivation curves, 442, 443f
representative models, 446t
probabilistic models, 448–450, 449f
secondary models, 447–448, 448f
optimization procedures, 433–434
parameter estimation, 456–457
research applications, 459–460, 459f, 460f
research summary, 461
validation methods, 457–459
Maxwell*Wagner polarization effects, 373–374, 374f
mayonnaise, 115t
mean absolute relative residual (MARR), 458
mean relative percentage residual (MRPR), 458
meats
infrared heating, 414t
irradiation processes, 177–178, 178t, 486
ohmic heating, 359
ozone processing, 237–243
melon juice
dense-phase carbon dioxide (DPCD), 294
pulsed electric field (PEF) processing, 72t, 86, 93–95
melons, 293–294
mercury lamps
background information, 186
characteristics, 197–199, 200t
design parameters, 191
emission spectra, 192f, 198f
Escherichia coli inactivation, 203f
methyl acrylate, 179t
methyl methacrylate, 179t
Mexico, 111t
microbial inactivation
dense-phase carbon dioxide (DPCD)
basic concepts, 263–264
carbon dioxide (CO_2) physicochemical properties, 276–277
carbon dioxide (CO_2) solubility
inactivation effects, 277–278
solubility measures, 278–279
thermodynamic models, 279–280
commercialization efforts
background information, 295–296
potential advantages, 295
potential difficulties, 296
depressurization rate, 280–281
inactivation mechanisms, 282–285, 283f
mixing speeds, 280
pressure and temperature parameters, 270–276, 272t, 275t
pressure cycling, 281–282
quality and nutritional parameter effects
enzyme inactivation, 289–293, 290t
inactivation and stability effects, 288–289
physicochemical and sensory effects, 293–294
research background, 285–294, 286t
spores, 271, 275t
high-pressure processing (HPP), 117–118
infrared heating
bacterial suspensions, 418–420, 419f, 421f
beer, 424–425
honey, 421–424, 424t
milk/milk products, 420–421, 423f
orange juice, 420–421, 422t
irradiation processes, 172–173, 178t
kinetic modeling, 433–472
achieved log-reduction assessments, 445–446
ad hoc hypotheses, 435–436
data collection guidelines, 439–440
data pre-processing guidelines, 441
experimental design
data collection guidelines, 436–441
dynamic conditions, 441
enzymatic and chemical data, 440
microbial data, 439–440
extrinsic parameters, 436, 437t
intrinsic parameters, 436, 438t
model structure characterization, 441–450
nonlinear kinetics, 435–436
optimization procedures, 433–434
primary models
basic concepts, 442–445
Class I curves, 443, 444, 446t
Class II curves, 444, 445, 446t
inactivation curves, 442, 443f
representative models, 446t
probabilistic models, 448–450, 449f
research applications, 459–460, 459f, 460f
secondary models, 447–448, 448f
static versus dynamic processes, 435
light pulse technologies
antimicrobial effects, 210–211
honey, 211
liquid sweeteners, 211
milk/milk products, 210–211

microbial inactivation (*Continued*)
photophysical effect, 204, 205f
photothermal effect, 203–204
microwave heating, 383
ohmic heating, 327–333, 330f, 331f
ozone processing
basic concepts, 225–226, 237–243
fluid foods, 227–228
inactivation mechanisms, 243–246
organic matter, 236, 237f
ozone concentration, 229–230, 231f, 232t
pH levels, 233–235
temperature effects, 230–233, 232t
voltage effects, 232f
pulsed electric field (PEF) processing
effectiveness, 68–69, 71–77, 72t
inactivation mechanisms, 71
ultrasound processing, 136, 141–146, 142t
ultraviolet (UV) light processing
background information, 185–186, 199–206
juices, 206–209, 206t
kinetic mechanisms, 204–206
milk/milk products, 210
photochemical effect, 199–203, 201f, 203f
microscopic ordering principle, 110–117
microwave heating, 369–410
basic concepts, 370, 370f
continuous flow-through systems, 41–43, 42f, 395, 399f
dielectric properties
basic concepts, 373–374
food products
cheese sauce, 380t
dielectric loss factors, 378–380, 379t, 381t
frequency effects, 374–375, 374f, 375f
moisture effects, 376–377
penetration depth, 379t, 381t
temperature/salt content effects, 375–376, 376f
product parameters, 372–377
extrinsic parameters, 437t
food component interactions
characteristics, 377–378
heat generation estimations, 378
penetration depth, 378, 379t, 381t
food processing
apple juice/apple cider, 394–395, 394f
apple puree, 390–392
aseptic processing, 395
avocado puree, 390–395
banana puree, 392
freeze drying, 380–396, 382t, 384f, 400f
fruit gels, 392
fruit juices, 387–390, 389f
heat transfer models, 395
honey, 384–387, 385t, 386f, 388f
industrial applications, 380–396, 382t
maple sap, 394
milk/milk products, 382–383
pasteurization, 395–396
time*temperature profiles, 392–394, 393f, 394f
vegetable soup, 392
intrinsic parameters, 438t
operation principles, 371–372, 372f
primary kinetic models, 446t
process control parameters
challenges
background information, 401–404
food composition changes, 402
material properties, 401
processing equipment, 402–404
costs, 396–401
microwave systems
basic concepts, 396–401
continuous flow-through systems, 399f
conveyorized cavity applicator, 398–401, 399f
freeze dryers, 400f
industrial microwave systems, 398–401
magnetrons, 397f, 398f
vacuum dryers, 400f
processing equipment
heat distribution and uniformity, 402–403
hybrid processes, 403
modeling and simulation tools, 403–404
performance factors, 402–404
research summary, 404
thermal processing methods, 369–371
wavelengths and frequencies, 370f
milk/milk products
dense-phase carbon dioxide (DPCD), 286t, 288–289, 290t
fluid rheology
dense-phase carbon dioxide (DPCD), 56–57
high-pressure processing (HPP), 47–51, 50f
pulsed electric fields (PEFs), 45–47
heat of compression value, 115t
high-pressure processing (HPP), 119t, 122t, 487–488

infrared heating, 420–421, 423f
light absorptivity, 192f
light pulse technologies, 210–211
microwave heating, 382–383
ohmic heating
electrical conductivity, 314t
microbial inactivation, 332
protein denaturation, 335–340
research background, 306–307
voltage gradient, 312
ozone processing, 237–243, 247–251
physical properties, 193–195, 194t, 195t
primary kinetic models, 446t
protein denaturation
microwave heating, 382–383
ohmic heating, 335–340
pulsed electric field (PEF) processing
effectiveness, 71–77, 72t
fluid rheology, 45–47
food-quality-related enzyme inactivation, 81t, 83–84
quality parameter effects, 86
research background, 63
shelf-life extension, 91–95, 92f
regulatory requirements, 488–489
ultrasound processing
microbial inactivation, 141–148, 151–152, 153t
quality parameter effects, 156
research background, 140–152
ultraviolet (UV) light processing, 185–186, 210, 212
mineral oil, 180t
minerals, 489
molds
dense-phase carbon dioxide (DPCD), 286t, 288–289
high-pressure processing (HPP), 117–118
irradiation processes, 172–173
ohmic heating, 329
ultraviolet (UV) light processing, 206–209, 206t
monoterpenes, 173–174
Most Probable Number (MPN) data collection technique, 439–440
MS2 coliphage
ozone processing, 242–243
photochemical effect, 199–203, 201f
Multiple Tube Method data collection technique, 439–440
mungbean starch suspension, 314t
muscadine grape juice, 286t, 288–289, 290t, 293–294
mushrooms, 359
Mycobacteria, 237–243
Mycobacterium avium subsp. paratuberculosis, 216–217
Mycobacterium fortuitum, 233–235, 238t
Mycobacterium terrae, 383
Mycobacterium tuberculosis, 193–195
mycotoxins, 185, 246–247
myrcene, 344
myricetin, 88

N

Naegleria gruberi, 238t
Naperian absorption coefficient, 190
naringenin, 88
near-infrared (NIR) radiation, 424–425
neural networks, 446t
neutraceuticals, 4
Newtonian fluid foods, 325
Newtonian fluids, 36–38, 37f
New Zealand, 111t
nisin, 78
nonenzymatic browning *See* browning
non-ionizing radiation *See* microwave heating; ultraviolet (UV) light processing
non-Newtonian fluid foods, 324–325
non-thermal technologies
hurdle technology, 5, 7–8
mathematical models, 433–472
process validation, 2
regulatory and legislative issues, 473–494
research background, 1–2
technology drivers
consumer acceptibility, 4
environmental impacts, 5
functional properties, 4
government regulations, 2–3
nutritional and sensory properties, 3
shelf-life extension, 3
Northern Ireland, 111t
Novel Foods Regulation, 476, 478–484
novel thermal and non-thermal technologies
hurdle technology, 5, 7–8
mathematical models, 433–472
process validation, 2
regulatory and legislative issues, 473–494
research background, 1–2
technology drivers
consumer acceptibility, 4
environmental impacts, 5
functional properties, 4
government regulations, 2–3

novel thermal and non-thermal technologies (*Continued*)
  nutritional and sensory properties, 3
  shelf-life extension, 3
nucleic acid, 244–246
nutrient broth, 272t, 275t
nutritional and sensory properties, 3
nuts, 414t
Nylon, 6, 179t
Nylon, 11, 179t

O

oat groats, 414t
ochratoxin A, 246–247
octanol, 344
ohmic heating, 305–368
  advantages, 310
  aseptic food processing, 310, 327–333
  basic concepts, 27–28, 305–306, 306f
  continuous flow-through systems, 40–41
  disadvantages, 310–311
  extrinsic parameters, 437t
  future outlook, 359
  intrinsic parameters, 438t
  microbial inactivation, 327–333, 330f, 331f
  operation principles, 306–311, 308f, 309f
  primary kinetic models, 446t
  process control parameters
    background information, 311–327
    current density, 312
    electrical conductivity, 312–320, 314t
    electrode types/electrochemical reactions, 321–323, 323f, 324f
    flow properties, 325–327, 326f
    food properties, 324–325
    power requirements, 311–312
    temperature, 320–321
    voltage gradient, 312, 349, 350f
  quality and nutritional parameter effects
    acidity, 334–335
    color changes, 345–348, 347t
    corrosion, 338
    enzyme inactivation, 348–352, 349t, 350f, 351t, 353t
    flavor characteristics, 344–345
    fouling processes, 335, 336f, 337f, 338f
    kinetic mechanisms, 333–358
    ohmic heater configuration, 339f
    phenols, 344–345
    proteins, 335–340, 338f
    rheological properties, 353–358, 354f, 355t, 356f, 357f
    sensory properties, 358
    soluble solid content, 334–335
    vitamins, 340–344, 341t, 343f, 391f
  research background, 306–307
  research summary, 359
  rheological properties, 57, 353–358, 354f, 355t, 356f, 357f
  solid*liquid systems, 44–45, 46f
Ohm's law, 307
oils
  dielectric properties, 378–380, 379t
  heat of compression value, 115t
  ozone processing, 56, 228
oleic acid, 180t
oligosaccharides, 143
olive oil
  heat of compression value, 115t
  ozone processing, 56, 228
onions, 414t
Optimal Experiment Design for Parameter Estimation (OED/PE), 439
orange juice
  antimicrobial treatments, 78
  cloud stability, 154–155, 250–251
  color changes, 152–156, 213, 345–348
  dense-phase carbon dioxide (DPCD), 56–57
    enzyme inactivation, 290t
    inactivation and stability effects, 288–289
    physicochemical and sensory effects, 293–294
    research background, 286t
    solubility measures, 279
  flavor changes, 173–174
  heat of compression value, 115t
  high-pressure processing (HPP), 119t, 122t, 127
  infrared heating, 420, 422f
  irradiation processes
    quality and nutritional effects, 173–174
    shelf-life extension, 172–173
  kinetic modeling, 446t, 459–460, 460f
  light absorptivity, 190
  microbial inactivation, 145
  microwave heating, 389
  ohmic heating
    color changes, 345–348, 347t
    electrical conductivity, 313, 314t, 318f
    electrode types/electrochemical reactions, 321–323
    enzyme inactivation, 352
    microbial inactivation, 330–331
    sensory properties, 358
    vitamin degradation, 340–344, 341t

ozone processing
color changes, 248–249
enzyme inactivation, 250–251
liquid food processing, 227–228
microbial inactivation, 237–243, 238t
organic matter, 236
physical properties, 250–251
quality parameter effects, 247–251, 248t
physical properties, 193–195, 194t, 195t
pulsed electric field (PEF) processing
effectiveness, 72t, 74–75
food-quality-related enzyme inactivation, 80–85, 81t
nutritional parameter effects, 87–91, 89t
quality parameter effects, 85–87, 93f
shelf-life extension, 91–95, 93f
treatment parameters, 67
ultrasound processing, 147–150, 148t
ultraviolet (UV) light processing
background and functional role, 185–186
flow dynamics, 197
microbial inactivation, 206–209, 206t, 207t
physical properties, 193–196, 194t, 195t
quality and nutritional effects, 211–213
reactor designs, 214–215
organic acids, 248–249
organic matter, 236, 237f
organoleptic foods, 4
Ostwald de Waele power law, 40–43
ozone processing, 225–262
efficacy-related factors
extrinsic parameters
flow rate, 229, 230t, 232f
ozone concentration, 229–230, 230t, 231f, 232t
solubility, 230–233
temperature, 230–233, 233f, 234f
voltage effects, 232f
intrinsic parameters
organic matter, 236, 237f
pH levels, 233–235
extrinsic parameters, 437t
food treatment effects, 55–56
generation mechanisms
basic concepts, 226–227
electrical (corona) discharge method, 226–227
electrochemical (cold plasma) method, 227
radiochemical ozone generation (RCOG), 227
ultraviolet (UV) radiation method, 227
intrinsic parameters, 438t
liquid food processing, 227–228
microbial inactivation
basic concepts, 225–226, 237–243
fluid foods, 227–228
inactivation mechanisms, 243–246
organic matter, 236, 237f
ozone concentration, 229–230, 231f, 232t
pH levels, 233–235
temperature effects, 230–233, 232t
voltage effects, 232f
mycotoxins, 246–247
nutritional parameter effects, 249–250
ozone characteristics and functional role, 225–226
primary kinetic models, 446t
quality parameter effects
bioactive compounds, 249–250
color changes, 248–249
enzyme inactivation, 250–251
sensory properties, 247–251, 248t
research summary, 251–252

P

packaging requirements
high-pressure processing (HPP), 15, 16f, 111–112
irradiation processes, 174–177, 175f, 176f, 179t, 180t
palmitic acid, 180t
parasites, 185–186, 227–228, 238t *See also* pathogenic microorganisms
particle image velocimetry (PIV), 326, 326f
passion fruit juice, 213
pasteurization
dense-phase carbon dioxide (DPCD), 288
far-infrared (FIR) radiation, 418–420, 419f, 421f
irradiation processes, 167–168
mathematical models, 434
microwave heating
food products, 395–396
industrial applications, 380–396, 382t
milk/milk products, 382–383
ohmic heating, 306–311
ultraviolet (UV) light processing, 193–196
pâté, 359
pathogenic microorganisms
high-pressure processing (HPP) effects, 10–17, 117–118
infrared heating, 420–421
irradiation processes, 172–173, 178t

pathogenic microorganisms (*Continued*)
- microwave heating, 383
- ohmic heating, 327–333
- ozone processing, 227–228, 237–243, 238t
- pulsed light technologies, 26–27
- ultrasound processing, 140–152
- ultraviolet (UV) light processing, 185–186, 193–196, 206t

patulin mycotoxin, 185, 246–247
peach juice
- dense-phase carbon dioxide (DPCD), 286t
- irradiation processes, 173–174
- ultraviolet (UV) light processing, 213

pêach puree, 314t, 319
peanut butter, 185
pea puree
- microwave heating, 42–43, 42f
- ohmic heating, 345–346, 347t, 352, 353t
- pulsed electric field (PEF) processing
  - effectiveness, 72t
  - food-quality-related enzyme inactivation, 81t, 84

pear juice
- dense-phase carbon dioxide (DPCD), 286t
- irradiation processes, 173–174
- pulsed electric field (PEF) processing, 72t
- regulatory requirements, 489

pears, 414t
peas, 414t
pectin
- irradiation effects, 53f
- ohmic heating, 314t
- ozone processing, 55–56
- ultrasound effects, 51–52

pectinase, 348–352, 349t
pectinerase, 290t
pectinesterase (PE)
- high-pressure processing (HPP), 118–126
- ohmic heating, 352
- ozone processing, 250–251

pectin methylesterase (PME)
- analytical measurement techniques, 440, 455, 455f, 460f
- cloud stability, 154–155
- dense-phase carbon dioxide (DPCD), 289–293
- high-pressure processing (HPP) effects, 119t, 126
- ohmic heating, 352
- pulsed electric field (PEF) processing, 45–47, 67, 80–85, 81t, 93f
- ultrasound processing, 147–150, 153t
- ultraviolet (UV) light processing, 213

Pediococcus damnosus, 76
Pediococcus sp., 390
pelargonidin-3-glucoside, 152–156
Peng*Robinson equation of state (EOS), 279–280
peppers, 359
pepsin, 151
peroxidase (POD)
- dense-phase carbon dioxide (DPCD), 289–293, 290t
- high-pressure processing (HPP) effects, 118–126, 119t
- infrared heating, 414t
- microwave heating, 392, 393f
- ohmic heating, 352, 353t
- ozone processing, 250–251
- pulsed electric field (PEF) processing, 67, 81t, 83–84, 93f
- ultrasound processing, 147, 149–150

peroxide *See* hydrogen peroxide (H_2O_2)
pest control, 178, 178t
petroleum wax, 180t
pH
- dense-phase carbon dioxide (DPCD), 282–285, 293–294
- fluid foods, 194t, 195t
- high-pressure processing (HPP), 49, 117
- intrinsic parameters, 438t
- milk/milk products
  - dense-phase carbon dioxide (DPCD), 56–57
  - high-pressure processing (HPP), 49
- ohmic heating, 334–335, 340
- ozone processing, 233–235, 250–251
- pulsed electric field (PEF) processing, 69–70, 85–87
- ultraviolet (UV) light processing, 213

phage T4/phage T7, 202, 244–246
phenols
- analytical measurement techniques, 440
- dense-phase carbon dioxide (DPCD), 293–294
- infrared heating, 414t
- microwave heating, 390
- ohmic heating, 344–345
- ozone processing, 250–251
- pulsed electric field (PEF) processing, 67–68, 69, 88, 95f
- ultrasound processing, 152–156

phosphatase, 118–126
phosphofructokinase-1, 244–246
photon emissions, 187–188
photophysical effect, 204, 205f
photoreactivation, 199

photothermal effect, 203–204
physiological saline, 272t, 275t
pineapple juice
  physical properties, 193–195, 194t
  ultraviolet (UV) light processing, 206–209, 206t
pineapples, 414t
pinene, 344
piridoxine, 126–128
Planck's law, 417–418
planktonic microorganisms, 228
platinized titanium electrodes, 321, 323f
Poland, 486
polioviruses
  microwave heating, 383
  ozone processing, 233–235, 238t, 244–246
pollen, 177–178
polyacrylamide gel electrophoresis (PAGE), 335–340
polyethylene film, 179t
polyethylene terephthalate, 179t
polygalacturonase (PG)
  high-pressure processing (HPP) effects, 119t, 126
  kinetic modeling, 459–460
  pulsed electric field (PEF) processing, 45–47, 80, 81t
  ultrasound processing, 153t
polyolefins, 179t
polyphenoloxidase (PPO)
  analytical measurement techniques, 440
  dense-phase carbon dioxide (DPCD), 289–293, 290t
  high-pressure processing (HPP), 14, 14f, 118–126
  high-pressure processing (HPP) effects, 119t
  kinetic modeling, 459–460
  kinetic rate constants, 351t
  microwave heating, 392, 393f
  ohmic heating, 348–352, 349t, 350f
  ozone processing, 249–251
  pulsed electric field (PEF) processing, 69, 81t
  ultrasound processing, 147–150
  ultraviolet (UV) light processing, 206–209, 213
polyphenols, 247–248
polypropylene, 180t
polysaccharides
  high-pressure processing (HPP), 118–126
  ozonation effects, 55–56
  ultrasound technologies, 51–52
polystyrene film, 179t
polyvinyl chloride, 179t
pomegranate juice
  high-pressure processing (HPP), 127
  ohmic heating, 344, 346, 347t, 357–358
pork, 177–178, 178t
Portugal, 111t
potassium stearate, 180t
potato dextrose broth, 272t
potatoes
  dense-phase carbon dioxide (DPCD), 290t, 292
  infrared heating, 414t
  ozone processing, 250–251
potatoes, mashed, 115t
potato starch suspension, 314t
poultry, 177–178, 178t, 486
power ultrasound technologies, 4 *See also* ultrasound technologies
pressure-assisted thermal processing (PATP), 117–118
pressure-assisted thermal sterilization (PATS), 117–118
pressure-transmitting vessels, 11, 12f
pressurized carbon dioxide (CO_2)
  carbon dioxide (CO_2) properties, 264–266, 265f, 276–277
  carbon dioxide (CO_2) solubility
    microbial inactivation, 277–278
    solubility measures, 278–279
    thermodynamic models, 279–280
  food preservation, 263–264
  microbial inactivation, 282–285, 283f
process validation, 2
propylene glycol
  heat of compression value, 115t
  high-pressure processing (HPP), 112
proteases
  dense-phase carbon dioxide (DPCD), 290t
  pulsed electric field (PEF) processing, 81t
  ultrasound processing, 151–152
proteins
  heat of compression value, 114, 115t
  high-pressure processing (HPP), 118–126
  microwave heating, 377–378
  ohmic heating, 335–340, 338f
  ozone processing, 243–244
Proteus vulgaris, 272t
protozoa, 225–228, 238t
Pseudomonas aeruginosa
  dense-phase carbon dioxide (DPCD), 272t
  microwave heating, 383
  ozone processing, 227–228

Pseudomonas fluorescens
dense-phase carbon dioxide (DPCD), 278
ozone processing, 237–243
pulsed electric field (PEF) processing, 72t, 76, 84–85
ultrasound processing, 151
Pseudomonas fragi, 237–243
Pseudomonas putida, 237–243
pseudoplastic fluids, 36–38, 37f
pulps *See also* juices; purees
irradiation processes, 172–173
ohmic heating, 312, 314t, 340–344, 341t
pulsed electric fields (PEFs), 63–108
basic concepts, 17–24
combination preservation strategies, 77–79, 79f
continuous flow-through systems, 43, 44f
electrohydrodynamic phenomena simulations, 18–19, 23f, 24f
extrinsic parameters, 437t
food-quality-related enzyme inactivation, 69, 80–85, 81t
functional foods, 4
health-related compounds, 69
intrinsic parameters, 438t
kinetic modeling, 435, 446t
microbial inactivation
effectiveness, 68–69, 71–77, 72t
inactivation mechanisms, 71
nutritional-related properties, 69, 87–91, 89t
operation principles, 64–65, 65f
process control parameters, 66
product parameters, 69–70
quality-related properties, 69, 85–87, 93f
research background, 63
research summary, 95–96
shelf-life extension, 91–95, 92f, 93f, 94f
treatment chamber geometries, 22–23, 22f
treatment effects, 45–47
treatment parameters, 66–68
pulsed light technologies *See* light pulse technologies; ultraviolet (UV) light processing
pulsed power energization, 189
purees
dense-phase carbon dioxide (DPCD), 279
high-pressure processing (HPP), 118, 119t, 122t
infrared heating, 414t
irradiation processes, 168
kinetic modeling, 459–460
microwave heating, 42–43, 42f, 390–395
ohmic heating, 312, 314t, 341t, 343–344, 347t, 353t
pulsed electric field (PEF) processing
effectiveness, 72t
food-quality-related enzyme inactivation, 81t, 84
purified tomatoes, 119t
purines, 244–246
pyrimidines, 202, 244–246
pyropheophytins, 348
pyruvate decarboxylase, 244–246

Q

quasi-chemical models, 446t
quercetin, 88, 89t
quince nectar, 357–358

R

radiochemical ozone generation (RCOG), 227
radiofrequency (RF) heating, 369, 370f
radura symbol, 169, 175f
raw milk *See* milk/milk products
Rayleigh-Bénard instability, 13
red currant juice, 173–174
Redlich*Kwong equation of state, 279–280
red wine
dielectric properties, 379–380
ultrasound processing, 152–156
regulatory and legislative issues, 473–494
European Union (EU)
consumer safety, 473–474, 477–478, 491–492
European Food Safety Agency (EFSA), 474–476
international standards, 490–491
law creation and enforcement
institutional role, 474–476
legislative characteristics, 476
supremacy factor, 476–477
novel food regime
application refusals, 480
Novel Foods Regulation, 476, 478–484
substantial equivalence concept, 481–482
update proposal draft, 482–484
regulatory framework, 477–478
regulatory requirements
fruit juices, 489
high-pressure processing (HPP), 487–488
irradiation processes, 484–487
labeling and health claims, 488–489

milk/milk products, 488–489
research background, 1–2
residence time (RTD), 196–197
resveratrol, 212
Rhodotorula rubra, 76
riboflavin, 87–91, 212
ribonucleic acid (RNA), 244–246
rice starch/rice flour, 353–358
rice wine, 111t
roasting processes, 380–396, 382t, 414t
root mean squared error (RMSE), 456
root-mean-squared residual (RMSR), 458
rotaviruses, 233–235, 238t
rubber hydrochloride film, 179t

S

Sabouraud dextrose broth, 272t
Saccharomyces cerevisiae
antimicrobial treatments, 78, 206–209
dense-phase carbon dioxide (DPCD)
carbon dioxide (CO_2) solubility, 278
depressurization rate, 280–281
microbial inactivation, 285
pressure and temperature parameters, 271–276, 272t
research background, 286t
light pulse technologies, 204, 205f
microwave heating, 390
ohmic heating, 329
pulsed electric field (PEF) processing, 69–70, 72t, 76
ultrasound processing, 143, 146f
ultraviolet (UV) light processing, 206–209, 206t
Saccharomyces uvarum, 76
salep, 53f
salmon, 115t
Salmonella anatum, 172–173
Salmonella choleraesuis, 210–211
Salmonella enteritidis
antimicrobial treatments, 78, 79f
irradiation processes, 172–173
ozone processing, 243
pulsed electric field (PEF) processing, 72t
ultraviolet (UV) light processing, 193–195, 202–203
Salmonella salford, 272t
Salmonella senftenberg, 69–70
Salmonella spp
irradiation processes, 486
ozone processing, 227–228, 238t
ultrasound processing, 151–152
ultraviolet (UV) light processing, 193–195
Salmonella typhimurium
dense-phase carbon dioxide (DPCD), 272t, 286t
ozone processing, 238t
pulsed electric field (PEF) processing, 72t
ultrasound processing, 147–148
salmonellosis, 167–168
salt solutions, 314t, 341t
sausages, 359
secalonic acid D, 246–247
sensory properties, 3
Serratia marcescens, 210–211
sesame seeds, 414t
shear flow, 36–38, 37f
shear thickening behavior, 36–38, 37f
shear thinning behavior, 36–38, 37f
shelf-life extension
irradiation processes, 172–173
pulsed electric field (PEF) processing, 91–95, 92f, 93f, 94f
technology drivers, 3
shellfish, 178t, 486
Sherwood number, 43, 44f
Shigella sonnei, 237–243, 238t
shrimp, brown, 290t, 292
sigmoidal-like models, 442, 443f, 446t
silicone oil, 112
skim milk
dense-phase carbon dioxide (DPCD), 286t
microwave heating, 395
ozone processing, 237–243, 247–251
pulsed electric field (PEF) processing
effectiveness, 72t, 74–75
food-quality-related enzyme inactivation, 81t
nutritional parameter effects, 89t
shelf-life extension, 91–95
thermal shock treatments, 77–79
ultrasound processing, 152, 153t
smoothies
high-pressure processing (HPP), 111t, 118
pulsed electric field (PEF) processing, 72t
ultrasound processing, 140–152
Soave*Redlich*Kwong (PSRK) equation of state, 279–280
sodium benzoate, 112
sodium citrate, 179t
sodium lauryl sulfate, 179t
sodium propionate, 180t
sonication, 136, 145f *See also* ultrasound technologies
sound waves, 370f

sour cherry juice, 313–317, 314t, 317f
soybean oil
  heat of compression value, 115t
  ozone processing, 56
soybeans, 290t
soymilk
  ohmic heating, 339–340
  pulsed electric field (PEF) processing
    effectiveness, 72t
    food-quality-related enzyme inactivation, 81t, 84
    nutritional parameter effects, 89t
    shelf-life extension, 91–95
spices, 177–178, 178t, 485–486
spinach
  infrared heating, 414t
  irradiation processes, 178t
  ohmic heating, 341t, 345–346, 347t
spiny lobsters, 290t, 292
spores
  dense-phase carbon dioxide (DPCD)
    depressurization rate, 281–282
    pressure and temperature parameters, 271, 275t
  far-infrared (FIR) radiation, 418–420
  high-pressure processing (HPP), 117–118
  light pulse technologies, 211
  ohmic heating, 327–333, 330f
  ozone processing
    fluid foods, 227–228
    inactivation mechanisms, 243–246
    microbial inactivation, 225–226, 237–243, 238t
    temperature effects, 232–233, 234f
  photochemical effect, 201f
  ultrasound processing, 145
  ultraviolet (UV) light processing, 185–186, 193–196
sprouts, 178, 178t
stable cavitation, 137
stainless steel electrodes, 321, 323f
standard type ice-cream mix, 312, 314t, 319, 325
Staphylococcus aureus
  dense-phase carbon dioxide (DPCD), 272t, 286t
  infrared heating, 420–421
  light pulse technologies, 204
  microwave heating, 383
  ozone processing, 227–228, 237–243
  pulsed electric field (PEF) processing, 72t, 74–75
  pulsed light technologies, 26–27, 210–211
  ultrasound processing, 145, 151–152
starch gelatinization, 314t, 320, 353–358
star-fruit honey, 384–387, 385t
stearic acid, 180t
Stefan*Boltzman's law, 417f, 418
sterile Ringer solution, 275t
sterilization, 380–396, 382t
sterilized water, 272t, 275t
strawberries
  dense-phase carbon dioxide (DPCD), 293–294
  infrared heating, 414t
  ozone processing, 249–250
strawberry*apple sauce, 314t, 318
strawberry filling, 314t, 318
strawberry juices
  color changes, 152–156
  high-pressure processing (HPP), 111t
  microwave heating, 390
  ozone processing, 248–250, 248t
  pulsed electric field (PEF) processing
    effectiveness, 72t
    fluid rheology, 45–47
    food-quality-related enzyme inactivation, 81t, 84
    nutritional parameter effects, 87–91, 89t
    quality parameter effects, 86
    shelf-life extension, 95f
    treatment parameters, 67
  ultrasound processing, 147–150
  ultraviolet (UV) light processing, 206–209, 206t, 213
strawberry pulp, 314t, 318, 340–344, 341t
strawberry topping, 314t, 318
Streptococcus faecalis, 237–243, 263–264
Streptococcus mutans, 146
Streptococcus spp., 151–152
Streptococcus thermophilus, 332
subsidiarity principle, 476
substantial equivalence concept, 481–482
sucrose
  dense-phase carbon dioxide (DPCD), 279
  light pulse technologies, 211
  microwave heating, 384–387, 385t
  physical properties, 194t, 195t, 195–196
sugars/sugar solutions
  microwave heating, 377–378, 384–387, 385t
  ohmic heating, 314t
  regulatory requirements, 489
sunflower oil, 228
superoxide dismutase, 244–246
surimi paste, 314t, 339
sweet potatoes, 414t
sweet potato wine, 177–178, 180–181

T

tamarind juice, 172, 174
tea
  infrared heating, 414t
  pulsed electric field (PEF) processing, 72t, 74–75
technology drivers
  consumer acceptibility, 4
  environmental impacts, 5
  functional properties, 4
  government regulations, 2–3
  nutritional and sensory properties, 3
  shelf-life extension, 3
tempering processes, 380–396, 382t
Tetrahymena pyriformis, 180–181
thawing processes, 414t
thermal boundary layer, 11–12
thermal pasteurization, 167–168
thermal shock treatments, 77–79, 143, 145f
thermal technologies
  hurdle technology, 5, 7–8
  mathematical models, 433–472
  process validation, 2
  regulatory and legislative issues, 473–494
  research background, 1–2
  technology drivers
    consumer acceptibility, 4
    environmental impacts, 5
    functional properties, 4
    government regulations, 2–3
    nutritional and sensory properties, 3
    shelf-life extension, 3
thermosonification, 145
thiamine, 87–91, 126–128
thixotropy, 38
thymine dimmers, 199–203
titanium electrodes, 321, 323f
tobacco mosaic virus (TMV), 244–246
tocopherol, 87–91
tofu
  heat of compression value, 115t
  ohmic heating, 339–340, 359
tomatoes, 119t
tomato juices
  antimicrobial treatments, 78
  fluid rheology
    pulsed electric fields (PEFs), 45–47
    ultrasound technologies, 51–52
  high-pressure processing (HPP), 122t
  irradiation processes, 173–174
  ohmic heating, 313, 314t
  ozone processing, 248t, 250–251
  physical properties, 194t
  pulsed electric field (PEF) processing
    effectiveness, 72t
    food-quality-related enzyme inactivation, 80–85, 81t
    nutritional parameter effects, 87–91, 89t
    quality parameter effects, 86
    shelf-life extension, 91–95, 94f
    treatment parameters, 67
  ultrasound processing, 147–150, 153t
  viscosity, 155
tomato paste, 52, 56–57, 359
tomato puree, 47, 119t, 122t, 126, 459–460
tomato salsa, 115t
tomato sauce
  microwave heating, 395
  ohmic heating, 314t
tomato soup, 314t, 332
tortillas, 414t
total color difference (TCD), 152–156
transient cavitation, 137
transmission electron microscopy (TEM), 143, 146f
trichothecene mycotoxins, 246–247
triethylene glycol, 180t
tropical fruit smoothie, 72t
tropical juices, 206–209, 206t, 213
Tryptic Soy Broth (TSB) model system, 229
Tyndallization effect, 330

U

ultra-high temperature (UHT) sterilization processes, 327–333
ultrasound technologies, 135–166
  basic concepts, 25–26
  batch systems, 44
  extrinsic parameters, 437t
  food processing, 135
  functional foods, 4
  generation mechanisms, 136–137, 138f
  intrinsic parameters, 438t
  liquid food sterilization
    enzyme inactivation, 147
    fruit juices, 147–150, 148t
    microbial inactivation, 141–146, 142t
    milk/milk products, 151–152, 153t
    research background, 140–152
  measurement techniques
    calorimetry, 139
    characteristics, 137–140
    chemical dosimetry, 139–140
  primary kinetic models, 446t

ultrasound technologies (*Continued*)
  quality parameter effects, 152–156
  research applications, 135–136
  research challenges, 156–157
  research summary, 157
  treatment effects, 51–52
ultraviolet (UV) light processing, 185–224
  background and functional role, 185–186
  extrinsic parameters, 437t
  food treatment effects, 54–55, 54f
  industrial scale-up
    challenges, 218–219
    future outlook, 218–219
    reactor designs
      annular reactors, 214–215, 214f
      coiled-tube reactors, 216–217, 216f
      dynamic mixers, 217–218, 217f
      static reactors, 216–217, 216f
      thin-film reactors, 214–215, 215f
  intrinsic parameters, 438t
  microbial inactivation
    background information, 185–186, 199–206
    juices, 206–209, 206t
    kinetic mechanisms, 204–206
    milk/milk products, 210
    photochemical effect, 199–203, 201f, 203f
  operation principles
    design parameters, 189–193, 192f
    gas discharges, 188–189
    generation mechanisms, 187–188
    lamp emission spectra, 192f
    light absorption, 189–193, 192f
    pulsed power energization, 189
    spectral characteristics, 187–193
  ozone generation, 227
  primary kinetic models, 446t
  process control parameters
    dose delivery, 196–197
    flow dynamics, 197
    fluid food parameters, 193–196, 194t
    lamp emission spectra, 198f, 200t
    light sources, 197–199, 200t
  quality and nutritional effects, 211–213
  research summary, 219
  wavelengths and frequencies, 370f
UNIFAC method, 279–280
United Kingdom, 486
United States
  high-pressure processing (HPP), 111t
  irradiation processes
    applications and dose limits, 178t
    background information, 177–179
    packaging requirements, 179t, 180t
  irradiation processes authorization, 486–487

V

vaccinia virus (VACV), 244–246
vacuum drying, 380–396, 382t, 400f
vanishing bulk viscosity, 8–10
vegetable juices
  high-pressure processing (HPP), 111t
  ohmic heating, 313
  physical properties, 194t
  ultrasound processing, 141–146
vegetable oils, 56, 228
vegetable parchments, 179t
vegetable purees, 341t, 343–344, 345
vegetable soup, 392
Venturi effect, 137
vesicular stomatitis Indiana virus (VSIV), 244–246
Vibrio cholerae, 178t
Vibrio parahaemolyticus, 332
Victoria grape must, 119t
vinylidene chloride, 179t
violaxanthin, 389
viruses
  high-pressure processing (HPP), 117–118
  microwave heating, 383
  ozone processing, 225–226, 233–235, 238t, 244–246
  ultraviolet (UV) light processing, 185–186
viscoplastic fluids, 36–38, 37f
viscosity
  fluid foods, 194t, 195, 195t
  pulsed electric field (PEF) processing, 69–70, 85–87
  ultrasound processing, 155
visible light, 370f
vitamin A, 87–91, 89t, 212
vitamin B, 89t, 126–128
vitamin C
  analytical measurement techniques, 440
  high-pressure processing (HPP), 126–128
  infrared heating, 391f, 420, 422t, 422f
  microwave heating, 389–390, 391f
  ohmic heating, 340–344, 343f, 391f
  pulsed electric field (PEF) processing, 67–68, 87–91, 89t
  ultraviolet (UV) light processing, 195–196, 212
vitamin D, 89t

vitamin E, 89t, 212
vitamins
  analytical measurement techniques, 440
  high-pressure processing (HPP), 118–128
  infrared heating, 391f, 420, 422t
  microwave heating, 389, 391f
  ohmic heating, 340–344, 341t, 343f, 391f
  pulsed electric field (PEF) processing, 87–91, 89t, 67–68
  regulatory requirements, 489
  ultraviolet (UV) light processing, 212
volumetric heating, 369

W

wastewater treatment, 225–226, 236, 238t, 339
water
  dense-phase carbon dioxide (DPCD)
    pressure and temperature parameters, 272t
    solubility measures, 279
  dielectric properties
    dielectric loss factors, 378–380, 379t
    frequency effects, 374–375, 374f
    moisture levels, 376–377
    temperature/salt content effects, 375–376, 376f
  heat of compression value, 115t
  high-pressure processing (HPP), 112
  ozone processing
    liquid food processing, 227–228
    microbial inactivation, 225–226, 237–243, 238t
    quality parameter effects, 247–251
  physical properties, 195t
  physicothermal properties, 113–114, 114t
water activity
  high-pressure processing (HPP), 117–118
  intrinsic parameters, 438t
  pulsed electric field (PEF) processing, 69–70
watermelon juices
  dense-phase carbon dioxide (DPCD), 286t
  physical properties, 193–195, 194t
  pulsed electric field (PEF) processing
    effectiveness, 72t
    fluid rheology, 45–47
    food-quality-related enzyme inactivation, 81t
    nutritional parameter effects, 87–88
    quality parameter effects, 86
    shelf-life extension, 93–95
wax-coated paperboard, 179t
Weibull*Hazard sensory shelf-life method, 358
Weibull type model, 445, 446t
whey
  dielectric properties, 379, 381t
  high-pressure processing (HPP), 119t, 122t
  light absorptivity, 192f
  ohmic heating, 312, 314t, 335–336, 354–355, 354f
  ozone processing, 237–243
  physical properties, 194t
  primary kinetic models, 446t
  rheological properties, 354f, 355t, 356f, 354–355
  ultrasound processing, 152
  ultraviolet (UV) light processing, 210, 214–215
white cabbage, 122t
Wien's displacement law, 418
wine
  dielectric properties, 379–380
  sweet potato wine, 177–178, 180–181
Wong and Sandler mixing rules (PRWS), 279–280
World Health Organization (WHO), 490
World Trade Organization (WTO), 490

X

xanthan gum
  infrared heating, 414t
  ohmic heating, 314t, 319, 335–336, 338f
X-rays
  characteristics, 169–171, 170t
  mode of action, 171–172
  packaging requirements, 174–177
  wavelengths and frequencies, 370f

Y

yams, 414t
Yarrowia lipolytica, 210
yeasts
  dense-phase carbon dioxide (DPCD)
    inactivation and stability effects, 288–289
    pressure and temperature parameters, 272t
    research background, 286t
  high-pressure processing (HPP), 117–118
  irradiation processes, 172–173
  microwave heating, 385–387, 385t, 386f
  ohmic heating, 329
  ozone processing, 227–228, 238t, 243–246

yeasts (*Continued*)
pulsed electric field (PEF) processing, 68–69, 71
ultrasound processing, 140–152, 145
ultraviolet (UV) light processing, 185–186, 206–209, 206t
Yersinia enterocolitica
light pulse technologies, 210–211
ozone processing, 227–228
ultrasound processing, 148–149
yoghurts
heat of compression value, 115t
high-pressure processing (HPP), 48–49, 50f
thermal shock treatments, 77–79
ultrasound processing, 151–152

Z

zearalenone (ZEN) mycotoxin, 246–247
zeaxanthin
high-pressure processing (HPP), 127
pulsed electric field (PEF) processing, 87–88
zinc salts, 390–395
z-value, 453
Zygosaccharomyces bailli
ohmic heating, 329
ozone processing, 227–228, 237–243